O9-ABH-683

Algebra 1 Instruction Manual

Published and distributed by Demme Learning

www.MathUSee.com

1-888-854-6284 or +1 717-283-1448 | www.demmelearning.com
Lancaster, Pennsylvania USA

ISBN 978-1-60826-031-7
Revision Code 0616

Printed in the United States of America by Bindery Associates LLC

For information regarding CPSIA on this printed material call: 1-888-854-6284 and provide reference #0616-07272016

Algebra 1

Instruction Manual

By Steven P. Demme

1·888·854·MATH (6284) - MathUSee.com
Sales@MathUSee.com

Algebra 1

Curriculum Sequence

∫ **Calculus**

cos **PreCalculus**
with Trigonometry

xy **Algebra 2**

Δ **Geometry**

x^2 **Algebra 1**

x **Pre-Algebra**

ζ **Zeta**
Decimals and Percents

ε **Epsilon**
Fractions

δ **Delta**
Division

γ **Gamma**
Multiplication

β **Beta**
Multiple-Digit Addition and Subtraction

α **Alpha**
Single-Digit Addition and Subtraction

P **Primer**
Introducing Math

Math-U-See is a complete, K-12 math curriculum that uses manipulatives to illustrate and teach math concepts. We strive toward "Building Understanding" by using a mastery-based approach suitable for all levels and learning preferences. While each book concentrates on a specific theme, other math topics are introduced where appropriate. Subsequent books continuously review and integrate topics and concepts presented in previous levels.

Where to Start

Because Math-U-See is mastery-based, students may start at any level. We use the Greek alphabet to show the sequence of concepts taught rather than the grade level. Go to MathUSee.com for more placement help.

Each level builds on previously learned skills to prepare a solid foundation so the student is then ready to apply these concepts to algebra and other upper-level courses.

Major concepts and skills for Algebra 1:

- Commutative, Associative, and Distributive Properties
- Order of operations
- Solving for an unknown
- Cartesian coordinate system
- Graphing lines, equations, and inequalities
- Solving simultaneous equations
- Factoring polynomials
- Significant digits and scientific notation
- Conic sections

Additional concepts and skills for Algebra 1:

- Unit multipliers
- Metric conversions
- Bases other than ten

Find more information and products at MathUSee.com

HONORS LESSON TOPICS

Here are the topics for the special challenge lessons included in the student text. You will find one honors page after the last systematic review page for each regular lesson. Instructions for the honors pages are included in the student text.

LESSON	TOPIC
01	Challenge word problems
02	More challenge word problems
03	Kinds of numbers; solving word problems with formulas
04	Interpreting graphs
05	Introduction to vectors
06	Plotting points on a Cartesian graph
07	Estimating slope
08	Practical application of slope
09	Word problems with different rates
10	Scatter diagrams
11	Using linear equations to solve challenging word problems
12	More linear word problems
13	Solving two or more inequalities by graphing; application with word problem involving sales
14	Word problems with equations; interpreting information from graphs
15	Word problems requiring simultaneous equations
16	Word problems involving the break-even point
17	Break-even point; linear equations
18	Word problems similar to coin and consecutive integer problems
19	Graphing word problems with exponential growth
20	More exponential word problems
21	Graphing exponential equations
22	Operations with exponents; graphing exponential equations
23	More exponential equations to graph
24	Area with polynomial multiplication; advanced polynomial addition and multiplication
25	Costs, revenue, and profit
26	More problems with cost, revenue, and profit
27	Polynomials that cannot be factored; graphing equations with negative exponents
28	Complex factoring
29	Velocity problems
30	Traditional measuring systems
31	More traditional measuring systems
32	Finding molecular mass; application of significant digits
33	Computer applications with base 16
34	Kepler's Third Law; astronomical units
35	Using the equation of a parabola to find maximum area

HOW TO USE

Welcome to *Algebra 1*. I believe you will have a positive experience with the unique Math·U·See approach to teaching math. These first few pages explain the essence of this methodology which has worked for thousands of students and teachers. I hope you will take five minutes and read through these steps carefully.

Since math builds upon previously studied concepts, the first step is to have your student take the readiness test found in the beginning of the student text. (The solutions to this test are at with the other solutions at the back of this book.) If the test reveals gaps in the student's understanding, please contact your trained Math·U·See representative and find out how to rebuild your student's math foundation. I am assuming a thorough grasp of the four basic operations (addition, subtraction, multiplication, and division), along with a mastery of fractions, decimals, percents, and pre-algebra skills.

If you are using the program properly and still need additional help, you may contact your authorized representative or visit Math·U·See online at MathUSee.com/support. —**Steve Demme**

The Goal of Math-U-See

The underlying assumption or premise of Math·U·See is that the reason we study math is to apply math in everyday situations. Our goal is to help produce confident problem solvers who enjoy the study of math. These are students who learn their math facts, rules, and formulas *and* are able to use this knowledge in solving word problems and real-life applications. Therefore, the study of math is much more than simply committing to memory a list of facts. It includes memorization, but it also encompasses learning underlying concepts that are critical to problem solving.

More than Memorization

Many people confuse memorization with understanding. Once while I was teaching seven junior high students, I asked how many pieces they would each receive if there were fourteen pieces. The students' response was, "What do we do: add, subtract, multiply, or divide?" Knowing *how* to divide is important; understanding *when* to divide is equally important.

THE SUGGESTED 4-STEP MATH-U-SEE APPROACH

In order to train students to be confident problem solvers, here are the four steps that I suggest you use to get the most from the Math·U·See curriculum:

Step 1. Preparation for the lesson.
Step 2. Presentation of the new topic.
Step 3. Practice for mastery.
Step 4. Progression after mastery.

Step 1. Preparation for the lesson.

Watch the DVD to learn the concept and see how to demonstrate this concept with the manipulatives when applicable. Study the written explanations and examples in the instruction manual. Many students watch the DVD along with their instructor. Older students in the secondary level who have taken responsibility to study math themselves will do well to watch the DVD and read through the instruction manual.

Step 2. Presentation of the new topic.

Now that you have studied the new topic, choose problems from the first lesson practice page to present the new concept to your students.

a. Build: Demonstrate how to use the manipulatives to solve the problem, if applicable. As students mature, they learn to think abstractly. However, we will still be using the manipulatives for some concepts in *Algebra 1*.

b. Write: Record the step-by-step solutions on paper as you work them through with the manipulatives.

c. Say: Explain the "why" and "what" of math as you build and write.

Do as many problems as you feel are necessary until the student is comfortable with the new material. One of the joys of teaching is hearing a student say, *"Now I get it!"* or *"Now I see it!"*

Step 3. Practice for mastery.

Using the examples and the lesson practice problems from the student text, have the students practice the new concept until they understand it. It is one thing for students to watch someone else do a problem; it is quite another to do the same problem themselves. Do enough examples together so that they can do them without assistance.

Do as many of the lesson practice pages as necessary (not all pages may be needed) until the student remembers the new material and gains understanding. Give special attention to word problems, which are designed to apply the concept being taught in the lesson. If your student needs more practice, look for the additional lesson practice pages under "Algebra 1 Updates" at MathUSee.com/support/downloads.

Step 4. Progression after mastery.

Once mastery of the new concept is demonstrated, proceed into the systematic review pages for that lesson. Mastery can be demonstrated by having each student teach the new material back to you. The goal is not to fill in worksheets, but to be able to teach back what has been learned.

The systematic review worksheets review the new material as well as provide practice of the math concepts previously studied. Remediate missed problems as they arise to ensure continued mastery.

After the last systematic review page in each lesson, you will find an "honors" lesson. These are optional, but highly recommended for students who will be taking advanced math or science courses. These challenging problems are a good way for all students to hone their problem-solving skills.

Proceed to the lesson tests. These were designed to be an assessment tool to help determine mastery, but they may also be used as extra worksheets. Your students will be ready for the next lesson only after demonstrating mastery of the new concept and continued mastery of concepts found in the systematic review worksheets.

Confucius is reputed to have said, "Tell me, I forget; show me, I understand; let me do it, I will remember." To which we add, **"Let me teach it and I will have achieved mastery!"**

Length of a Lesson

So how long should a lesson take? This will vary from student to student and from topic to topic. You may spend a day on a new topic, or you may spend several days. There are so many factors that influence this process that it is impossible to predict the length of time from one lesson to another. I have spent three days on a lesson, and I have also invested three weeks in a lesson. This occurred in the same book with the same student. If you move from lesson to lesson too quickly without the student demonstrating mastery, he will become overwhelmed and discouraged as he is exposed to more new material without having learned the previous topics. But if you move too slowly, your student may become bored and lose interest in math. I believe that as you regularly spend time working along with your student, you will sense when is the right time to take the lesson test and progress through the book.

By following the four steps outlined above, you will have a much greater opportunity to succeed. Math must be taught sequentially, as it builds line upon line and precept upon precept on previously learned material. I hope you will try this methodology and move at your student's pace. As you do, I think you will be helping to create a confident problem solver who enjoys the study of math.

ONGOING SUPPORT AND ADDITIONAL RESOURCES

Welcome to the Math-U-See Family!

Now that you have invested in your children's education, I would like to tell you about the resources that are available to you. Allow me to introduce you to our staff, our ever improving website, the Math·U·See blog, our free e-mail newsletter, and other online resources.

Many of our customer service representatives have been with us for over 10 years. What makes them unique is their desire to serve and their expertise. They are able to answer your questions, place your student(s) in the appropriate level, and provide knowledgeable support throughout the school year.

Come to your local curriculum fair where you can meet us face-to-face, see the latest products, attend a workshop, meet other MUS users at the booth, and be refreshed. We are at most curriculum fairs and events. To find the fair nearest you, click on "Events" under "E-Sources."

The **Website**, at MathUSee.com, is continually being updated and improved. It has many excellent tools to enhance your teaching and provide more practice for your student(s).

Math-U-See Blog

Interesting insights and up-to-date information appear regularly on the Math·U·See Blog. It features updates, rep highlights, fun pictures, and stories from other users. Visit us to get the latest scoop on what is happening.

Email Newsletter

For the latest news and practical teaching tips, sign up online for the free Math·U·See e-mail newsletter. Each month you will receive an e-mail with a teaching tip from Steve as well as the latest news from the website. It's short, beneficial, and fun. Sign up today!

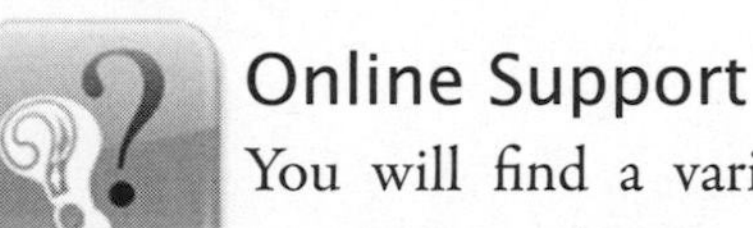

Online Support

You will find a variety of helpful tools on our website, including corrections, placement tests, answers to questions, and support options.

For Specific Math Help

When you have watched the DVD and read the instruction manual and still have a question, we are here to help. Call us or click the contact link. Our trained staff are available to answer a question or walk you through a specific lesson.

Feedback

Send us an e-mail by clicking the contact link. We are here to serve you and help you teach math. Ask a question, leave a comment, or tell us how you and your student are doing with Math·U·See.

Our hope and prayer is that you and your students will be equipped to have a successful experience with math!

Blessings,

Steve

Steve Demme

LESSON 1

Commutative and Associative Properties

The first few lessons of *Algebra 1* are a quick review of concepts taught in Math-U-See *Pre-Algebra*. Please take your time with these lessons, as understanding these concepts is critical for the rest of the course. There is a readiness test in the beginning of the student text. Please use this test to see whether your student is ready to begin *Algebra 1*.

Commutative Property

We have been using the commutative property since the addition and multiplication curriculum. I think of it as the "commute" ative property. Let's say you commute to work seven miles. When you come home, it is still seven miles. Regardless of which direction you are going, it is the same distance. This is true for addition and multiplication. Changing the order, or direction, does not affect the sum (addition) or the product (multiplication). Study the following examples, and notice that the commutative property is not applicable to subtraction or division, where order and direction are critical.

Example 1

$5 + 7 = 12$ $\quad$ $7 + 5 = 12$ $\quad$ $12 = 12$

Example 2

$5 \times 7 = 35$ $\quad$ $7 \times 5 = 35$ $\quad$ $35 = 35$

Example 3

$7 - 5 = (+2)$ $\quad$ $5 - 7 = (-2)$ $\quad$ $(+2) \neq (-2)$

Example 4

$7 \div 5 = 7/5$ $\quad$ $5 \div 7 = 5/7$ $\quad$ $7/5 \neq 5/7$

Associative Property

I remember the associative property by thinking of whom I associate, or am grouped, with. Read through the following examples and notice how the grouping by parentheses affects the answer.

Example 5

$$\begin{aligned} (3 + 5) + 7 &= 3 + (5 + 7) \\ 8 + 7 &= 3 + 12 \\ 15 &= 15 \end{aligned}$$

Example 6

$$\begin{aligned} (4 \times 5) \times 6 &= 4 \times (5 \times 6) \\ 20 \times 6 &= 4 \times 30 \\ 120 &= 120 \end{aligned}$$

Example 7

$$\begin{aligned} (10 - 7) - 4 &= 10 - (7 - 4) \\ 3 - 4 &= 10 - 3 \\ -1 &\neq 7 \end{aligned}$$

Example 8

$$\begin{aligned} (8 \div 4) \div 2 &= 8 \div (4 \div 2) \\ 2 \div 2 &= 8 \div 2 \\ 1 &\neq 4 \end{aligned}$$

As with the commutative property, the associative property is true for addition and multiplication, but not for subtraction and division.

Operations with negative numbers

Lesson Practice 1A in the student text has an important review of negative numbers. Notice especially the examples of negative numbers with exponents. Watch for more Quick Reviews as you work through the student text.

LESSON 2

Order of Operations and Absolute Value

Levels of Math

When solving an algebraic equation with several different operations (addition, subtraction, multiplication, division, and exponents), there is an order that determines which operation to do first. This order is related to the four levels of math.

Level 1 - Counting
Level 2 - Fast counting, or instant counting, is adding, and the inverse of adding is subtracting.
Level 3 - Fast adding of the same number is multiplying, and the inverse of multiplying is dividing.
Level 4 - Fast multiplying of the same number is exponents, and their inverse, roots.

When systematically solving an equation, go in order with the highest level first and then sequentially to the lower levels. First exponents, then multiplying, and then adding. Before you get to the operations themselves, make sure that all the parentheses have been taken care of first.

Order of Operations

A fun way to remember this is to think of your. . . **PARA**chute **EX**pert **M**y **D**ear Aunt Sally.

Parenthesis - Parachute
Level 4) **E**xponents - Expert
Level 3) **M**ultiply and **D**ivide - My Dear
Level 2) **A**dd and **S**ubtract - Aunt Sally

When you have two operations on the same level such as adding and subtracting, proceed from left to right in the natural order of the equation. This is illustrated in the third line of the first example.

Example 1

$(4 \times 3) + 4^2 - 6 =$	**PARA**chute - parentheses
$12 + 4^2 - 6 =$	**EX**pert - exponents
$12 + 16 - 6 = 22$	**A**dding and **S**ubstracting from left to right

Sometimes parentheses are added to the expressions to organize and group the same kinds of things.

Example 2

First, group terms according to their kind.

$$5A + 3 + 4A + 2 - 5 - A + 1$$
$$(5A + 4A - A) + (3 + 2 - 5 + 1)$$
$$8A + 1$$

Absolute Value

Absolute value signs make the final result of the operations between them positive. When deciding on the order of operations, treat absolute value bars like parentheses.

$|-3| = 3$ The absolute value bars make negative three positive.

$|3| = 3$ The absolute value bars leave positive three unchanged

$-|-3| = -3$ The absolute value bars make negative three positive. Then the negative sign takes effect, and the answer is a negative three.

$-|3| = -3$ The absolute value bars leave positive three unchanged. Then the negative sign takes effect, and the answer is a negative three.

LESSON 3

Solve for Unknown with One Variable

Using an Unknown

Algebra is used when we don't know a number, and so we use a letter instead.

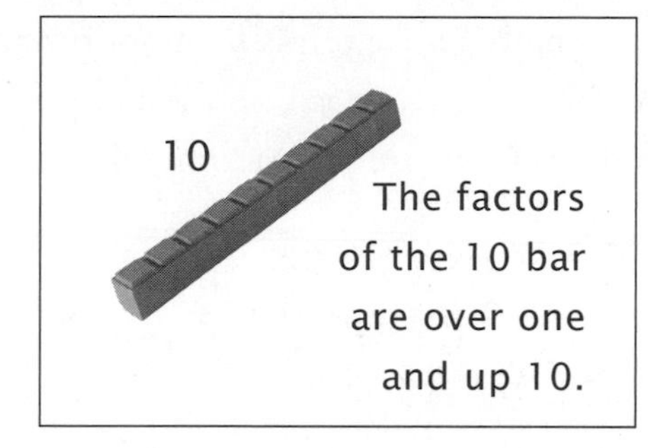

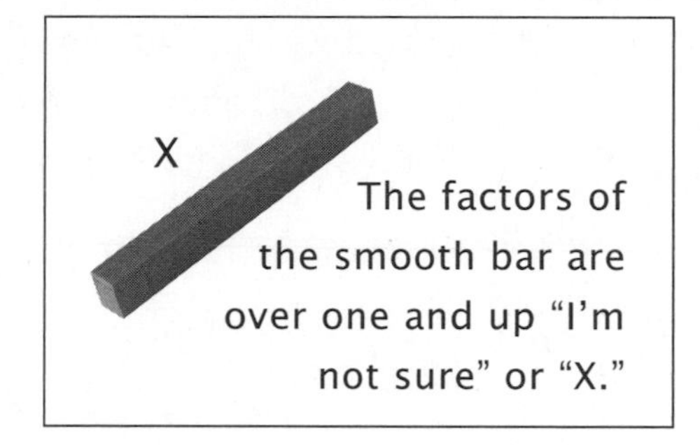

To compare or combine, your numbers must be the same kind.

Expanding an Equation

The key concept in solving for an unknown is remembering that the equals sign (=) means that both sides of the equation are the same and must remain the same. If we add something to one side, we must add it to the other side to maintain their equality or sameness. If we multiply one side by two, we must multiply the other side by two to maintain their equality.

In the following example, we will start with an easy equation and make it look complicated by adding the same thing to both sides.

Example 1

$$
\begin{array}{c}
X = 2 \\
\underline{+X \quad +X} \\
2X = X + 2 \\
\underline{+3 \quad +3} \\
2X + 3 = X + 2 + 3 \\
\underline{+2X \quad +2X} \\
2X + 2X + 3 = X + 2 + 3 + 2X \\
\underline{-1 \quad -1} \\
2X + 2X + 3 - 1 = X + 2 + 3 + 2X - 1 \\
\underline{-X \quad -X} \\
2X + 2X + 3 - 1 - X = X + 2 + 3 + 2X - 1 - X
\end{array}
$$

Solving an Equation

Now we will go backwards and solve for the unknown "X." First, group things according to their kind with parentheses. Then, combine like terms to simplify the equation. Next, subtract the same thing from both sides instead of adding, since subtraction is the opposite of addition. In the example, we are subtracting "Xs" first, and then numbers.

Example 2

$$
\begin{array}{c}
2X + 2X + 3 - 1 - X = X + 2 + 3 + 2X - 1 - X \\
(2X + 2X - X) + (3 - 1) = (X + 2X - X) + (2 + 3 - 1) \\
3X + 2 = 2X + 4 \\
\overline{-2X \quad -2X} \\
X + 2 = 4 \\
\overline{-2 \quad -2} \\
X = 2
\end{array}
$$

Using Manipulatives

Now let's do the same kind of problem with the manipulatives. We'll need the X bars that we make by snapping the smooth inserts into the backs of the tens. The blue inserts represent positive X, and the gray inserts represent negative X (–X).

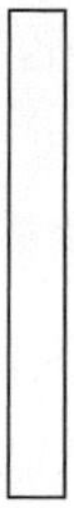

blue inserts = X  **gray inserts = –X**

Example 3

3X + 2 = 2X + 4

–2X –2X

X + 2 = 4

–2 –2

X = 2

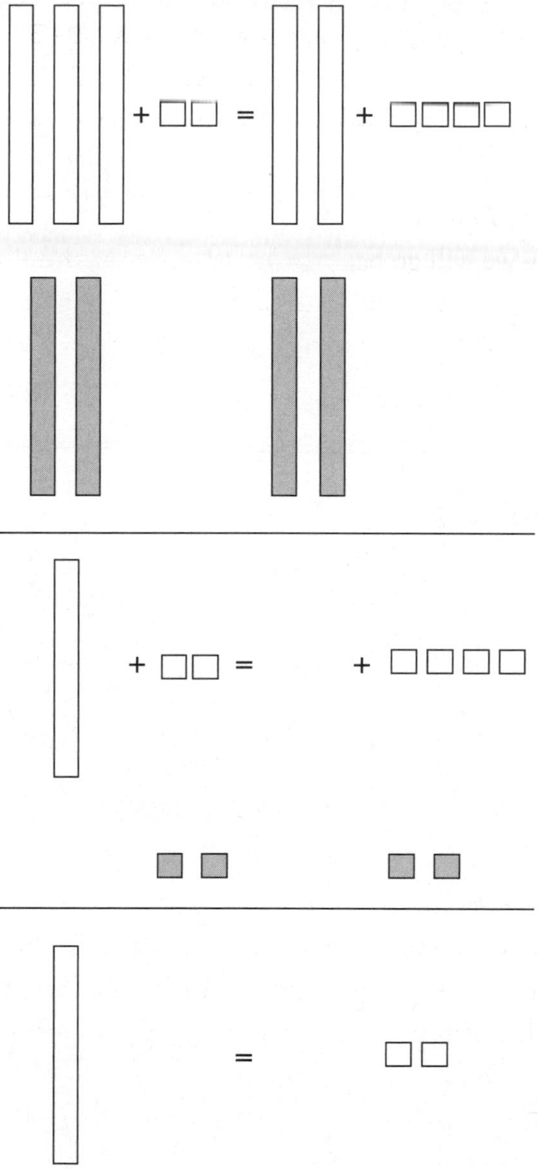

Solving with Multiplication and Division

We can multiply both sides of an equation by the same, or common, factor.

Example 4

$$2 = 2$$
$$2 \times 5 = 2 \times 5$$
$$10 = 10$$

$$1 \text{ apple} = 4 \text{ oz}$$
$$\text{multiplying both sides by 5 yields}$$
$$5 \text{ apples} = 5(4 \text{ oz}) \text{ or } 20 \text{ oz}$$

Division is like multiplication since they are inverses or opposites. We can also divide both sides of an equation by the same factor.

Example 5

$$5 \text{ apples} = 20 \text{ oz}$$
$$\text{dividing both sides by 5 yields}$$
$$1 \text{ apple} = 4 \text{ oz}$$

LESSON 4

Distributive Property

In the following picture and corresponding equation, notice how two is distributed across the five and the three, and of course, the eight (or 5 + 3) that we began with.

Example 1

$(2)(8) = (2)(5) + (2)(3)$

$\rightarrow \uparrow \quad \rightarrow \uparrow \quad \rightarrow \uparrow$

$(2)(5+3) = (2)(5) + (2)(3)$

$\rightarrow \uparrow \quad \rightarrow \uparrow \quad \rightarrow \uparrow$

$16 = 10 + 6$

$16 = 16$

↑ 8

↑ 3

↑ 5

2 →

2 →

$(2)(5+3) = (2)(5) + (2)(3)$

2 times 5, plus 3, is the same as 2 times 5, plus 2 times 3.

Example 2

$(8)(D+3) = (8)(D) + (8)(3) = 8D + 24$

This can be very helpful in multiplying larger numbers.

Example 3

$(4)(12) = (4)(10 + 2) = (4)(10) + (4)(2) = 40 + 8 = 48$

What we did was turn the larger product into two smaller "parts." Instead of multiplying times 12, we made 12 into 10 + 2. Then we multiplied by four, using the distributive property. These smaller parts, or smaller products, are called *partial products.*

Inverse of Distributive Property

You can also go backwards using the same principle. In the following equation, what was distributed between A + B? $(\quad)(A + B) = 3A + 3B$. As you read this problem, think, "What times A + B is equal to 3A + 3B?" The answer is three. In this example, we had to find the missing three. In example 4, we know the three but must find what goes inside the parentheses.

Example 4

$3(\quad + \quad) = 3A + 3B$ » $3(A + B) = 3A + 3B$

The answer is A + B.

Now we'll take it one step further. If we are given half of the equation, can we work backwards to find the other half?

Example 5

$7X + 7M = (\quad)(\quad + \quad)$.

The answer is $7X + 7M = (7)(X + M)$.

This is often referred to as finding the *common factor,* in this case seven, and factoring it out of the equation. It is simply the distributive property in reverse.

Throughout the rest of the book, we will be using both of these skills. Sometimes we will distribute, and other times we will divide through by a common factor.

LESSON 5

Number Lines and Cartesian Coordinates

A number line is used to graphically show real-number solutions to a problem. It looks like this:

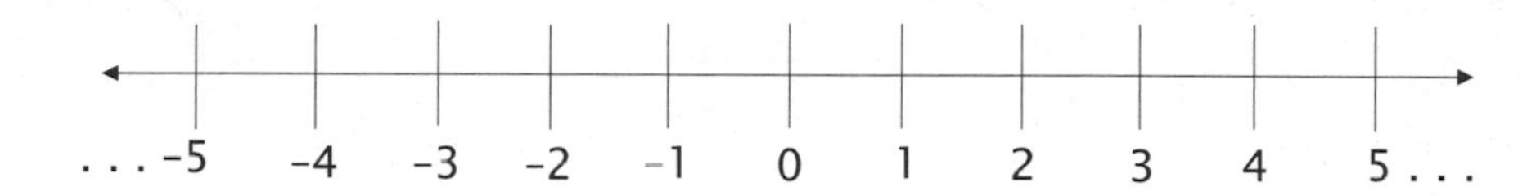

Zero is in the middle of the number line. All the negative numbers extend infinitely to the left. Similarly, the positive numbers extend infinitely to the right.

Notice that even though only integers are labeled on the line above, you could show fractions or irrational numbers as well.

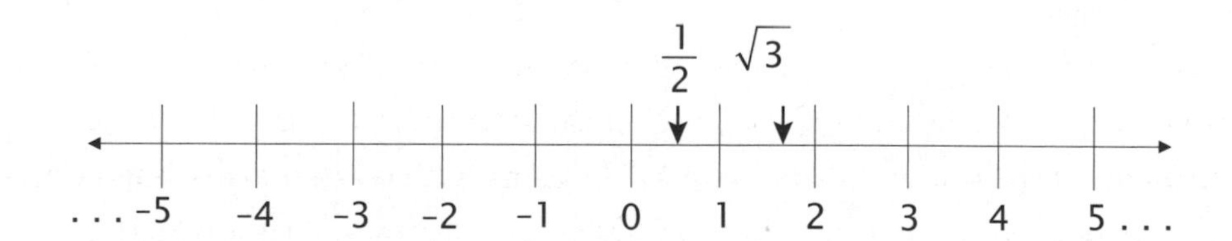

Example 1

Find all even numbers between 1 and 7. Plot your answer on a real-number line.

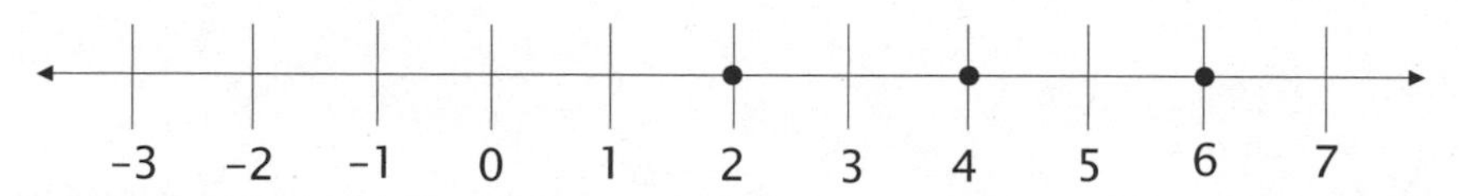

Filled-in circles are used to designate a solution to the problem.

Example 2

Find all numbers between -3 and 1, including the endpoints -3 and 1. Plot your answer on a real-number line.

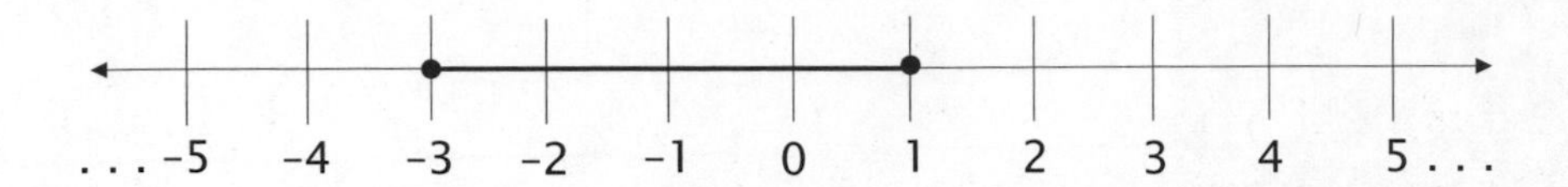

This time we have an infinite range of answers. Therefore, we circle and fill in the endpoints because they are included in the solution; and then shade in all the numbers in between them.

Example 3

Find all the numbers that are greater than -2. Show your answer on a real-number line.

This time we have left −2 as an open circle because it is not a part of the solution. Only numbers greater than −2 comprise the solution. Since the answer includes the numbers that extend infinitely to the right, a shaded arrow is used.

Example 4

Find all $X \geq \frac{5}{2}$.

Here, 5/2 = 2 1/2 is shown as filled in. Answers extend infinitely to the right as shown. We are showing here that the real number line can be used to describe the values of the variable, X. Since the answer includes the numbers that extend infinitely to the right, a shaded arrow is used as well.

Example 5

Show on a number line where $A < \sqrt{2}$.

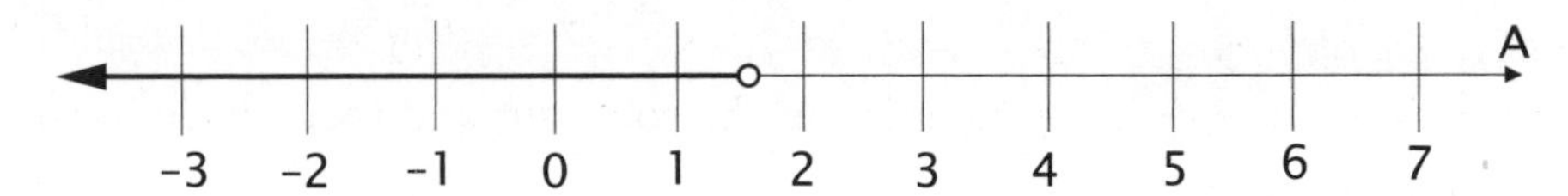

We are showing the possible values of A on this number line.

$\sqrt{2} \approx 1.4$ (Note that 1.4 is an approximate value.)

In the next part of this lesson, we are going to be using a graph of two number lines perpendicular to each other at point "0." The horizontal number line is labeled as the *X-axis,* and the vertical number line is referred to as the *Y-axis.*

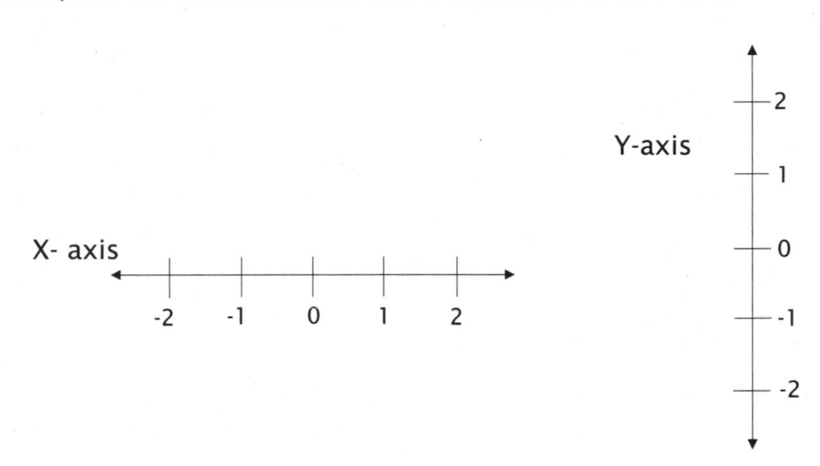

The X-axis starts at zero and moves to the right for positive numbers and to the left for negative numbers. The Y-axis starts at zero and moves up for positive numbers and down for negative numbers. If you think of the Cartesian graph explained below as two number lines, it should help you in understanding it.

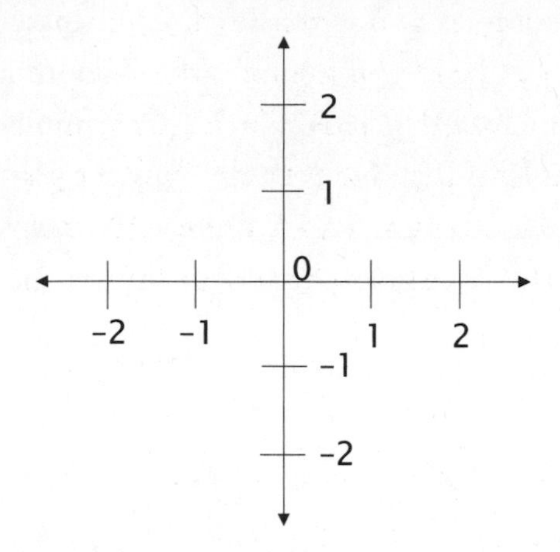

Analytic Geometry

Since lesson 3, we've solved equations. This is known as algebra. Another subject that concerns points, lines, planes, and angles is *geometry.* For hundreds of years, these two subjects were treated as separate entities. In the early 1600s a young man in his teens, René Descartes, discovered how to "show" algebra geometrically, resulting in what is called "*analytic* (algebra) *geometry.*" Math·U·See owes a great deal to this mathematician since our attempt is to "show" all aspects of math with visual, concrete objects. The following coordinate system is named after him. It is the (Des)*cartesian coordinate system.*

Cartesian Coordinate System

In graphing equations, there are over and up coordinates. The over, or horizontal, coordinate is called the *X-coordinate,* and the up, or vertical, coordinate is called the *Y-coordinate.* Think of playing the game Battleship, which has horizontal and vertical coordinates. Instead of numbers and letters, each coordinate is a number. The first number given is the X-coordinate; the second is the Y-coordinate.

Each axis is a number line with positive and negative numbers. Figure 1 shows the two perpendicular axes as lines that intersect, dividing the plane into four *quadrants.* These quadrants are numbered counterclockwise with Roman numerals. Figure 2 on the next page represents the same graph but does not have the numbers written.

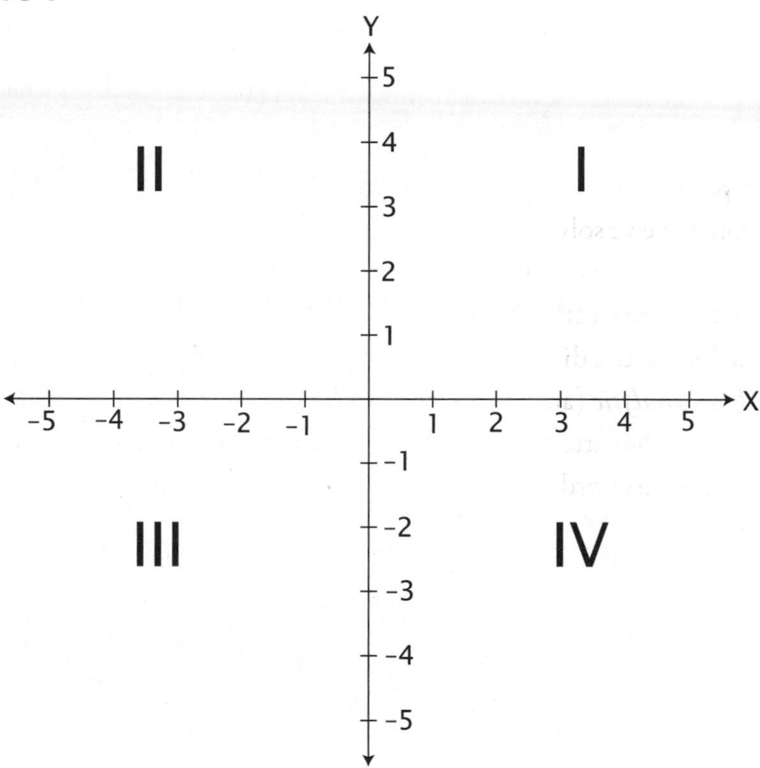

Figure 2

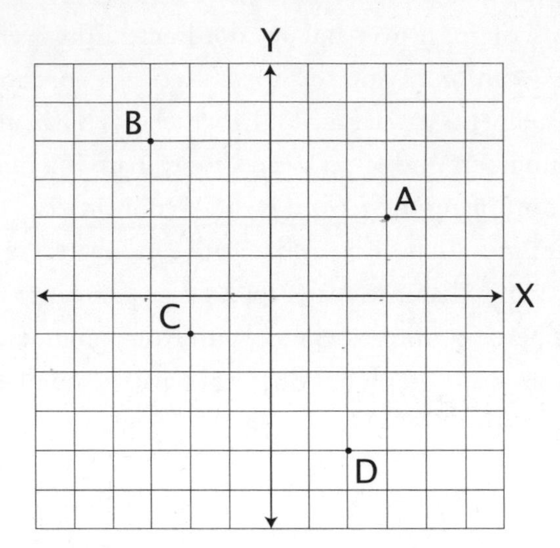

The *point* where the two lines intersect, (0, 0), is called the *origin*. When counting over and up to find the coordinates of a point, always begin at the origin.

Graphing Points

Let's graph a few points and label them as we plot them on the graph.

A = (3, 2). This means over three to the right because it is positive, and up two since two is also positive.

B = (–3, 4). This means over three to the left because it is negative, and up four (positive).

C = (–2, –1). This means over to the left two and down one (negative).

D = (2, –4). This means over to the right two and down four.

LESSON 6

Graphing a Line

Word Problem 1

Let's pretend you are baking bread for a school or church bake sale. Your mom wants to take pictures of this marathon bake-o-rama for the family album. Right before you start, you bake two loaves to make sure that you have all the necessary ingredients and to test the oven to see that it is working properly. You figure that with everything taken into consideration, you'll be able to bake three loaves per hour.

Bright and early the next morning, Mom arrives with the camera and takes the first picture of the two loaves sitting on the counter. She will return and take a snapshot once every hour. When she returns after the first hour, you have five loaves of bread: the two you began with and three that just came out of the oven. In two hours, you have eight loaves: the two you began with and three plus three for the two hours you've been baking. In three hours, there are 11 tasty brown loaves (2 + 3 + 3 + 3). How many would you have in 10 hours? You would have 32 loaves (three each hour, or 3 x 10, plus the two you started with).

When we record this data in two columns, it looks like this:

Hour	Bread
0	$(3\times 0)+2=2$
1	$(3\times 1)+2=5$
2	$(3\times 2)+2=8$
3	$(3\times 3)+2=11$
10	$(3\times 10)+2=32$

Graphing the Problem

Now we can interpret the data in our table with an equation. Bread equals three per hour, plus two ($B = 3H + 2$). Each hour we have a new set of data.

The data, or points to be graphed, are:

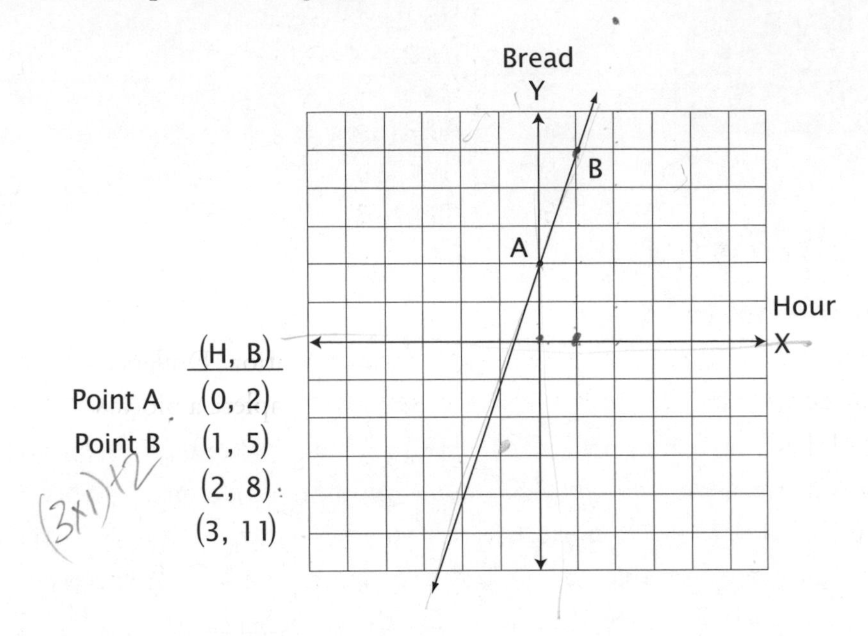

Due to the size of the graph, we can only plot the first two points. They are points A and B. Connecting the points with a straight line gives us our first line graphed with Cartesian coordinates.

Word Problem 2

Sometimes problems begin with someone owing, or in debt, or in the hole. This will be represented by negative numbers. If I owe you five dollars, this appears on the graph as -5. In the next example, Bobby owed three (negative three) dollars to Barbara. He earns one (positive one) dollar for every block set he assembles. How much money will he have after assembling six block sets and then paying back his debt to Barbara? When he begins, without having put together any block sets; he has no money, in fact, he is in the hole three dollars. This is shown in the data as -3. And since he didn't have a chance to begin working yet, the money from the block sets is 1×0, or 0.

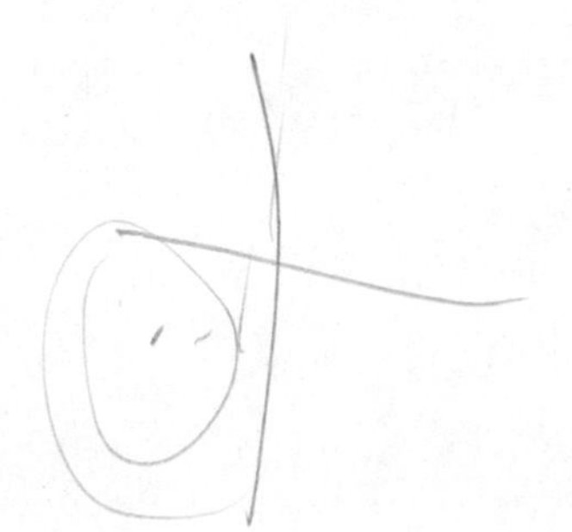

When we record this data in two columns, it looks like this:

	Block Sets	Dollars
when he starts:	0	$(1 \times 0) - 3 = -3$
	1	$(1 \times 1) - 3 = -2$
	2	$(1 \times 2) - 3 = -1$
	3	$(1 \times 3) - 3 = \ 0$
	4	$(1 \times 4) - 3 = +1$
	5	$(1 \times 5) - 3 = +2$
	6	$(1 \times 6) - 3 = +3$

Graphing the Problem

We can interpret the data in our table with an equation. Dollars equals one per block set minus three ($D = 1B - 3$). Each time we complete a block set, we have a new set of data.

The data, or points to be graphed, are:

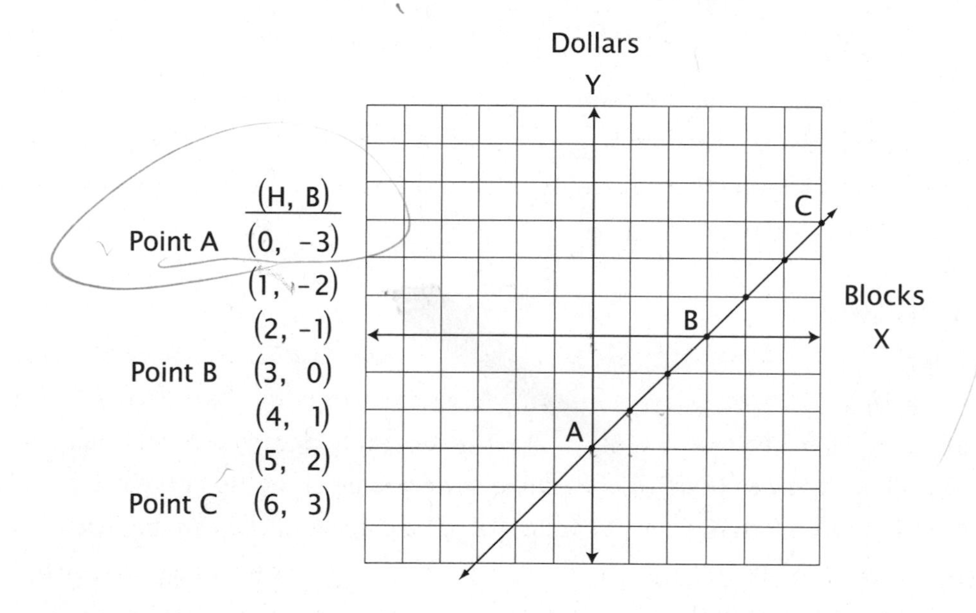

Points A, B, and C are shown as three points on the graph. Connecting the points with a straight line shows us the data drawn as a line and graphed with Cartesian coordinates.

LESSON 7

Slope-Intercept Formula

Here are two new words that describe lines– slope and intercept. The *slope* is given by *m* (a mountain has slope and starts with *m*), and *intercept* is indicated with the letter *b* (a bee that intercepts your leg with his stinger). The *slope-intercept formula* is Y = mX + b. In the equation Y = 2X + 3, the 2 is the slope and 3 is the intercept.

Slope

Slope is the $\frac{\text{up dimension}}{\text{over dimension}}$ or the rise over the run $\left(\frac{\text{rise}}{\text{run}}\right)$. In our example of the bread-baking, for every one hour (over), we were able to bake three loaves (up). So for every hour, we move over to the right one space and up three spaces. We continue to do this, and when we connect two or more points, we have a line that "slopes" up. We describe the slope as $\frac{3 \text{ up}}{1 \text{ over}}$, or 3.

Figure 1

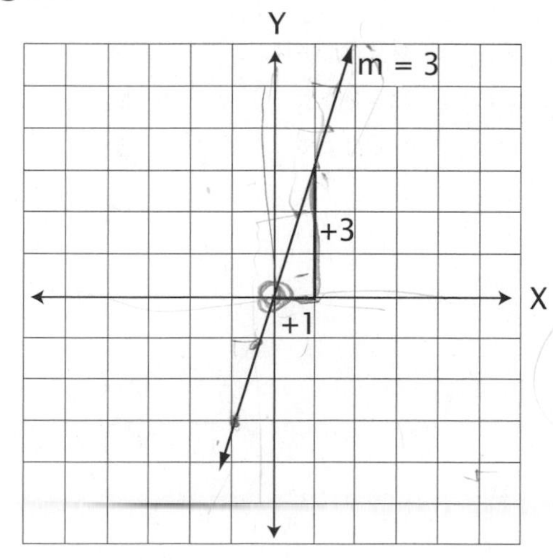

$$\frac{\text{Rise}}{\text{Run}} = \frac{+3}{+1} = +3 \text{ or } 3$$

m = 3

If we were able to bake five loaves for every one hour, we move over to the right one space and up five spaces. The slope is steeper. The rise is five and the run is one. So the slope is $\frac{5 \text{ up}}{1 \text{ over}}$, or 5.

Figure 2

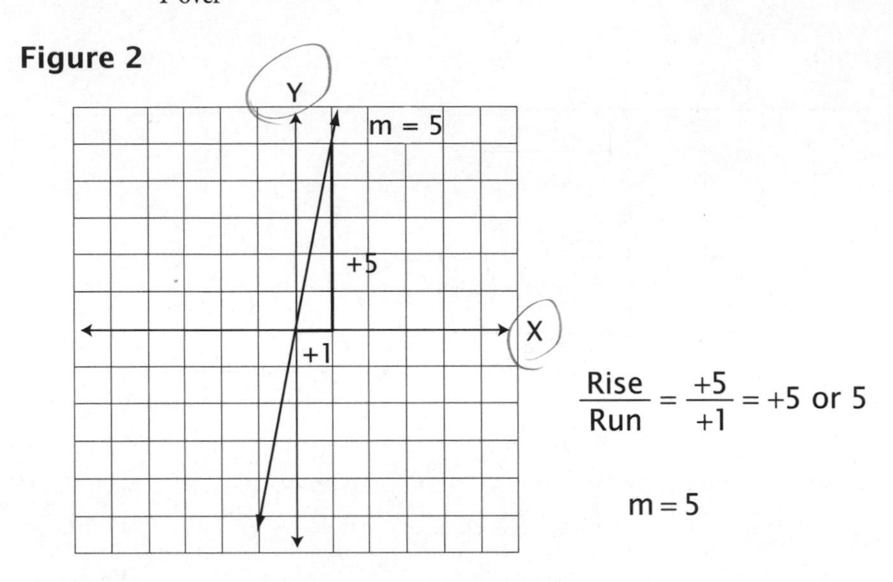

If we bake one loaf for every one hour, we move over to the right one space and up one space. The slope is not as steep as three loaves or five loaves per hour. The rise is one and the run is one. The slope is $\frac{1 \text{ up}}{1 \text{ over}}$, or 1.

Figure 3

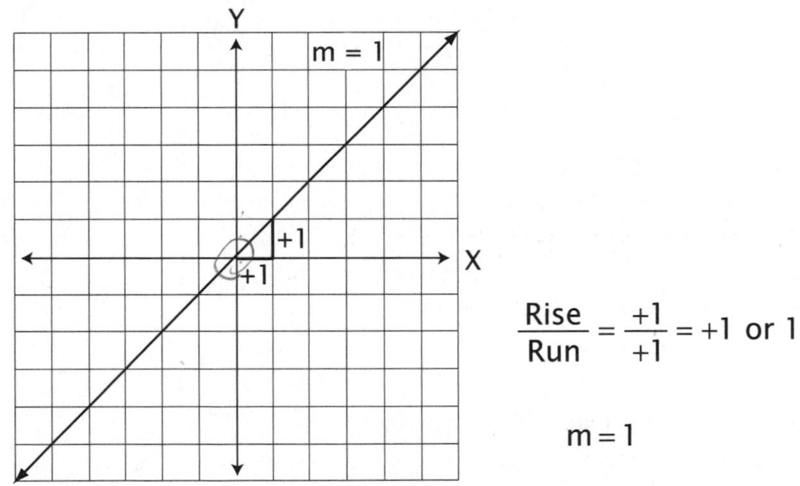

If we bake one loaf and it takes four hours, we move over to the right four spaces and up one space. The slope is not nearly as steep as three loaves per hour. The rise is one and the run is four. The slope is $\frac{1 \text{ up}}{4 \text{ over}}$, or $\frac{1}{4}$.

Figure 4

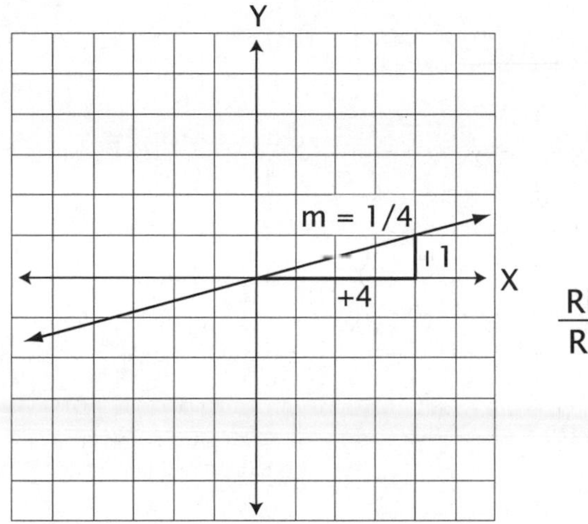

$$\frac{\text{Rise}}{\text{Run}} = \frac{+1}{+4} \text{ or } \frac{1}{4}$$

$$m = \frac{1}{4}$$

If we bake three loaves and it takes five hours, we move over to the right five spaces and up three spaces. The rise is three and the run is five. The slope is $\frac{3 \text{ up}}{5 \text{ over}}$, or $\frac{3}{5}$.

Figure 5

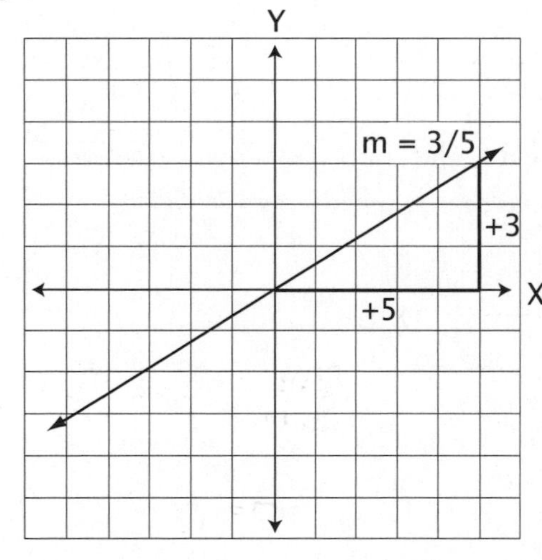

$$\frac{\text{Rise}}{\text{Run}} = \frac{+3}{+5} = \text{ or } \frac{3}{5}$$

$$m = \frac{3}{5}$$

Negative Slope

You can also have negative slopes. An example would be the business man who loses two dollars each day. The slope is over one and down two (the opposite of up because it is minus). It will look like the line in figure 6: $Y = -2X + 0$, or $Y = -2X$.

Figure 6

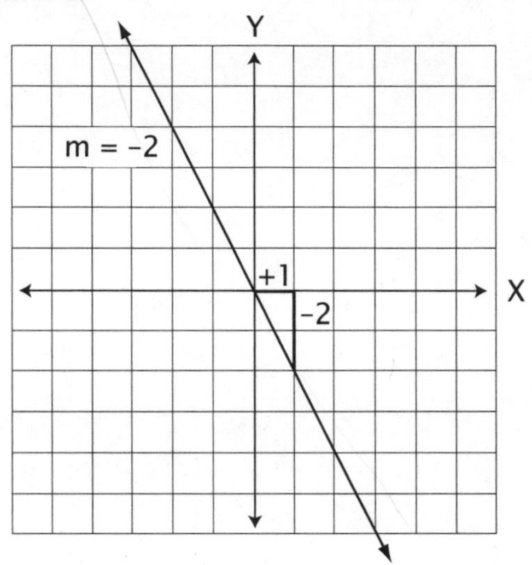

Days	Money
1	2
2	4
3	6

$$\frac{\text{Rise}}{\text{Run}} = \frac{-2}{+1} = \text{or } -2$$

$$m = -2$$

We move over to the right one space and up (really down) two spaces. The rise is negative two and the run is positive one. The slope is $\frac{-2 \text{ down}}{1 \text{ over}}$, or $-\frac{2}{1}$ or -2.

Figure 7 on the next page is an example of losing one dollar per day:
$Y = -X + 0$, or $Y = -X$.

We move over to the right one space and up (really down) one space. The rise is negative one and the run is positive one. The slope is $\frac{-1 \text{ down}}{1 \text{ over}}$, or $-\frac{1}{1}$, or -1. Notice that $-1X$ is the same as $-X$.

Figure 7

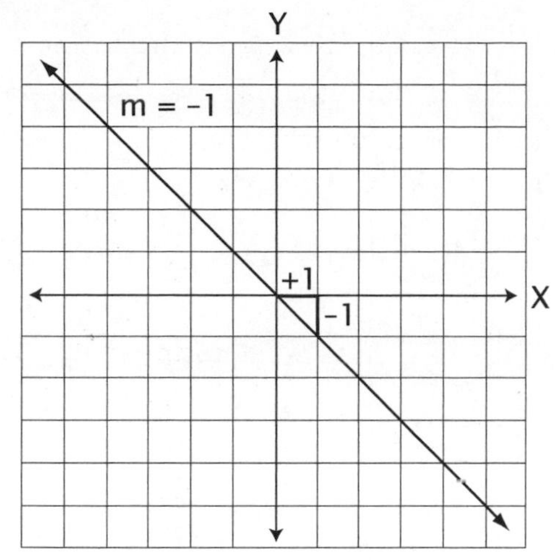

$$\frac{\text{Rise}}{\text{Run}} = \frac{-1}{+1} \text{ or } -1$$

$$m = -1$$

Negative Fraction Slope

You can also have *negative fraction slopes.* An example would be the business man who loses two dollars every three days. The slope is over three and down two (the opposite of up because it is minus).

Figure 8

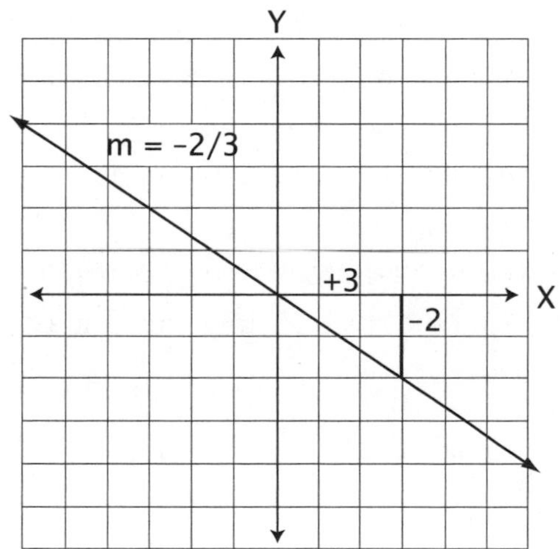

Days	Money
3	2
6	4

$$m = \frac{-2}{+3} = -\frac{2}{3}$$

We move over to the right three spaces and up (really down) two spaces. The rise is negative two and the run is positive three. The slope is $\frac{\text{2 down}}{\text{3 over}}$, or $\frac{-2}{+3}$, or $-\frac{2}{3}$.

Figure 9

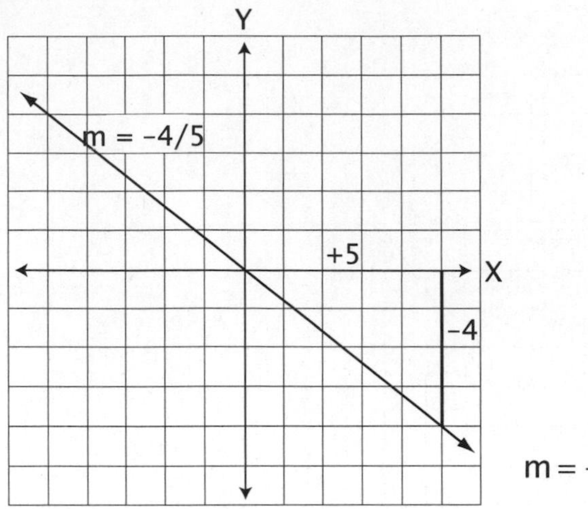

$$m = \frac{-4}{+5} = -\frac{4}{5}$$

If the slope is $^{-4}/_{5}$, move over to the right five spaces and up (down) four spaces. The rise is negative four and the run is positive five. The slope is $\frac{4 \text{ down}}{5 \text{ over}}$, or $^{-4}/_{+5}$, $^{-4}/_{5}$.

Finding the Slope of a Line

If you encounter the graph of a line, you can determine the slope by **making a right triangle between any two points on the line. Here is how to find the slope of line *w* if it is not already given:**

First choose two points that are on the grid or at the intersection of a Y-coordinate and an X-coordinate. In figure 10 on the next page, I chose two points: (2, 4) and (0, 1).

Then choose what direction you want to move. I chose to move away from the origin starting at the point (0, 1) and moving towards (2, 4). By making a right triangle, we can find the up dimension or the rise and the over dimension or the run. In figure 10 the rise is positive three and the run is positive two. The slope is +3/+2 (rise over run) or 3/2.

Figure 10

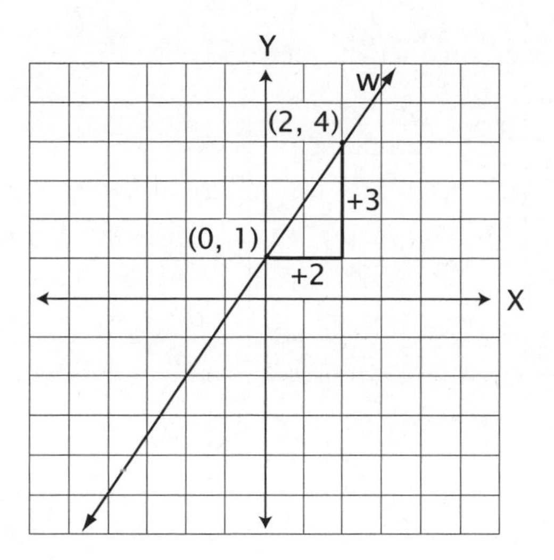

We can also find the slope by moving towards the origin from point (2, 4) towards (0, 2). In figure 11, I drew the triangle above the line and moved to the left for the run and down for the rise. Moving to the left on the Cartesian graph for the X-coordinates means the run will be a negative number. Moving down for the rise, or Y-coordinates, also yields a negative number. Both of these are negative values. The rise is negative three and the run is negative two. The slope is –3/–2 (rise over run) or 3/2.

So whether you move down the line or up the line, you still end up with a positive slope of 3/2.

Figure 11

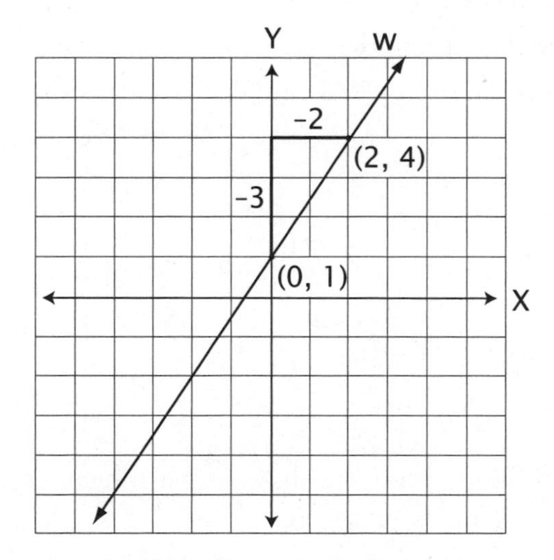

Intercepts

There are an infinite number of lines that have the same slope. Determining where a line crosses, or intercepts, the Y-axis narrows it down to one specific line.

In figure 12 there are three lines. Notice that each of these lines has the same slope. It is 2/1 or 2. What makes each line different is where it intercepts the Y-axis. The point where the line intercepts the Y-axis is called the Y-intercept.

Figure 12

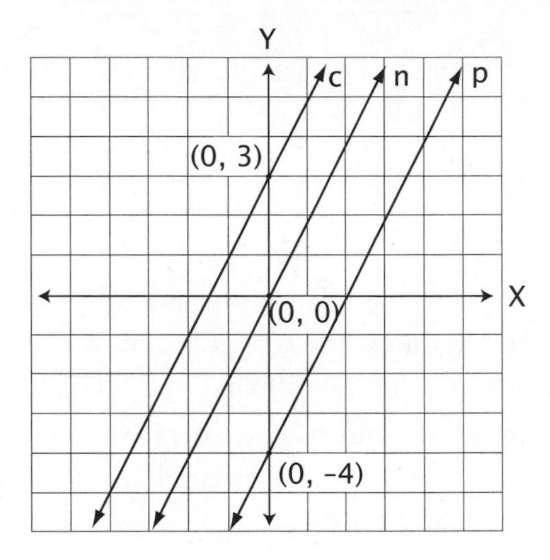

The intercept of line c is the point (0, 3).
All along the Y-axis, the coordinate for X is 0.
The point (0, 3) tells us that $X = 0$ and $Y = 3$.
So, our Y-intercept is found whenever X is 0, and in this case the intercept is 3.
Here is the intercept for each line:

line c: (0, 3)
line n: (0, 0)
line p: (0, –4)

The slope of each line is 2, so the slope–intercept formulas for the lines are:

line c: $Y = 2X + 3$
line n: $Y = 2X + 0$ or $Y = 2X$
line p: $Y = 2X - 4$.

If an equation is given with no b term, we can assume that b is zero. This means that the Y-intercept of the line is also zero. See line n above.

We looked at the graph of each line, and it was easy to see the value of the intercept. If you have an equation of a line in the slope-intercept form, you can find the intercept algebraically by inserting "0" as the value of "X," because when the line intercepts the Y-axis, X is always equal to zero. In line *c* above, $Y = 2X + 3$. If you insert the value "0" for X, then $Y = 2(0) + 3$, or $Y = 3$, and the intercept is 3. In line *p* above, $Y = 2X - 4$. When you insert the value "0" for X, then $Y = 2(0) - 4$, or $Y = -4$, and the intercept is -4.

Figure 13

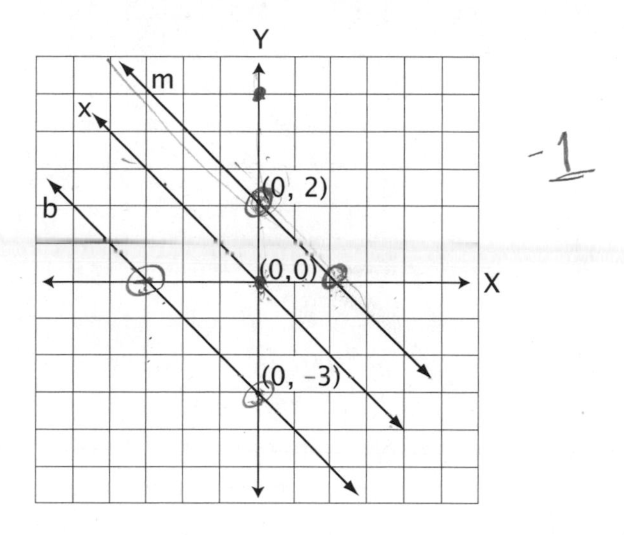

The intercept of line *m* is the point (0, 2).
All along the Y-axis, the coordinate for X is 0.
The point (0, 2) tells us that when $X = 0$, then $Y = 2$.
So, our Y-intercept is found whenever X is 0, and in this case the intercept is 2.
Here is the intercept for each line.

line *m*: (0, 2)
line *x*: (0, 0)
line *b*: (0, –3)

The slope of each line is –1, so the slope-intercept formulas for the lines are:

line *m*: $Y = -X + 2$
line *x*: $Y = -X + 0$ or $Y = -X$
line *b*: $Y = -X - 3$

Figure 14

Find the slope and intercept of line *d*.

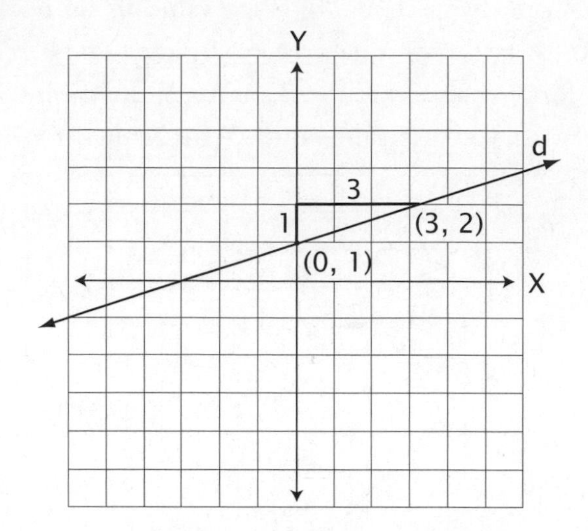

First choose two points that are on the grid or at the intersection of a Y-coordinate and an X-coordinate. In figure 14, I chose two points: (3, 2) and (0, 1).

I then chose to move away from the origin, starting at the point (0, 1) and moving towards (3, 2). In making a right triangle, I found the up dimension, or rise, to be positive one and the over dimension, or run, to be positive three. So the slope is $+1/+3$ (rise over run), or $1/3$. Replacing m with $1/3$, we get $Y = 1/3\,X$.

We can see that the intercept is (0, 1), so the slope-intercept formula representing line *d* is $Y = 1/3\,X + 1$.

Figure 15

Find the slope and intercept of line *e*.

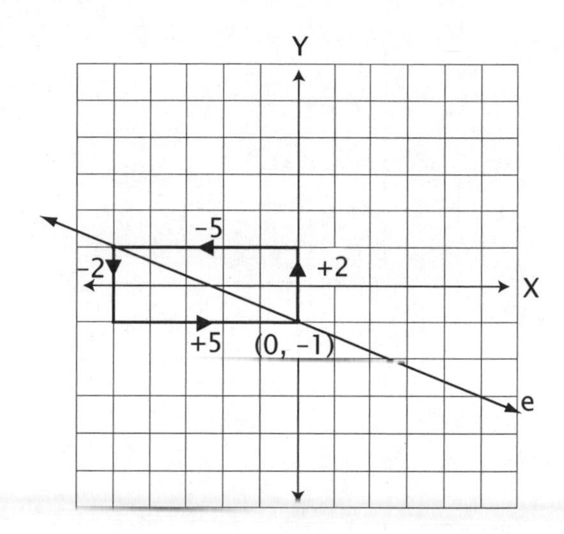

First choose two points that are on the grid or at the intersection of a Y-coordinate and an X-coordinate. In figure 15, I chose two points: (–5, 1) and (0, –1). I then chose to move up, starting at the point (0, –1) and moving towards (–5, 1). In making a right triangle, we find the up dimension, or rise, to be positive two and the over dimension, or run, to be negative five. So the slope is +2/–5 (rise over run) or –2/5. Replacing *m* with –2/5 yields Y = –2/5 X.

We can see that the intercept is (0, –1). So the slope-intercept formula representing line *e* is Y = –2/5 X – 1.

Notice that we could also have moved from (–5, 1) towards (0, –1) to find the slope. Then the rise would have been down two, or negative two, and the run would have been a positive five. In that scenario, the slope would have been –2 over +5, or –2/+5, which is equal to –2/5. Either way, the slope is the same.

Figure 16

Find the slope and intercept of line *k*.

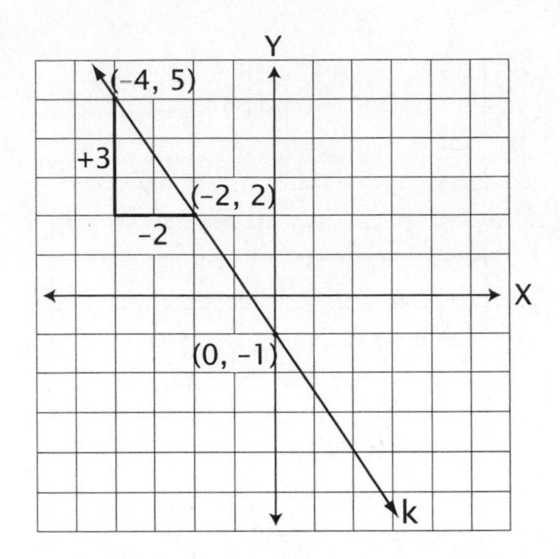

First choose two points that are on the grid or at the intersection of a Y-coordinate and an X-coordinate. In figure 16, I chose two points: (–2, 2) and (–4, 5).

I then chose to move up, starting at the point (–2, 2) and moving towards (–4, 5). In making a right triangle, I found the up dimension, or rise, to be positive three and the over dimension, or run, to be negative two. So the slope is +3/–2 (rise over run) or –3/2, and replacing m with –3/2 yields Y = –3/2 X.

We can see that the intercept is (0, –1). So the slope-intercept formula representing line *k* is Y = –3/2 X – 1.

Figure 17

Find the slope and intercept of line *v*.

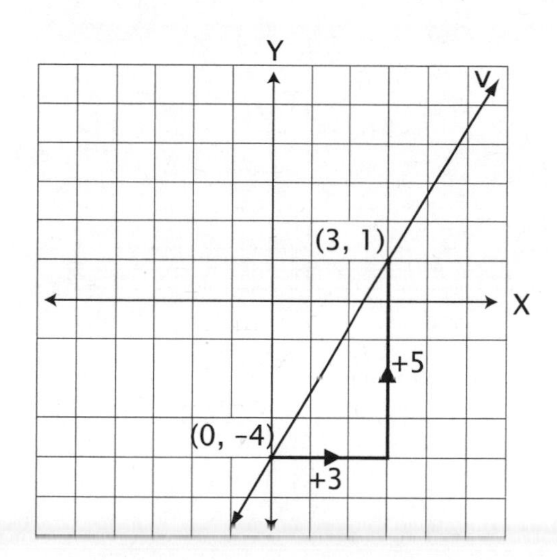

First choose two points that are on the grid or at the intersection of a Y-coordinate and an X-coordinate. In figure 17, I chose two points: (0, –4) and (3, 1).

I then chose to move away from the point (0, –4) and move towards (3, 1). In making a right triangle, I found the up dimension, or rise, to be positive five and the over dimension, or run, to be positive three. So the slope is +5/+3 (rise over run), or 5/3, and replacing m with 5/3 yields Y = 5/3 X.

We can see that the intercept is (0, –4). So the slope-intercept formula representing line v is Y = 5/3 X – 4.

Notice that we could also have moved from (3, 1) towards (0, –4) to find the slope. Then the rise would have been down five or negative five, and the run would have been to the left three or a negative three. In that scenario the slope would have been –5 over –3, or –5/–3, which is equal to 5/3. Either way the slope is the same.

$$\frac{+5}{+3} = \frac{-5}{-3} = \frac{5}{3}$$

LESSON 8

Graphing a Line
from the Slope-Intercept Formula

The inverse of finding the slope–intercept formula of a line is graphing a line when you are given the slope–intercept formula (Y = mX + b). In figure 1, we are given the formula Y = 3X + 2.

Figure 1

Graph Y = 3X + 2.

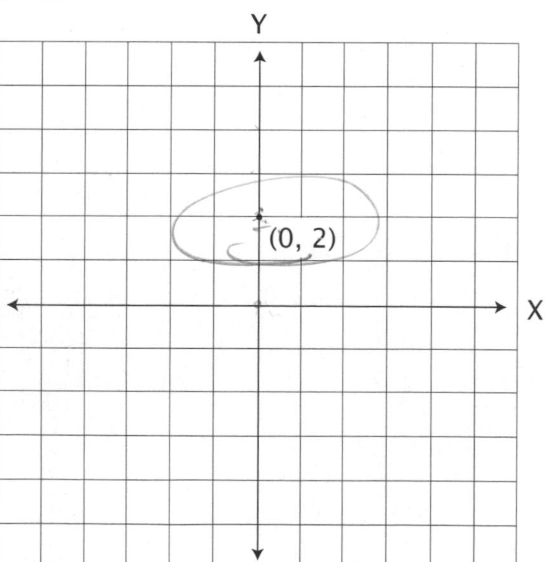

We'll begin by locating the intercept. Looking at the graph, it appears to be at the point (0, 2). We can solve for *b* by using algebra. Since the coordinate of X is zero at the intercept, we can make X = 0 in Y = 3X + 2.

Substituting for X: $Y = 3(0) + 2$

$Y = 0 + 2$

$Y = 2$

So if $X = 0$, then $Y = 2$, and the intercept is $(0, 2)$.

That's the point we located on the graph on the previous page.

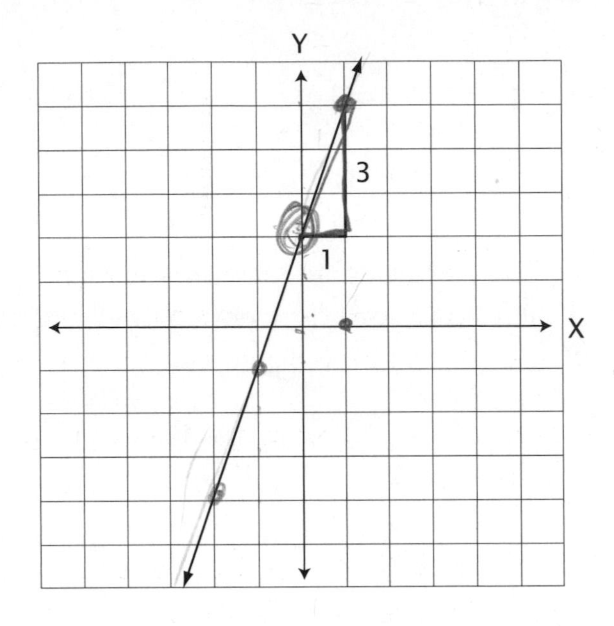

Now let's examine the slope. Since the slope is 3 or +3, then the slope is positive and leaning toward the right.

We know the slope is the coefficient of X. It is 3, which is the same as +3, or +3/+1, and has a rise of positive three and a run of positive one.

Beginning at the intercept (0, 2), we draw a triangle with a run (over) positive one and a rise (up) positive three. This brings us to the point (1, 5).

If we connect the two points, we have a line representing $Y = 3X + 2$.

Figure 2
Graph Y = -3X - 4.

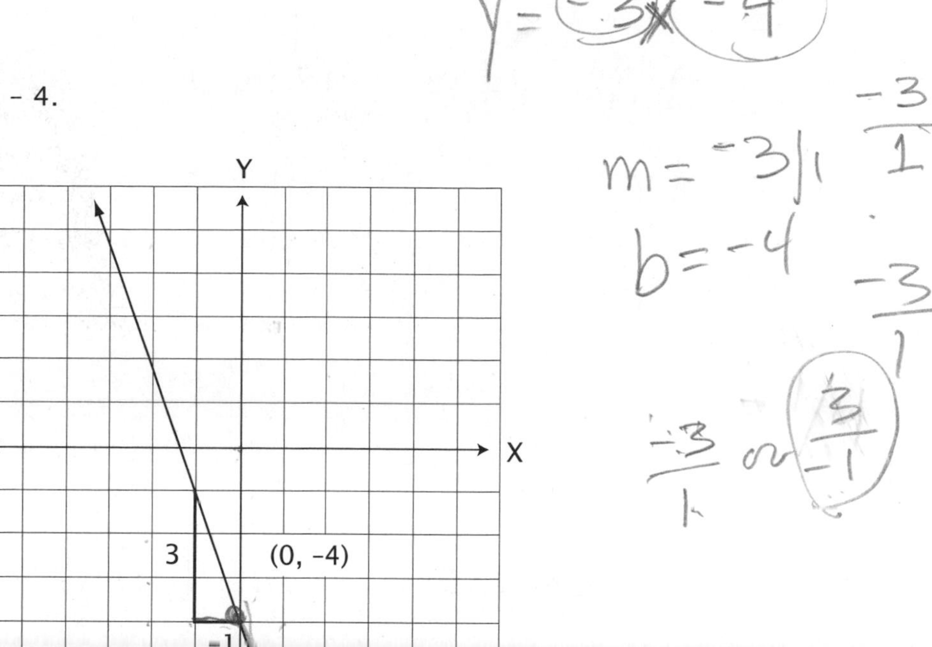

We begin by examining the slope. Since the slope is -3, then the slope is going to be negative and leaning toward the left.

To find the intercept, we make X equal to zero.

Substituting this value into the equation:
$$Y = -3(0) - 4$$
$$Y = 0 - 4$$
$$Y = -4$$

When $X = 0$, then $Y = -4$, and the intercept is $(0, -4)$.

Or, we could have used the information from the slope-intercept formula. $Y = mX + b$; where *m* is the slope, and *b* is the intercept. In $Y = -3X - 4$, the intercept, or *b*, is -4.

In $Y = -3X - 4$, the slope, or *m*, is the coefficient of X and is -3. This could be $-3/+1$ or $+3/-1$.

Beginning at $(0, -4)$, we chose to construct a triangle with a run of negative one and a rise of positive three. After drawing a line through the points $(0, -4)$ and $(-1, -1)$, we have a graphic representation of the line described as $Y = -3X - 4$.

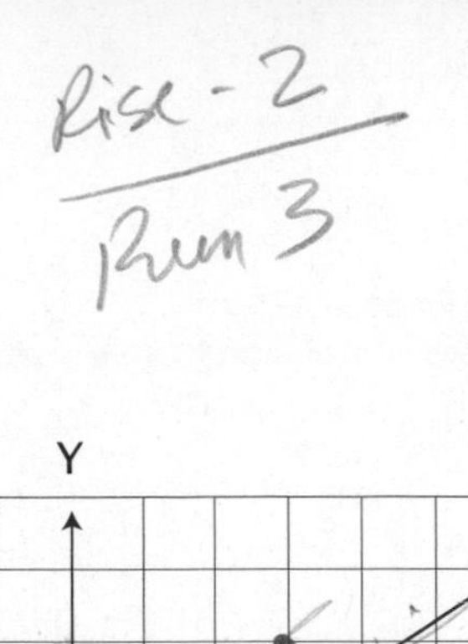

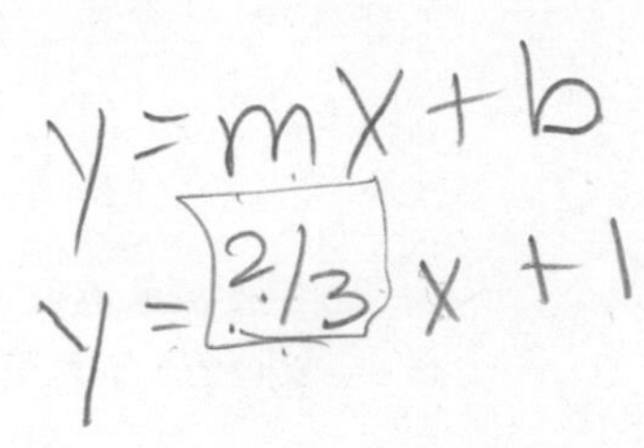

Figure 3

Graph Y = 2/3 X + 1.

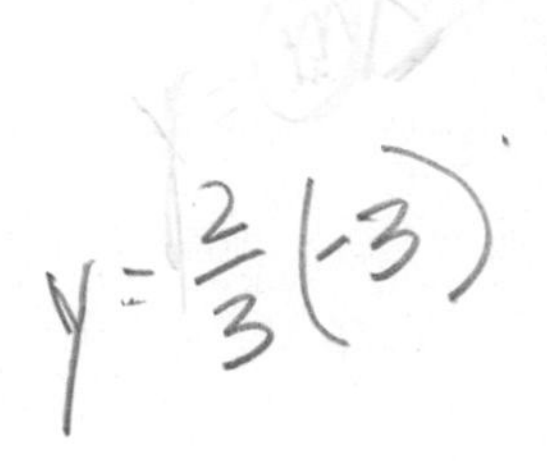

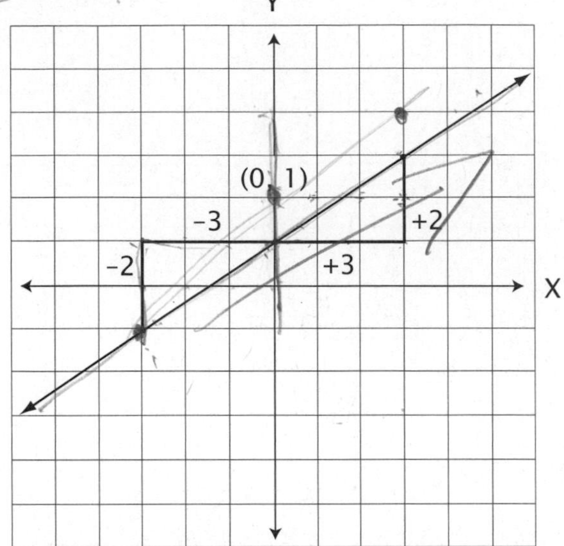

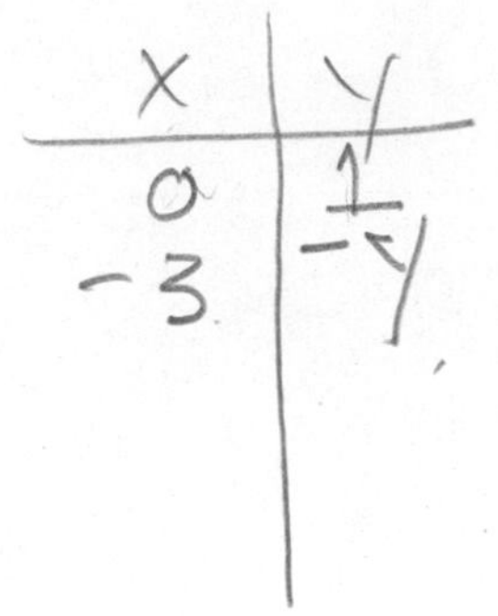

We begin by examining the slope. Since the slope is 2/3, or +2/3, then the slope is going to be positive and leaning toward the right.

Find the intercept by making X equal to zero, or by using the slope-intercept formula. The intercept is one, and we plot the point (0, 1).

We can see that the slope is 2/3, since it is the coefficient or what is multiplied times X. Beginning at (0, 1), we construct a triangle with a run of positive three and a rise of positive two.

Notice that we could also make a triangle with a run of negative three and a rise of negative two.

$$\frac{2}{3} = \frac{+2}{+3} = \frac{-2}{-3}$$

After connecting the points (0, 1) and (−3, −1), or the points (0, 1) and (3, 3), we have a graphic representation of the line described as Y = 2/3 X + 1.

Figure 4
Graph Y = 2X - 1.

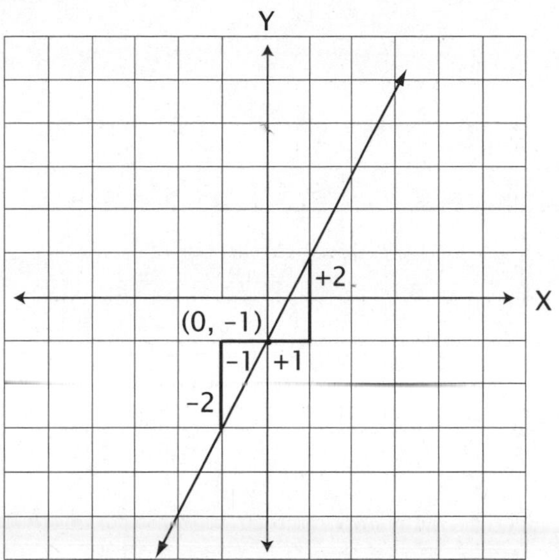

We begin by examining the slope. Since the slope is 2, or positive two, then the slope is going to be positive and leaning toward the right.

To find the intercept, we make X equal to zero. Substituting this value into the equation:

$$Y = 2(0) - 1$$
$$Y = 0 - 1$$
$$Y = -1$$

When X = 0, then Y = –1, and the intercept is (0, –1).

Or, we could have used the information from the slope-intercept formula. Y = mX + b, where *m* is the slope and *b* is the intercept. In Y = 2X – 1, the intercept, or *b*, is –1

In Y = 2X – 1, the slope, or *m*, is the coefficient of X, and is 2. This could be +2/+1 or –2/–1.

Beginning at (0, –1), we construct a triangle with a run of positive one and a rise of positive two.

Notice that we could also make a triangle with a run of negative one and a rise of negative two.

After drawing a line through the points (0, –1) and (+1, +1), or the points (0, –1) and (–1, –3), we have a graphic representation of the line described as Y = 2X – 1.

Figure 5
Graph Y = –4/5 X.

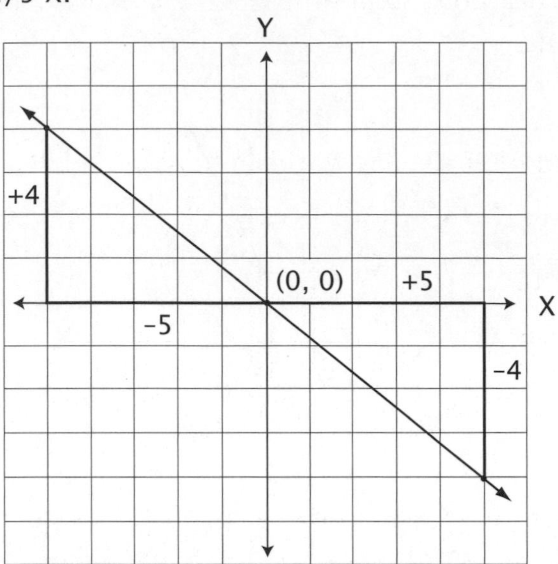

We begin by examining the slope. Since the slope is –4/5, the slope is negative and leaning toward the left.

Find the intercept by making X equal to zero or by using the slope-intercept formula. The intercept is zero, and we plot the point (0, 0). This line will go through (0, 0) or the origin.

We can see the slope is –4/5 since it is the coefficient of X. This could be +4/–5 or –4/+5.

Beginning at (0, 0), we construct a triangle with a run of positive five and a rise of negative four.

Notice that we could also make a triangle with a run of negative five and a rise of positive four.

After connecting the points (0, 0) and (–5, +4), or the points (0, 0) and (+5, –4), we have a graphic representation of the line described as Y = –4/5 X.

Horizontal and Vertical Lines

There are lines that don't seem to have any slope at all. They may be either horizontal or vertical.

Figure 6

Find the equation of the line for each horizontal line.

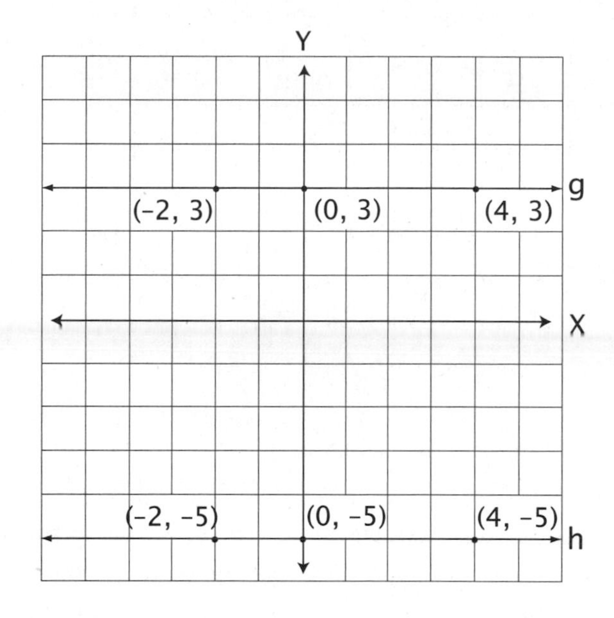

Notice line *g*. Observe that every point along that line has one coordinate that is the same.

All three points have a Y-coordinate of 3. In fact, every single point along this line has a Y-coordinate of 3.

The slope of any line is the rise over the run. In the case of line *g*, moving from the point (0, 3) to the point at (4, 3), the rise is zero and the run is positive four. The slope of line *g* is 0/4 or $0 \div 4$, which is 0.

If we use the slope-intercept formula, we have $Y = 0X + 3$. This simplifies to $Y = 3$, the same equation given above. So the equation of this line is $Y = 3$.

What is the equation of line *h*? If you said $Y = -5$, you are correct.

Figure 7
Find the equation of the line for each vertical line.

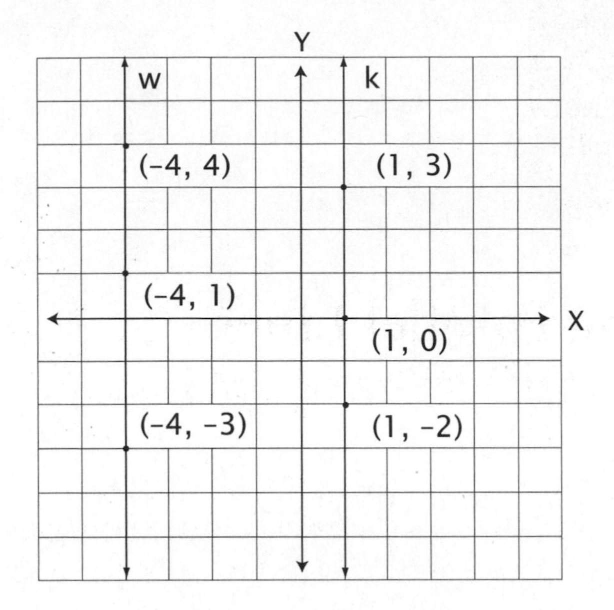

Now that you know horizontal lines, you could surmise that vertical lines will be X = some number, and you would be correct.

Figure 5 has two lines, *k* and *w*. Notice that the three points on each line have the same X-coordinate. In fact, every single point along these lines also has the same X-coordinate.

The equation of line *k* is X = 1.

The equation of line *w* is X = −4.

The slope of line *k* is rise over run, which is three over zero, or 3/0, between the points (1, 0) and (1, 3). However, you cannot divide a number by zero. We say that 3/0, or 3 ÷ 0, is *undefined*. There is no answer which when multiplied by zero will yield three.

To summarize: The slope of a horizontal line is always zero, and the slope of a vertical line is always undefined.

LESSON 9

Graphing Parallel Lines

and the Standard Equation of a Line

Parallel Lines

Two or more lines that have the same slope and different intercepts are *parallel.* Recall that parallel lines are defined as two lines in the same plane that never intersect or touch. We've talked about the fact that there are an infinite number of lines that have the same slope. The intercept distinguishes one from the other. In figure 1, notice that all the lines have a slope of 2/3, and thus all are parallel. Only the intercepts are different for each line.

Figure 1

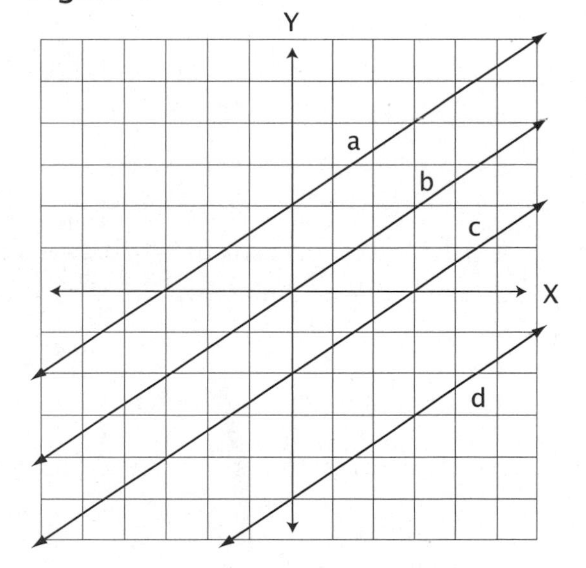

line *a*: $Y = 2/3\ X + 2$
line *b*: $Y = 2/3\ X + 0$,
or $Y = 2/3\ X$
line *c*: $Y = 2/3\ X - 2$
line *d*: $Y = 2/3\ X - 5$

Figure 2

line *e*: Y = –3/4 X + 2
line *f*: Y = –3/4 X – 1
line *g*: Y = –3/4 X – 4

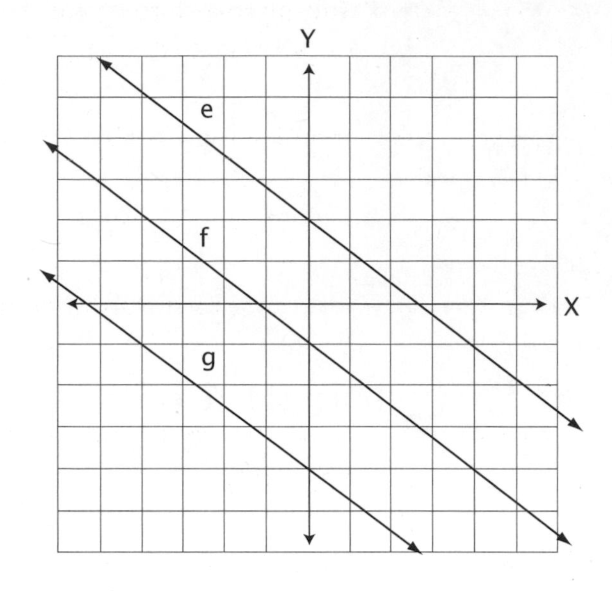

Equation of a Line

As I mentioned in the DVD, another way of describing a line besides the slope-intercept formula or form, is the *standard form of the equation of a line.* (Sometimes this is referred to simply as the "equation of a line" or "standard form.") Instead of Y = mX + b, it is defined as AX + BY = C. In the slope-intercept form, we want the coefficient of Y to be one (and thus not seen). In the standard form of the equation of a line, the coefficient of X is A (instead of *m*), the coefficient of Y is B (instead of one), and both variables are written on the left-hand side of the equation.

While the slope-intercept form may have fractional coefficients, we will leave our standard form with whole-number coefficients. Look at the next few examples, and notice the differences and similarities.

Example 1

Change the slope-intercept form to the standard form of the equation of a line.

Y = 4/5 X + 2	Slope-intercept form, where the coefficient of Y is 1, the coefficient of X is 4/5, and b = 2.
5Y = 4X + 10	Multiply both sides by 5.
–4X + 5Y = 10	Subtract 4X from both sides.
	A = –4, B = 5, C = 10

$4X - 5Y = -10$ Multiply by -1 to make the coefficient of X positive.
$A = 4, B = -5, C = -10$

In example 1, either of the last two lines is correct as long as the equation is in the form $AX + BY = C$.

Example 2

Change the slope-intercept form to the standard form of the equation of a line.

$Y = -2X - 1$ Slope-intercept form, where the coefficient of Y is 1, the coefficient of X is -2 and $b = -1$.

$2X + Y = -1$ Add 2X to both sides.
$A = 2, B = 1, C = -1$

Example 3

Change the slope-intercept form to the standard form of the equation of a line.

$Y = 5/3\ X$ Slope-intercept form: the coefficient of Y is 1, the coefficient of X is 5/3 and $b = 0$.

$3Y = 5X$ Multiply both sides by 3.

$-5X + 3Y = 0$ Subtract 5X from both sides.
$A = -5, B = 3, C = 0$

Example 4

Change the standard form of the equation of a line to the slope-intercept form.

$2X + 3Y = 6$ Standard equation of a line, where $A = 2$, $B = 3$, and $C = 6$.

$3Y = -2X + 6$ Subtract 2X from both sides.

$Y = -2/3\ X + 2$ Divide both sides by 3. In the slope-intercept form, the coefficient of Y is 1, the coefficient of X is -2/3, and the intercept, or *b*, is 2.

Example 5

Change the standard form of the equation of a line to the slope-intercept form.

$-X - 2Y = 4$	standard equation of a line In the standard form, $A = -1$, $B = -2$, and $C = 4$. $AX + BY = C$ $-1 \quad -2 \quad 4$
$-2Y = X + 4$	Add X to both sides.
$Y = -1/2\ X - 2$	Divide both sides by −2. In the slope-intercept form, the coefficient of Y is 1, the coefficient of X is −1/2, and the intercept, or *b*, is −2.

Example 6

Change the standard form of the equation of a line to the slope-intercept form.

$-3X + 4Y = -8$	standard equation of a line In the standard form, $A = -3$, $B = 4$, and $C = -8$. $AX + BY = C$ $-3 \quad 4 \quad -8$
$4Y = 3X - 8$	Add 3X to both sides.
$Y = 3/4\ X - 2$	Divide both sides by 4. In the slope-intercept form, the coefficient of Y is 1, the coefficient of X is 3/4, and the intercept, or *b*, is −2.

Writing an Equation

Now let's put together all we've learned about plotting points, finding the slope, and describing the line with the slope-intercept form ($Y = mX + b$). Then we can describe a line using the standard form of an equation of a line ($AX + BY = C$).

Example 7

Plot the points (-1, 4) and (1, 0), and then find the slope-intercept formula and the standard form of an equation of a line.

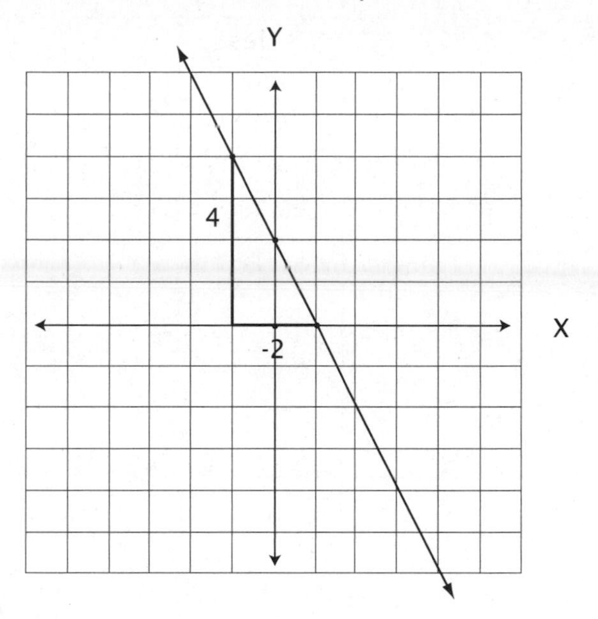

Evaluating the slope of the line, we can see it will be negative since it is leaning backward.

Then we make a right triangle to find the slope. The rise (up distance) is four, and the run is negative two, since it is to the left.

$$m = \frac{\text{rise}}{\text{run}} = \frac{4}{-2} = -2$$

Connecting the points reveals an intercept of positive two, so $b = 2$.

The slope–intercept form is $Y = -2X + 2$.

The standard form is found by adding $+2X$ to both sides, giving us $2X + Y = 2$.

Example 8

Plot the points (-3, -3) and (3, 1), and then find the slope-intercept formula and the standard form of an equation of a line.

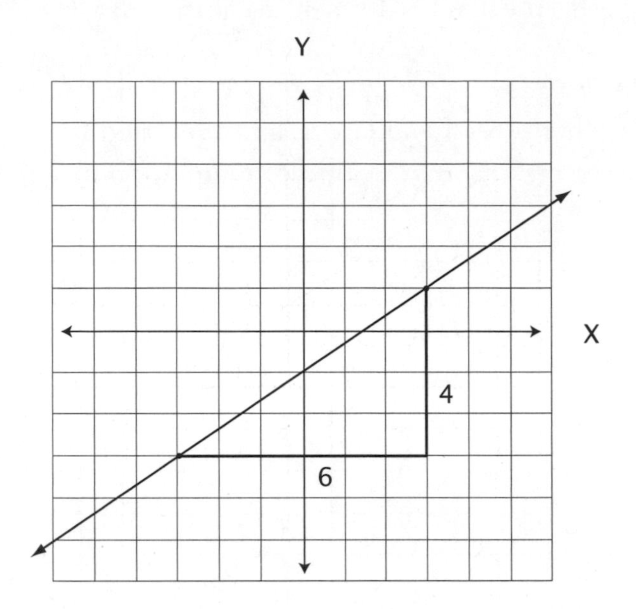

Evaluating the slope of the line, we can see it will be positive since it is falling forward.

Then we make a right triangle to find the slope. The rise (up distance) is four, and the run is six, so the slope $m = \frac{\text{rise}}{\text{run}} = \frac{4}{6} = \frac{2}{3}$.

Connecting the points reveals an intercept of negative one, so $b = -1$.

The slope–intercept form is $Y = 2/3\,X - 1$.

The standard form is found by subtracting $2/3\,X$ from both sides and multiplying all the elements by three: $3(-2/3\,X + Y = -1)$.

The standard form can be written as $-2X + 3Y = -3$ or $2X - 3Y = 3$.

Example 9

Plot the points (-6, 4) and (3, -2), and then find the slope-intercept formula and the standard form of an equation of a line.

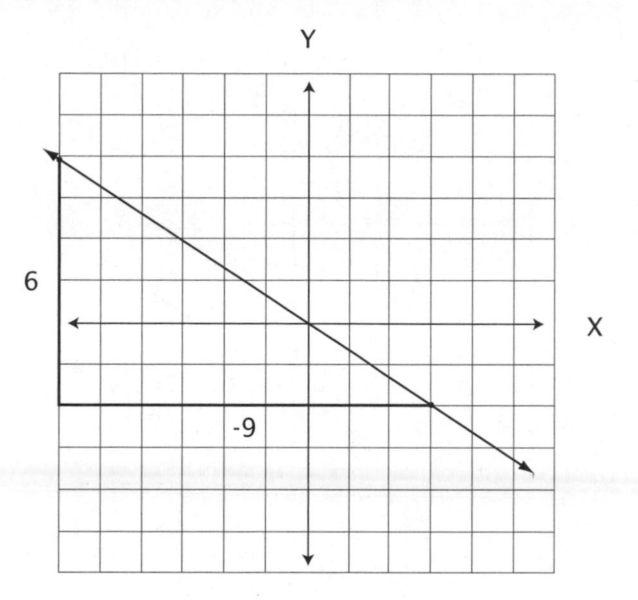

Evaluating the slope of the linc, we can see it will be negative since it is falling backward.

Then we make a right triangle to find the slope. The rise is six, and the run is negative nine (to the left), so the slope $m = \frac{\text{rise}}{\text{run}} = \frac{6}{-9} = \frac{2}{-3}$.

Connecting the points reveals an intercept of zero, so b = 0.

The slope-intercept form is Y = −2/3 X + 0 or Y = −2/3 X.

The standard form is found by adding 2/3 X to both sides, and multiplying all the elements by three: 3(2/3 X + Y = 0).

The standard form of the equation is 2X + 3Y = 0.

Example 10

Plot the points (0, 5) and (-2, -3), and then find the slope-intercept formula and the standard form of the equation of a line.

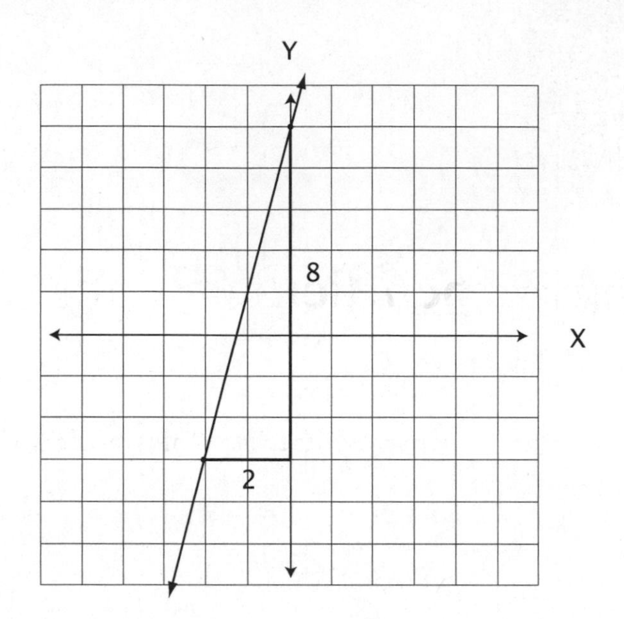

Evaluating the slope of the line, we can see it will be positive since it is falling forward.

Then we make a right triangle to find the slope. The rise (up distance) is eight, and the run is two, so the slope $m = \frac{\text{rise}}{\text{run}} = \frac{8}{2} = \frac{4}{1} = 4$.

Connecting the points reveals an intercept of five, so $b = 5$.

The slope-intercept form is $Y = 4X + 5$.

The standard form is found by subtracting $4X$ from both sides.

The standard form of the equation is $-4X + Y = 5$.

LESSON 10

Graphing Perpendicular Lines

In the previous lesson, we saw that the lines like *a*, *b*, and *c* in figure 1 are parallel and have the same slope.

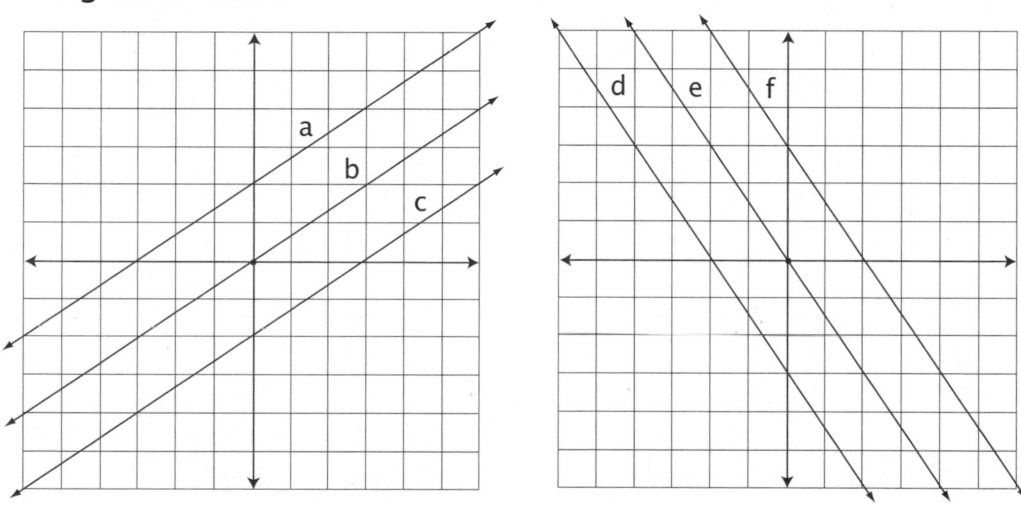

In figure 2, all the lines are parallel as well. Are the slopes in figure 2 the same? Yes, the slopes are identical, but they are expressed in different ways. What is the slope of the lines *d*, *e*, and *f* in figure 2? Do you see a relationship between the slope in figure 1 and the slope in figure 2?

Perpendicular Lines

Lines that intersect each other in the same plane and form right angles are called *perpendicular* lines.

Figure 3

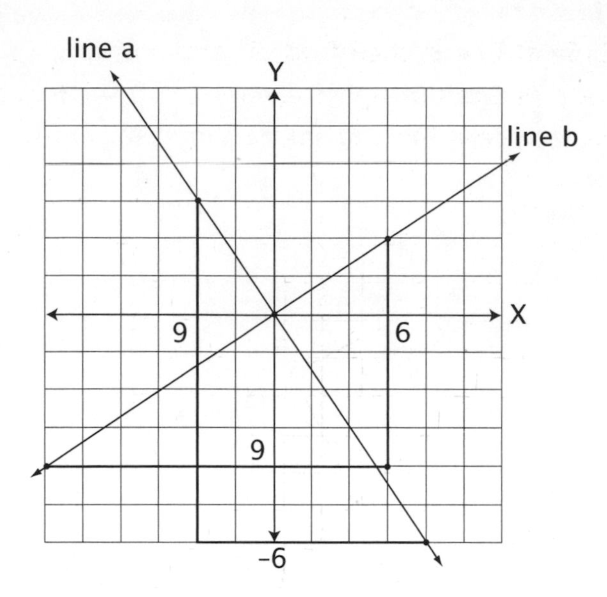

Slopes of Perpendicular Lines

In figure 3, notice the slopes of lines a and b. The slope of line a is $-\frac{9}{6} = -\frac{3}{2}$. The slope of line b is $\frac{6}{9} = \frac{2}{3}$. Also observe that line a is perpendicular to line b, or line $a \perp$ line b.

Now look at the slopes and see whether there is a relationship between $\frac{2}{3}$ and $-\frac{3}{2}$. Did you figure out that $-\frac{3}{2}$ is the negative reciprocal of $\frac{2}{3}$? This is true for all perpendicular lines.

Example 1

Find the slope of line *c*, then find the slope of a line perpendicular to *c*. Now draw this line through the point (1, 2) and label it line *d*. What is the slope-intercept form of the equation for line *d*?

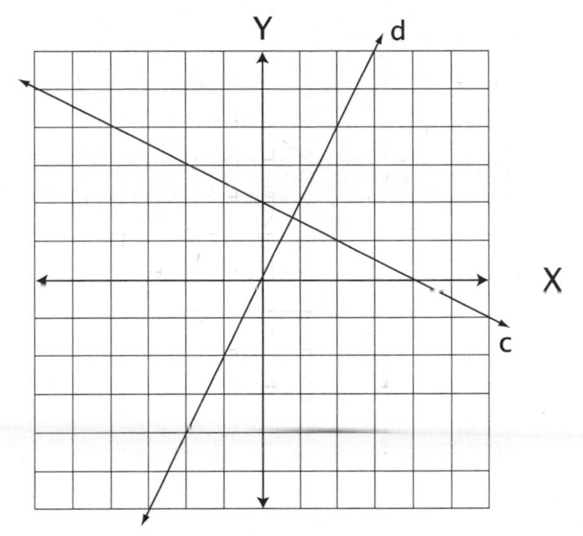

First we find the slope of line *c*, which is $-\frac{1}{2}$.

The slope of a line perpendicular to line *c* is $-\left(-\frac{2}{1}\right)$ or 2.

Two, or $\frac{2}{1}$, is the negative reciprocal of $-\frac{1}{2}$.

From the graph, we see the Y-intercept of line *d* is zero, so the slope-intercept form for line *d* is $Y = 2X + 0$.

Example 2

Find the slope of line *f*. Then find the slope of a line perpendicular to *f* through the point (0, –2) and label it line *g*. What are the slope-intercept form and the standard form of the equation for line *g*?

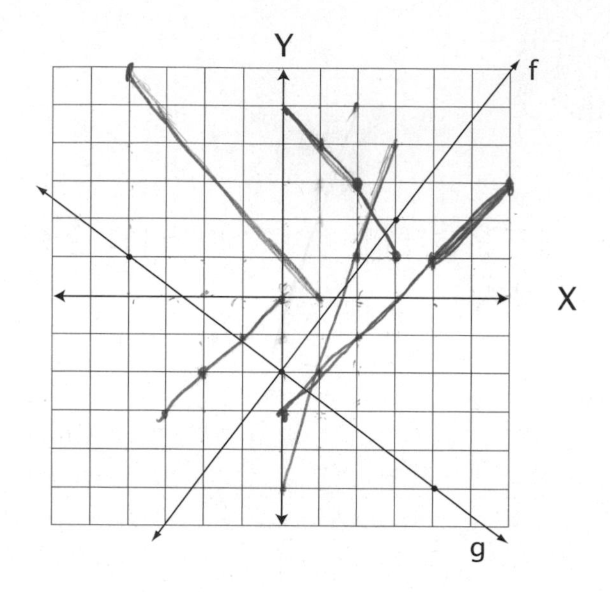

First we find the slope of line *f*, which is $\frac{4}{3}$.

The slope of a line perpendicular to line *f* is $-\left(\frac{3}{4}\right)$.

$-\frac{3}{4}$ is the negative reciprocal of $\frac{4}{3}$.

From the graph, we see the Y-intercept of line *g* is negative two, so the slope-intercept form for line *g* is $Y = -3/4\,X - 2$.

Multiply each term by four, and then rearrange the terms to find the standard form of the equation of the line, which is $3X + 4Y = -8$.

LESSON 11

Finding the Slope-Intercept Formula
with Different Givens

Slope and One Point

The slope-intercept formula, or form, is very helpful in drawing a line on a Cartesian coordinate graph. Find the intercept and draw a point. Then move from that point to another point by counting over and up according to the slope. Once you have two points, you can connect them and draw your line.

In this lesson, we are given slightly different information, but we still need to find the slope and the intercept in order to draw an accurate line and have an exact equation.

If you were given the slope and another point on the line, you would need to find the intercept. By graphing the information given, you can estimate the intercept, which is helpful, but you still need to work further to arrive at an exact solution.

Example 1

Given the slope and one point, find the slope-intercept form of the equation of the line.

Given: Slope = 3, through the point (1, 2) on the same line.

Drawing this:

Y

X

We start at the point (1, 2), and move over one and up three (slope = 3/1 or 3). Connecting these points, the line appears to intercept the Y-axis at point (0, –1). We have a pretty good idea that our intercept is negative one.

To find out for sure, we'll substitute what we know from the givens into the slope-intercept form.

$Y = mX + b$

$Y = (3)X + b$ Substituting 3 for the slope *m*.

Now we'll substitute the point (1, 2), where X = 1 and Y = 2, into the equation. Make sure you understand this, as it is the new and essential part of solving this equation to find *b*.

$(2) = 3(1) + b$

$2 = 3 + b$

$-1 = b$ So our intercept is –1.

Substituting for *m* and *b*, the slope-intercept form is $Y = 3X - 1$.

Two Points – A step removed from this is if you are given only two points, and you have to find the intercept and slope. See example 2 below.

Example 2

Given two points, find the slope-intercept form of the equation of the line.

Given: point 1 is (1, 1) and point 2 is (-1, 5).

Plot the points, and estimate the intercept and the slope.

To find the slope, draw a right triangle from point 1 to point 2. You can see that the over dimension of the rectangle is two to the left or negative two. The up dimension is four. The slope is 4/–2 or –2. Now we are at the same place as in example 1. We have the slope, and then we choose one of the given points to find the intercept. For example 2, I chose point 1 (1, 1).

$Y = -2X + b$ The slope *m* is -2.

$(1) = -2(1) + b$ Substituting (1, 1).

$1 = -2 + b$

$3 = b$ So our intercept is 3. The equation is $Y = -2X + 3$.

Computing the Slope

Another way to find the slope, without plotting the points and drawing the triangle, is by finding the differences between the two X-coordinates and the two Y-coordinates. In example 3, we drew lines from the points to the X- and Y-axes to illustrate these differences. Notice the length of the sides of the triangle.

Example 3

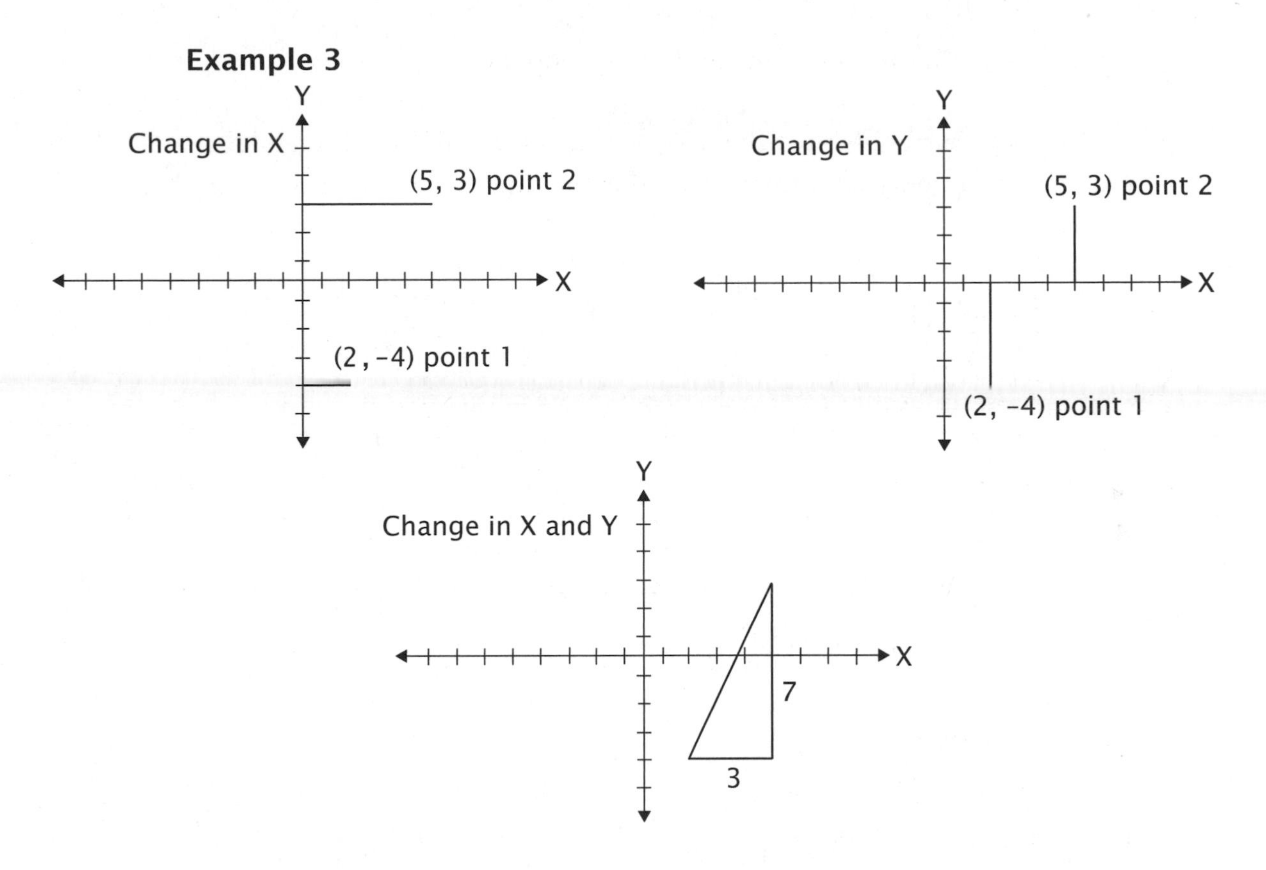

The over dimension of the right triangle is three. We could also have figured this by subtracting: five (the X-coordinate of point 2) minus two (the X-coordinate of point 1) is three, because $5 - 2 = 3$.

The up dimension is seven. We could also find this by subtracting: three (Y-coordinate of point 2) minus a negative four (the Y-coordinate of point 1) is seven, because $3 - (-4) = 7$. To construct a formula for determining slope using this principle, we can begin by labeling points 1 and 2 using subscripts. *Subscripts* are little numbers below the line that help us to identify points but do not affect the value of the number as exponents do. Therefore point 1 is written as (X_1, Y_1) and point 2 as (X_2, Y_2).

Our formula for calculating slope is on the next page.

To find the change, you subtract the coordinates. (Δ means "the change in.") So the change in the Y coordinates would be found by subtracting: Y_2 minus Y_1. Using the new symbol, we would write the "change in Y" as $\Delta Y = Y_2 - Y_1$.

$$m = \frac{\text{rise}}{\text{run}} = \frac{Y}{X} = \frac{Y_2 - Y_1}{X_2 - X_1} = \frac{(5)-(-4)}{(5)-(2)} = \frac{7}{3}$$

↗ ↗ ↖ ↖

pt. 2 pt.1 pt. 2 pt.1

So the slope of the line in example 3 is 7/3.

Go back to example 2, and use this method of subtracting the points to figure the slope. It doesn't make any difference which point is labeled point 1 or point 2, as long as you are consistent in applying the subscripts and plugging the values into the formula. Be careful that you don't mix up the subscripts like this:

$$\frac{Y_1 - Y_2}{X_2 - X_1}$$

Example 2 (solved with the formula)

Given two points, find the slope-intercept form of the equation of a line.

Given: point 1 is (1, 1) and point 2 is (-1, 5).

Plot the points, and then estimate the intercept and the slope.

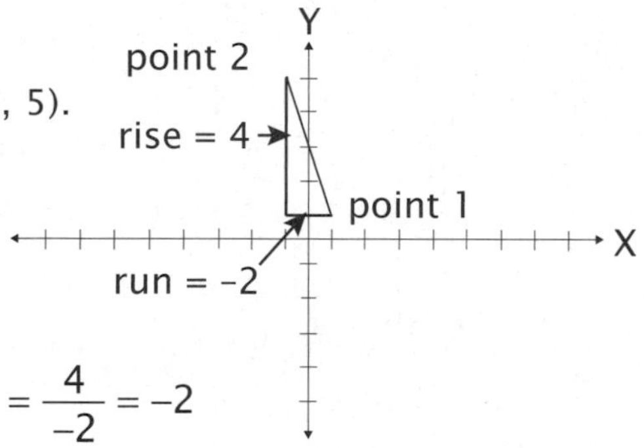

point 2

rise = 4

point 1

run = -2

Y

X

$$m = \frac{\text{rise}}{\text{run}} = \frac{Y}{X} = \frac{Y_2 - Y_1}{X_2 - X_1} = \frac{(5)-(1)}{(-1)-(1)} = \frac{4}{-2} = -2$$

↗ ↗ ↖ ↖

pt. 2 pt.1 pt. 2 pt.1

So the slope of the line in example 2 is 4/−2 or −2.

LESSON 12

Graphing Inequalities

Kinds of Inequalities

Up to this point, whenever we graphed a line, the equations expressed equalities, such as $Y = 2X - 3$. But there are also other situations, known as *inequalities*, where Y may be greater than (>), greater than or equal to (≥), less than (<), or less than or equal to (≤) another term. Based on the line $Y = 2X - 3$, here are the possible inequalities: $Y > 2X - 3$, $Y \geq 2X - 3$, $Y < 2X - 3$, $Y \leq 2X - 3$.

Let's consider the line $Y = 2X - 3$. In terms of points satisfying the equation, there are three areas to consider: points to the left of the line, points on the line, and points to the right of the line. We'll examine three sample points, one in each of these areas, and see whether they agree or disagree with our equation. Let's choose the points (0, 0), (2, 1), and (4, 0). After plotting them and seeing that each of the three areas is represented, we'll substitute each of the points in the equation.

Figure 1

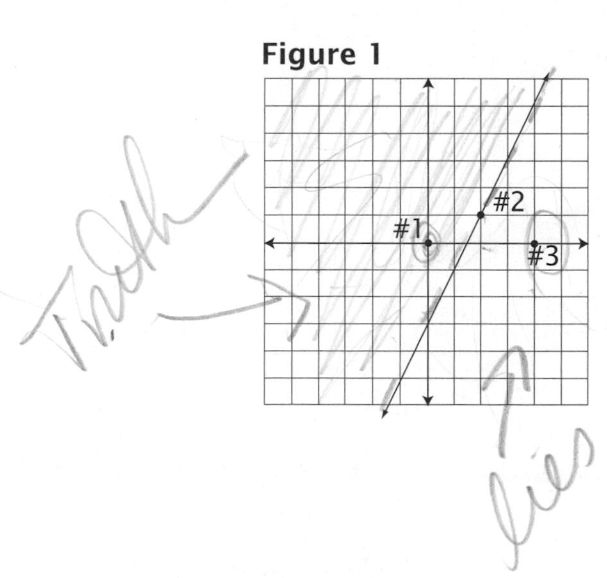

#1 $Y > 2X - 3$
$(0) > 2(0) - 3$
$0 > -3$
This is true!

#2 $Y > 2X - 3$
$(1) > 2(2) - 3$
$1 > 1$
This is not true!

#3 $Y > 2X - 3$
$(0) > 2(4) - 3$
$0 > 5$
This is not true!

Since point #1 (on previous page) provides a true solution, then all the points in the shaded area also satisfy the equation and give a true solution. So we shade this area. Point #3 does not satisfy the equation, so we leave the area to the right of the line alone. Point #2, representing the line itself, also does not yield a true solution, so we draw the line as a dotted line. The answer is the shaded area.

Figure 2

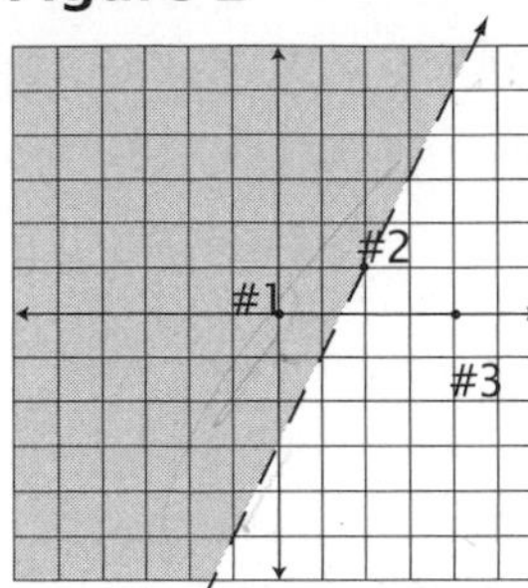

Figure 3

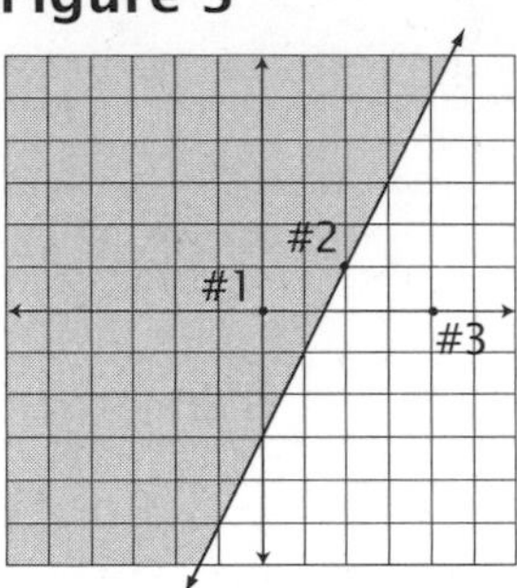

Figure 3 is the graph of $Y \geq 2X - 3$. This is exactly like the previous graph, except it includes the line. This is a combination of $Y > 2X - 3$ (the previous graph) and $Y = 2X - 3$ (which is the line itself).

Figure 4 is the graph of $Y < 2X - 3$. This is the opposite of figure 2, and since it is less than, not less than and equal to, the line is a dotted line. Use the origin (0, 0) as the test point: $(0) < 2(0) - 3$ or $0 < -3$, which is not true. So the shading is to the right of the line.

Figure 4

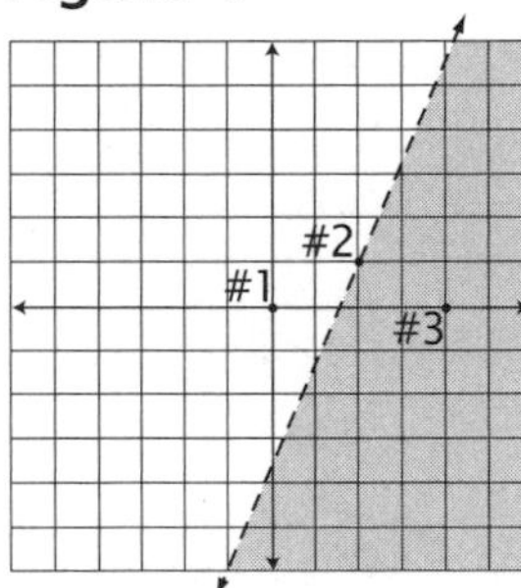

Figure 5

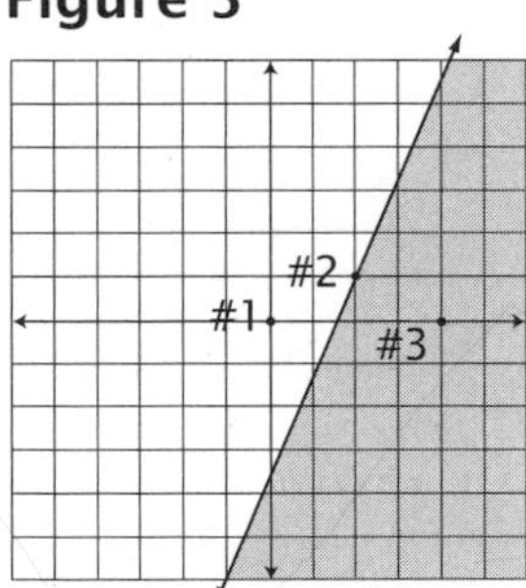

Figure 5 is the graph of $Y \leq 2X - 3$, which is exactly like the previous graph except that it includes the line. This is a combination of $Y < 2X - 3$ (the previous graph) and $Y = 2X - 3$ (which is the line itself).

Inequalities and Negative Numbers

When you have an inequality like $-2Y \geq 3X + 6$ that has a negative Y that you want to be positive in order to be in the slope-intercept form, multiplying or dividing by a negative number changes the inequality sign. When you have an inequality, you can multiply or divide by a positive number without affecting the equation. Adding or subtracting something from both sides does not change the sign, but multiplying or dividing by a negative number is a different situation.

Notice the following equations with real numbers to see how this works.

Example 1

$8 = 8$ is true.

When both sides are multiplied by positive two, the result is $16 = 16$, which is also true.

When both sides are multiplied by negative two, the result is $-16 = -16$, which is still true.

Example 2

$5 > -3$ is true.

When both sides are multiplied by positive two, the result is $10 > -6$, which is also true.

When both sides are multiplied by negative two, the result is $-10 > 6$, which is not true.

To make it true, change the inequality sign, and then $-10 < 6$, which is now true.

Example 3

$1 < 2$ is true.

When both sides are multiplied by positive six, the result is $6 < 12$, which is also true.

When both sides are multiplied by negative six, the result is $-6 < -12$, which is not true.

To make it true, change the inequality sign, and then $-6 > -12$, which is now true.

So if you have $-2Y \geq 3X + 6$, divide both sides by a negative two (or multiply both sides by a negative one-half), and the result is $Y \leq -3/2\,X - 3$. See figure 6 on the next page for a graph of this. The equals sign is not affected by multiplying by a negative number, as we saw in example 1.

Figure 6

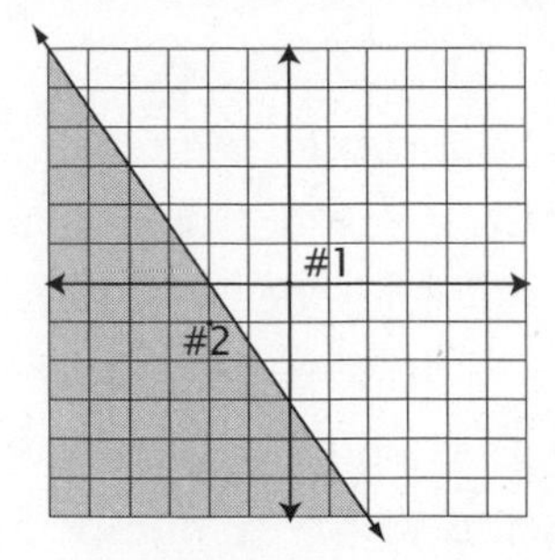

Let's consider the graph for $-2Y \geq 3X + 6$. In order to put the equation in the slope-intercept form, we must divide both sides by a negative two (or multiply both sides by a negative one-half). Since we are dividing or multiplying by a negative number, we must reverse the sign of the inequality.

$$-2Y \geq 3X + 6$$
$$Y \leq -3/2\ X - 3$$

Draw a line with a slope of $-3/2$ and a Y-intercept of -3. The equation combines an inequality and an equality, so we will use a solid line.

Now examine two points, one on either side of the line, to see whether they agree or disagree with the inequality. The points chosen here are (0, 0) and (−2, −1). You may check your points in either inequality, but be sure that you have the inequality sign pointing the correct way! With the original inequality:

#1	#2
$-2Y \geq 3X + 6$	$-2Y \geq 3X + 6$
$-2(0) \geq 3(0) + 6$	$-2(-1) \geq 3(-2) + 6$
$0 \geq +6$	$2 \geq -6 + 6$
	$2 \geq 0$
This is not true!	This is true!

Since #2 is true, we will shade the area under the line. Here are the same points checked with the simplified inequality:

#1	#2
$Y \leq -\frac{3}{2}X - 3$	$Y \leq -\frac{3}{2}X - 3$
$(0) \leq -\frac{3}{2}(0) - 3$	$(-1) \leq -\frac{3}{2}(-2) - 3$
$0 \leq -3$	$-1 \leq 3 - 3$
	$-1 \leq 0$
This is not true!	This is true!

We get the same results as with the original inequality.

LESSON 13

Solving Simultaneous Equations
By Graphing

Graphing Equations

Every line is made up of an infinite number of points. That's why arrows are drawn on the ends of a line to show that a line proceeds in both directions. Each point on that line satisfies or fits into the particular equation of that specific line.

In the line drawn in figure 1, $1X + Y = 3$ or $Y = -1X + 3$, we know -1 is the slope and 3 is the intercept. Each point on this line (I've picked three, marked A, B, and C) "works" when placed in the equation.

The same is true for figure 2 and the points marked D, E, and F for the equation $Y = 2X - 3$; each point "works" when placed in the equation.

Figure 1

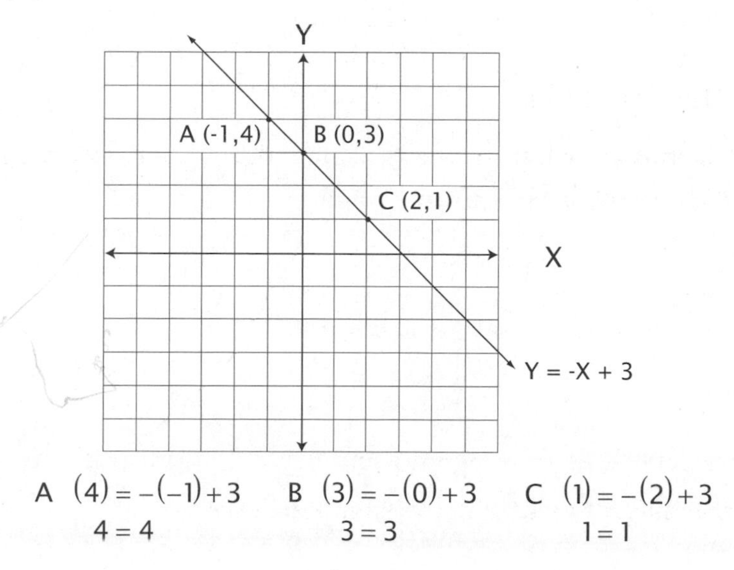

Figure 3

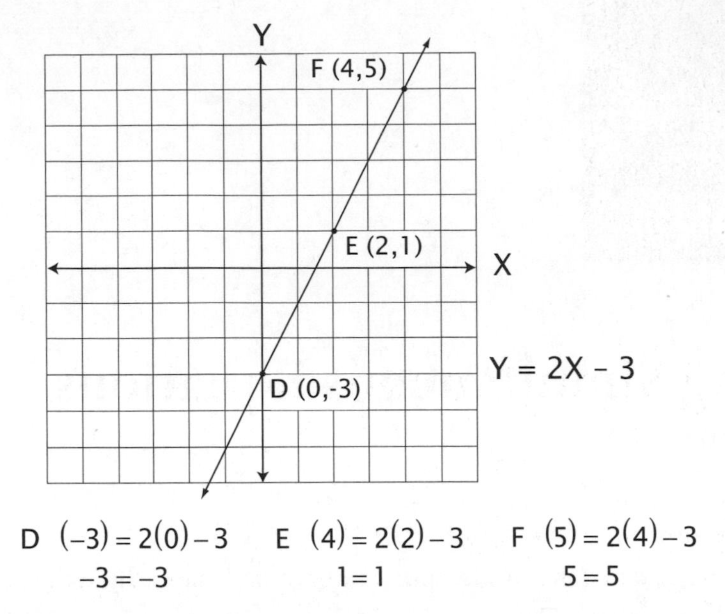

Graphing Simultaneous Equations

When two lines intersect, there is only one point that is present in both lines. This is the one point that "works" in both equations. An infinite number of points work in each line, but only one point works for both. This is the point of intersection.

Figure 3

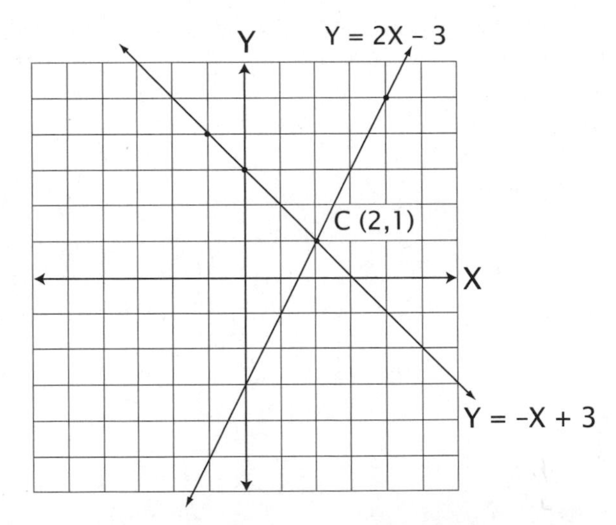

Putting the two lines from figure 1 and figure 2 together, we notice that the point of intersection is (2, 1). This is point C from figure 1 and point E from figure 2.

LESSON 14

Solving Simultaneous Equations
By Substitution

Finding Solutions Algebraically

In lesson 13, we graphed the two lines on one graph in order to find the solution—the one point that satisfied both equations. This is a geometric solution to the problem. In this lesson, we're searching for the same point, but we'll solve it algebraically.

Substitution

The first method used to solve these equations simultaneously is called substitution. Our strategy for solving by this method is to find a way to get an equation with only one variable. After solving for the value of the variable, substitute this in either of the original equations to find the value of the other variable. Look at example 1 below.

Example 1

Line 1: $Y - X = 0$ Line 2: $Y - 3X = -4$

$$Y = 3X - 4$$

Since $Y = 3X - 4$ in line 2, we can substitute $(3X - 4)$ for Y in line 1.

$$(3X - 4) - X = 0$$
$$3X - 4 - X = 0$$
$$2X - 4 = 0$$
$$2X = 4$$
$$X = 2$$

Now we know the value of X, and we can substitute it in either equation to find that Y = 2.

$$\begin{array}{ll} Y - X = 0 & \text{or } Y - (2) = -4 \\ Y - (2) = 0 & \quad Y - 6 = -4 \\ Y = 2 & \quad Y = 2 \end{array}$$

The answer, or solution, is (2, 2).

Example 2

Line 1: $Y = X - 1$ Line 2: $Y = 3X + 1$

Since $Y = X - 1$ in line 1, we can substitute $(X - 1)$ for Y in line 2.

$$\begin{aligned} (X - 1) &= 3X + 1 \\ X - 1 &= 3X + 1 \\ -2 &= 2X \\ -1 &= X \end{aligned}$$

Now we know the value of X, and we can substitute it in either equation to find that:

$$\begin{array}{ll} Y = X - 1 & \text{or } Y = 3X + 1 \\ Y = (-1) - 1 & \quad Y = 3(-1) + 1 \\ Y = -2 & \quad Y = -3 + 1 \\ & \quad Y = -2 \end{array}$$

The answer, or solution, is (-1, -2).

LESSON 15

Solving Simultaneous Equations

By Elimination

Adding Equations

Another method for solving simultaneous equations algebraically is called *elimination*. This method produces the same answer as if we had used substitution, but there are certain types of problems where elimination is more efficient.

Notice this simple equation:	$3 = 3$
Another similar equation:	$4 = 4$
If we add these two equations,	$3 + 4 = 3 + 4$
we find another "equal-ation."	$7 = 7$

So if two equations are equal, then we can add them to produce another "equal-ation" or equation.

Example 1

equation 1	$X + Y = 3$
equation 2	$\underline{2X - Y = 3}$
adding	$3X \quad = 6$
dividing by 3	$X \quad = 2$

Notice the Ys were eliminated, leaving only one variable: X. Since $X = 2$, we can substitute 2 for X in equation 1 (or equation 2 if you prefer) and solve for Y.

equation 1	$X + Y = 3$
substituting 2 for X	$(2) + Y = 3$
subtracting 2	$Y = 1$

Elimination

Solving these two equations at the same time, or "simultaneously," produces the point (2, 1) as the solution. Both equations have two variables. In order to solve an equation, we have to eliminate one of the variables so we can solve for the remaining one. Then we use this answer to find the value of the first variable, which we had previously eliminated.

Example 2

equation 1	$Y - X = 0$
equation 2	$Y - 3X = 4$
adding produces	$2Y - 4X = 4$

But there are still two variables, as neither of the variables was eliminated.

We need to multiply one of the equations by some number which will allow one of the variables to be eliminated. Multiply equation 1 by negative one. Then when we add the two equations, the Ys will be eliminated.

equation 1	$-Y + X = 0$
equation 2	$Y - 3X = 4$
adding now produces	$-2X = 4$
dividing by -2	$X = -2$

We now know that $X = -2$, but what is Y? Place this value for X in either equation to find the value of Y. Let's substitute $X = -2$ in both equations.

$-Y + X = 0$	$Y - 3X = 4$
$-Y + (-2) = 0$	$Y - 3(-2) = 4$
$-Y = 2$	$Y + 6 = 4$
$Y = -2$	$Y = -2$

Example 2 can also be solved by eliminating the X. See example 3.

Example 3

equation 1 (multiplied by −3)	$-3(Y - X = 0)$
equation 2	$Y - 3X = 4$
new version of equation 1	$-3Y + 3X = 0$
adding now produces	$-2Y = 4$
dividing by −2	$Y = -2$

Substituting Y = −2 in either equation yields X = −2, so the solution is (−2, −2), the same as in example 2.

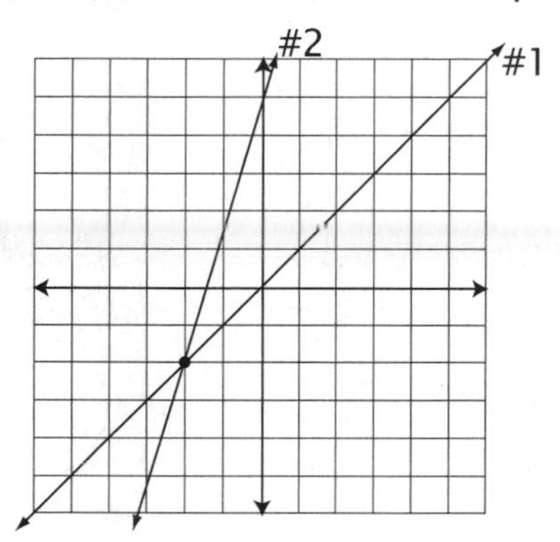

Example 4

equation 1	equation 2
$Y = -X - 3$ or $Y + X = -3$	$Y = 2X + 3$ or $Y - 2X = 3$
multiplied by 2: $2Y + 2X = -6$	

add new equatons 1 and 2

$$\begin{aligned} 2Y + 2X &= -6 \\ Y - 2X &= 3 \\ \hline 3Y &= -3 \\ Y &= -1 \end{aligned}$$

Substituting Y = −1 in either equation produces X = −2. The solution is (−2, −1).

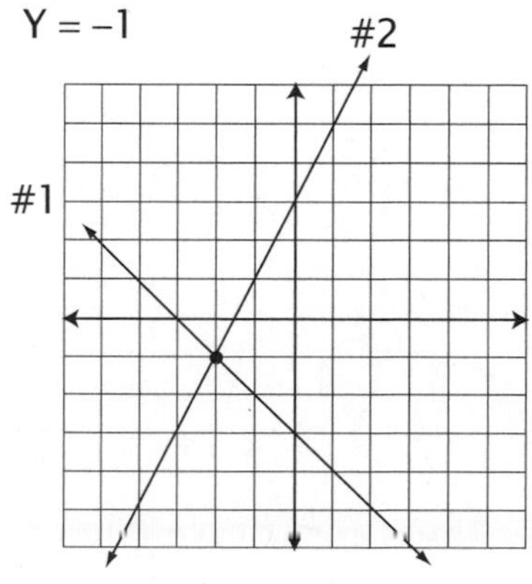

LESSON 16

Coin Problems

We can use what we've learned about solving simultaneous equations and apply it to some interesting coin problems. Did you ever wonder how to find out how many of each kind of coin there are in someone's pocket, given the amount of money and the number of coins? Here is how you do it.

Example 1

I have seven coins in my pocket. They are all either dimes or nickels. The value of the coins is $.55. How many dimes and nickels do I have? There are two equations, one representing the number of coins (how many) and the other representing the dollar amount of the coins (how much). Each may be represented by an equation.

How many: Nickels plus Dimes equals seven, or $N + D = 7$

What kind: Nickels (.05) plus Dimes (.10) equals .55, or $.05N + .10D = .55$

Using what we know about the least common multiple (LCM), we can multiply the second equation by 100, transforming it into $5N + 10D = 55$. Putting our two equations together yields:

$$\begin{array}{rr} (N+D=7)\text{ times }(-5)= & -5N-5D=-35 \\ 5N+10D=55 & \underline{5N+10D=\ \ 55} \\ & 5D=\ \ 20 \\ & D=\ \ \ 4 \end{array}$$

If $D = 4$ and $N + D = 7$, then $N = 3$.

Checking the answer to make sure, four dimes is $.40 and three nickels is $.15, which adds up to $.55.

The key to remembering the two equations is "count" and "amount." "Count" describes how many of each kind, and "amount" describes what kind. This is the same idea we learned when studying place value: the digits represent how many of something, and the place value tells us what kind.

Example 2

I have 11 coins in my pocket. They are all either dimes or nickels. The value of the coins is $.70. How many dimes and how many nickels do I have?

How many: Nickels plus Dimes equals 11, or $N + D = 11$

What kind: Nickels (.05) plus Dimes (.10) equals .70, or $.05N + .10D = .70$

$$\begin{array}{lrr} (N+D=11) \text{ times } (-10) = & & -10N-10D=-110 \\ & 5N+10D=70 & 5N+10D=\ \ 70 \\ \hline & & -5N \quad =-40 \\ & & N=\quad 8 \end{array}$$

If $N = 8$ and $N + D = 11$, then $D = 3$.

Example 3

I have 12 coins in my pocket. They are all either pennies or nickels. The value of the coins is $.32. How many pennies and how many nickels do I have?

How many: Pennies plus Nickels equals 12, or $P + N = 12$.

What kind: Pennies (.01) plus Nickels (.05) equals .32, or $.01P + .05N = .32$

$$\begin{array}{lrr} (P+N=12) \text{ times } (-1) = & & -P-N=-12 \\ & 1P+5N=32 & P+5N=\ \ 32 \\ \hline & & 4N=\ \ 20 \\ & & N=\quad 5 \end{array}$$

If $N = 5$ and $P + N = 12$, then $P = 7$.

LESSON 17

Consecutive Integers

Integers include whole numbers and their negative counterparts. Examples of *consecutive* integers are 2, 3, 4 or 10, 11, 12. They begin with the smallest and increase by one. Consecutive *even* integers begin with a smallest number which is even, and increase by two, as in 6, 8, 10 or 22, 24, 26. Consecutive *odd* integers begin with a smallest number which is odd, and also increase by two, as in 7, 9, 11 or 33, 35, 37.

Integers may also be negative. Three consecutive even integers could be –14, –12, –10, the smallest being –14 and the largest –10.

Representing these relationships with algebra would look like this:

Consecutive integers: $N, N + 1, N + 2$

If $N = 5$, then $N + 1 = 6$ and $N + 2 = 7$, so 5, 6, 7.

Consecutive even integers: $N, N + 2, N + 4$

If $N = 12$, then $N + 2 = 14$ and $N + 4 = 16$, so 12, 14, 16.

Consecutive odd integers: $N, N + 2, N + 4$

If $N = 23$, then $N + 2 = 25$ and $N + 4 = 27$, so 23, 25, 27.

Example 1

Find three consecutive integers where three times the first integer, plus two times the third integer, is equal to 29.

$3(\text{first}) + 2(\text{third}) = 29$, and $N = \text{first}$, $N+1 = \text{second}$, $N+2 = \text{third}$

$$3N + 2(N+2) = 29$$

$$3N + 2N + 4 = 29 \qquad N = 5, \text{ so } N+1 = 6 \text{ and } N+2 = 7$$

$$5N + 4 = 29$$

$$5N = 25$$

$$N = 5$$

The solution (three consecutive integers) is 5, 6, 7.

To Check: $3(5)+2(7)=29$
$15+14=29$ It checks!

Example 2

Find three consecutive even integers such that two times the first integer, plus two times the second integer, is equal to six times the third.

$2(\text{first})+2(\text{second})=6(\text{third})$, and $N=\text{first}$, $N+2=\text{second}$, $N+4=\text{third}$

$$2N+2(N+2)=6(N+4)$$
$$2N+2N+4=6N+24$$
$$4N+4=6N+24$$
$$-2N=20$$
$$N=-10$$

$N=-10$, so $N+2=-8$ and $N+4=-6$

Three consecutive even integers are -10, -8, -6.

To Check: $2(-10)+2(-8)=6(-6)$
$-20-16=-36$ It checks!

Example 3

Find three consecutive odd integers such that five times the first integer, plus two times the second integer, is equal to three times the third integer, plus four.

$5(\text{first})+2(\text{second})=3(\text{third})+4$, and $N=\text{first}$, $N+2=\text{second}$, $N+4=\text{third}$

$$5N+2(N+2)=3(N+4)+4$$
$$5N+2N+4=3N+12+4$$
$$7N+4=3N+16$$
$$4N=12$$
$$N=3$$

$N=3$, so $N+2=5$ and $N+4=7$

Three consecutive even integers are 3, 5, 7.

To Check: $5(3)+2(5)=3(7)+4$
$15+10=25$ It checks!

LESSON 18

Multiplication and Division
with Exponents

Squares

In figure 1, notice that the over factor is 13 and the up factor is 13.

Thirteen is used as a factor two times, and we write that as 13^2. The 2 is the exponent. We verbalize this as "thirteen squared." The *factors* are 13, and the *product* is 169.

Figure 1

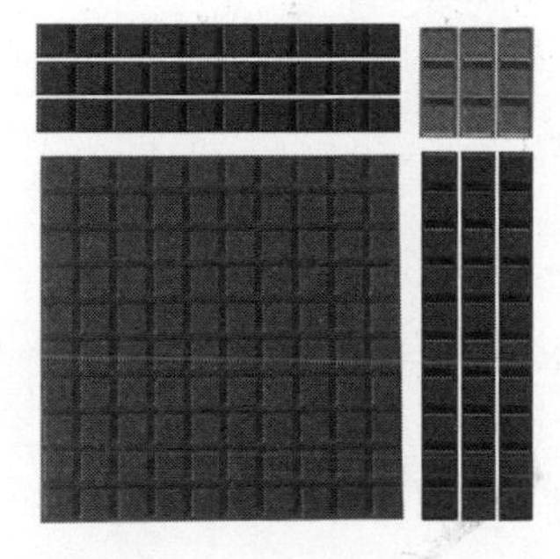

Figure 2

This is 4 over and 4 up. It is 4 used as a factor twice, 4^2, or 4 squared. The factors are 4, and the product is 16.

Square Roots

The opposite, or inverse of squares, is when you are given the product and you build a square to find the square's factors.

Example 1

Find the factors of a square, or the square's factors, of 196. Here we are given the product, or area (196), and we need to build a square and find its factors.

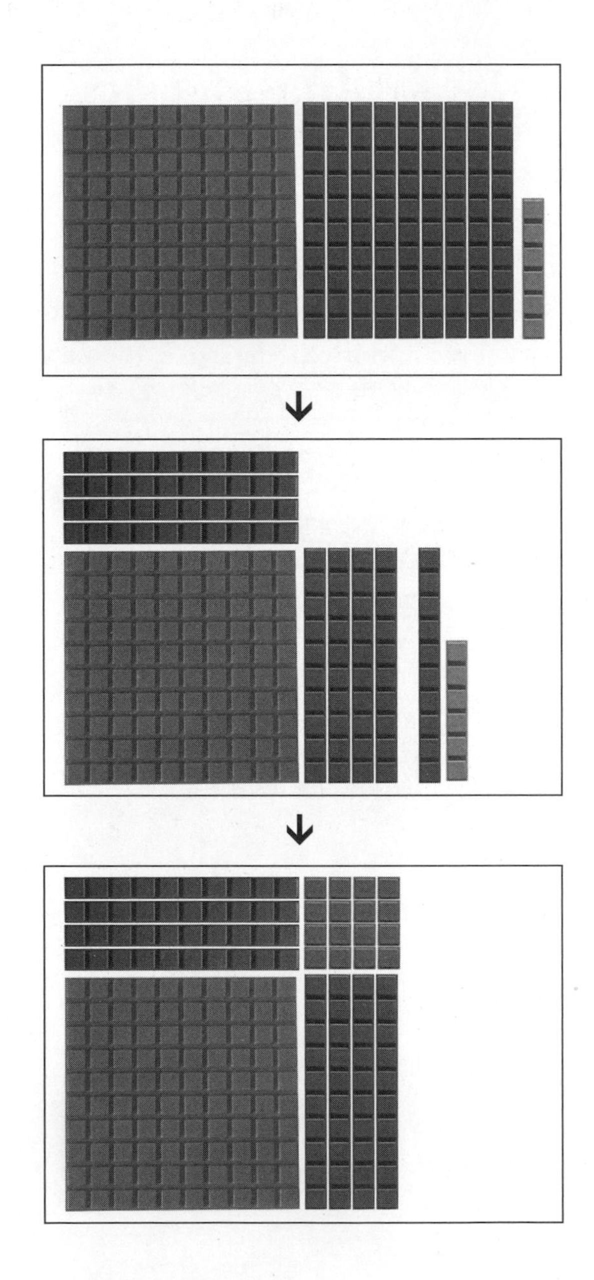

The square's factor is 14. We refer to the square's factor as the square's root or the *square root*. This may also be referred to as a *radical* or a *radical expression*.

$$\sqrt{196} = 14$$

The square root of a positive number can also be a negative number, because a negative times a negative is also a positive.

$$\sqrt{196} = 14 \text{ or } -14 \text{ since } (-14)(-14) = +196.$$

$\sqrt{}$ When you see this symbol, say to yourself, "What number times itself is equal to . . . ?" In the above example, you are asking, "What number times itself is equal to 196?" The answer is either +14 or –14. To distinguish, $\sqrt{196}$ will mean 14, and $-\sqrt{196}$ will mean –14.

Multiplication with Exponents

So far, most of this is review. But when multiplying and dividing numbers with the same base, there are some interesting patterns revealed in the exponents. Consider two to the third power times two to the fourth power. Two is the *base*, and three and four are the *exponents*.

$$2^3 \times 2^4 = (2^1 \times 2^1 \times 2^1) \times (2^1 \times 2^1 \times 2^1 \times 2^1) = 2^7$$
$$2^3 \times 2^4 = 2^7 = 128 \text{ or } 8 \times 16 = 128$$

Another example:

$$3^2 \cdot 3^3 = (3^1 \cdot 3^1) \cdot (3^1 \cdot 3^1 \cdot 3^1) = 3^5$$
$$3^2 \cdot 3^3 = 3^5 = 243 \text{ or } 9 \cdot 27 = 243$$

Using algebra, we move from real numbers to the generalization:

$$3^2 \cdot 3^3 = 3^5 \text{ or } 3^A \cdot 3^B = 3^{A+B} \text{ or } X^A \cdot X^B = X^{A+B}$$

When two numbers with the same base are being multiplied, you add the exponents. Study the examples and the algebraic representation to make sure this rule is understood before proceeding.

Division of Exponents

The opposite of multiplying is dividing, and the inverse of adding is subtracting. Notice the pattern in the exponents when two numbers with the same base are divided.

$$2^5 \div 2^2 = \frac{2^5}{2^2} = \frac{\cancel{2} \cdot \cancel{2} \cdot 2 \cdot 2 \cdot 2}{\cancel{2} \cdot \cancel{2}} = 2^3$$

Because division is the opposite of multiplication, the exponents are subtracted in division. Using algebra, we arrive at the following generalization:

$$X^A \div X^B = X^{A-B}$$

We could solve the previous example as follows:

$$2^5 \div 2^2 = 2^{5-2} = 2^3$$

Consider this problem:

$$2^2 \div 2^5 = \frac{2^2}{2^5} = \frac{\cancel{2} \cdot \cancel{2}}{\cancel{2} \cdot \cancel{2} \cdot 2 \cdot 2 \cdot 2} = 2^{-3}$$

$$\text{or } 2^2 \div 2^5 = 2^{2-5} = 2^{-3}$$

See the next lesson for more on negative exponents.

Summary

We have discovered how to add exponents when multiplying two numbers with the same base and how to subtract exponents when dividing two numbers with the same base. These operations are summed up as follows:

$$X^A \cdot X^B = X^{A+B} \quad \frac{X^A}{X^B} = X^A \div X^B = X^{A-B}$$

LESSON 19

Exponents

Negative and Raising to a Power

Negative Exponents

Using what we have learned so far, we can apply this knowledge to negative exponents.

$$2^2 \div 2^5 = \frac{2^2}{2^5} = \frac{\cancel{2} \cdot \cancel{2}}{\cancel{2} \cdot \cancel{2} \cdot 2 \cdot 2 \cdot 2} = \frac{1}{2^3}$$

$$2^2 \div 2^5 = 2^{2-5} \qquad = 2^{-3}$$

If you change the place of an exponent from numerator to denominator or vice versa, then you must also change the sign from negative to positive or from positive to negative. Conversely, if you want to change the sign, you must change the place.

$$\frac{1}{2^5} = 2^{-5} \text{ and } \frac{1}{2^{-5}} = 2^5$$

Using what we know, we can prove that anything with a zero exponent is equal to one.

Start here.
↓

$$1 = \frac{100}{100} = \frac{10^2}{10^2} = 10^{2-2} = 10^0 \qquad \text{Therefore } 10^0 = 1.$$

So $X^0 = 1$. $^0 = 1$.

Summary

We have discovered that if we change the sign of an exponent, we also change its place from numerator to denominator or from denominator to numerator.

$$X^A = \frac{1}{X^{-A}} \text{ or } X^{-A} = \frac{1}{X^A}$$

Using this information, we deduced that $X^0 = 1$.

Raising an Exponent to a Power

We know that $(5^4)^3 = (5^4)(5^4)(5^4) = (5^{4+4+4}) = 5^{12}$. We could also write this as: $(5^4)^3 = 5^{12}$. This is fast adding of the fours or four counted three times.

When raising an exponent to another power, or exponent, you fast add, or multiply, the exponents. In the example, we either add $4 + 4 + 4 = 12$, or say that $4 \times 3 = 12$.

Here is another example:

$$(10{,}000)^3 = (10^4)^3 = 10^{4 \cdot 3} = 10^{12}$$

In algebraic terms:

$$(X^A)^B = X^{AB}$$

Simplifying an Expression Using Exponents

We may use what we know about exponents and apply this to letters, or variables as well. In the ensuing example, we first group all of the variables by kind. Then, since we are multiplying numbers with the same bases (or same letters in this case), we may add the exponents.

$$M^1B^2\,B^0M^{-4}B^{-5}M^8$$

Combine like terms by putting all of the Ms and Bs side by side.

$$M^1M^{-4}M^8\,B^2\,B^0B^{-5}$$

When we multiply two numbers or letters, we add the exponents. The final result is written below.

$$M^5\,B^{-3}$$

Simplifying an Expression using Negative Exponents

We may use what we have learned about negative exponents when simplifying an algebraic expression, as in the following example. First we move all of the terms into the numerator by changing the sign of each of the exponents in the denominator.

Remember: when you change the place (from the denominator to the numerator), change the sign.

$$\frac{F^4K^3K^{-1}}{K^2F^{-5}K^6} = \frac{F^4K^3K^{-1}\,K^{-2}\,F^5K^{-6}}{1}$$

Then we can combine like terms by putting together all of the Fs and Ks. Also, we don't need to put the expression over one, as the meaning is the same without it.

$$F^4F^5\,K^3K^{-1}K^{-2}K^{-6}$$

When we multiply two or more like numbers or letters, we add the exponents. The final result is written below.

$$F^9\,K^{-6}$$

LESSON 20

Addition and Multiplication of Polynomials

Base 10 and Base X

Recall the factors of each of the pieces in base 10. The unit block (green) is 1 x 1.

The 10 bar (blue) is 1 x 10, and the 100 square (red) is 10 x 10. Each of these pieces may also be expressed in terms of exponents: 1 x 1 = 1, which is 1^0; 1 x 10 = 10, which is 10^1; 10 x 10 = 100, which is 10^2. Below is the number 156 shown with the blocks and expressed in different ways.

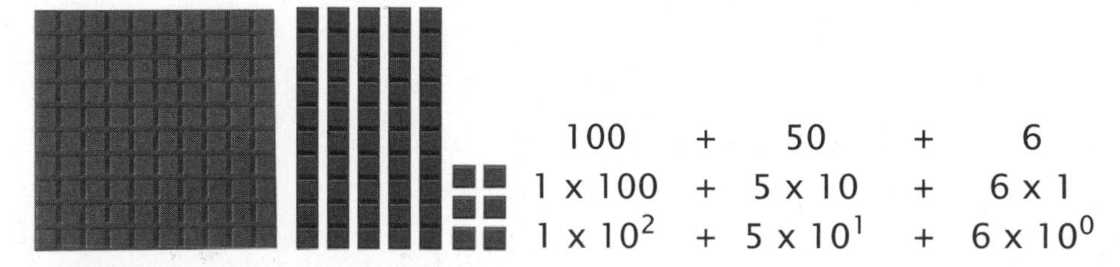

In the decimal system, every value is based on 10. The decimal system is referred to as base 10.

In algebra, the unit bar is still one by one. The smooth blue piece that snaps into the back of the 10 bar is one by X, and the smooth red piece that snaps into the back of the 100 square is X by X. Each of these pieces may also be expressed in terms of exponents: $1 \cdot 1 = 1$, which is 1^0, or X^0 (which is the same thing since both are equal to one); $1 \cdot X = X$, which is X^1; and $X \cdot X = X^2$.

On the next page is the polynomial $X^2 + 5X + 6$ shown with the blocks and expressed in different ways.

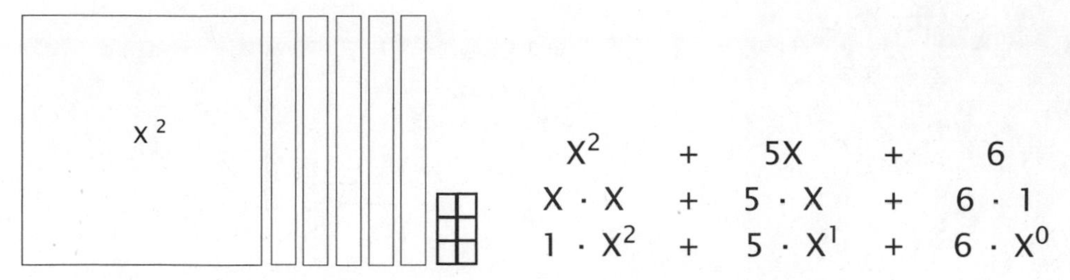

In algebra, every value is based on X. Algebra is arithmetic in base X.

Kinds of Polynomials

Polynomial derives from polys (many) and nomen (name), so literally it means "many names." If a polynomial has three components, it is called a *trinomial* (tri- meaning "three"). A *binomial* (bi- meaning "two") has two parts. A *monomial* (mono- meaning "one") has one part.

In the next few sections, whenever you feel the need to reassure yourself that you are on the right track, simply change the equation from base X to base 10, and redo the problem.

On the next two pages, operations that you are familiar with, such as addition and multiplication, will now be performed with polynomials in base X instead of base 10. Take your time and remember the connection with what you already know. Someone has said, "Algebra is not difficult, just different."

Addition of Polynomials

When adding or subtracting polynomials, remember that "to combine, they must be the same kind." Units may be added (or subtracted) with other units, Xs with Xs, X^2s with X^2s, etc. Since we don't know what the value is for X, all the addition and subtraction is done in the coefficients. Read through the following examples for clarity. The gray inserts are used to show –X.

Example 1

$$\begin{array}{r} X^2+2X+\ \ 4 \\ +\,X^2+5X+\ \ 6 \\ \hline 2X^2+7X+10 \end{array}$$

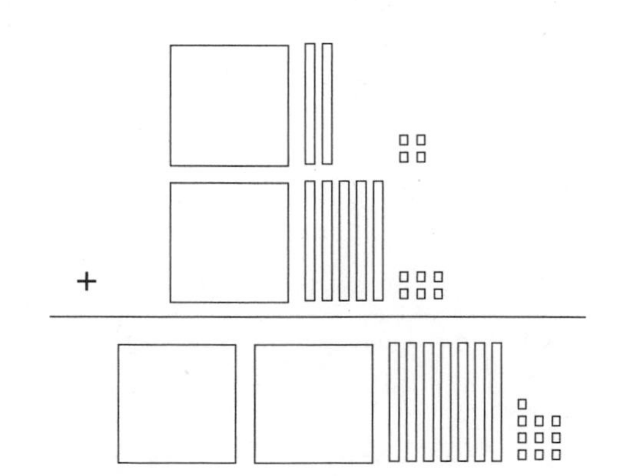

Example 2

$$\begin{array}{r} X^2 - X - 4 \\ + X^2 + 5X + 6 \\ \hline 2X^2 + 4X + 2 \end{array}$$

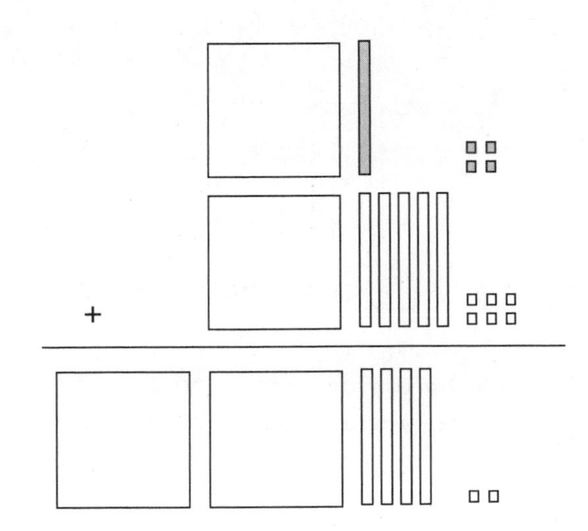

Multiplication of Polynomials

When we multiply two binomials, the result is a trinomial. I like to use the same format for multiplying a binomial as for multiplying any double-digit number in the decimal system.

Let's look at a problem in the decimal system using expanded notation.

Example 3

$$\begin{array}{r} 13 \rightarrow \\ \times 12 \uparrow \\ \hline \end{array} = \begin{array}{r} 10 + 3 \\ \times 10 + 2 \\ \hline 20 + 6 \\ 100 + 30 \\ \hline 100 + 50 + 6 \end{array}$$

If this were in base X instead of base 10, it would look like this:

$$\begin{array}{r} X + 3 \\ \times X + 2 \\ \hline 2X + 6 \\ X^2 + 3X \\ \hline X^2 + 5X + 6 \end{array}$$

The area, or product, of this rectangle is $X^2 + 5X + 6$. Do you see it?

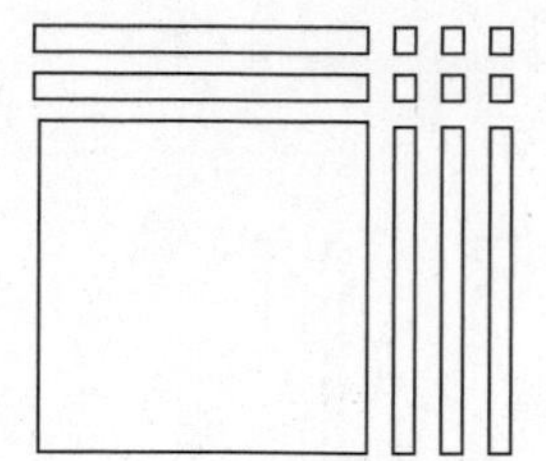

In this rectangle, we cover up most of it to reveal the factors, which are $(X + 3)$ over and $(X + 2)$ up.

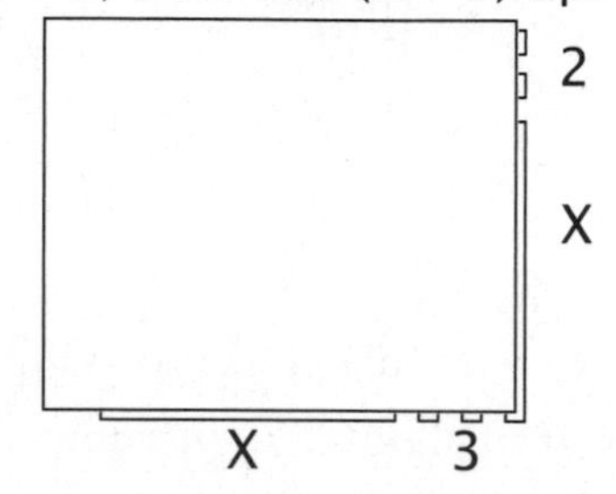

The written equivalent of the picture looks just like double-digit multiplication, which it is.

$$\begin{array}{r} X+3 \ \rightarrow \\ \times\, X+2 \ \uparrow \\ \hline 2X+6 \\ X^2+3X \quad \\ \hline X^2+5X+6 \end{array}$$

Example 4

$$\begin{array}{r} 23 \ \rightarrow \\ \times\, 13 \ \uparrow \\ \hline \end{array} = \begin{array}{r} 20+3 \\ \times\, 20+3 \\ \hline 60+9 \\ 200+30 \quad \\ \hline 200+90+9 \end{array}$$

If this were in base X instead of base 10, it would look like this:

$$\begin{array}{r} 2X+3 \\ \times\, X+3 \\ \hline 6X+9 \\ 2X^2+3X \quad \\ \hline 2X^2+9X+9 \end{array}$$

The area, or product, of this rectangle is $2X^2 + 9X + 9$. Do you see it?

In this rectangle, we cover up most of it to reveal the factors, which are $(2X + 3)$ over and $(X + 3)$ up.

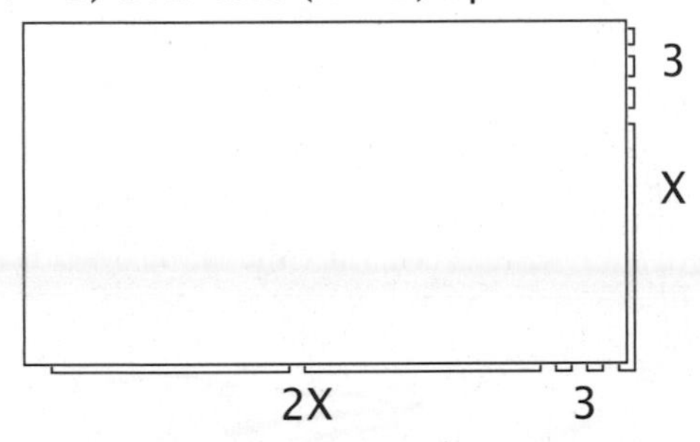

The written equivalent of the picture looks just like double-digit multiplication, which it is.

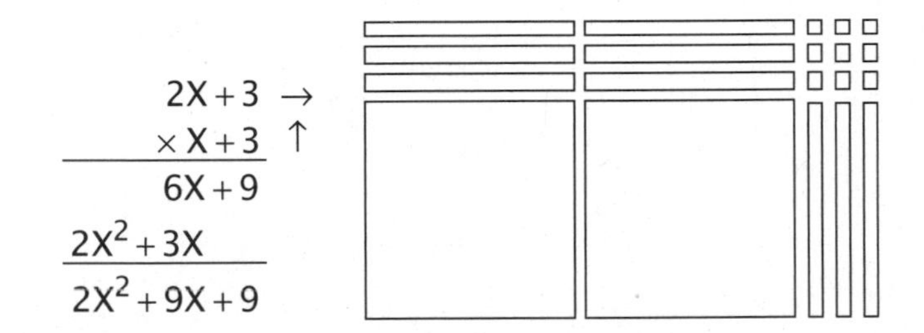

You can also multiply binomials that include negative numbers. To show $-X$, use the gray inserts. The addition identity tells us that $X + (-X) = 0$.

+ = 0

X -X

Example 5

$$\begin{array}{r} X-3 \;\rightarrow \\ \times\, X+2 \;\uparrow \\ \hline 2X-6 \\ X^2-3X \\ \hline X^2-\;X-6 \end{array}$$

X + 2

X - 3

Here it is in base 10:

$$\begin{array}{rl} 10-3 & =7 \\ \times\, 10+2 & =12 \\ \hline 20-6 & \;84 \\ 100-30 & \;\nwarrow\!\searrow \\ \hline 100-10-6 & =84 \end{array}$$

Look at the next two examples carefully.

Example 6

(X - 2)(X + 4)

$$\begin{array}{r} X-2 \;\rightarrow \\ \times\, X+4 \;\uparrow \\ \hline 4X-8 \\ X^2-2X \\ \hline X^2-2X-8 \end{array}$$

Example 7

(X - 1)(X - 5)

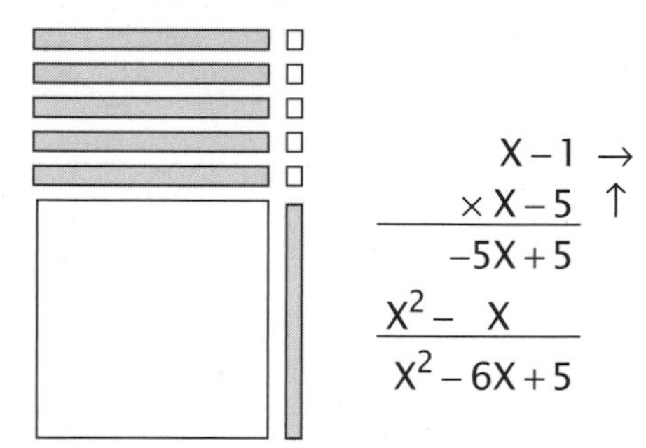

$$\begin{array}{r} X-1 \;\rightarrow \\ \times\, X-5 \;\uparrow \\ \hline -5X+5 \\ X^2-\;X \\ \hline X^2-6X+5 \end{array}$$

LESSON 21

Factor Polynomials

Factoring Trinomials

We will be finding the factors of $X^2 + 7X + 12$ using the blocks with the algebra inserts snapped into the backs. Finding the factors of a trinomial is the opposite of multiplying two binomials to find the product, which is a trinomial. In lesson 20, you were given the factors, and you were to find the product. Now, you are given the product and are asked to find the factors.

Example 1

First, build $X^2 + 7X + 12$. This is the product, which is given.
Next, build a rectangle using all the blocks. Then find the factors by reading the lengths of the over dimension and the up dimension.

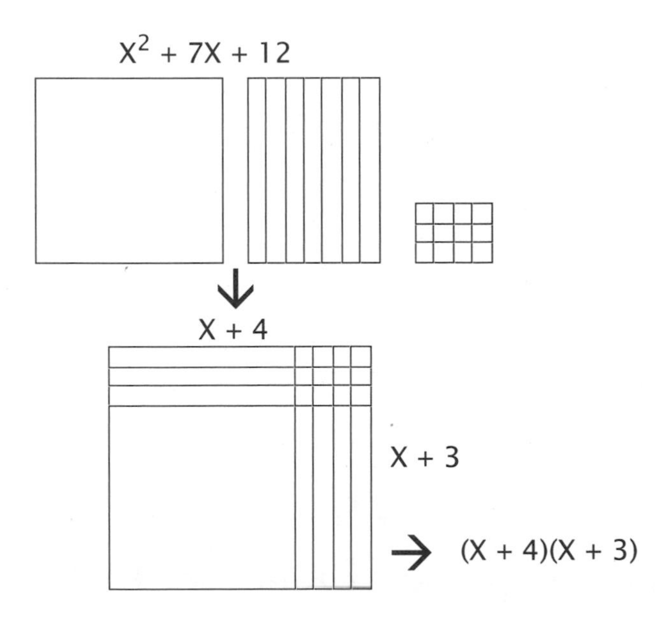

Example 2

Now find the factors of $X^2 + 8X + 12$. Represent the trinomial with the manipulatives, build a rectangle, and then read the factors.

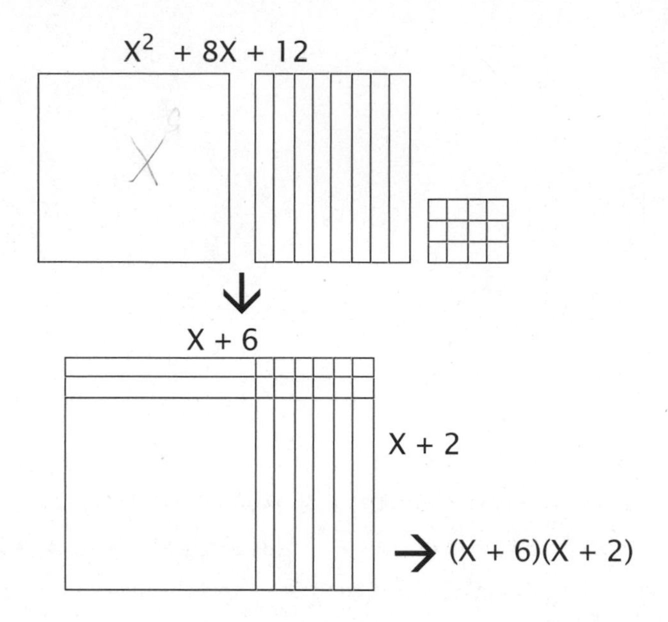

In examples 1 and 2, notice the relationship between the last term (12), the middle term (7X or 8X), and the factors.

$$X^2 + 7X + 12 = (X + 4)(X + 3)$$

The last term is found by multiplying: 3 x 4.
The middle term is found by adding: 3X + 4X.

$$X^2 + 8X + 12 = (X + 6)(X + 2)$$

The last term is found by multiplying: 6 x 2.
The middle term is found by adding: 6X + 2X.

The factors of the last term are the addends of the middle term. This always works when the coefficient of X^2 is one.

$X^2 + 13X + 12 = (X + 1)(X + 12)$
The last term is found by multiplying: 1 + 12.
The middle term is found by adding: 1X + 12X.

More on Multiplying Polynomials

Polynomials may be multiplied vertically (A) or horizontally (B) using the distributive property. Study examples 3 and 4 on the next two pages.

Example 3

A.

$$\begin{array}{r} 2X+3 \\ \times X+2 \\ \hline 4X+6 \\ 2X^2+3X \\ \hline 2X^2+7X+6 \end{array}$$

B. $(X+2)(2X+3) = (X)(2X+3)+(2)(2X+3) = (2X^2+3X)+(4X+6)$

$= 2X^2+7X+6$

When multiplying horizontally, there are four *partial products* just as before, but they are arrived at using a formula called FOIL: F–First, O–Outside, I–Inside, L–Last. Each letter corresponds to a partial product.

F In $(X + 2)(2X + 3)$, $X \cdot 2X$ is the **F**irst term times the **F**irst term: $2X^2$

O In $(X + 2)(2X + 3)$, $X \cdot 3$ is the **O**utside term times the **O**utside term: $3X$

I In $(X + 2)(2X + 3)$, $2 \cdot 2X$ is the **I**nside term times the **I**nside term: $4X$

L In $(X + 2)(2X + 3)$, $2 \cdot 3$ is the **L**ast term times the **L**ast term: 6

$2X^2 + 3X + 4X + 6$

$2X^2 + 7X + 6$

Example 4

A.

$$\begin{array}{r} X+\ 3 \\ \times X+\ 4 \\ \hline 4X+12 \\ X^2+3X \\ \hline X^2+7X+12 \end{array}$$

B. $(X+4)(X+3) = (X)(X+3)+(4)(X+3) = (X^2+3X)+(4X+12)$

$= X^2+7X+12$

F In (X + 4)(X + 3), $X \cdot X$ is the **F**irst term times the **F**irst term: X^2

O In (X + 4)(X + 3), $X \cdot 3$ is the **O**utside term times the **O**utside term: $3X$

I In (X + 4)(X + 3), $4 \cdot X$ is the **I**nside term times the **I**nside term: $4X$

L In (X + 4)(X + 3), $4 \cdot 3$ is the **L**ast term times the **L**ast term: 12

$X^2 + 3X + 4X + 12$

$X^2 + 7X + 12$

LESSON 22

Factoring Trinomials with Coefficients

Place Value

In the decimal system, each digit written is assigned a value dependent upon its "place" in the numeral. For example, in the numeral 423, the 4 is assigned a "value" of 100. The digit 4 tells us only "how many"; where we place it tells us "what kind" (4 equals how many; 100 equals what kind). This distinction, so subtly displayed in our decimal system, is very clear in algebraic terminology. Since algebra is merely arithmetic in base X, we can convert the above example (423) into an algebraic polynomial as follows:

$$423_{10} = 4 \times 100 + 2 \times 10 + 3 \times 1$$
$$423_{10} = 4 \times 10^2 + 2 \times 10 + 3$$
$$423_X = 4X^2 + 2X + 3$$

In a polynomial, coefficients are the numbers, and variables (X^3, X^2, X^1, etc.) are the place values. The coefficient tells "how many" and the variable tells "what kind." In 7X, the 7 is the coefficient. In $3X^2$, the 3 is the coefficient.

In lesson 21, we learned that when factoring a trinomial, the factors of the last term are the addends of the middle term. This statement applies only when the coefficient of X^2 is one. In this lesson, we will learn how to find the factors of a trinomial when the coefficient of X^2 is not one.

Example 1

Find the factors of $2X^2 + 7X + 6$.

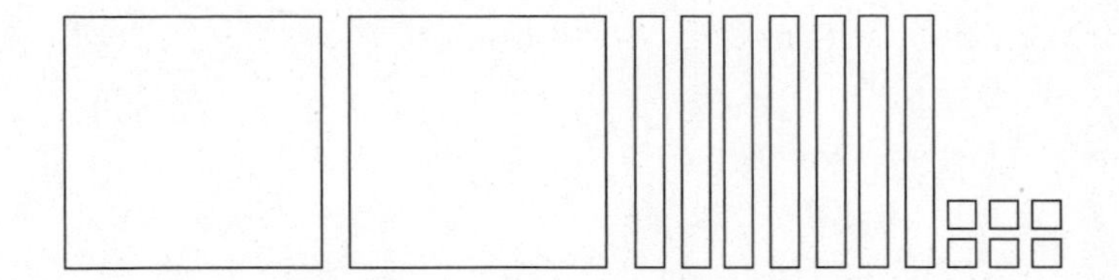

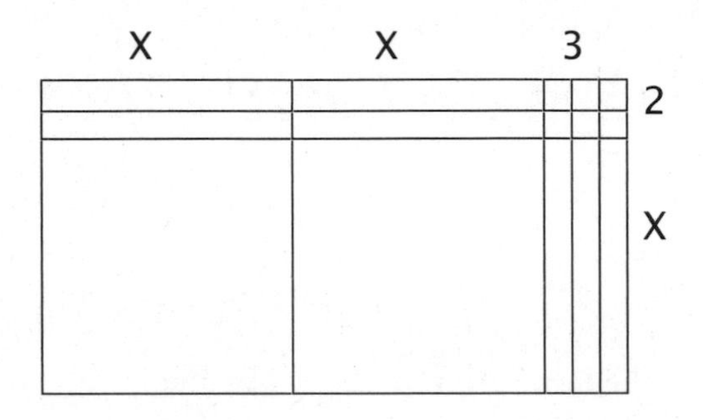

The factors are $(2X + 3)(X + 2)$.

Example 2

Find the factors of $2X^2 + 5X + 3$.

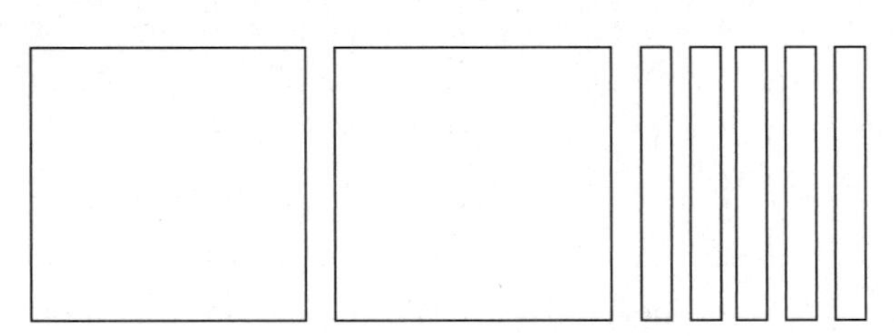

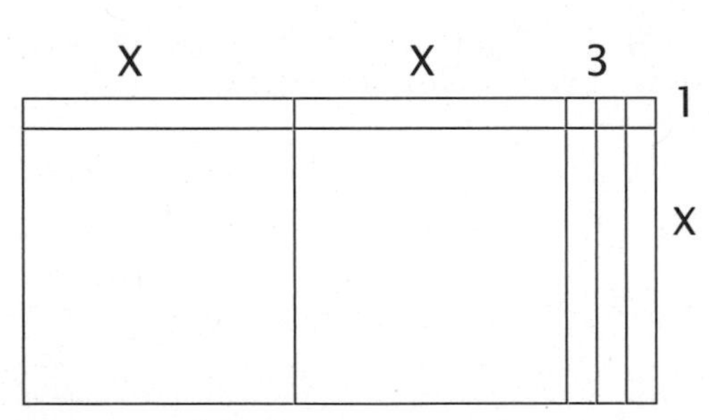

The factors are $(2X + 3)(X + 1)$.

Example 3

Find the factors of $3X^2 + 11X + 6$.

The factors of 6 could be 1 x 6 or 2 x 3. We decide to try 2 x 3.

$(3X + 2)(X + 3)$

$3X^2 + 9X + 2X + 6 = 3X^2 + 11X + 6$. So our choice was correct.

Now try using 3 x 2 instead of 2 x 3.

$(3X + 3)(X + 2)$

$3X^2 + 6X + 3X + 6 = 3X^2 + 9X + 6$. This is not the polynomial we started with!

Here is a trial with 6 x 1.

$(3X + 6)(X + 1)$

$3X^2 + 3X + 6X + 6 = 3X^2 + 9X + 6$. This is not the desired result.

Now try 1 x 6.

$(3X + 1)(X + 6)$

$3X^2 + 18X + X + 6 = 3X^2 + 19X + 6$.

This doesn't work either. Only the first trial gives the correct combination of factors.

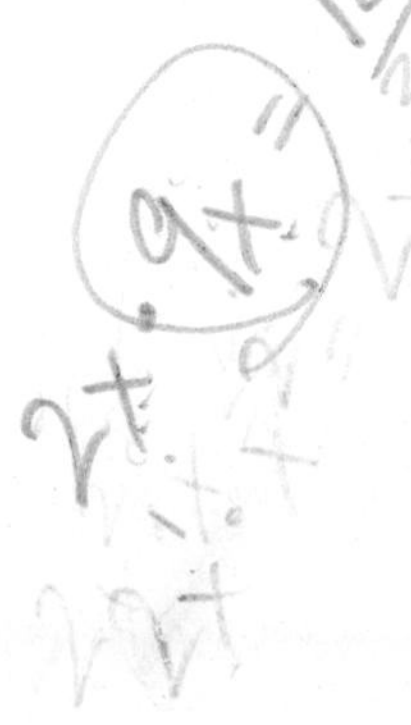

Coefficients and the GCF

When factoring polynomials, always check to see if there is a greatest common factor (GCF) that can be factored out first. A GCF may contain coefficients, variables, or both. In $2X^3 + 18X^2 + 36X$, each term has a common factor of two. Each term also has a common factor of X. We can factor 2X out of each term, yielding $2X(X^2 + 9X + 18)$. Then we can factor the second term. The final result is $2X(X + 6)(X + 3)$.

LESSON 23

Factoring Trinomials
with Negative Numbers

Negative Last Term

Recall that the shaded bars are negative X (or –X). They are represented by gray algebra inserts snapped into the backs of the blue ten bars. The shaded unit bars in the picture are unit pieces placed upside down (the hollow side is showing) to represent negative numbers.

Example 1

Find the factors of $X^2 - 2X - 8$.

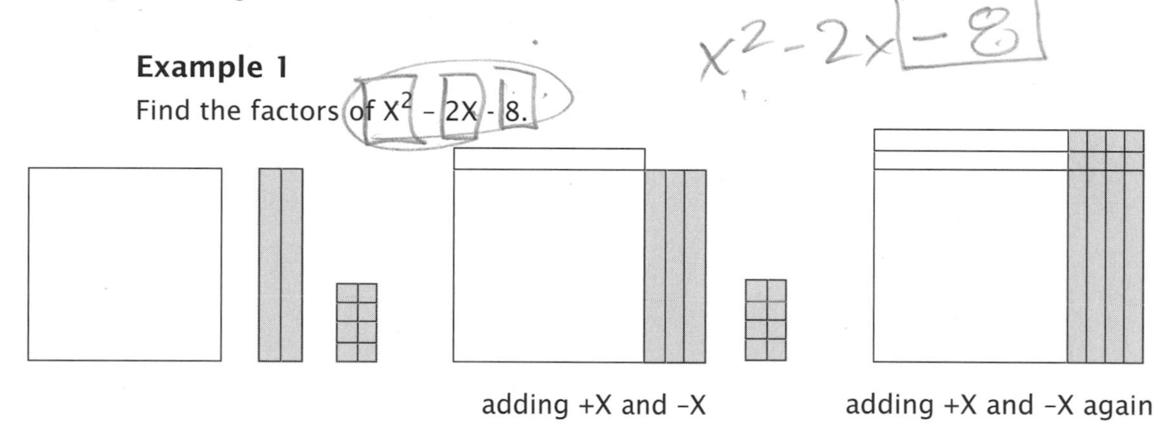

Adding Zero

To find the factors of $X^2 - 2X - 8$, we must build a rectangle. But this is not possible with the manipulatives that are given, so we have to add something else to the manipulatives. We know that zero plus anything is still that same thing. So we will add zero (nothing) to our attempted rectangle by adding a positive X and a negative X, or a gray insert and a blue insert. This still doesn't allow us enough pieces to build a rectangle, so we'll add zero again $(+X - X)$. This works!

We still have $X^2 - 2X - 8$, but now it is in a different form: $X^2 - 4X + 2X - 8$. The factors are $(X - 4)(X + 2)$.

Check your answer by multiplying.

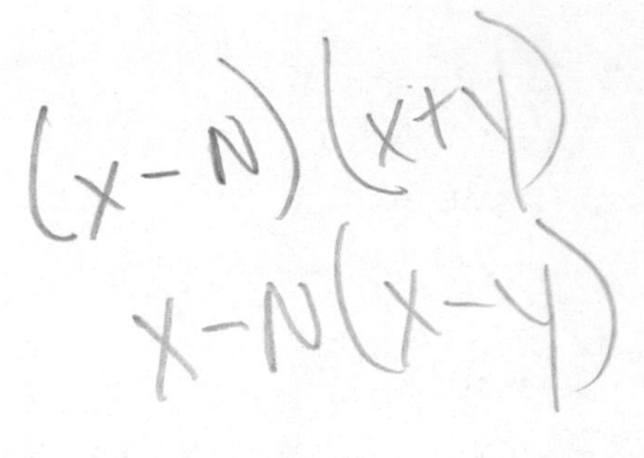

$$\begin{array}{r} X-4 \\ \times\, X+2 \\ \hline 2X-8 \\ X^2-4X \\ \hline X^2-2X-8 \end{array}$$

If the last term of the trinomial is negative, then one of its factors must be negative as well. The middle term of the trinomial will be the same sign as the sign of the larger of the last terms in the factors.

Positive Last Term

In the last couple of lessons in the student text, you were given problems to multiply, such as (X − 3)(X − 2). I hope you built these with the blocks and the algebra inserts. If you did, you discovered that the units (the final terms) in these cases were positive because you multiplied a negative by a negative, which gave you a positive. But the Xs (the middle terms) were all negative. This is illustrated in the following example.

Example 2

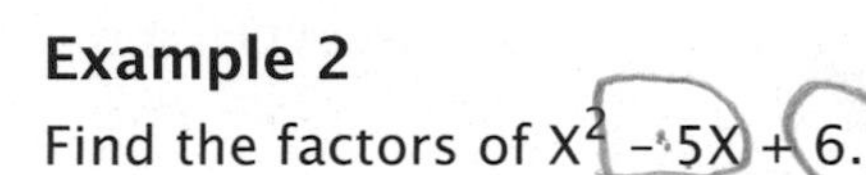

Find the factors of $X^2 - 5X + 6$.

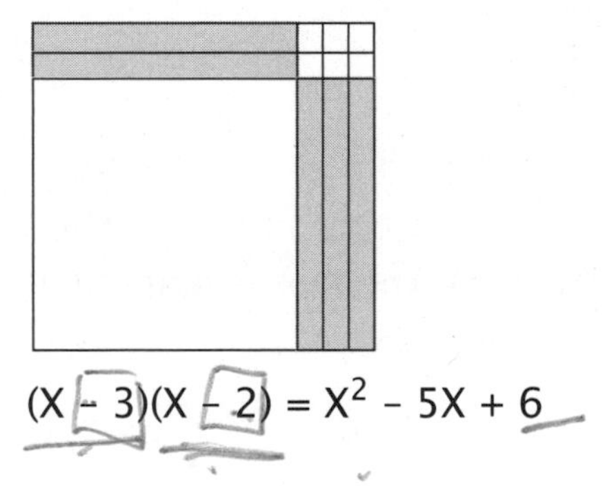

$(X - 3)(X - 2) = X^2 - 5X + 6$

If the last term in the trinomial is positive and the middle term is negative, then the last term in both factors will be negative, as in example 2. If the last term and the middle term are both positive in the trinomial, then the last term in both of the factors will be positive.

Study these examples:

Example 3

$X^2 - 5X + 6 = (X - 2)(X - 3) \quad X^2 + 5X + 6 = (X + 2)(X + 3)$

Example 4

$X^2 - 6X + 8 = (X - 2)(X - 4) \quad X^2 + 6X + 8 = (X + 2)(X + 4)$

Example 5

$X^2 - 4X + 3 = (X - 1)(X - 3) \quad X^2 + 4X + 3 = (X + 1)(X + 3)$

Remember that if the last term of the trinomial is negative, then one of its factors must be negative as well. The middle term of the trinomial will be the same sign as the sign of the larger of the last terms in the factors.

Example 6

$(X - 2)(X + 5) = X^2 + 3X - 10 \quad (X - 7)(X + 3) = X^2 - 4X - 21$

Coefficients of X^2

Problems with factoring trinomials are quickly solved when the coefficient of X^2 is one. When there are other coefficients involved, this process is complicated.

Example 7

$2X^2 - X - 6 = (2X + 3)(X - 2)$

The three in the first factor is positive, but the middle term of the trinomial ends up being negative, because the negative two for the second factor is multiplied by 2X to get (–4X). The four is larger than the three in (+3X), so the middle term is negative:

$-4 + 3 = -1$ and $-4X + 3X = -X$

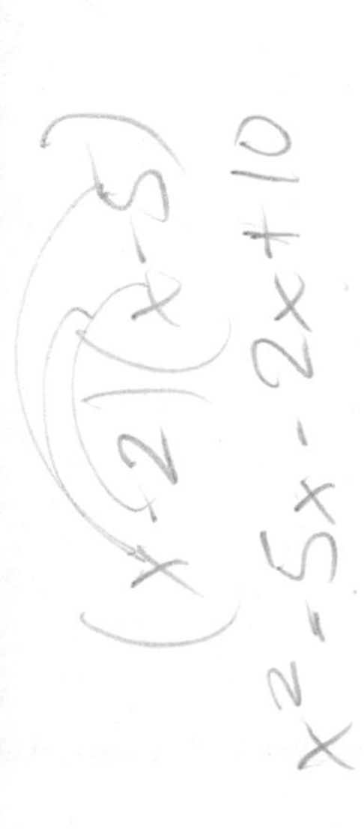

Negative Coefficients of X^2

You may be asking why you never see negative X^2. When given a factoring problem such as $-2X^2 - 7X - 6$, you first factor out a negative one to give you $-(2X^2 + 7X + 6)$. Further factoring produces $-(2X + 3)(X + 2)$.

So whenever you are faced with a negative coefficient of X^2, factor out the negative one and proceed as before. Example 8 is the same as example 7 except for the negative sign.

Example 8

$-2X^2 + X + 6 = -(2X^2 - X - 6) = -(2X + 3)(X - 2)$

LESSON 24

Square Roots and Dividing Polynomials

Square Roots

The opposite of squaring a number is finding the square root of a number. Instead of being given the factors and solving to find the product, we are given the product and have to solve to find the factors, or roots. In example 1, we will find the square root of $X^2 + 4X + 4$.

Example 1

$\sqrt{X^2 + 4X + 4}$

Looking at this, think, "What factor times itself equals $X^2 + 4X + 4$?" or, "What are the factors of a square that is the product of $X^2 + 4X + 4$?" or, "What is the square root of $X^2 + 4X + 4$?"

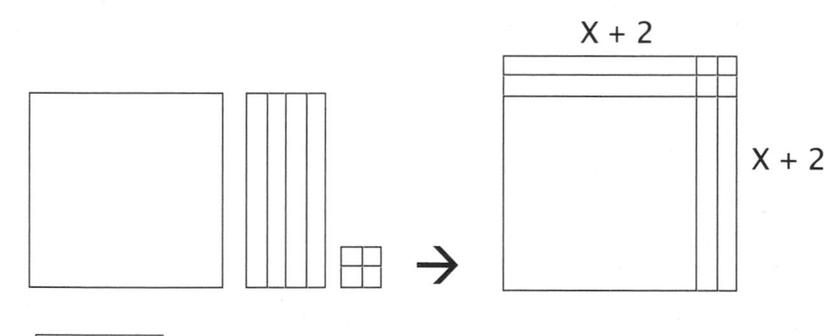

so $\sqrt{X^2 + 4X + 4} = (X + 2)$

Division of Polynomials

Next we will look at how to divide polynomials.

Example 2

Looking at the X in X + 3, ask, "What times X equals X^2?" The answer is X, so we place this in the X place above 7X.

$$\begin{array}{r} X \\ \boxed{X}+3\,\overline{)\,X^2+7X+13} \\ \underline{X^2+3X} \end{array}$$

We ask the above question because our goal is to get X^2 in the product. When we subtract the product from X^2 + 7X + 13, we have eliminated the X^2 and can continue with the division process.

$$\begin{array}{r} X \\ \boxed{X}+3\,\overline{)\,X^2+7X+13} \\ \underline{-(X^2+3X)} \\ 4X+13 \end{array}$$

Subtract X^2 + 3X.

Notice that the subtraction sign applies to both X^2 and 3X.

$$\begin{array}{r} X+\ 4 \\ \boxed{X}+3\,\overline{)\,X^2+7X+13} \\ \underline{-(X^2+3X)} \\ 4X+13 \\ \underline{-(4X+12)} \\ 1 \end{array}$$

After bringing down the 13, we have 4X + 13. Look at the 4X and ask, "What times X equals 4X?" The answer is 4, and we place this above the 13 in the units place. Multiplying X + 3 by 4 produces 4X + 12.

Subtract 4X + 12, leaving a remainder of one.

To check this, multiply.

$$\begin{array}{r} X+\ 4 \\ \underline{\times X+\ 3} \\ 3X+12 \\ \underline{X^2+4X } \\ X^2+7X+12 \end{array}$$

Then add one, which was the remainder.

$X^2 + 7X + 12 + 1 = X^2 + 7X + 13$

Example 3

$$\begin{array}{r} 2X \\ \boxed{2X}+1 \,\overline{\big)\, 4X^2 - 4X - 3} \\ \underline{4X^2 + 2X} \end{array}$$

Looking at the 2X in 2X + 1, ask, "What times 2X equals $4X^2$?"

The answer is 2X, so we place this in the X place above 4X.

$$\begin{array}{r} 2X \\ \boxed{2X}+1 \,\overline{\big)\, 4X^2 - 4X - 3} \\ \underline{-(4X^2 + 2X)} \\ -6X - 3 \end{array}$$

Subtract $4X^2 + 2X$.

Notice that the subtraction sign applies to both $4X^2$ and 2X.

$$\begin{array}{r} 2X - 3 \\ \boxed{2X}+1 \,\overline{\big)\, 4X^2 - 4X - 3} \\ \underline{-(4X^2 + 2X)} \\ -6X - 3 \\ \underline{-(-6X - 3)} \end{array}$$

After bringing down -3, we have −6X − 3. Look at the -6X and ask, "What times 2X equals -6X?" The answer is −3. Place this above the 3 in the units place.

Multiplying 2X + 1 by −3 produces −6X − 3. Subtracting (−6X − 3) gives the same result as adding (6X + 3). The remainder is zero.

Check by multiplication.

$$\begin{array}{r} 2X - 3 \\ \underline{\times\ 2X + 1} \\ 2X - 3 \\ \underline{4X^2 - 6X } \\ 4X^2 + 4X - 3 \end{array}$$

LESSON 25

Difference of Two Squares
and Oriental Squares

Difference of Two Squares

This skill is essential in some factoring problems.

Example 1

Find the factors of $X^2 - 9$.

Notice that both X^2 and 9 are squares and $X^2 - 9$ is equal to $X^2 - 3^2$.

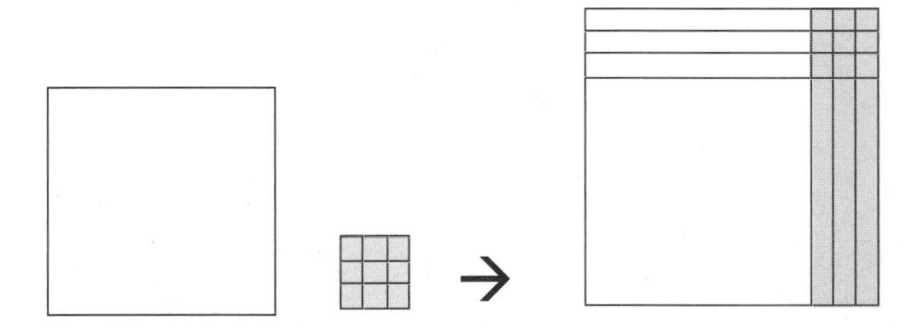

By adding +3X and −3X (which equals zero), we may now build a rectangle using all the pieces.

So the factors of $X^2 - 9$ are $(X + 3)(X - 3)$.

$$(X + 3)(X - 3) = X^2 - 3X + 3X - 9 = X^2 - 9$$

Example 2

Find the factors of $X^2 - 4$. Notice that both X^2 and 4 are squares and $X^2 - 4$ is equal to $X^2 - 2^2$.

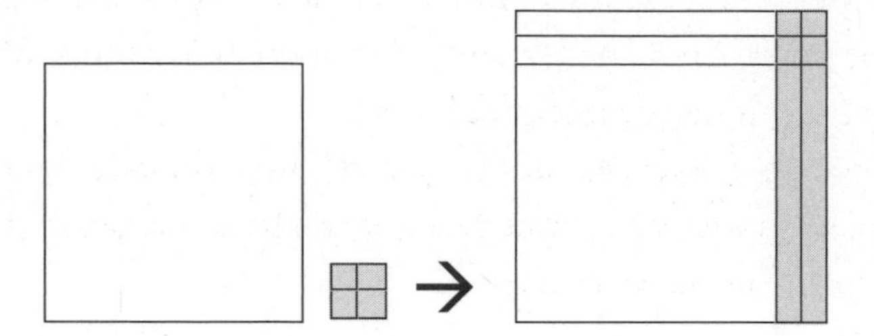

By adding +2X and -2X (which equals zero), we may now build a rectangle using all the pieces.

So the factors of $X^2 - 4$ are $(X + 2)(X - 2)$.

$$(X + 2)(X - 2) = X^2 - 2X + 2X - 4 = X^2 - 4$$

Algebraically, we may represent this pattern as $A^2 - B^2 = (A + B)(A - B)$.

Oriental Squares

This is a slick trick taught to students in the Far East (I've been told) for multiplying squares that end in five and have factors such as 15, 35, 65 . . .

Example 3

Build 15 x 15, or 15^2. It should look like this the picture on the left.

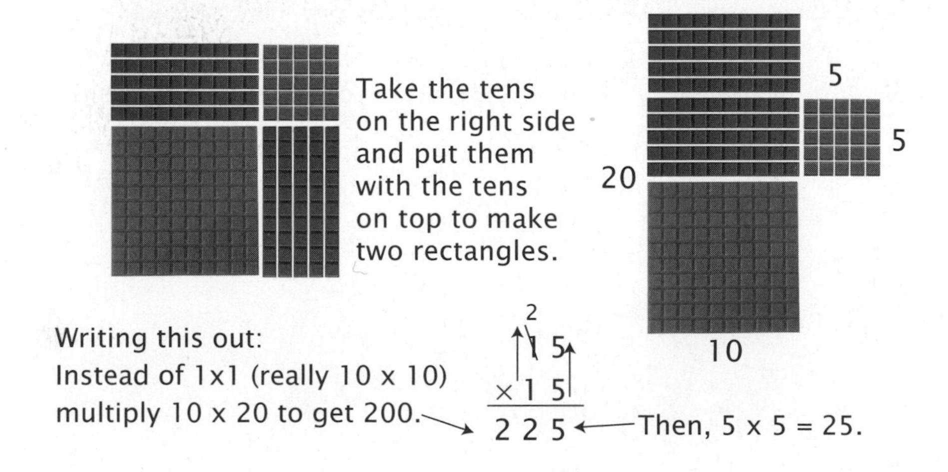

Study example 3. What we are doing is eliminating the middle term and making this a simple two-step problem instead of a four-step problem. In the units place you multiply 5 x 5 = 25. Then in the tens place, instead of 1 x 1 (really 10 x 10), you multiply 1 x 2 (really 10 x 20), because the hundreds are made up of the tens. We don't have to multiply five units times one ten and one ten times five units, because we took the tens and made one hundred out of them.

Look at the picture to make sure you see this extra hundred made out of the tens. This is now one ten times two tens (notice that I crossed out one ten in the top number and increased it from one ten to two tens).

Example 4

Build 25 x 25, or 25 squared.

Move the tens from the right to the top of the other tens, so we have two rectangles, and thus two problems.

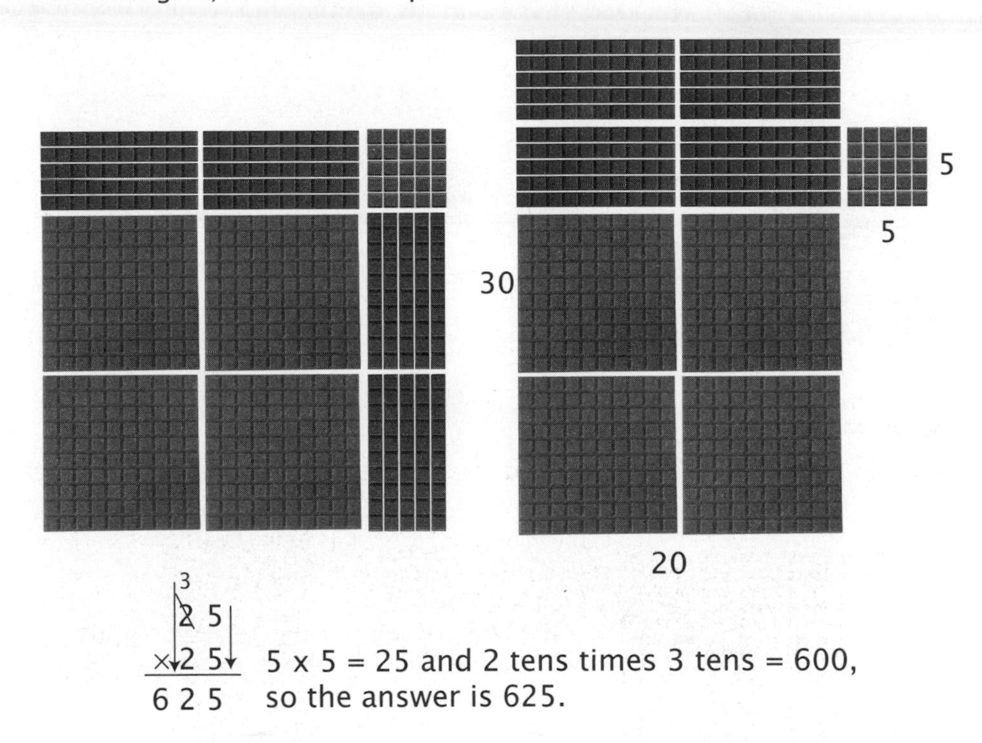

5 x 5 = 25 and 2 tens times 3 tens = 600, so the answer is 625.

If the digits in the tens place are the same, and the digits in the units place are both five, then this works.

Here are a few more problems. The 5 x 5 = 25 is obvious. Where the blocks come in handy is showing us why we add one more to the tens place in the top number before multiplying the tens.

Example 5

$$\begin{array}{r} 45 \\ \times\, 45 \\ \hline 2025 \end{array}$$

5 x 5 in the units place gives us 25.
Instead of 4 x 4 (40 x 40) in the tens place,
we multiply 4 x 5 (40 x 50) to get 20 (2,000).

Example 6

$$\begin{array}{r} 75 \\ \times\, 75 \\ \hline 5625 \end{array}$$

5 x 5 in the units place gives us 25.
Instead of 7 x 7 in the tens place,
we multiply 7 x 8 to get 56. So, the answer is 5,625.

Now what about 27 x 23? Build this problem. If you bring all the tens on top again, won't this make another hundred? Yes. We can prove it using the difference of two squares.

In 27 x 23, the formula indicates the answer should be 621. First we'll express 27 and 23 in relation to 25.

Example 7

$$\begin{array}{r} 27 \\ \times\, 23 \\ \hline 621 \end{array} \quad \begin{array}{r} 27 \\ \times\, 23 \\ \hline \end{array} \begin{array}{l} \rightarrow (25+2) \\ \rightarrow (25-2) \\ \end{array}$$

Now we'll use what we know about the difference of two squares.

Remember that $A^2 - B^2 = (A + B)(A - B)$.

The converse is also true. $(A + B)(A - B) = A^2 - B^2$.
So $(25 + 2)(25 - 2) = 25^2 - 2^2 = 625 - 4 = 621$.

So as long as the numbers in the units place add up to 10, and the numbers in the tens place are the same, this formula works.

LESSON 26

Repeated Factoring of Polynomials

The binomial X^4 - 16 is the same as $(X^2)(X^2) - (4)(4)$, or $(X^2)^2 - (4)^2$. This is the difference of two squares. The resultant factors will be $(X^2 - 4)(X^2 + 4)$. Notice that we are not done yet, as $X^2 - 4$ is also the difference of two squares $(X^2 - 2^2)$, and its factors are $(X - 2)(X + 2)$. Putting it all together:

$$X^4 - 16 = (X^2)^2 - (4)^2 = (X^2 + 4)(X^2 - 4) = (X^2 + 4)(X - 2)(X + 2)$$

It might help to relate this to a factor tree. The first set of factors of 30 is 5 x 6. But six isn't prime and has its own factors, which are 2 x 3. The final factors of 30 are 5 x 2 x 3.

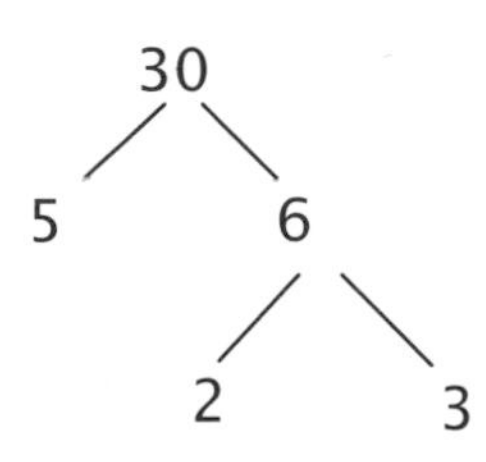

The final factors are 5 x 2 x 3.

X^4 - 16

$(X^2 + 4)$ $(X^2 - 4)$

$(X - 2)(X + 2)$

The final factors are $(X^2 + 4)(X - 2)(X + 2)$.

Example 1

Factor X^3 - 81X completely.

Look for a common factor, whether in the coefficients or in the variables. In the variables, we have a common factor of X.

$$X^3 - 81X = X(X^2 - 81) = X(X^2 - 9^2) = X(X - 9)(X + 9)$$

Example 2

Factor $X^4 - 81$ completely.

$$X^4 - 81 = (X^2)^2 - (9)^2 = (X^2 + 9)(X^2 - 9) = (X^2 + 9)(X - 3)(X + 3)$$

Example 3

Factor $3X^3 - 12X$ completely.

In the coefficients, we have a common factor of 3, and in the variables, we have a common factor of X.

$$3X^3 - 12X = 3X(X^2 - 4) = 3X(X + 2)(X - 2)$$

Example 4

Factor $2X^3 + 12X^2 + 18X$ completely.

In the coefficients, we have a common factor of 2, and in the variables, we have a common factor of X.

$$2X^3 + 12X^2 + 18X = 2X(X^2 + 6X + 9) = 2X(X + 3)(X + 3)$$

$2x^2 + 3x = 2$

$2x^2 + 3x - 2 = 0$

$(2x - 1)(x + 2) = 0$

$2x - 1 = 0 \quad x + 2 = 0$

$x = 1/2 \quad x = -2$

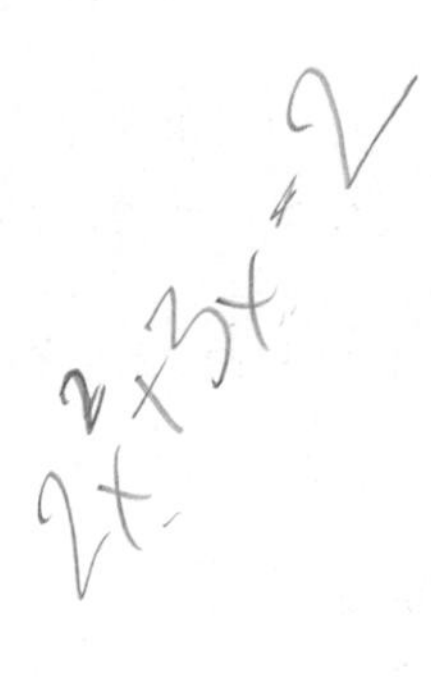

LESSON 27

Solving Equations with Factoring

Solving for X

In this lesson, we'll use what we've learned about factoring to solve equations and find the value of X.

Example 1

$2X^2 + 3X = 2$ First subtract 2 from both sides.

$2X^2 + 3X - 2 = 0$ Now find the factors.

$(2X - 1)(X + 2) = 0$ Think of these two factors as A and B.

If $A \times B = 0$, then either A or B or both have to be equal to zero.

The options are:

$0 \times B = 0$

$A \times 0 = 0$

$0 \times 0 = 0$ So either $(2x - 1) = 0$ or $(X + 2) = 0$.

$2X - 1 = 0$ $\quad$ $X + 2 = 0$

$2X = 1$ $\quad$ $X = -2$

$X = \frac{1}{2}$

The solutions are $X = \frac{1}{2}$ or $X = -2$.

Let's check these in the original equation.

$2\left(\frac{1}{2}\right)^2 + 3\left(\frac{1}{2}\right) = 2$ and $2(-2)^2 + 3(-2) = 2$ They both work, thus validating our solutions.

$2\left(\frac{1}{4}\right) + 3\left(\frac{1}{2}\right) = 2$ $\quad$ $2(4) + (-6) = 2$

$\frac{1}{2} + \frac{3}{2} = 2$ $\quad$ $8 - 6 = 2$

$2 = 2$ ✓ $\quad$ $2 = 2$ ✓

Example 2

$$2X^2 - X - 3 = 12$$ Subtract 12 from both sides.

$$2X^2 - X - 15 = 0$$ Now find the factors.

$$(2X+5)(X-3) = 0$$ Think of these two factors as A and B.

$$(2X+5) = 0 \text{ or } (X-3) = 0$$

If $2X+5=0$, then $X = -2\frac{1}{2}$.

If $X-3=0$, then $X=+3$.

The solutions are $X = -2\frac{1}{2}$ or 3.

Let's check these in the original equation. We used the decimal -2.5 instead of the fraction $-2\frac{1}{2}$ to make it easier to calculate.

$$2(-2.5)^2 - (-2.5) - 3 = 12 \quad \text{and} \quad 2(+3)^2 - (+3) - 3 = 12$$
$$2(6.25) + 2.5 - 3 = 12 \qquad 2(9) - 3 - 3 = 12$$
$$12 = 12 \checkmark \qquad 12 = 12 \checkmark$$

Example 3

$$3X^3 - 27X = 0$$ Factor out 3X.

$$3X(X^2 - 9) = 0$$ Now find the factors of $(X^2 - 9)$.

$$3X(X-3)(X+3) = 0$$ Set each of the factors equal to 0 and solve.

If $3X = 0$, then $X = 0$.

If $X - 3 = 0$, then $X = 3$.

If $(X+3) = 0$, then $X = -3$.

Let's check these in the original equation.

$$3X^2 - 27X = 0 \qquad 3X^2 - 27X = 0 \qquad 3X^2 - 27X = 0$$
$$3(0^3) - 27(0) = 0 \qquad 3(3^3) - 27(3) = 0 \qquad 3\left[(-3)^3\right] - 27(-3) = 0$$
$$0 = 0 \checkmark \qquad 3(27) - 27(3) = 0 \qquad 3(-27) - 27(-3) = 0$$
$$81 - 81 = 0 \qquad -81 + 81 = 0$$
$$0 = 0 \checkmark \qquad 0 = 0 \checkmark$$

Some polynomials cannot be factored. We will learn how to solve equations with unfactorable polynomials in *Algebra 2.*

LESSON 28

Unit Multipliers

To change inches to feet, inches to yards, ounces to pounds, pounds to tons, etc., you need to learn only two new skills. The first is how to make one, and the second is how to divide so as to produce the correct unit of measure. Those skills are admittedly rather vague, but they do sound easy, and they are. Let's view some examples to see how these two new skills function.

Change two feet to inches. First we write two feet as a fraction, with feet in the numerator.

Example 1

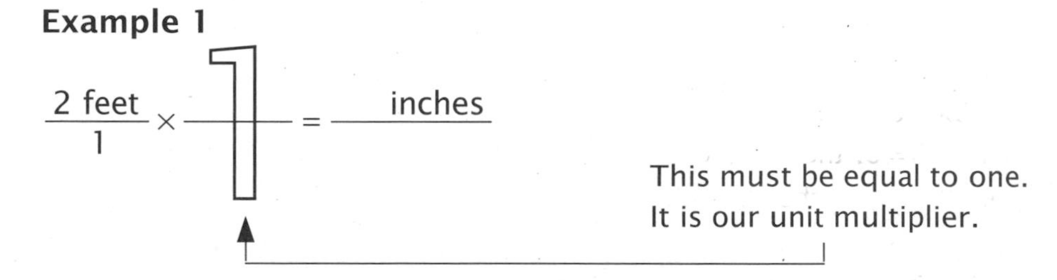

Making One - Skill #1

We must make one so that when we multiply one unit times two feet, we still have two feet. The solution may be more pieces and a different form, but it must be still equal to two feet. So we ask ourselves, "What has inches and feet in it that is equal to one?" There are two possibilities:

$$\frac{1 \text{ foot}}{12 \text{ inches}} = 1 \text{ or } \frac{12 \text{ inches}}{1 \text{ foot}} = 1$$

Both equal one because the numerator and denominator are different ways of expressing the same thing and are identical in value. These fractions are referred to as *unit multipliers.*

Either of these fractions could be multiplied by two feet without changing the value of two feet because both fractions are equal to one, and one times anything is still one. Now we come to skill #2.

Choosing the Correct Multiplier - Skill #2

Which one of these unit multipliers, when multiplied by two feet, will leave only inches, the desired unit of measure? Let's try each and see which gives the desired result.

Here we still have feet in the answer, but we just want inches.

$$\frac{2\text{ feet}}{1} \times \frac{1\text{ foot}}{12\text{ inches}} = \frac{2\text{ feet}^2}{12\text{ inches}}$$

Here we have only inches in the answer, which is what we want.

$$\frac{2\text{ }\cancel{\text{feet}}}{1} \times \frac{12\text{ inches}}{1\text{ }\cancel{\text{foot}}} = \frac{24\text{ inches}}{1}$$

Notice that the feet are canceled in the second problem as a result of having feet in the numerator of the first fraction and feet in the denominator of the second fraction. The key here is not the numbers, but the unit of measure, or kind—in this example the feet and the inches. Since we begin with feet in the numerator, skill #2 is to make sure there is feet in the denominator of our unit multiplier. And since we want inches in the numerator at the end, we must have inches in the numerator of our unit multiplier. Here are two more examples.

Example 2

Change 64 ounces to pounds.

$$\frac{64\text{ }\cancel{\text{oz}}}{1} \times \frac{1\text{ pound}}{16\text{ }\cancel{\text{oz}}} = \frac{4\text{ pounds}}{1}$$

Example 3

Change 1/2 yard to inches.

$$\frac{1\text{ }\cancel{\text{yard}}}{2} \times \frac{36\text{ in}}{1\text{ }\cancel{\text{yard}}} = \frac{18\text{ inches}}{1}$$

LESSON 29

Square Unit Multipliers

This lesson builds on the previous one. What is new is that with square units, we need to divide (cancel) **twice**, since squares are numbers that are used as a factor twice. Cubes need to be multiplied or divided three times.

Square Units

Square units are divided, or canceled, twice.

Example 1

How many square inches are in one square foot?
Or how many in^2 are in one ft^2?

$$1 \text{ square foot} = 1 \text{ ft} \times 1 \text{ ft} = 1 \text{ ft}^2$$

$$1 \text{ square inch} = 1 \text{ in} \times 1 \text{ in} = 1 \text{ in}^2$$

$$\frac{1 \text{ ft}}{1} \times \frac{1 \text{ ft}}{1} \times \frac{?}{?} \times \frac{?}{?} = \frac{? \text{ in}^2}{1}$$

$$\frac{1 \cancel{\text{ft}}}{1} \times \frac{1 \cancel{\text{ft}}}{1} \times \frac{12 \text{ in}}{1 \cancel{\text{ft}}} \times \frac{12 \text{ in}}{1 \cancel{\text{ft}}} = \frac{144 \text{ in}^2}{1}$$

Cubic Units

Cubic units are divided, or canceled, three times.

Example 2

Change two (cubic) ft^3 to in^3.

$$\frac{2 \cancel{\text{ft}}}{1} \times \frac{1 \cancel{\text{ft}}}{1} \times \frac{1 \cancel{\text{ft}}}{1} \times \frac{12 \text{ in}}{1 \cancel{\text{ft}}} \times \frac{12 \text{ in}}{1 \cancel{\text{ft}}} \times \frac{12 \text{ in}}{1 \cancel{\text{ft}}} = \frac{3{,}456 \text{ in}^3}{1}$$

More Than One Multiplier

Sometimes you need to use more than one unit multiplier in the same problem. In example 3, watch as one gallon is changed to cups.

Example 3

$$\frac{1\ \cancel{\text{gal}}}{1} \times \frac{4\ \cancel{\text{quarts}}}{1\ \cancel{\text{gal}}} \times \frac{2\ \cancel{\text{pints}}}{1\ \cancel{\text{quart}}} \times \frac{2\ \text{cups}}{1\ \cancel{\text{pint}}} = \frac{16\ \text{cups}}{1}$$

Some Useful Measures

You should memorize the following measures.

An *acre* is 43,560 square feet.

A *cord* of wood is 4 x 4 x 8 feet, or 128 cubic feet.

4 ft
4 ft
8 ft

A *yard of carpet* is a square yard, or 3 feet by 3 feet.

1 yd
1 yd
3 ft
3 ft

A *yard of concrete* refers to a cubic yard, or 3 ft by 3 ft by 3 ft, or 1 yd by 1 yd by 1 yd.

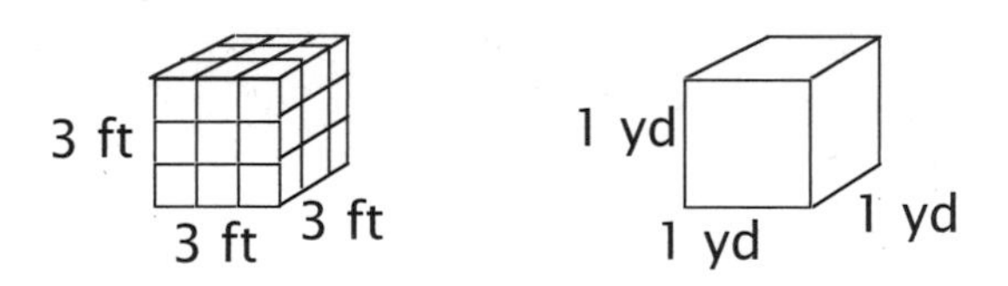

LESSON 30

Metric Conversions

There are many metric conversions available for use when converting from metric to U. S. customary measure and vice versa. These are shown in the following table. I've listed the ones I feel are most important. The conversions below are approximate and have been rounded for ease in calculations. Because the conversions have been rounded, you may get slightly different answers, depending on whether you choose the table on the left or the right for your calculations. The easiest way to calculate conversions is to make the denominator in the unit multiplier one instead of a decimal. This way you multiply instead of divide a decimal. Notice examples 2A and 2B, which illustrate this.

U. S. Customary to Metric

1 inch ≈ 2.5 centimeters
1 yard ≈ .9 meter
1 mile ≈ 1.6 kilometers
1 ounce ≈ 28 grams
1 pound ≈ .45 kilogram
1 quart ≈ .95 liter

Metric to U. S. Customary

1 centimeter ≈ .4 inch
1 meter ≈ 1.1 yards
1 kilometer ≈ .62 mile
1 gram ≈ .035 ounce
1 kilogram ≈ 2.2 pounds
1 liter ≈ 1.06 quarts

Example 1
Change 34 inches to centimeters.

$$\frac{34 \not{\text{in}}}{1} \times \frac{2.5 \text{ cm}}{1 \not{\text{in}}} = \frac{85 \text{ cm}}{1}$$

Example 2A

Change 5 liters to quarts.

$$\frac{5 \text{ liters}}{1} \times \frac{1.06 \text{ quarts}}{1 \text{ liter}} = \frac{5.3 \text{ quarts}}{1}$$

Example 2B

Change 5 liters to quarts.

$$\frac{5 \text{ liters}}{1} \times \frac{1 \text{ quart}}{.95 \text{ liter}} = \frac{5.26 \text{ quarts}}{1}$$

It seems easier to multiply by 1.06 than to divide by .95. Notice the slightly different answers.

LESSON 31

Fractional Exponents

Square Root

The square root of nine is three. This is written as $\sqrt{9} = 3$. We can also write this using exponents. Remember that we write nine squared as 9^2. This could also be written as $9^{\frac{2}{1}}$, since 2/1 is the same as 2. What is the opposite of nine squared? It is the square root of nine. What is the opposite of $9^{\frac{2}{1}}$? Hint: remember the reciprocal.

The answer is $9^{\frac{1}{2}}$. So $9^{\frac{1}{2}}$ is the way we represent the square root using exponents. In $3^{\frac{4}{1}}$, the 4 indicates $3 \cdot 3 \cdot 3 \cdot 3$. The number in the numerator of the exponent tells how many times the number three (the base) is used as a factor.

Example 1

$7^{\frac{2}{1}} = 49$ $\quad$ $25^{\frac{1}{2}} = 5$

Example 2

$9^{\frac{2}{1}} = 81$ $\quad$ $121^{\frac{1}{2}} = 11$

Example 3

$5^{\frac{2}{1}} = 25$ $\quad$ $36^{\frac{1}{2}} = 6$

Cube Root

In $8^{\frac{1}{3}}$, we ask, "What times itself three times equals eight?" $2 \cdot 2 \cdot 2 = 8$, so the *cube root* of eight is two.

When the symbol for a radical $\sqrt{}$ is used, the 2 is taken for granted, or understood. If the 2 were used, it would look like this: $\sqrt[2]{}$. When other roots are to be sought, the appropriate number must be placed outside the symbol. The cube root looks like this: $\sqrt[3]{\ }$. Consider the following examples.

Example 4

$\sqrt[3]{27} = 3$ because $3 \cdot 3 \cdot 3 = 27$, or $27^{\frac{1}{3}} = 3$.

Example 5

$\sqrt[2]{64} = 8$ because $8 \cdot 8 = 64$, or $64^{\frac{1}{2}} = 8$.

Example 6

$\sqrt[3]{64} = 4$ because $4 \cdot 4 \cdot 4 = 64$, or $64^{\frac{1}{3}} = 4$.

Combination Problems

You can also get real tricky and have different numbers in the numerator and the denominator. In the following example, first find the root (the denominator), and then raise to the power (the numerator). The number in the denominator of the exponent indicates how many times the root will be multiplied by itself to equal the number in the base.

Example 7

$8^{\frac{4}{3}} = \left(\sqrt[3]{8}\right)^4$ is the same as $\left(8^{\frac{1}{3}}\right)^4 = 2^4 = 16$.

The cube root of 8 is 2, and 2 to the fourth power is 16. We solve this problem by breaking it down into two separate problems.

Example 8

$9^{\frac{3}{2}} = \left(\sqrt[2]{9}\right)^3$ is the same as $\left(9^{\frac{1}{2}}\right)^3 = 3^3 = 27$.

The square root of 9 is 3, and 3 to the third power is 27.

Now let's put together what we know about multiplying two numbers with the same bases (if two numbers have the same base, you can add the exponents) and rewrite a number using exponents.

Example 9

$9 \cdot 3^3 = 3^2 \cdot 3^3 = 3^5$

We change 9 to 3^2, and since both terms have the same base, the exponents are added.

Example 10

$(8^3)(4^5) = (2^3)^3(2^2)^5 = 2^9 \cdot 2^{10} = 2^{19}$

When any doubt persists, write the problem out the long way.

$$(2\cdot 2\cdot 2)^3(2\cdot 2)^5 = (2\cdot 2\cdot 2)(2\cdot 2\cdot 2)(2\cdot 2\cdot 2)(2\cdot 2)(2\cdot 2)(2\cdot 2)(2\cdot 2)(2\cdot 2)$$
$$= 2^{19}$$

LESSON 32

Significant Digits and Scientific Notation

Measurement

Undergirding the entire discussion of significant digits is the fact that measurements are approximate. Whether you are reading a ruler or a digital scale in your bathroom, measurements are not completely accurate. However, some are more accurate than others, and that is what we have to learn to recognize and take into consideration. Accuracy is how close you get to the reality of what you are trying to measure. Precision describes this reality, and how many significant digits you use tells how precise you are. Numbers that are not measures, but are specific absolute numbers, tell how many and are not to be treated in the same way. If I say that my shoe is 13.125 inches long, that is an approximate measure. But if I asked how long two of my shoes would be end to end, then the two is not a measurement or an approximation; it is the whole number two.

Significant Digits

The digits used in describing a measure are called significant digits. The first postulate is that all digits, except zero, are significant. In the number 13.125, all digits are significant, and so we say it has five significant digits. (You may see this abbreviated as SD.)

Sometimes, however, zeros are significant, and the hard part about significant digits is deciding which zeros are significant and which are place holders. Common sense will help more than a formula, so let's think this through. The number 6,007 has four significant digits because the zeros are between the 6 and the 7, which are significant digits. All the digits in 6,007 are significant. The number 600 (which could be 591 or 604 rounded to the nearest hundred) has two zeros which are just holding the units and tens places, so this number has only one significant digit. This is true for decimal numbers as well. Consider .0006, which has one significant digit and three place holders. This is pretty straightforward, but it gets tricky when you have a number like .060. Now how many significant digits are there? The answer is two, because the zero to the right of the 6 shows that the measurement is accurate to the thousandths place. The number 60.0 has three significant digits because the zeros are showing precision to the tenths place. We can sum up these thoughts with three more postulates.

A. If the zero is between numbers, it is a significant digit.
B. If the zero is a place holder, which means between the decimal point and the number, it is not a significant digit.
C. If the zero is to the right of a decimal point and is expressing precision, it is a significant digit.

Significant Digits with Addition and Subtraction

Since you add and subtract the same kind, for example tenths added to tenths, and hundredths subtracted from hundredths, then it follows that an answer can only be as accurate as the least precise component of the problem. For example, if you are adding 4.1 + 2.333, then you must round your answer to the tenths place, because tenths are less precise than thousandths. 4.1 + 2.333 = 6.433, which is rounded to 6.4. It is a question of place value and the preciseness of the addends.

Significant Digits with Multiplication and Division

In these operations, it is the number of significant digits that is the key. For example, let's multiply 41 by 2.333. We first ask which number has the fewest number of significant digits. The answer is 41, which has two significant digits. Therefore, while 95.653 is the answer, this measurement can be trusted to only two significant digits and must be rounded to 96.

For division, consider the problem 9.5653 divided by 2.33. Since 2.33 has the fewest number of significant digits, then the answer should also have three significant digits. Instead of an answer of 4.10527897, we round to 4.11, which has three significant digits.

Remember that in multiplication and division, if one of the factors is a whole number and not a measurement, then it does not fit the same template. Recall the shoe problem: 2 x 13.125 inches. Since two is a whole number, the answer is left as 26.250. If I were dividing 12,549 pumpkin seeds among three farmers, my answer would not have one significant digit like three because the three is a whole number, and not an approximation. The number of seeds per farmer would be 4,183 (not 4,000).

Scientific Notation

Scientific notation is used in science to solve equations that use very large and/or very small numbers. If we were asked to compute 200 times the distance from the earth to the sun (93 million, or 93,000,000 miles) using the normal method of multiplying, it would require a good deal of paper and pencil work, and lots of zeroes. (The distance is an estimate, since the orbit of the sun is an ellipse.)

Example 1

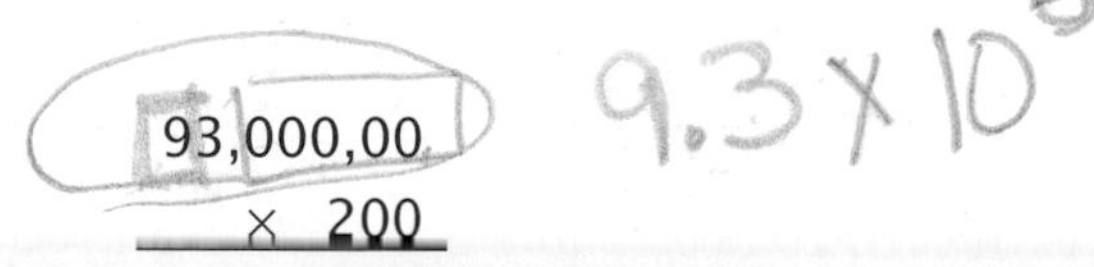

Exponential Notation

Scientific notation provides an easier and more efficient method of computing. It is closely related to exponential notation, which separates the digit from the place value. In example 1, ninety million, or 90,000,000, would be shown in exponential notation as nine times 10,000,000. The 10,000,000 is expressed with exponents as 10 to the seventh power. Three million is expressed as three times 10 to the sixth power. 93,000,000 in exponential notation is $9 \times 10^7 + 3 \times 10^6$

Writing Scientific Notation

In scientific notation, you keep the 9 and the 3 together instead of separating them, and place the decimal point so the 9 is in the units place. The 9 and 3 are the significant digits (SD) and are called the *mantissa*. The exponent is figured from the number in the units place in the mantissa. It's almost as if you forget the 3 when choosing the exponent. The exponent is referred to as the *characteristic*. In example 1, the mantissa has two significant digits and is 9.3, and the characteristic is 7.

93,000,000 in scientific notation is 9.3×10^7. Let's express a few more numbers using scientific notation:

$135{,}000 = 1.35 \times 10^5$. It has 3 significant digits (SD). The mantissa is 1.35 and the characteristic is 5.

$62{,}700{,}000 = 6.27 \times 10^7$. It has 3 SD. The mantissa is 6.27 and the characteristic is 7.

$8{,}900 = 8.9 \times 10^3$. It has 2 SD. The mantissa is 8.9 and the characteristic is 3.

$50 = 5 \times 10^1$. It has 1 SD. The mantissa is 5 and the characteristic is 1.

Computing with Scientific Notation

To solve our original problem of 200 times the distance from the sun to the earth, multiply the mantissas and add the exponents. Remember that 9.3 is a measurement and 200 is not a measurement. So the answer will be rounded to two SD.

$$\left(9.3\times10^7\right)\left(2\times10^2\right)=\left(9.3\times2\right)\left(10^7\times10^2\right)=\left(18.6\right)\left(10^9\right)=\left(1.86\times10^1\right)\left(10^9\right)$$
$$=1.86\times10^{10}=1.9\times10^{10}$$

Small Numbers

You can also show very small decimal numbers with scientific notation. Remember that you may only have one number in the units place and the rest must be decimals. Let's begin with .000547.

Example 2

$.000547 = 5.47 \times 10^{-4}$ (three SD)

Let's solve some more problems. Try these first; then check the solutions Assume all numbers to be measurements.

Example 3

$30{,}000{,}000 \times .000023$

$(3\times10^7)(2.3\times10^{-5}) = (3\times2.3)(10^7\times10^{-5}) = (6.9)(10^2) = 7\times10^2$ (one SD)

Example 4

$93{,}000{,}000 \times .000547$

$(9.3\times10^7)(5.47\times10^{-4}) = (9.3\times5.47)(10^7\times10^{-4}) = (50.871)(10^3)$

$= (5.0871\times10^1)(10^3) = 5.0871\times10^4$

$= 5.1\times10^4$ (two SD)

Example 5

$280{,}000{,}000 \div .0007$

$(2.8 \times 10^8) \div (7 \times 10^{-4}) = (2.8 \div 7)(10^8 \div 10^{-4}) = (.4)(10^{12})$

$= (.4 \times 10^{-1})(10^{12}) = 4 \times 10^{11}$

The least number of significant digits in the first line of the solution is one, and the answer has one significant digit as well.

LESSON 33

Bases Other than Ten

Bases

In the decimal system, every place value is based on 10. In algebra, every value is based on X. There are other bases. We'll examine base 2, base 4, base 8, and base 12.

Base 2, which is used in calculators, uses only two digits, or numbers: 0 and 1.
Base 4 uses only four digits: 0, 1, 2, and 3.
Base 8 uses only eight digits: 0, 1, 2, 3, 4, 5, 6 and 7.

Base 12 uses 12 digits. Since we know of only 10 digits in base 10, we have to create two more, A and B, to represent 10 and 11. So the digits are 0, 1, 2, 3, 4, 5, 6, 7, 8, 9, A, and B.

Look at table 1 on the next page to see how to count in the different bases.

Subscripts

Most of us have never operated in any other base save base 10. To tell when another base is being used, *subscripts* (*sub* means under) are used to denote the base. In base 10, these are understood, but they could be written with a subscript of 10. We would write 143 in base 10 as 143_{10}.

Base 10	Base 2	Base 4	Base 8	Base 12
0	0	0	0	0
1	1	1	1	1
2	10	2	2	2
3	11	3	3	3
4	100	10	4	4
5	101	11	5	5
6	110	12	6	6
7	111	13	7	7
8	1,000	20	10	8
9	1,001	21	11	9
10	1,010	22	12	A
11	1,011	23	13	B
12	1,100	30	14	10
13	1,101	31	15	11
14	1,110	32	16	12
15	1,111	33	17	13
16	10,000	100	20	14

Notice that $16_{10} = 10{,}000_2 = 100_4 = 20_8 = 14_{12}$

Changing from a Different Base to Base 10

When changing one base to a different base, the key is the place value. Recall exponential notation, where we separate the number (how many) from the place value (what kind). Exponential notation helps to distinguish the number that tells how many from the place value.

In base 10, we had Decimal Street. The numbers in base 10 go from zero to nine. The place values in base 10 are based on 10: $10^0, 10^1, 10^2, 10^3 \ldots$

156 in base 10 is written as $1 \times 10^2 + 5 \times 10^1 + 6 \times 10^0$. It looks like this:

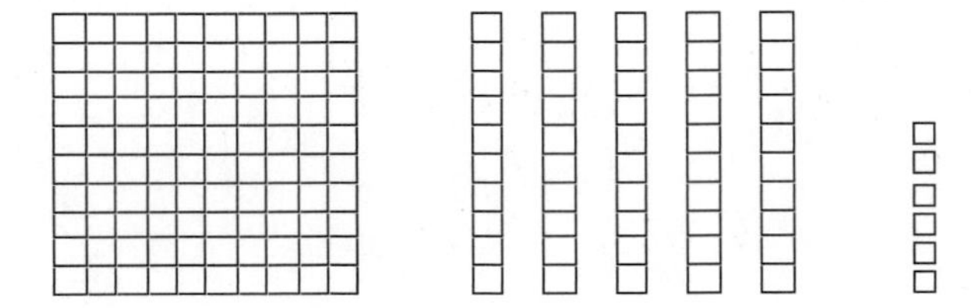

In base 4, we have Quad Street. The numbers in base 4 go from zero to three. The place values in base 4 are based on 4: $4^0, 4^1, 4^2, 4^3 \ldots$

312 in base 4 is written as $3 \times 4^2 + 1 \times 4^1 + 2 \times 4^0$. On Quad Street it looks like this:

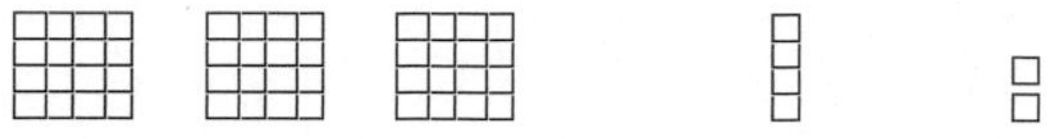

Example 1

Let's change from 312 in base 4 to some number in base 10. What you are asking is how many 10^2s there are, how many 10^1s there are, how many 10^0s there are, etc. To find out, follow these steps in order:

1. Write 312_4 in exponential notation.
2. Change the place values from exponents to number form.
3. Then multiply each number by the place value.
4. Add them all together.

$$3 \times 4^2 + 1 \times 4^1 + 2 \times 4^0$$
$$3 \times 16 + 1 \times 4 + 2 \times 1$$
$$48 + 4 + 2$$
$$54$$
$$312_4 = 54_{10}$$

Example 2

Let's change from 5,8B2 in base 12 to some number in base 10. What you are asking is how many 10^2s there are, how many 10^1s there are, how many 10^0s there are, etc. To find out, follow these steps in order:

1. Write 5,8B2 base 12 in exponential notation.
2. Change the place values from exponents to number form.
3. Then multiply each number by the place value.
4. Add them all together.

$$5 \times 12^3 + 8 \times 12^2 + B \times 12^1 + 2 \times 12^0$$
$$5 \times 1{,}728 + 8 \times 144 + 11 \times 12 + 2 \times 1$$
$$8{,}640 + 1{,}152 + 132 + 2$$
$$9{,}926_{10}$$
$$5{,}8B2_{12} = 9{,}926_{10}$$

Changing from Base 10 to a Different Base

Now we will learn how to go in the opposite direction.

Example 3

Let's change from 54 in base 10 to some number in base 4. What you are asking is how many 4^3s (sixty-fours) there are, how many 4^2s (sixteens) there are, how many 4^1s (fours) there are, how many 4^0s (ones) there are. To find out, begin dividing 54 by the values in base 4 and work your way from the largest value (sixteen in this case) to the smallest value (this will be one).

$$\begin{array}{r} 0 \\ 64\overline{)54} \\ \underline{0} \\ 54 \\ (4^3) \end{array} \nearrow \begin{array}{r} 3 \\ 16\overline{)54} \\ \underline{48} \\ 6 \\ (4^2) \end{array} \nearrow \begin{array}{r} 1 \\ 4\overline{)6} \\ \underline{4} \\ 2 \\ (4^1) \end{array} \nearrow \begin{array}{r} 2 \\ 1\overline{)2} \\ \underline{2} \\ 0 \\ (4^0) \end{array}$$

$$54_{10} = 3 \times 4^2 + 1 \times 4^1 + 2 \times 4^0 = 312_4$$

Example 4

Change $9{,}926_{10}$ to base 12.

$$\begin{array}{r} 5 \\ 1{,}728\overline{)9{,}926} \\ \underline{8{,}640} \\ 1{,}286 \\ (12^3) \end{array} \nearrow \begin{array}{r} 8 \\ 144\overline{)1{,}286} \\ \underline{1{,}152} \\ 134 \\ (12^2) \end{array} \nearrow \begin{array}{r} 11 \text{ or B} \\ 12\overline{)134} \\ \underline{132} \\ 2 \\ (12^1) \end{array} \nearrow \begin{array}{r} 2 \\ 1\overline{)2} \\ \underline{2} \\ 0 \\ (12^0) \end{array}$$

$$9{,}926_{10} = 5 \times 12^3 + 8 \times 12^2 + B \times 12^1 + 2 \times 12^0 = 5{,}8B2_{12}$$

LESSON 34

Graphing a Circle and an Ellipse

You know that $Y = mX + b$ and $AX + BY = C$ represent equations of a line. In this lesson, we have two new equations that represent curves.

Graphing a Circle

The basic equation for a circle is $X^2 + Y^2 = R^2$, where R is the radius of the circle.

Example 1

Graph $X^2 + Y^2 = 3^2$

$X^2 + Y^2 = 9$

X	Y
0	+3
0	–3
+3	0
–3	0

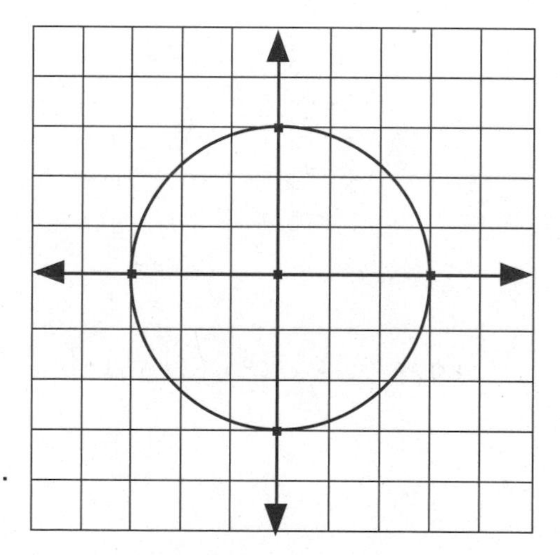

If $X = 0$: $(0)^2 + Y^2 = 9$

$\sqrt{Y^2} = \pm\sqrt{9}$

$Y = \pm 3$

The points are (0, 3) and (0, –3).

If $Y = 0$: $X^2 + (0)^2 = 9$

$\sqrt{X^2} = \pm\sqrt{9}$

$X = \pm 3$

The points are (3, 0) and (–3, 0).

You see that the radius of the circle is 3, and our equation is $X^2 + Y^2 = R^2$. (3 = R or radius).

Example 2

Graph $(X-1)^2+(Y-2)^2=3^2$

X	Y
1	+5
1	−1
+4	2
−2	2

If $X=1$: $(Y-2)^2=9$

$$Y^2-4Y+4=9$$
$$Y^2-4Y-5=0$$
$$(Y-5)(Y+1)=0$$

$Y-5=0$ $\quad$ $Y+1=0$

$Y=5$ $\quad$ $Y=-1$

$Y=5,\ -1$

The points are $(1,\ 5)$ and $(1,\ -1)$.

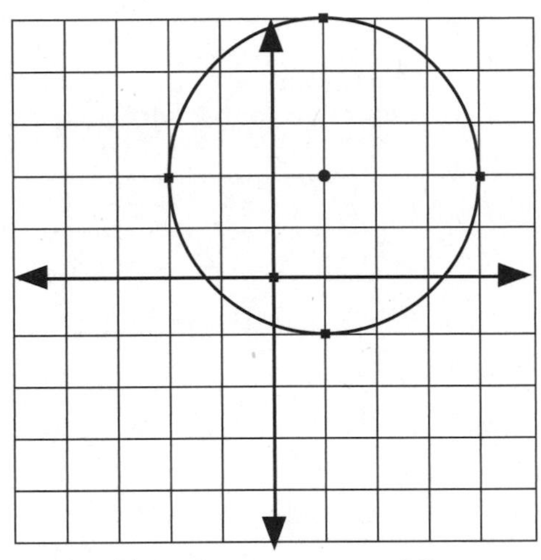

If $Y=2$: $(X-1)^2=9$

$$X^2-2X+1=9$$
$$X^2-2X-8=0$$
$$(X-4)(X+2)=0$$

$X-4=0$ $\quad$ $X+2=0$

$X=4$ $\quad$ $X=-2$

$X=4,\ -2$

The points are $(4,\ 2)$ and $(-2,\ 2)$.

So (1, 2) is the center, and 3 is the radius. A formula for the center (h, k) and the radius (R) is $(X-h)^2+(Y-k)^2=R^2$.

The coefficients of the equation of a circle are equal.

Graphing an Ellipse

An ellipse is a "stretched," or elongated, circle. The coefficients of the variables in the equation of an ellipse are not equal.

Example 3

Graph $9X^2 + 4Y^2 = 36$.

X	Y
0	+3
0	–3
+2	0
–2	0

Ellipse

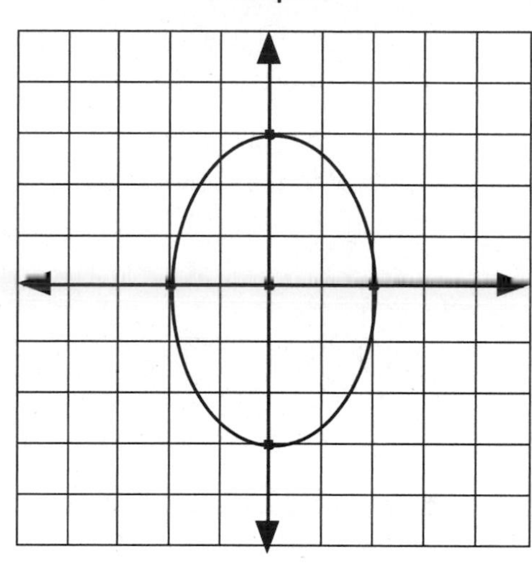

If $X = 0$: $9(0)^2 + 4Y^2 = 36$

$$4Y^2 = 36$$
$$Y^2 = 9$$
$$\sqrt{Y^2} = \sqrt{9}$$
$$Y = \pm 3$$

The points are $(0, 3)$ and $(0, -3)$.

If $Y = 0$: $9X^2 + 4(0)^2 = 36$

$$9X^2 = 36$$
$$X^2 = 4$$
$$\sqrt{X^2} = \sqrt{4}$$
$$X = \pm 2$$

The points are $(2, 0)$ and $(-2, 0)$.

The equation is $AX^2 + BY^2 = C$.

When given the equation for a circle or an ellipse, always see if you can simplify first by dividing each term by the greatest common factor (GCF).

Conic Sections

A circle and an ellipse can each be shown as a plane intersecting a cone. These are called *conic sections.*

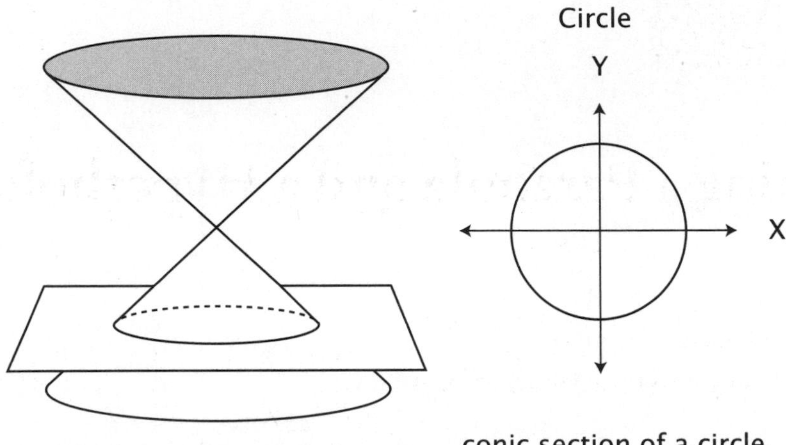

conic section of a circle

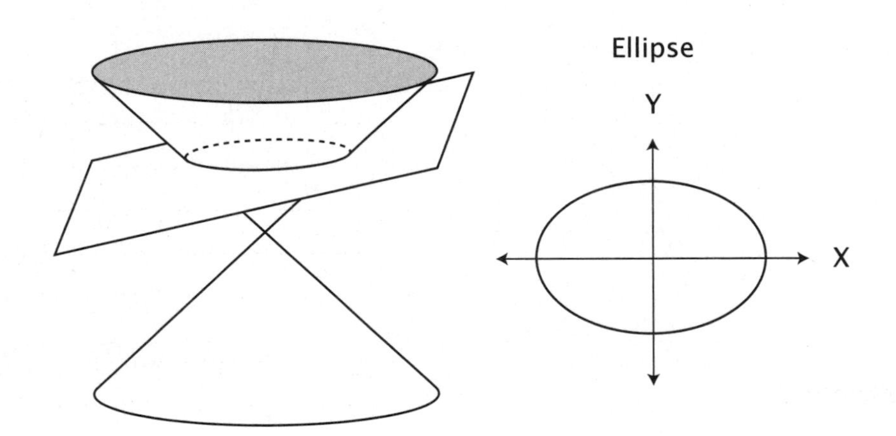

conic section of an ellipse

LESSON 35

Graphing a Parabola and a Hyperbola

Graphing a Parabola

Here is an introduction to the graph of a parabola.

Example 1

Graph $Y = X^2$

X	Y
0	0
1	1
-1	1
2	4
-2	4

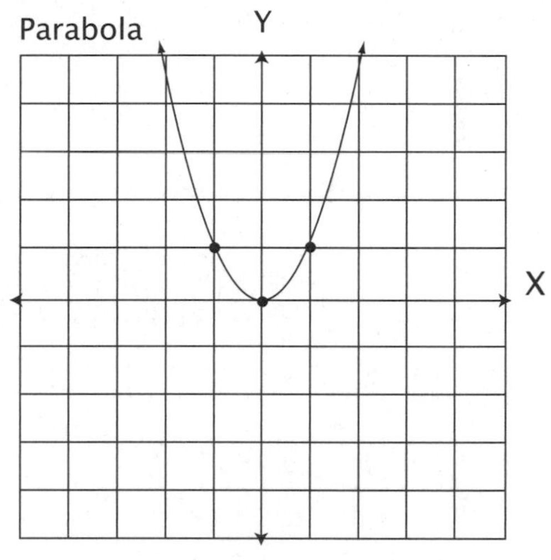

Example 2

Graph $Y = 2X^2$

X	Y
0	0
1	2
-1	2
2	8
-2	8

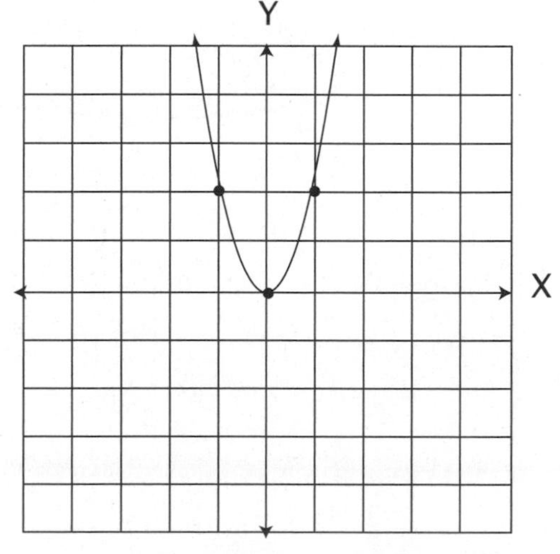

Example 3

Graph $Y = 1/4\ X^2$

$1/4\ (0)^2 = 0$

$1/4\ (\pm 1)^2 = 1/4$

$1/4\ (\pm 2)^2 = 1$

X	Y
0	0
1	1/4
-1	1/4
2	1
-2	1

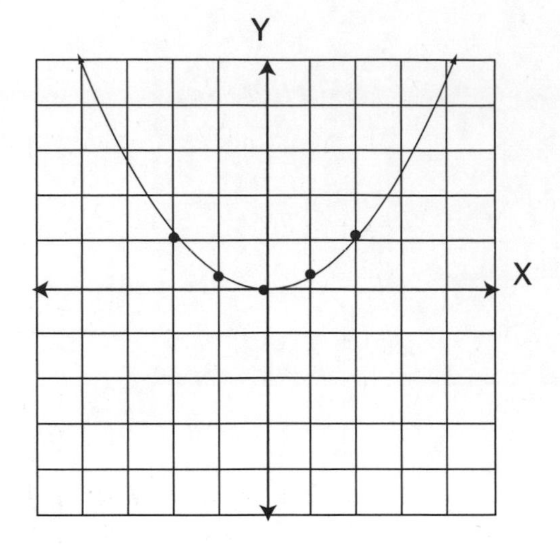

Example 4

Graph $Y = -X^2$

X	Y
0	0
1	-1
-1	-1
2	-4
-2	-4

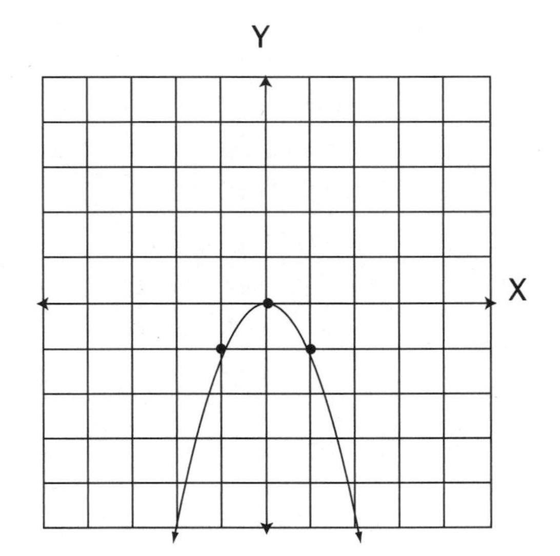

Example 5

Graph $Y = X^2 + 2$

X	Y
0	2
1	3
-1	3
2	6
-2	6

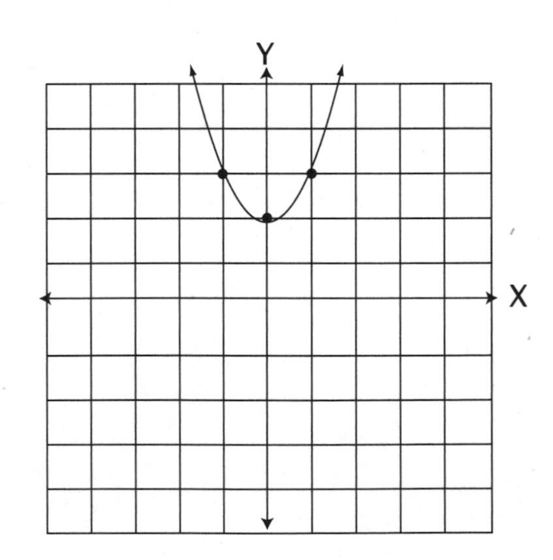

Graphing a Hyperbola

The hyperbola is another graph you should recognize.

Example 6

Graph $XY = 6$

X	Y
1	6
2	3
3	2
6	1
-1	-6
-2	-3
-3	-2
-6	-1

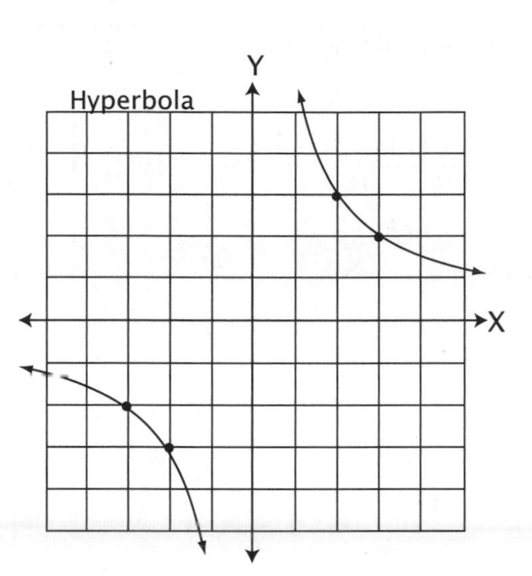

The value of X or Y can never be zero in this case. What value of X gives $X(0) = 6$? There is no possible value. We say X is *undefined* for zero.

Example 7

Graph $XY = -6$

X	Y
1	-6
2	-3
3	-2
6	-1
-1	6
-2	3
-3	2
-6	1

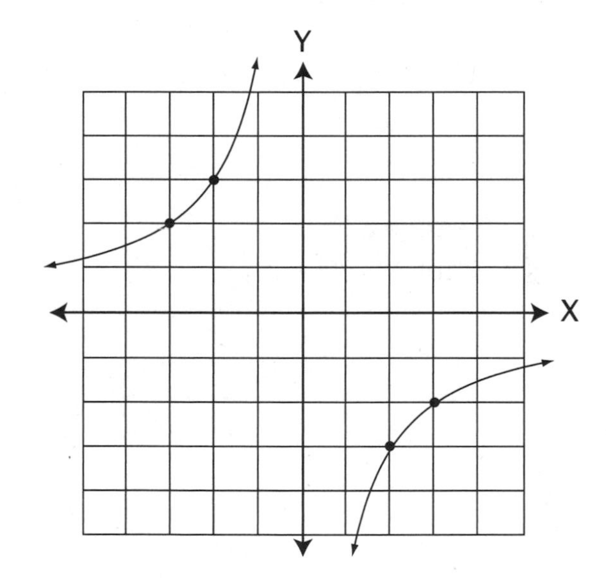

Conic Sections

A parabola and a hyperbola can also be shown as conic sections.

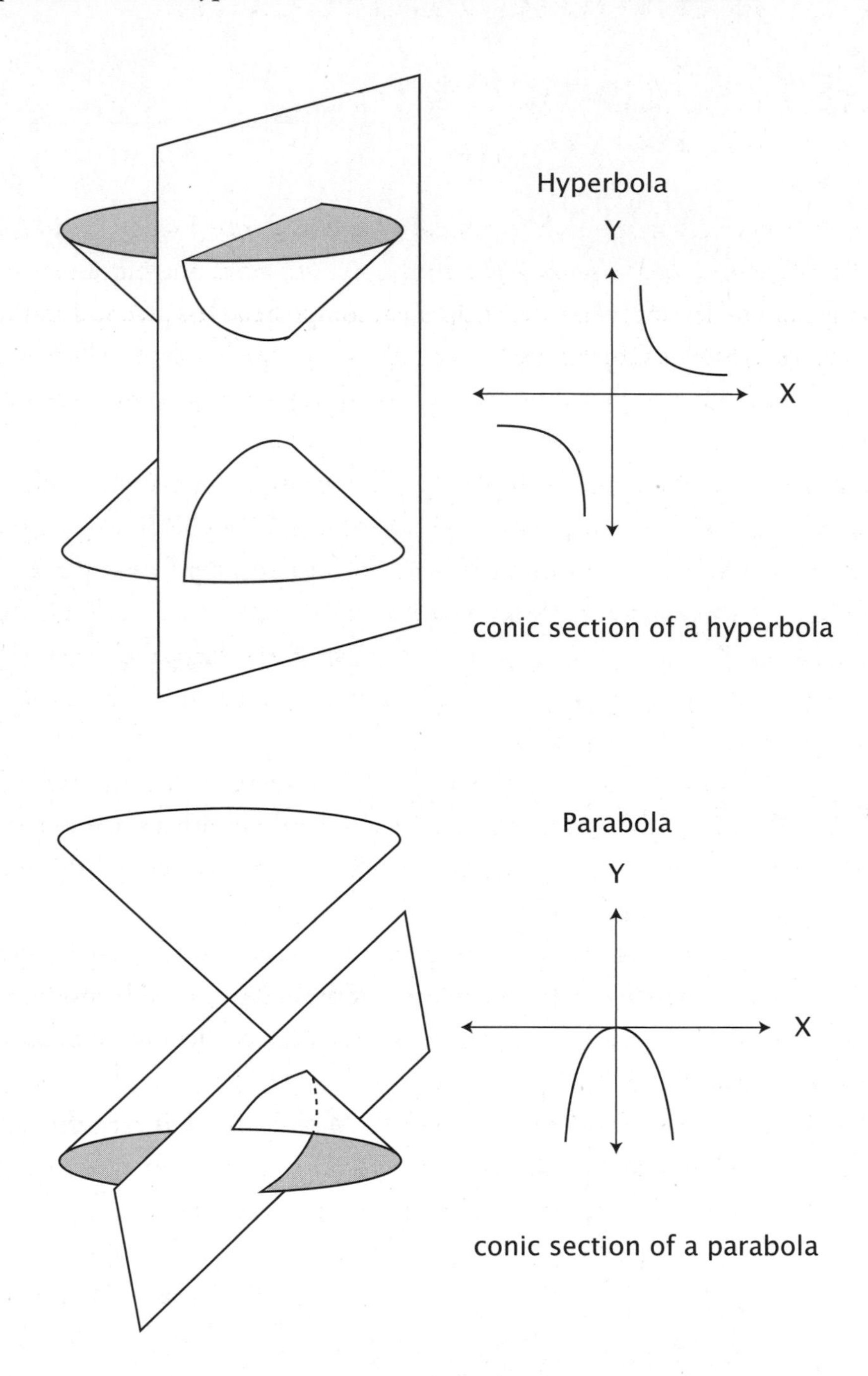

When Should A Student Study Geometry?

I have often been asked about the secondary math sequence. I recommend *Geometry* between *Algebra 1* and *Algebra 2*. I want to give students one more year for their thinking skills to develop, which will help them comprehend less practical and concrete math topics. Abstract subjects are better studied by older students whose reasoning ability has had a chance to mature. The ability to employ logic seems to be related to age and maturity.

Geometry is very visual with plenty of applications in everyday life. Angles, perimeter, area, volume, line graphs, and other subjects can be drawn and pictured. For this reason, the first two-thirds of Math-U-See *Geometry* focus on the concrete material with two-column proofs strategically placed at the end of the book. By taking *Algebra 2,* which is mostly theoretical, after a year of geometry, the ability to reason has one more year to develop. My hope is that the extra year will help the student grasp abstract concepts more readily.

Another consideration is the SAT exam. The material on this exam focuses primarily on material covered in *Algebra 1* and *Geometry* with a small portion from *Algebra 2.* If a student needs to take this exam early in high school, this would be another factor in your decision.

I am aware of the potential for students to forget algebra while studying geometry. This is a concern I have heard from parents. I addressed this need by inserting many review units from *Algebra 1* throughout the *Geometry* curriculum in order to keep these algebra skills sharp.

This is my rationale for the way I designed Math-U-See, but even though this is my preference, I know that students differ in their capabilities. So, I have written out other possible sequences that would also fit with our materials.

—Steve Demme

RECOMMENDED SEQUENCE FOR SECONDARY MATH

Algebra 1—Geometry—Algebra 2—PreCalculus with Trigonometry If you are taking one course per year it would look like this:

Year 1 – *Algebra 1*
Year 2 – *Geometry*
Year 3 – *Algebra 2*
Year 4 – *PreCalculus with Trigonometry*

ALTERNATIVE 1

Take *Algebra 1* and *Geometry* simultaneously. When you have completed these, take *Algebra 2,* followed by *PreCalculus with Trigonometry*. If you choose this option, you will notice that algebra review lessons are systematically sprinkled throughout *Geometry.* Skip these problems or do them as your knowledge of algebra increases.

Year 1 – *Algebra 1* and *Geometry*
Year 2 – *Algebra 2*
Year 3 – *PreCalculus with Trigonometry*

ALTERNATIVE 2

Take *Algebra 1,* and then follow this by taking *Geometry* and *Algebra 2* at the same time. The following year, you will be prepared for *PreCalculus with Trigonometry.* If you decide on this course of action, finish *Geometry* lesson 18 before beginning *Algebra 2* lesson 22. Also, some of the tests for *Algebra 2* (which assumes you have finished geometry) have questions that deal with geometry. Do them if you can, or skip them until you finish the geometry course.

Year 1 – *Algebra 1*
Year 2 – *Algebra 2* and *Geometry*
Year 3 – *PreCalculus with Trigonometry*

ALGEBRA 1 READINESS TEST SOLUTIONS

1. $\frac{1}{2}$ of $36 = 18$

2. $\frac{2}{3}$ of $12 = 8$

3. $\frac{7}{8}$ of $56 = 49$

4. $\frac{2}{5} = \frac{4}{10} = \frac{6}{15} = \frac{8}{20}$

5. $\frac{3}{7} = \frac{6}{14} = \frac{9}{21} = \frac{12}{28}$

6. $\frac{1}{2} + \frac{3}{4} + \frac{5}{8} =$

   $\frac{4}{8} + \frac{6}{8} + \frac{5}{8} = \frac{15}{8} = 1\frac{7}{8}$

7. $\frac{4}{5} - \frac{2}{3} = \frac{12}{15} - \frac{10}{15} = \frac{2}{15}$

8. $\frac{1}{5} \div \frac{1}{6} = \frac{1}{5} \times \frac{6}{1} = \frac{6}{5} = 1\frac{1}{5}$

9. $4\frac{2}{3} \div \frac{7}{18} = \frac{\cancel{14}^{2}}{\cancel{3}} \times \frac{\cancel{18}^{6}}{\cancel{7}} = 12$

10. $\frac{3}{5} \times 2\frac{1}{4} \times 4\frac{1}{3} =$

    $\frac{3}{5} \times \frac{9}{4} \times \frac{13}{3} = \frac{117}{20} = 5\frac{17}{20}$

11. $$\begin{array}{r} 4\frac{1}{2} = 3\frac{9}{6} \\ -3\frac{2}{3} = -3\frac{4}{6} \\ \hline \frac{5}{6} \end{array}$$

12. $\frac{47}{7}$

13. $$\begin{array}{r} 8.63 \\ -1.85 \\ \hline 6.78 \end{array}$$

14. $$\begin{array}{r} 7.0 \\ +6.38 \\ \hline 13.38 \end{array}$$

15. $$\begin{array}{r} 21.052 \\ -\ .486 \\ \hline 20.567 \end{array}$$

16. $$\begin{array}{r} 4.29 \\ \times\ .5 \\ \hline 2.145 \end{array}$$

17. $$\begin{array}{r} 2.7 \\ \times\ 3. \\ \hline 8.1 \end{array}$$

18. $$\begin{array}{r} .005 \\ \times\ .08 \\ \hline .00040 \end{array}$$ or .0004

19. $5\overline{)16.6}$ = 3.32

20. $.04\overline{).033}$ = .825 rounds to .83

21. $11\overline{)8.}$ = .727 rounds to .73

22. 6% = .06

23. 45% = .45

24. $\frac{6}{10} = .6 = 60\%$

25. $\frac{1}{4} = .25 = 25\%$

26. $(-7) + (-24) = -31$

27. $(-6) \times (-14) = 84$

28. $(10) - (-5) = 15$

29. $(-36) \div (9) = -4$

30. $-1^2 = -(1)(1) = -1$

31. $-(5)^3 = -(5)(5)(5) = -125$

32. $(-5)^2 = (-5)(-5) = 25$

33. $\left(-\frac{2}{5}\right)^2 = \left(-\frac{2}{5}\right)\left(-\frac{2}{5}\right) = \frac{4}{25}$

34. $2 \times 10^3 + 7 \times 10^2 + 1 \times 10^1 + 6 \times 10^0 + 8 \times \frac{1}{10^1} =$

$$2 \times 1000 + 7 \times 100 + 1 \times 10 + 6 \times 1 + 8 \times .1 =$$

$$2000 + 700 + 10 + 6 + .8 = 2{,}716.8$$

35. $\sqrt{81} = 9$

36. $\sqrt{25} = 5$

37. $\sqrt{X^2} = X$

38. 14

39. 8

40. 40

41. 18

42. $2 \times 2 \times 2 \times 7 = 56$

43. $5 \times 5 \times 3 = 75$

44. 7

45. $\frac{1}{4}$

Student Solutions

Lesson Practice 1A

1. done
2. done
3. $(6)(-5) = -30$
4. $(-8)(-5) = 40$
5. $8 + 6 = 6 + 8$ (commutative property)
   $14 = 14$ true
6. $5 \times 9 = 9 \times 5$ (commutative property)
   $45 = 45$ true
7. $8 - 4 = 4 - 8$
   $4 = -4$ false
8. $36 \div 4 = 4 \div 36$
   $9 = \frac{4}{36}$ or $\frac{1}{9}$ false
9. $(2 + 9) + 8 = 2 + (9 + 8)$ (associative property)
   $11 + 8 = 2 + 17$
   $19 = 19$ true
10. $(4 \times 5) \times 6 = 4 \times (5 \times 6)$ (associative property)
    $20 \times 6 = 4 \times 30$
    $120 = 120$ true
11. $(11 - 4) - 2 = 11 - (4 - 2)$
    $7 - 2 = 11 - 2$
    $5 \neq 9$ false
12. $(9 \div 3) \div 3 = 9 \div (3 \div 3)$
    $3 \div 3 = 9 \div 1$
    $1 \neq 9$ false
13. false; see #7
14. true; see #9
15. true; see #6 and #10

Lesson Practice 1B

1. $(-3) + (-10) = -13$
2. $(-3) - (+10) = (-3) + (-10) = -13$
3. $(6) - (-5) = (6) + (+5) = 11$
4. $(-8) - (-5) = (-8) + (+5) = -3$
5. $5D - 6C + 8D - 3C + B =$
   $5D + 8D - 6C - 3C + B = B - 9C + 13D$
6. $2A + B - A + 3B =$
   $2A - A + B + 3B = A + 4B$
7. $5Q + 3C - C + Q + 4Q - 5C =$
   $5Q + Q + 4Q + 3C - C - 5C = -3C + 10Q$
8. $20 + 5X - 6Y + Y + 2X + X - 9 =$
   $20 - 9 + 5X + 2X + X - 6Y + Y = 8X - 5Y + 11$
9. $2X + 2 - X + 2X =$
   $2X - X + 2X + 2 = 3X + 2$
10. $3Y - 1 + 2Y - 1 - 4Y =$
    $3Y + 2Y - 4Y - 1 - 1 = Y - 2$
11. $5A - 6B - 3B + 10A - 8 =$
    $5A + 10A - 6B - 3B - 8 = 15A - 9B - 8$
12. $18X - 5Y - 9X + Y =$
    $18X - 9X - 5Y + Y = 9X - 4Y$
13. false; see 1A #12
14. true; see 1A #6
15. false; see 1A #11

Systematic Review 1C

1. $4Q + 2C - 2C - 2Q - 3C =$
   $4Q - 2Q + 2C - 2C - 3C = -3C + 2Q$
2. $-5M - 7 + 3M - 4 + 5 =$
   $-5M + 3M - 7 - 4 + 5 = -2M - 6$
3. $2A - 3B + 4C - A + B + C =$
   $2A - A - 3B + B + 4C + C = A - 2B + 5C$
4. $4A - 5 - 2A + 7 - 1 =$
   $4A - 2A - 5 + 7 - 1 = 2A + 1$
5. $4X - 3Y - 6Y + 10X - 5 =$
   $4X + 10X - 3Y - 6Y - 5 = 14X - 9Y - 5$
6. $15X - 4Y - 6X + Y =$
   $15X - 6X - 4Y + Y = 9X - 3Y$
7. $15X + 6X - 4Y - 5Y - 14X + 10 =$
   $15X + 6X - 14X - 4Y - 5Y + 10 = 7X - 9Y + 10$
8. $3A - 4B + 6A + 7B + 8 =$
   $3A + 6A - 4B + 7B + 8 = 9A + 3B + 8$
9. $(-3)(5) = -15$
10. $(-81) \div (-9) = 9$
11. $4 \div (-2) = -2$

12. $(-5)^2 = (-5)(-5) = 25$

13. $4 + (-2) = 4 - 2 = 2$

14. $-4^2 = -(4 \times 4) = -16$

15. $\frac{1}{\cancel{4}} \times \frac{\cancel{7}}{11} \times \frac{\cancel{4}}{\cancel{7}} = \frac{1}{11}$

16. $\frac{1}{2} \times \frac{5}{6} \times \frac{11}{12} = \frac{55}{144}$

17. $\frac{1}{3} \div \frac{4}{5} = \frac{5}{15} \div \frac{12}{15} = \frac{5 \div 12}{15 \div 15} = \frac{5 \div 12}{1} = \frac{5}{12}$

18. $7\frac{1}{2} \div 2\frac{4}{7} = \frac{15}{2} \div \frac{18}{7} = \frac{105}{14} \div \frac{36}{14}$
$= \frac{105 \div 36}{14 \div 14} = \frac{105 \div 36}{1}$
$= \frac{105}{36} = 2\frac{33}{36} = 2\frac{11}{12}$

19. $\frac{1}{3} \div \frac{4}{5} = \frac{1}{3} \times \frac{5}{4} = \frac{5}{12}$

20. $7\frac{1}{2} \div 2\frac{4}{7} = \frac{15}{2} \div \frac{18}{7} = \frac{15}{2} \times \frac{7}{18}$
$= \frac{105}{36} = 2\frac{33}{36} = 2\frac{11}{12}$

Systematic Review 1D

1. $2A - 3B + 4A + 4B - 5A =$
$(2A + 4A - 5A) + (-3B + 4B) = A + B$

2. $18X + 5X - 6Y - 8Y - 11X + 10Y =$
$(18X + 5X - 11X) + (-6Y - 8Y + 10Y) = 12X - 4Y$

3. $4A - 4B + 16A + 7B + 18 =$
$(4A + 16A) + (-4B + 7B) + 18 = 20A + 3B + 18$

4. $-5X + 3 + 8X - 4 =$
$(-5X + 8X) + (3 - 4) = 3X - 1$

5. $8K - 6 + 3K - 2K + 3 =$
$(8K + 3K - 2K) + (3 - 6) = 9K - 3$

6. $10C - 3C - 9D + 3D - C =$
$(10C - 3C - C) + (-9D + 3D) = 6C - 6D$

7. $13A - 8Z - 2A - 12Z =$
$(13A - 2A) + (-8Z - 12Z) = 11A - 20Z$

8. $7D - 4D - 4 + 5D + 8 - 7D =$
$(7D - 4D + 5D - 7D) + (-4 + 8) = D + 4$

9. $(-3)^2 = (-3)(-3) = 9$

10. $-3^3 = -(3)(3)(3) = -27$

11. $(-6)(-2) = +12$

12. $(-4) - (-3) = (-4) + (+3) = -1$

13. $\frac{\cancel{4}^{1}}{{}_{1}\cancel{5}} \times \frac{1}{2} \times \frac{\cancel{5}^{1}}{\cancel{8}_{2}} = \frac{1}{4}$

14. $\frac{1}{{}_{1}\cancel{2}} \times \frac{\cancel{6}^{\cancel{3}^{1}}}{7} \times \frac{2}{\cancel{3}_{1}} = \frac{2}{7}$

15. $\frac{5}{8} \div \frac{1}{7} = \frac{35}{56} \div \frac{8}{56} = \frac{35}{8} = 4\frac{3}{8}$

16. $\frac{5}{8} \div \frac{1}{7} = \frac{5}{8} \times \frac{7}{1} = \frac{35}{8} = 4\frac{3}{8}$

17.
```
      28
     /  \
    2    14
        /  \
       2    7
```
$2 \times 2 \times 7$

18.
```
      42
     /  \
    2    21
        /  \
       3    7
```
$2 \times 3 \times 7$

19.

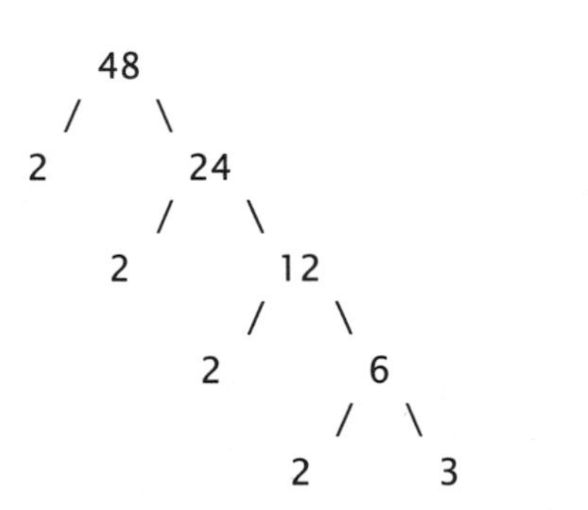

$2 \times 2 \times 2 \times 2 \times 3$

20.

```
        100
       /   \
      2     50
           /  \
          2    25
              /  \
             5    5
```

$2\times2\times5\times5$

Systematic Review 1E

1. $\left(1^2+2^2\right)+3^2=1^2+\left(2^2+3^2\right)$

$(1+4)+9=1+(4+9)$

$5+9=1+13$

$14=14$

2. yes

3. $\left[(81\div9)\div3\right]\neq\left[81\div(9\div3)\right]$

$(9\div3)\neq(81\div3)$

$3\neq27$

4. no

5. $3\times4\times3=4\times3\times3$

$36=36$

6. yes

7. $125-15-4\neq15-4-125$

$110-4\neq11-125$

$106\neq-114$

8. no

9. $\frac{1}{4}\times\frac{^1\cancel{3}}{_1\cancel{5}}\times\frac{\cancel{5}^1}{\cancel{3}_1}=\frac{1}{4}$

10. $\frac{^1\cancel{11}}{_1\cancel{2}\cancel{6}}\times\frac{\cancel{3}^1}{\cancel{11}_1}\times\frac{\cancel{4}^2}{7}=\frac{2}{7}$

11. $\frac{7}{4}\div\frac{7}{8}=\frac{56}{32}\div\frac{28}{32}=\frac{56}{28}=2$

12. $\frac{7}{4}\div\frac{7}{8}=\frac{^1\cancel{7}}{_1\cancel{4}}\times\frac{\cancel{8}^2}{\cancel{7}_1}=2$

13. $2\times2\times2\times2$

```
        16
       /  \
      2    8
          / \
         2   4
            / \
           2   2
```

14. $2\times3\times3\times3$

```
        54
       /  \
      2    27
          /  \
         3    9
             / \
            3   3
```

15.

```
        72
       /  \
      2    36
          /  \
         2    18
             /  \
            2    9
                / \
               3   3
```

$2\times2\times2\times3\times3$

16. $2\times2\times3\times3$

```
        36
       /  \
      2    18
          /  \
         2    9
             / \
            3   3
```

17. $\frac{24}{36}=\frac{2\times\cancel{12}}{3\times\cancel{12}}=\frac{2}{3}$ 12 is GCF

18. $\frac{10}{25}=\frac{2\times\cancel{5}}{5\times\cancel{5}}=\frac{2}{5}$ 5 is GCF

19. $\frac{30}{45}=\frac{2\times\cancel{15}}{3\times\cancel{15}}=\frac{2}{3}$ 15 is GCF

20. $\frac{32}{56}=\frac{4\times\cancel{8}}{7\times\cancel{8}}=\frac{4}{7}$ 8 is GCF

Lesson Practice 2A

1. $\frac{3}{8} = \frac{3}{8}$ LCM is 8
   $+\frac{1}{4} = \frac{2}{8}$
   $\frac{5}{8}$

2. $\frac{5}{6} = \frac{25}{30}$ LCM is 30
   $-\frac{3}{10} = \frac{9}{30}$
   $\frac{16}{30} = \frac{8}{15}$

3. $\frac{2}{3} = \frac{10}{15}$ LCM is 15
   $+\frac{4}{5} = \frac{12}{15}$
   $\frac{22}{15} = 1\frac{7}{15}$

4. $5 \cdot 6 + 4^2 = 5 \cdot 6 + 16 = 30 + 16 = 46$
5. $9 \cdot 4^2 - 19 = 9 \cdot 16 - 19$
   $= 144 - 19 = 125$
6. $6^2 \cdot 8 \div 2 = 36 \cdot 8 \div 2 = 288 \div 2 = 144$
7. $12 \cdot 3 + 4^2 - 8 = 12 \cdot 3 + 16 - 8 =$
   $36 + 16 - 8 = 52 - 8 = 44$
8. $18 \div 2 \cdot 5 + 6 = 9 \cdot 5 + 6 = 45 + 6 = 51$
9. $(-3)^2 + (8 + 3^2) = (-3)(-3) + (8 + 9)$
   $= 9 + 17 = 26$
10. $8 + 32 \div 4 - 2^2 = 8 + 32 \div 4 - 4 =$
    $8 + 8 - 4 = 16 - 4 = 12$
11. $3A - 3B + 5A + 4B + 7 =$
    $(3A + 5A) + (-3B + 4B) + 7 = 8A + B + 7$
12. $|5 \cdot 6^2| = |5 \cdot 36| = |180| = 180$
13. $|18 + 2^3| = |18 + 8| = |26| = 26$
14. $|3^2 - 8^2| = |9 - 64| = |-55| = 55$
15. $|4^2 - 2^2| = |16 - 4| = |12| = 12$

Lesson Practice 2B

1. $16 = 2 \times 2 \times 2 \times 2$
   $18 = 2 \times 3 \times 3$
   $\text{LCM} = 2 \times 2 \times 2 \times 2 \times 3 \times 3 = 144$
2. $10 = 2 \times 5$
   $14 = 2 \times 7$
   $\text{LCM} = 2 \times 5 \times 7 = 70$
3. $24 = 2 \times 2 \times 2 \times 3$
   $50 = 2 \times 5 \times 5$
   $\text{LCM} = 2 \times 2 \times 2 \times 3 \times 5 \times 5 = 600$
4. $4 \cdot 8 + 3^2 = (4 \cdot 8) + 9 = 32 + 9 = 41$
5. $10 \cdot 4^2 - 25 = (10 \cdot 16) - 25 = 160 - 25 = 135$
6. $7^2 - 9 \div 2 = 49 - (9 \div 2) = 49 - 4.5 = 44.5$
7. $18 \cdot 2 + 5^2 - 11 = (18 \cdot 2) + 25 - 11$
   $= 36 + 25 - 11 = 50$
8. $15 \div 3 \cdot 8 + 10 = (5 \cdot 8) + 10 = 40 + 10 = 50$
9. $(-5)^2 + (9 + 4^2) = 25 + (9 + 16) = 25 + 25 = 50$
10. $9^2 + 48 \div 12 - 3^3 = 81 + (48 \div 12) - 27$
    $= 81 + 4 - 27 = 58$
11. $|4^2 - 9| + (8 \div 4)^2 = |4^2 - 9| + (2)^2$
    $= |16 - 9| + 4 = 7 + 4 = 11$
12. $|3^2 - 5^2| - (15 \div 3)^3 + 18 = |3^2 - 5^2| - (5)^3 + 18$
    $= |9 - 25| - 125 + 18$
    $= |-16| - 125 + 18$
    $= 16 - 125 + 18 = -91$
13. $|10^2 - 5^2| + |-8 + 2^2| = |100 - 25| + |-8 + 4|$
    $= |75| + |-4| = 75 + 4 = 79$
14. $|18 - 36| + (|3 - 5^2| - 15)^2$
    $= |18 - 36| + (|3 - 25| - 15)^2$
    $= |-18| + (|-22| - 15)^2$
    $= 18 + (22 - 15)^2$
    $= 18 + 7^2 = 18 + 49 = 67$
15. $|(-10)^2 - 9| - |2^4 - 5^2| = |100 - 9| - |16 - 25|$
    $= |91| - |-9| = 91 - 9 = 82$

Systematic Review 2C

1. $4 \cdot 7 + 3^2 = (4 \cdot 7) + 9 = 37$
2. $5^2 + 8 \div 2 = 25 + 8 \div 2 = 25 + 4 = 29$
3. $12^2 \times (2+3) - 4 = (144 \times 5) - 4$
   $= 720 - 4 = 716$
4. $9 \times 1^2 - 8 = (9 \times 1) - 8 = 9 - 8 = 1$
5. $14 \div 2 - 1 \times 6 = 7 - 6 = 1$
6. $6 + 28 \div 7 - 4^2 = 6 + 4 - 16 = -6$
7. $(-3)^2 \div 9 + 6 = 9 \div 9 + 6 = 1 + 6 = 7$
8. $|6 \div (-2)| \times 5 + 3^2 = |-3| \times 5 + 9$
   $= 3 \times 5 + 9 = 15 + 9 = 24$
9. $\frac{\cancel{3}^1}{{}_2\cancel{4}\cancel{8}} \times \frac{\cancel{2}^1}{5} \times \frac{\cancel{2}^1}{\cancel{3}_1} = \frac{1}{10}$
10. $\frac{1}{\cancel{2}} \times \frac{\cancel{2}}{\cancel{3}} \times \frac{\cancel{3}}{\cancel{4}} \times \frac{\cancel{4}}{5} = \frac{1}{5}$
11. 64 → 2, 32; 32 → 2, 16; 16 → 2, 8; 8 → 2, 4; 4 → 2, 2
    $2 \times 2 \times 2 \times 2 \times 2 \times 2$
12. 81 → 3, 27; 27 → 3, 9; 9 → 3, 3
    $3 \times 3 \times 3 \times 3$
13. $\frac{32}{48} = \frac{2 \times \cancel{16}}{3 \times \cancel{16}} = \frac{2}{3}$ 16 is GCF
14. $24 = 2 \times 2 \times 2 \times 3$; $36 = 2 \times 2 \times 3 \times 3$
    $LCM = 2 \times 2 \times 2 \times 3 \times 3 = 72$
15. $\frac{2}{3} \div \frac{2}{7} = \frac{14}{21} \div \frac{6}{21} = \frac{14}{6} = 2\frac{2}{6} = 2\frac{1}{3}$
16. $\frac{2}{3} \div \frac{2}{7} = \frac{2}{3} \times \frac{7}{2} = \frac{14}{6} = 2\frac{1}{3}$
17. $.7 \times .3 = .21$
    (because $1/10 \times 1/10 = 1/100$)
18. $2.4 \times 1.2 = 2.88$ (see note for #17)
19. $1.3 \times 2.1 = 2.73$ (see note for #17)
    or:
    $$\begin{array}{r} 1.3 \\ \underline{2.1} \\ 1\,3 \\ \underline{2\,6\;\,} \\ 2.7\,3 \end{array}$$
    (two decimal places in answer)
20. $.4 \times 3.2 = 1.28$

Systematic Review 2D

1. $-4^2 + (7-3)^2 - |-2| = -16 + (4)^2 - 2$
   $= -16 + 16 - 2 = -2$
2. $4(10-3) - 5(6) + 8 \div 2 = 4(7) - 30 + 4$
   $= 28 - 30 + 4 = 2$
3. $-19 - (7)(-2) + 6^2 = -19 - (-14) + 36$
   $= -19 + 14 + 36 = 31$
4. $-(A - B) + A - B = (-A + B) + A - B$
   $= (-A + A) + (B - B) = 0$
5. $11^2 \div 4 + \frac{2}{3} = 121 \div 4 + \frac{2}{3} = \frac{121}{4} + \frac{2}{3}$
   $= \frac{363}{12} + \frac{8}{12} = \frac{371}{12} = 30\frac{11}{12}$
6. $5 \times 3 + 4^2 - 7 + (-8 \div 4) = 5 \times 3 + 16 - 7 + (-2)$
   $= 15 + 16 - 7 - 2 = 22$
7. $-5^2 + (-5)^2 = -(5 \times 5) + (-5)(-5) = -25 + 25 = 0$
8. $|(9^2 \div 9) \div 3| = |(81 \div 9) \div 3| = |9 \div 3| = 3$
9. $\frac{\cancel{2}^1}{5} \times \frac{\cancel{7}^1}{\cancel{8}\cancel{2}_1} \times \frac{\cancel{4}}{\cancel{7}} = \frac{1}{5}$
10. $\frac{5}{24} = \frac{20}{96}$ LCM is 96
    $+\frac{9}{32} = \frac{27}{96}$
    $\frac{47}{96}$

11. $(3\times4)\times6=3\times(4\times6)$
$12\times6=3\times24$
$72=72$

12. yes, see #11

13. $10-(8-6)\neq(10-8)-6$
$10-2\neq2-6$
$8\neq-4$

14. no, see #13

15. $\frac{12}{7}\div\frac{7}{4}=\frac{48}{28}\div\frac{49}{28}=\frac{48}{49}$

16. $\frac{12}{7}\div\frac{7}{4}=\frac{12}{7}\times\frac{4}{7}=\frac{48}{49}$

17.

```
        38.33
.06 ) 2.3000
      180
       50
       48
        20
        18
         20
         18
```

18.

```
       5
.5 ) 2.5
     2 5
       0
```

19.

```
       50
.05 ) 2.50
      2 5
        0
```

20.

```
        .2
5.3 ) 1.06
      1 06
         0
```

Systematic Review 2E

1. $-3+2^3-8+7^2=-3+8-8+49=46$

2. $(5\times6)\div3=30\div3=10$

3. $\left[(10+3)^2-9\right]\div20=\left[13^2-9\right]\div20$
$=\left[169-9\right]\div20$
$=160\div20=8$

4. $A+B+2A-3B=(A+2A)+(B-3B)=3A-2B$

5. $(42\div6-2)\times11=(7-2)\times11=5\times11=55$

6. $8+45\div9+3=8+5+3=16$

7. $(-4)^2+(5)^2-3^2=16+25-9=32$

8. $(192\div8)\times4-|67-200|=24\times4-|-133|$
$=96-133=-37$

9. $\frac{\overset{5}{\cancel{10}}}{3}\times\frac{7}{4}\times\frac{7}{\underset{6}{\cancel{12}}}=\frac{245}{72}=3\frac{29}{72}$

10. $\frac{3}{7}+\frac{11}{13}=\frac{3}{7}\times\frac{13}{13}+\frac{11}{13}\times\frac{7}{7}$
$=\frac{39}{91}+\frac{77}{91}=\frac{116}{91}=1\frac{25}{91}$

Using this method always yields a common denominator. In some cases, you will have to reduce after finding the answer.

11. $\frac{30}{54}=\frac{5\times\cancel{6}}{9\times\cancel{6}}=\frac{5}{9}$

6 is GCF

12. $10=10$
$100=10\times10$
$LCM=10\times10=100$
(LCM may also be found using prime factors)

13. $6+2+9=2+6+9$
$8+9=8+9$
$17=17$

14. yes; see #13

15. $\frac{37}{8}\div\frac{11}{4}=\frac{37}{8}\div\frac{22}{8}=\frac{37}{22}=1\frac{15}{22}$

(Either method may be used for dividing fractions)

16.

```
        .45
3.1 ) 1.395
      1 24
        155
        155
```

17. $\frac{14}{3}\div\frac{5}{4}=\frac{56}{12}\div\frac{15}{12}=\frac{56}{15}=3\frac{11}{15}$

18. $.4\overline{).0016}$ = $.004$

$\cup \quad \cup \quad \underline{16}$

19. $.4\overline{)1.2}$ = 3 groups of $.40

$\underline{16}$

20. $6\overline{)1.44}$ = $.24 per person

$\underline{12}$

24

Lesson Practice 3A

1.
$$-5A+3+8A-4=9+3-1$$
$$(-5A+8A)+(3-4)=11$$
$$3A+(-1)=11$$
$$\underline{\quad +1 \quad +1}$$
$$3A=12$$
$$A=4$$

Check:
$$-5(4)+3+8(4)-4=9+3-1$$
$$-20+3+32-4=9+3-1$$
$$11=11$$

2.
$$3B-B+7+4B=43$$
$$6B+7=43$$
$$\underline{\quad -7 \quad -7}$$
$$\frac{6B}{6}=\frac{36}{6}$$
$$B=6$$

Check:
$$3(6)-(6)+7+4(6)=43$$
$$18-6+7+24=43$$
$$43=43$$

3.
$$-4Y-6+7Y+3+Y=17$$
$$4Y-3=17$$
$$\underline{\quad +3 \quad +3}$$
$$\frac{4Y}{4}=\frac{20}{4}$$
$$Y=5$$

Check:
$$-4(5)-6+7(5)+3+(5)=17$$
$$-20-6+35+3+5=17$$
$$17=17$$

4.
$$5Q+3Q-6+2Q=(2+3)+9$$
$$10Q-6=14$$
$$\underline{\quad +6 \quad +6}$$
$$\frac{10Q}{10}=\frac{20}{10}$$
$$Q=2$$

Check:
$$5(2)+3(2)-6+2(2)=(2+3)+9$$
$$10+6-6+4=5+9$$
$$14=14$$

5.
$$6K-5+4K-K+2=12\cdot 2$$
$$9K-3=24$$
$$\underline{\quad +3 \quad +3}$$
$$\frac{9K}{9}=\frac{27}{9}$$
$$K=3$$

Check:
$$6(3)-5+4(3)-(3)+2=12\cdot 2$$
$$18-5+12-3+2=24$$
$$24=24$$

6.
$$5C-2C-8+7-C=3\cdot 4+1$$
$$2C-1=13$$
$$\underline{\quad +1 \quad +1}$$
$$\frac{2C}{2}=\frac{14}{2}$$
$$C=7$$

Check:
$$5(7)-2(7)-8+7-(7)=3\cdot 4+1$$
$$35-14-8+7-7=12+1$$
$$13=13$$

7.
$$4A+6=2A+12$$
$$4A-2A=12-6$$
$$\frac{2A}{2}=\frac{6}{2}$$
$$A=3$$

Check:
$$4(3)+6=2(3)+12$$
$$12+6=6+12$$
$$18=18$$

8.
$$10B-2B+3=5B+21$$
$$8B-5B=21-3$$
$$\frac{3B}{3}=\frac{18}{3}$$
$$B=6$$

Check:
$$10(6)-2(6)+3=5(6)+21$$
$$60-12+3=30+21$$
$$51=51$$

9. $6C-8+3C=7C-2C+12$
$9C-5C=12+8$
$\frac{4C}{4}=\frac{20}{4}$
$C=5$
Check: $6(5)-8+3(5)=7(5)-2(5)+12$
$30-8+15=35-10+12$
$37=37$

10. $6D-10=-2D-34$
$6D+2D=-34+10$
$\frac{8D}{8}=\frac{-24}{8}$
$D=-3$
Check: $6(-3)-10=-2(-3)-34$
$-18-10=6-34$
$-28=-28$

11. $-3A-3-6A+10A+5=10$
$A+2=10$
$\underline{-2\quad -2}$
$A=\ 8$
Check: $-3(8)-3-6(8)+10(8)+5=10$
$-24-3-48+80+5=10$
$10=10$

12. $-5B-B+4+10B-7=7\cdot 11$
$4B-3=77$
$\underline{+3\quad +3}$
$\frac{4B}{4}=\frac{80}{4}$
$B=20$
Check: $-5(20)-(20)+4+10(20)-7=7\cdot 11$
$-100-20+4+200-7=77$
$77=77$

13. $-4R+7R-3+5R=10^2-7$
$8R=100-7+3$
$\frac{8R}{8}=\frac{96}{8}$
$R=12$
Check: $-4(12)+7(12)-3+5(12)=10^2-7$
$-48+84-3+60=100-7$
$93=93$

14. $-7Q+8-6+5Q=3\cdot 5-7$
$-2Q+2=8$
$\underline{-2\quad -2}$
$\frac{-2Q}{-2}=\frac{6}{-2}$
$Q=-3$
Check: $-7(-3)+8-6+5(-3)=3\cdot 5-7$
$21+8-6-15=15-7$
$8=8$

Lesson Practice 3B

1. $-3A-5+4A-6+2A=19$
$3A-11=19$
$\underline{+11\quad +11}$
$\frac{3A}{3}=\frac{30}{3}$
$A=10$
Check: $-3(10)-5+4(10)-6+2(10)=19$
$-30-5+40-6+20=19$
$19=19$

2. $8B-6+5B-3-3B=41$
$10B-9=41$
$\underline{+9\quad +9}$
$\frac{10B}{10}=\frac{50}{10}$
$B=5$
Check: $8(5)-6+5(5)-3-3(5)=41$
$40-6+25-3-15=41$
$41=41$

3. $-5Y+3-6Y+2Y+4=13$
$-9Y+7=13$
$\underline{-7\quad -7}$
$\frac{-9Y}{-9}=\frac{6}{-9}$
$Y=-\frac{2}{3}$

$$\text{Check: } -5\left(-\frac{2}{3}\right)+3-6\left(-\frac{2}{3}\right)+2\left(-\frac{2}{3}\right)+4=13$$
$$\frac{10}{3}+3+\frac{12}{3}-\frac{4}{3}+4=13$$
$$\frac{18}{3}+7=13$$
$$6+7=13$$
$$13=13$$

4. $8Q-Q+7-4-3Q=7+4\times10$
$$4Q+3=47$$
$$-3 \quad -3$$
$$\frac{4Q}{4}=\frac{44}{4}$$
$$Q=11$$
$$\text{Check: } 8(11)-(11)+7-4-3(11)=7+4\times10$$
$$88-11+7-4-33=47$$
$$47=47$$

5. $8M-4M-6-3+5M=8^2-1$
$$9M-9=63$$
$$+9 \quad +9$$
$$\frac{9M}{9}=\frac{72}{9}$$
$$M=8$$
$$\text{Check: } 8(8)-4(8)-6-3+5(8)=8^2-1$$
$$64-32-6-3+40=63$$
$$63=63$$

6. $7C-4C+5-8+C=5^2+4$
$$4C-3=29$$
$$+3 \quad +3$$
$$\frac{4C}{4}=\frac{32}{4}$$
$$C=8$$
$$\text{Check: } 7(8)-4(8)+5-8+(8)=5^2+4$$
$$56-32+5-8+8=5^2+4$$
$$29=29$$

7. $11A-4A-18=2A+A+10$
$$7A-3A=10+18$$
$$\frac{4A}{4}=\frac{28}{4}$$
$$A=7$$
$$\text{Check: } 11(7)-4(7)-18=2(7)+(7)+10$$
$$77-28-18=14+7+10$$
$$31=31$$

8. $2B-10B-15+5=8B-40-4B-6$
$$-8B-10=4B-46$$
$$-8B-4B=-46+10$$
$$\frac{-12B}{-12}=\frac{-36}{-12}$$
$$B=3$$
$$\text{Check: } 2(3)-10(3)-15+5=8(3)-40-4(3)-6$$
$$6-30-15+5=24-40-12-6$$
$$-34=-34$$

9. $3C-6+2C=10C-2C+6$
$$5C-6=8C+6$$
$$5C-8C=6+6$$
$$\frac{-3C}{-3}=\frac{12}{-3}$$
$$C=-4$$
$$\text{Check: } 3(-4)-6+2(-4)=10(-4)-2(-4)+6$$
$$-12-6-8=-40+8+6$$
$$-26=-26$$

10. $2D-8-5D=-3D-2D+6$
$$-3D-8=-5D+6$$
$$-3D+5D=6+8$$
$$\frac{2D}{2}=\frac{14}{2}$$
$$D=7$$
$$\text{Check: } 2(7)-8-5(7)=-3(7)-2(7)+6$$
$$14-8-35=-21-14+6$$
$$-29=-29$$

11. $8K-6+3K-2K+3=4\times33$
$$9K-3=132$$
$$+3 \quad +3$$
$$\frac{9K}{9}=\frac{135}{9}$$
$$K=15$$
$$\text{Check: } 8(15)-6+3(15)-2(15)+3=4\times33$$
$$120-6+45-30+3=132$$
$$132=132$$

12. $B+B+B+6=6B+5-2B+9$
$$3B+6=4B+14$$
$$6-14=4B-3B$$
$$B=-8$$

Check:
$$(-8)+(-8)+(-8)+6=6(-8)+5-2(-8)+9$$
$$-24+6=-48+5+16+9$$
$$-18=-18$$

13.
$$\begin{aligned}
-2C+12 &= 2C-6+6C-12\\
-2C+12 &= 8C-18\\
-2C-8C &= -18-12\\
\frac{-10C}{-10} &= \frac{-30}{-10}\\
C &= 3
\end{aligned}$$

Check:
$$\begin{aligned}
-2(3)+12 &= 2(3)-6+6(3)-12\\
-6+12 &= 6-6+18-12\\
6 &= 6
\end{aligned}$$

14.
$$\begin{aligned}
10X-3X-9+3-X &= 51\div 3+1\\
6X-6 &= 18\\
+6 \quad & \quad +6\\
\frac{6X}{6} &= \frac{24}{6}\\
X &= 4
\end{aligned}$$

Check:
$$\begin{aligned}
10(4)-3(4)-9+3-(4) &= 51\div 3+1\\
40-12-9+3-4 &= 17+1\\
18 &= 18
\end{aligned}$$

Systematic Review 3C

1.
$$\begin{aligned}
X+3 &= 9\\
X+3-3 &= 9-3\\
X &= 6
\end{aligned}$$

2.
$$\begin{aligned}
X+6 &= 10\\
X+6-6 &= 10-6\\
X &= 4
\end{aligned}$$

3.
$$\begin{aligned}
2X+5 &= 11\\
2X+5-5 &= 11-5\\
2X &= 6\\
X &= 3
\end{aligned}$$

4.
$$\begin{aligned}
4Q-2 &= 10\\
4Q-2+2 &= 10+2\\
4Q &= 12\\
Q &= 3
\end{aligned}$$

5.
$$\begin{aligned}
4X+2 &= 2X+8\\
4X+2-2 &= 2X+8-2\\
4X &= 2X+6\\
4X-2X &= 2X-2X+6\\
2X &= 6\\
X &= 3
\end{aligned}$$

6.
$$\begin{aligned}
3Y+5 &= 2Y+7\\
3Y+5-5 &= 2Y+7-5\\
3Y &= 2Y+2\\
3Y-2Y &= 2Y-2Y+2\\
Y &= 2
\end{aligned}$$

7.
$$\begin{aligned}
Q+4 &= 3Q-6\\
Q+4-Q &= 3Q-Q-6\\
4 &= 2Q-6\\
4+6 &= 2Q-6+6\\
10 &= 2Q\\
Q &= 5
\end{aligned}$$

8.
$$\begin{aligned}
2R+8 &= 3R-2\\
2R-2R+8 &= 3R-2R-2\\
8 &= R-2\\
8+2 &= R-2+2\\
R &= 10
\end{aligned}$$

9.
$$\begin{aligned}
9-3 &< |4-11|\\
6 &< |-7|\\
6 &< 7
\end{aligned}$$

10.
$$\begin{aligned}
|1-2-3| &< |2\times 3|\\
|-4| &< |6|\\
4 &< 6
\end{aligned}$$

11.
$$\begin{aligned}
&(-3)\times 4+6^2\times(-3)+5^2=\\
&-3\times 4+36\times(-3)+25=\\
&-12+(-108)+25=-120+25=-95
\end{aligned}$$

12.
$$\begin{aligned}
&(14-9+2^2)-(3\div 6\times 2^2)=\\
&(14-9+4)-(3\div 6\times 4)=\\
&9-\left(\frac{3}{6}\times 4\right)=\\
&9-\frac{12}{6}=9-2=7
\end{aligned}$$

13. $\frac{4}{3}\times\frac{6}{10}\div\frac{2}{3}=\frac{\cancel{4}^{2}}{\cancel{3}_{1}}\times\frac{\cancel{6}^{\cancel{2}^{1}}}{\cancel{10}_{5}}\times\frac{3}{\cancel{2}_{1}}=\frac{6}{5}=1\frac{1}{5}$

14.
$$\begin{array}{r}
.1\,7\\
\times\ .8\\
\hline
5\\
8\,6\\
\hline
.1\,3\,6
\end{array}$$

(Three decimal places in answer)

15. $(-8)(-7)=56$

16. $(-4)^2=(-4)(-4)=16$

17. $2 = 2;\ 3 = 3;\ 4 = 2 \times 2;$ so $\text{LCM} = 2 \times 2 \times 3 = 12$

$$^{6}\cancel{(12)}\frac{1}{\cancel{2}} + {}^{4}\cancel{(12)}\frac{2}{\cancel{3}} = {}^{3}\cancel{(12)}\frac{1}{\cancel{4}}X$$

It is not necessary to write in "1" when dividing terms, unless you wish.

$6 + 8 = 3X;\ X = 4\frac{2}{3}$

18. $2 = 2;\ 5 = 5;\ 4 = 2 \times 2;$ so $\text{LCM} = 2 \times 2 \times 5 = 20$

$$^{4}\cancel{(20)}\frac{3}{\cancel{5}}X + {}^{5}\cancel{(20)}\frac{3}{\cancel{4}} = {}^{10}\cancel{(20)}\frac{3}{\cancel{2}}$$

$12X + 15 = 30;\ X = 1\frac{1}{4}$

19. $3 = 3;\ 5 = 5;\ 9 = 3 \times 3;$ so $\text{LCM} = 3 \times 3 \times 5 = 45$

$$^{5}\cancel{(45)}\frac{1}{\cancel{9}}X + {}^{15}\cancel{(45)}\frac{2}{\cancel{3}} = {}^{9}\cancel{(45)}\frac{1}{\cancel{5}}$$

$5X + 30 = 9;\ 5X = -21;\ X = -4\frac{1}{5}$

20. $5 = 5,\ 4 = 2 \times 2,\ 8 = 2 \times 2 \times 2$

so $\text{LCM} = 2 \times 2 \times 2 \times 5 = 40$

$$^{5}\cancel{(40)}\frac{3}{\cancel{8}} - {}^{8}\cancel{(40)}\frac{1}{\cancel{5}}X = {}^{10}\cancel{(40)}\frac{3}{\cancel{4}}$$

$15 - 8X = 30;\ -8X = 15;\ X = -1\frac{7}{8}$

Systematic Review 3D

1.
$$\begin{aligned} Y - 3 &= 10 \\ Y - 3 + 3 &= 10 + 3 \\ Y &= 13 \end{aligned}$$

2.
$$\begin{aligned} 2B - 5 &= 13 \\ 2B &= 18 \\ 2B \div 2 &= 18 \div 2 \\ B &= 9 \end{aligned}$$

3.
$$\begin{aligned} 3C + 6 &= -9 \\ 3C &= -15 \\ 3C \div 3 &= -15 \div 3 \\ C &= -5 \end{aligned}$$

4.
$$\begin{aligned} 2D - 5 &= 1 \\ 2D &= 6 \\ 2D \div 2 &= 6 \div 2 \\ D &= 3 \end{aligned}$$

5.
$$\begin{aligned} 4E - 3 &= -3 \\ 4E &= 0 \\ E &= 0 \end{aligned}$$

6.
$$\begin{aligned} 3X + 8 &= -2X - 2 \\ 3X + 2X &= -2X + 2X - 10 \\ 5X &= -10 \\ 5X \div 5 &= -10 \div 5 \\ X &= -2 \end{aligned}$$

7.
$$\begin{aligned} 2Y - 2 &= 3Y - 6 \\ 2Y - 3Y &= -6 + 2 \\ -Y &= -4 \\ (-1)(-Y) &= (-1)(-4) \\ Y &= 4 \end{aligned}$$

8.
$$\begin{aligned} Z + 8 &= 2Z + 18 \\ Z - 2Z &= 18 - 8 \\ -Z &= 10 \\ (-1)(-Z) &= (-1)(10) \\ Z &= -10 \end{aligned}$$

9.
$$\begin{aligned} |3 \times 2 \times (-2)| &> 24 \div (-3) \\ |-12| &> -8 \\ 12 &> -8 \end{aligned}$$

10.
$$\begin{aligned} |17 - 3 - 20| &< |7 + 0 + 1| \\ |-6| &< |8| \\ 6 &< 8 \end{aligned}$$

11.
$$\begin{aligned} &\left[(6 - 2) \times 5^2 - 10\right] \div 5^2 = \\ &[4 \times 25 - 10] \div 25 = \\ &[100 - 10] \div 25 = \\ &90 \div 25 = 3\tfrac{3}{5} \text{ or } 3.6 \end{aligned}$$

12.
$$\begin{aligned} &(-7 - 6)^2 - (4 + 5 - 3)^2 = \\ &(-13)^2 - 6^2 = 169 - 36 = 133 \end{aligned}$$

13. $\frac{5}{6} \times \frac{3}{7} \div \frac{2}{3} = \frac{5}{{}_{2}\cancel{6}} \times \frac{\cancel{3}}{7} \times \frac{3}{2} = \frac{15}{28}$

14.

```
        14.
 12. ) 168.
       12
        48
        48
```

15. $2 = 2;\ 5 = 5;\ 10 = 2 \times 5;\ \text{LCM} = 2 \times 5 = 10$

16.
$$^{2}\cancel{(10)}\frac{6}{\cancel{5}}X + {}^{1}\cancel{(10)}\frac{7}{\cancel{10}} = {}^{5}\cancel{(10)}\frac{5}{\cancel{2}}X$$

$$\begin{aligned} 12X + 7 &= 25X \\ 7 &= 13X \\ 7 \div 13 &= 13X \div 13 \\ \frac{7}{13} &= X \end{aligned}$$

17. $100 = 10 \times 10;\ 1000 = 10 \times 10 \times 10$

$\text{LCM} = 10 \times 10 \times 10 = 1000$

$$1000(.83) + 1000(.04X) = 1000(.325)$$
$$830 + 40X = 325$$
$$40X = -505$$
$$X = \frac{-505}{40}$$
$$= -12\frac{5}{8} \text{ or } -12.625$$

18. $10 = 10;\ 100 = 10 \times 10;$

$\text{LCM} = 10 \times 10 = 100$

$$100(.18) + 100(.2X) = 100(.17)$$
$$18 + 20X = 17$$
$$20X = -1$$
$$X = \frac{-1}{20} \text{ or } -.05$$

19. $10 = 10;\ 100 = 10 \times 10;$

$\text{LCM} = 10 \times 10 = 100$

$$100(.8X) + 100(1.3) = 100(7) + 100(.24)$$
$$80X + 130 = 700 + 24$$
$$80X = 594$$
$$X = \frac{594}{80} = 7\frac{17}{40} \text{ or } 7.425$$

20. $10 = 10;\ 100 = 10 \times 10;$

$\text{LCM} = 10 \times 10 = 100$

$$100(8.2) - 100(4) = 100(.08X)$$
$$820 - 400 = 8X$$
$$420 = 8X$$
$$X = \frac{420}{8} = 52\frac{1}{2} \text{ or } 52.5$$

Systematic Review 3E

1. $-2X + 7 + 3X - 4 = 10 - 1$
$$X + 3 = 9$$
$$X = 6$$

2. $3Y + 8 - 2 - 2Y = 9 - 4 + 5$
$$Y + 6 = 10$$
$$Y = 4$$

3. $2X - 2 + 7 + X - X = 6 + 6 - 1$
$$2X + 5 = 11$$
$$2X = 6$$
$$X = 3$$

4. $-2B + 3 + 5B + 1 = 2(3 + 2) + 9$
$$3B + 4 = 2(5) + 9$$
$$3B = 19 - 4$$
$$3B = 15$$
$$B = 5$$

5. $3Q - 2 + Q = 3(2 + 2) - 2$
$$4Q - 2 = 3(4) - 2$$
$$4Q = 12 - 2 + 2$$
$$4Q = 12$$
$$Q = 3$$

6. $5X + 5 - X - 3 = 3X - X + 4(2)$
$$4X + 2 = 2X + 8$$
$$4X - 2X = 8 - 2$$
$$2X = 6$$
$$X = 3$$

7. $2Y - 4 + Y + 9 = -2Y - 4 + 4Y + 11$
$$3Y + 5 = 2Y + 7$$
$$3Y - 2Y = 7 - 5$$
$$Y = 2$$

8. $-4Q + 2 + 5Q + 2 = 3Q - 6$
$$Q + 4 = 3Q - 6$$
$$4 + 6 = 3Q - Q$$
$$10 = 2Q$$
$$5 = Q$$

9. $(7 - 3)^2 \times |3 - 7| = (4)^2 \times |-4| = 16 \times 4 = 64$

10. $8 + (5 + 4)^2 \times 2 + 11^2 = 8 + 9^2 \times 2 + 121$
$$= 8 + 81 \times 2 + 121$$
$$= 8 + 162 + 121 = 291$$

11. $(4 \times 8 - 6 + 3^2) + (3 - 6 - 7^2 \times 3 + 4) =$
$$(4 \times 8 - 6 + 9) + (3 - 6 - 49 \times 3 + 4) =$$
$$(32 - 6 + 9) + (3 - 6 - 147 + 4) =$$
$$35 + (-146) = -111$$

12. $(15 - 6 + 8^2 + 3 \div 3) - (10 + 9^2 - 40 \div 8) =$
$$(15 - 6 + 64 + 3 \div 3) - (10 + 81 - 40 \div 8) =$$
$$(15 - 6 + 64 + 1) - (10 + 81 - 5) =$$
$$74 - 86 = -12$$

13. $\frac{3}{4} \times \frac{8}{3} \div \frac{2}{1} = \frac{\cancel{3}}{\cancel{4}} \times \frac{\cancel{8}^{4}}{\cancel{3}} \times \frac{1}{\cancel{2}} = 1$

14.
$$\begin{array}{r} 1.7 \\ \underline{.8} \\ 5 \\ \underline{86} \\ 1.36 \end{array}$$
(two decimal places in answer)

15. $(-19)(6) = -114$

16. $-6^2 = -(6^2) = -(6)(6) = -36$

17. $-[-(-6)] = -[+6] = -6$

18. $-7-(-3) = -7+3 = -4$

19. $3 = 3 \times 1$; $6 = 2 \times 3$; $8 = 2 \times 2 \times 2$

$\text{LCM} = 2 \times 2 \times 2 \times 3 = 24$

$$^{3}(\cancel{24})\frac{7}{\cancel{8}} + {}^{8}(\cancel{24})\frac{2}{\cancel{3}}X = {}^{4}(\cancel{24})\frac{1}{\cancel{6}}$$
$$21 + 16X = 4$$
$$16X = -17$$
$$X = -1\frac{1}{16}$$

20. $10 = 10$; $100 = 10 \times 10$;

$\text{LCM} = 10 \times 10 = 100$

$$100(.03X) - 100(.6) = 100(.75)$$
$$3X - 60 = 75$$
$$3X = 135$$
$$X = 45$$

Lesson Practice 4A

1. $5(4+3) = 5(4)+5(3)$
2. $6(2+3+1) = 6(2)+6(3)+6(1)$
3. $7(A+B) = 7A+7B$
4. $3(4C+3B) = 3(4C)+3(3B)$
5. $5(2X+3Y-3+4X) = 5(2X)+5(3Y)-5(3)+5(4X)$
6. $8(A+3B+8+4A) = 8(A)+8(3B)+8(8)+8(4A)$
7. $6X+6Y = 6(X+Y)$
8. $8A+16B = 8(A+2B)$
9. $14X+21Y = 7(2X+3Y)$
10. $-2M-6N = -2(M+3N)$
11. $6B+18C = 6(B+3C)$
12. $15X+10A = 5(3X+2A)$
13. $5X+15 = 45$

    $5(X+3) = 5(9)$ divide out the 5s:

    $X+3 = 9$

    $X = 6$

14. $10X+16 = 26$

    $2(5X+8) = 2(13)$ divide out the 2s:

    $5X+8 = 13$

    $5X = 5$

    $X = 1$

15. $13Y-26+39Y = 52$

    $13(Y-2+3Y) = 13(4)$ divide out the 13s:

    $4Y-2 = 4$

    $4Y = 6$

    $Y = \frac{6}{4} = 1\frac{1}{2}$

16. $8A-10-6A = 14$

    $2(4A-5-3A) = 2(7)$ divide out the 2s:

    $A-5 = 7$

    $A = 12$

17. $12X+21 = 30$

    $3(4X+7) = 3(10)$ divide out the 3s:

    $4X+7 = 10$

    $4X = 3$

    $X = \frac{3}{4}$

18. $8X-28 = 12$

    $4(2X-7) = 4(3)$ divide out the 4s:

    $2X-7 = 3$

    $2X = 10$

    $X = 5$

Lesson Practice 4B

1. $8(5+2) = 8(5)+8(2)$
2. $5(4-3+2) = 5(4)-5(3)+5(2)$
3. $9(C+D) = 9(C)+9(D)$
4. $5(2C+4D) = 5(2C)+5(4D)$
5. $3(X+Y+4X) = 3(X)+3(Y)+3(4X)$
6. $-2(3X+2Y+Y) = (-2)(3X)+(-2)(2Y)+(-2)(Y)$
7. $8X+12Y = 4(2X+3Y)$
8. $-7X-21Y = 7(-X-3Y)$ or $-7(X+3Y)$
9. $18A+24B = 6(3A+4B)$
10. $8X+10 = 16$

    $2(4X+5) = 2(8)$

11. $6A+3 = 15$

    $3(2A+1) = 3(5)$

12. $8A+10 = 20$

    $2(4A+5) = 2(10)$

13.
$$8X+32=40$$
$$8(X+4)=8(5)$$
$$X+4=5$$
$$X=1$$

14.
$$18Y+27=45$$
$$9(2Y+3)=9(5)$$
$$2Y+3=5$$
$$2Y=2;\ Y=1$$

15.
$$15X-10+5X=25$$
$$5(3X-2+X)=5(5)$$
$$4X-2=5$$
$$4X=7$$
$$X=\frac{7}{4}=1\frac{3}{4}$$

16.
$$9C-6-12C=18$$
$$3(3C-2-4C)=3(6)$$
$$-C-2=6$$
$$-C=8$$
$$C=-8$$

17.
$$14M-42+56M=28$$
$$14(M-3+4M)=14(2)$$
$$5M-3=2$$
$$5M=5$$
$$M=1$$

18.
$$6A-16-4A=20$$
$$2(3A-8-2A)=2(10)$$
$$A-8=10$$
$$A=18$$

Systematic Review 4C

1. $4(A+B+3)=4A+4B+12$
2. $5(X-Y+6+Z)=5X-5Y+30+5Z$
3. $3(2Q-4+3T+7)=6Q-12+9T+21$
4. $2(2X+3Y-5)=4X+6Y-10$
5. $15Y+30X=10;\ 5(3Y+6X)=5(2)$
6. $12Q+6Y=15;\ 3(4Q+2Y)=3(5)$
7. $24Q+18Y=30;\ 6(4Q+3Y)=6(5)$
8. $36A-14B=10;\ 2(18A-7B)=2(5)$

9.
$$3-9<|4+1^2|$$
$$-6<|4+1|$$
$$-6<|5|$$
$$-6<5$$

10.
$$4X-16=24$$
$$4(X-4)=4(6)$$
$$X-4=6$$
$$X=10$$

11.
$$30-42Y=18$$
$$6(5-7Y)=6(3)$$
$$5-7Y=3$$
$$-7Y=3-5$$
$$-7Y=-2$$
$$Y=\frac{-2}{-7}=\frac{2}{7}$$

12.
$$-24+56=16Q$$
$$8(-3+7)=8(2Q)$$
$$4=2Q$$
$$Q=2$$

13.
$$-36=72A+45$$
$$9(-4)=9(8A+5)$$
$$-4=8A+5$$
$$-9=8A$$
$$\frac{-9}{8}=A=-1\frac{1}{8}$$

14. $10=10\times1;\ 100=10\times10;$
$LCM=10\times10=100$

15.
$$100(.2X)-100(.03)=100(.97)$$
$$20X-3=97$$
$$20X=100$$
$$X=5$$

16. $3=3\times1;\ 4=2\times2;\ 6=2\times3;$
$LCM=2\times2\times3=12$

17.
$$^{3}(\cancel{12})\frac{3}{\cancel{4}}+^{4}(\cancel{12})\frac{1}{\cancel{3}}Q=^{2}(\cancel{12})\frac{5}{\cancel{6}}$$
$$9+4Q=10$$
$$4Q=1$$
$$Q=\frac{1}{4}$$

18. $10=10\times1;\ 100=10\times10;$
$LCM=10\times10=100$

19.
$$100(-.7A)+100(.8A)=100(.12)$$
$$-70A+80A=12$$
$$10A=12$$
$$A=\frac{12}{10}$$
$$=1\frac{1}{5}\text{ or }1.2$$

20.

$$\begin{array}{r} 18.9 \\ 4\overline{)\,75.6} \\ \underline{4} \\ 35 \\ \underline{32} \\ 36 \\ \underline{36} \end{array}$$

Systematic Review 4D

1. $3(A-B-2)=3A-3B-6$
2. $5(3A-9+2A)=15A-45+10A$
3. $Q(X+3)=QX+Q3$ or $QX+3Q$
4. $-(-A-B+2C)=A+B-2C$
5. $10X-25Y=40;\ 5(2X-5Y)=5(8)$
6. $24A+12B=36;\ 12(2A+B)=12(3)$
7. $-14Q-21D=-42$
   $-7(2Q+3D)=-7(6)$
8. $3X+4XY=7X;$
   $X(3+4Y)=X(7)$ or $7X$
9. $$\begin{aligned} 22X+33&=44 \\ 11(2X+3)&=11(4) \\ 2X+3&=4 \\ 2X&=1 \\ X&=\frac{1}{2} \end{aligned}$$
10. $$\begin{aligned} 7Q-15&=9-5Q \\ 7Q+5Q&=9+15 \\ 12Q&=24 \\ Q&=2 \end{aligned}$$
11. $$\begin{aligned} 30Y-10&=10 \\ 10(3Y-1)&=10(1) \\ 3Y-1&=1 \\ 3Y&=2 \\ Y&=\frac{2}{3} \end{aligned}$$
12. $$\begin{aligned} 56B-49&=28 \\ 7(8B-7)&=7(4) \\ 8B-7&=4 \\ 8B&=11 \\ B&=\frac{11}{8}=1\frac{3}{8} \end{aligned}$$
13. $10=10\times1;\ 100=10\times10;$
    $LCM=10\times10=100$
14. $$\begin{aligned} 100(.3X)-100(1.2)&=100(.34) \\ 30X-120&=34 \\ 30X&=154 \\ X&=\frac{154}{30}=5\frac{2}{15}\text{ or }5.1\overline{3} \end{aligned}$$
15. $4=2\times2;\ 6=2\times3;\ 10=2\times5;$
    $LCM=2\times2\times3\times5=60$
16. $$\begin{aligned} {}^{15}(\cancel{60})\left(-\frac{3}{\cancel{4}}\right)+{}^{10}(\cancel{60})\frac{1}{\cancel{6}}R&={}^{6}(\cancel{60})\frac{7}{\cancel{10}} \\ -45+10R&=42 \\ 10R&=87 \\ R&=\frac{87}{10} \\ R&=8\frac{7}{10}\text{ or }8.7 \end{aligned}$$
17. $$\begin{array}{r} 75. \\ .05\overline{)\,3.75} \\ \underline{3\,5} \\ 25 \\ \underline{25} \end{array}$$ gum balls
18. $\frac{1}{4}=\frac{25}{100}=.25=25\%$
19. $40\%=.40=\frac{40}{100}=\frac{2}{5}$
20. $125\%=1.25=\frac{125}{100}=1\frac{1}{4}$

Systematic Review 4E

1. $-2(Q+2R-3E)=-2Q-4R+6E$
2. $A^2(3+B)=3A^2+A^2B$
3. $-X(Y+2+M)=-XY-2X-MX$
4. $-4(A^2+B^2+C^2)=-4A^2-4B^2-4C^2$
5. $4A-16B=-18;\ 2(2A-8B)=2(-9)$
6. $20A-40D=100;\ 20(A-2D)=20(5)$
7. $6Q+12G=3;\ 3(2Q+4G)=3(1)$
8. $-5R-15T=-20;\ -5(R+3T)=-5(4)$
9. $\frac{5}{6}\times\frac{4}{1}\div\frac{5}{2}=\frac{\cancel{5}}{3\cancel{6}}\times\frac{4}{1}\times\frac{\cancel{2}}{\cancel{5}}=\frac{4}{3}=1\frac{1}{3}$
10. $$\begin{aligned} -8&=-10C-14 \\ -2(4)&=-2(5C+7) \\ 4&=5C+7 \\ -3&=5C \\ C&=\frac{-3}{5} \end{aligned}$$

11.
$$15 = -45M - 30$$
$$15(1) = 15(-3M - 2)$$
$$1 = -3M - 2$$
$$3 = -3M$$
$$M = -1$$

12.
$$40 + 64 = 48N$$
$$8(5 + 8) = 8(6N)$$
$$13 = 6N$$
$$N = 2\frac{1}{6}$$

13.
$$63 = 35 - 7P$$
$$7(9) = 7(5 - P)$$
$$9 = 5 - P$$
$$4 = -P$$
$$P = -4$$

14. $10 = 10 \times 1$; $1000 = 10 \times 10 \times 10$;
$LCM = 10 \times 10 \times 10 = 1000$

15.
$$1000(.5Y) - 1000(.3) = 1000(.002)$$
$$500Y - 300 = 2$$
$$500Y = 302$$
$$Y = \frac{302}{500}$$
$$Y = \frac{151}{250} \text{ or } .604$$

16. $3 = 3 \times 1$; $4 = 2 \times 2$; $12 = 2 \times 2 \times 3$;
$LCM = 2 \times 2 \times 3 = 12$

17.
$$^{4}(\cancel{12})\frac{11}{\cancel{3}} + {}^{1}(\cancel{12})\frac{5}{\cancel{12}}K = {}^{3}(\cancel{12})\left(-\frac{5}{\cancel{4}}\right)$$
$$44 + 5K = -15$$
$$5K = -59$$
$$K = \frac{-59}{5}$$
$$K = -11\frac{4}{5} \text{ or } -11.8$$

18. $\frac{3}{4} = \frac{75}{100} = .75 = 75\%$

19. $20\% = .20 = \frac{20}{100} = \frac{1}{5}$

20. $380\% = 3.80 = \frac{380}{100} = 3\frac{4}{5}$

Lesson Practice 5A

1. (–2, 3)
2. 2
3. (–4,–2)
4. 3
5. (2,–2)
6. 4
7. (2, 3)
8. 1
9. (–1,–5)
10. 3
11. see graph
12. 2
13. see graph
14. 1
15. see graph
16. 4 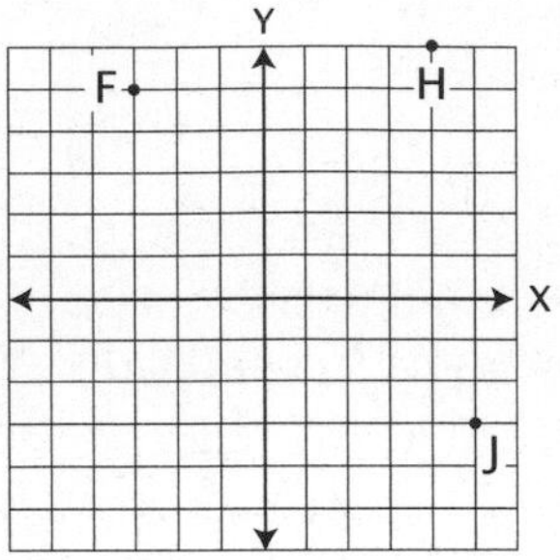
17. geometrically
18. positive, negative
19. the same X coordinate
20. X, 5

Y X F H J

21. -5 -4 -3 -2 -1 0 1 2 3 4 5

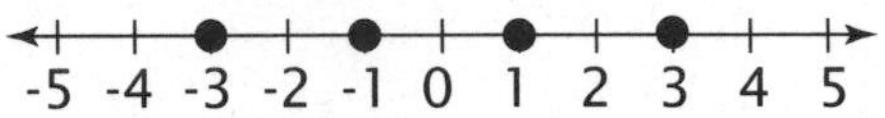

22. -5 -4 -3 -2 -1 0 1 2 3 4 5

23. -5 -4 -3 -2 -1 0 1 2 3 4 5

24. -5 -4 -3 -2 -1 0 1 2 3 4 5

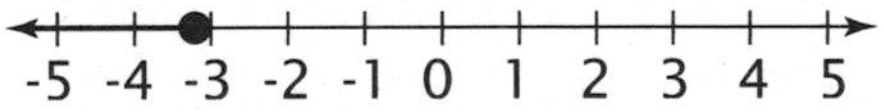

Lesson Practice 5B

1. (2, 3)
2. 1
3. (–1,–3)
4. 3
5. (2,–2)
6. 4
7. (–2, 1)
8. 2
9. (5,–5)
10. 4

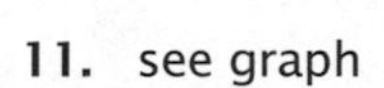

11. see graph
12. 3
13. see graph
14. 1
15. see graph
16. 4
17. (0, 0)

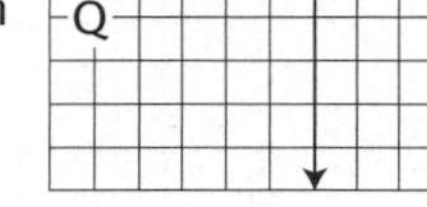

18. negative, negative
19. the same Y coordinate
20. Y, −2
21. −5 −4 −3 −2 −1 0 1 2 3 4 5
22. −5 −4 −3 −2 −1 0 1 2 3 4 5
23. −5 −4 −3 −2 −1 0 1 2 3 4 5
24. −5 −4 −3 −2 −1 0 1 2 3 4 5

Systematic Review 5C

1. (5,4)
2. (2,6)
3. (−2,1)
4. see graph
5. see graph
6. see graph
7. Descartes

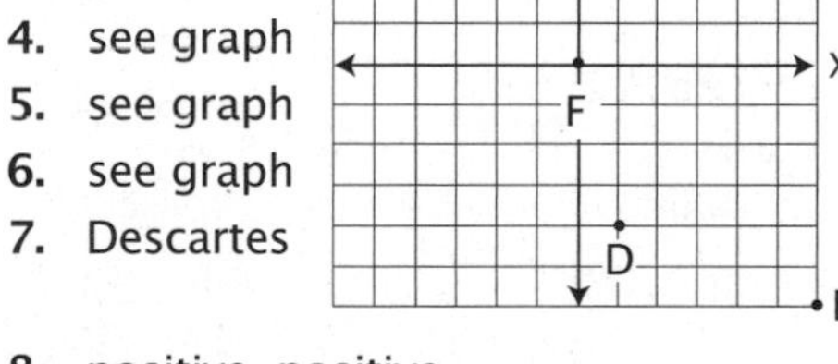

8. positive, positive
9. Y, X
10. origin
11. $100(.05X)+100(.12X)=100(.85)$
$$5X+12X=85$$
$$17X=85$$
$$X=5$$
12. $-72+8Y=32$
$$8Y=104$$
$$Y=13$$
13. $7(-B+2+7-1)=13+3B+5B$
$$7(-B+8)=13+8B$$
$$-7B+56=13+8B$$
$$43=15B$$
$$B=\frac{43}{15}$$
$$B=2\frac{13}{15}$$
14. $-4(P-6)+2P=|5-3+6|$
$$-4P+24+2P=8$$
$$-2P+24=8$$
$$-2P=-16$$
$$P=8$$
15. $3=3;\ 4=2\times2;\ 7=7;\ LCM=2\times2\times3\times7=84$
$${}^{12}(\cancel{84})\frac{18}{\cancel{7}}-{}^{21}(\cancel{84})\frac{1}{\cancel{4}}Q={}^{28}(\cancel{84})\frac{-17}{\cancel{3}}$$
$$216-21Q=-476$$
$$-21Q=-692$$
$$Q=\frac{-692}{-21}$$
$$Q=32\frac{20}{21}$$
16. $100(.3X)-100(.06X)=100(1.25)$
$$30X-6X=125$$
$$24X=125$$
$$X=\frac{125}{24}$$
$$X=5\frac{5}{24} \text{ or } X\approx5.21$$
17.
```
          116
         /   \
      2        58
              /  \
            2      29
```
$2\times2\times29$
18.
```
        36
       /  \
     2     18
          /  \
        2      9
              / \
             3   3
```
$2\times2\times3\times3$
19. $B+A$
20. $A+(B+C)$

Systematic Review 5D

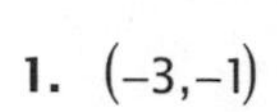

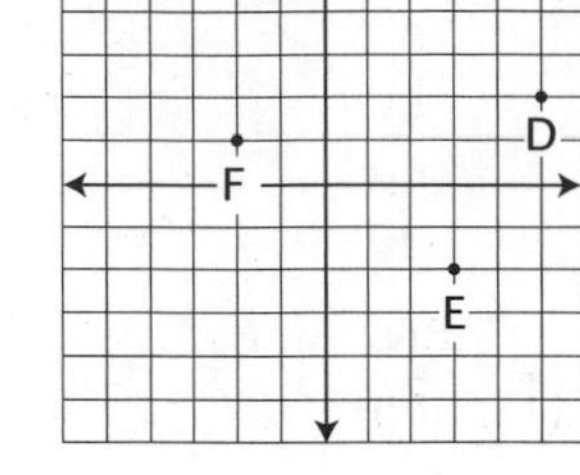

1. $(-3,-1)$
2. $(0,-4)$
3. $(-4,2)$
4. see graph
5. see graph
6. see graph
7. cartesian
8. negative, positive
9. same X coordinate
10. X, 3
11. $10(-1.3)+10(2.7)=10(.2Y)$
    $-13+27=2Y$
    $14=2Y$
    $Y=7$
12. $17Q-14XQ=11Q$
    $Q(17-14X)=Q(11)$
    $17-14X=11$
    $-14X=-6$
    $X=\frac{-6}{-14}$
    $X=\frac{3}{7}$
13. $D(3-7)-12=0$
    $3D-7D-12=0$
    $-4D=12$
    $D=\frac{12}{-4}$
    $D=-3$
14. $(6^2\div 9)\times 2-9Y=8(Y-4+9)$
    $(36\div 9)\times 2-9Y=8(Y+5)$
    $4\times 2-9Y=8Y+40$
    $8-9Y=8Y+40$
    $-32=17Y$
    $\frac{-32}{17}=Y$
    $Y=-1\frac{15}{17}$
15. $2=2;\ 4=2\times 2;\ 7=7$
    $LCM=2\times 2\times 7=28$
    ${}^{14}(\cancel{28})\frac{9}{\cancel{2}}={}^{7}(\cancel{28})\frac{5}{\cancel{4}}R+{}^{4}(\cancel{28})\frac{17}{\cancel{7}}$
    $126=35R+68$
    $58=35R$
    $\frac{58}{35}=R$
    $R=1\frac{23}{35}$
16. $100(.35P)+100(3.2)=100(-4P)$
    $35P+320=-400P$
    $435P=-320$
    $P=\frac{-320}{435}$
    $P=\frac{-64}{87}$ or $P\approx -.74$
17. $75\%=.75=\frac{75}{100}=\frac{3}{4}$
18. $113\%=1.13=\frac{113}{100}=1\frac{13}{100}$
19. $\frac{2}{5}=\frac{40}{100}=.40=40\%$
20. AB + AB

Systematic Review 5E

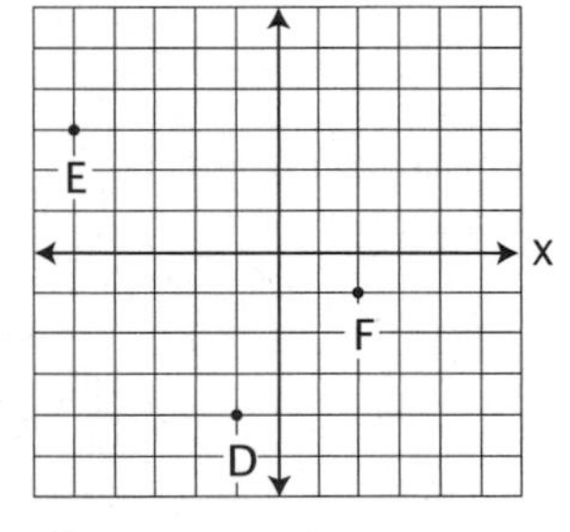

1. $(3,3)$
2. $(4,-2)$
3. $(-5,5)$
4. see graph
5. see graph
6. see graph
7. analytic
8. negative, negative
9. same X coordinate
10. X, –2
11. $100(1.08V)=100(.7)-100(.24)$
    $108V=70-24$
    $108V=46$
    $V=\frac{46}{108}=\frac{23}{54}$

12.
$$9X^2M = 10X^2 - 19X^2$$
$$X^2(9M) = X^2(10-19)$$
$$9M = 10-19$$
$$9M = -9$$
$$M = \frac{-9}{9}$$
$$M = -1$$

13.
$$(11-4)^2 \div 7 - |3-9| = 14(R+3R-2R+1)$$
$$7^2 \div 7 - |-6| = 14(2R+1)$$
$$49 \div 7 - 6 = 28R + 14$$
$$7 - 6 = 28R + 14$$
$$1 = 28R + 14$$
$$-13 = 28R$$
$$R = \frac{-13}{28}$$

14.
$$6[8-(Y+4)] = 3[(10+1)^2 - (7-5+4)]$$
$$6[8-Y-4] = 3[11^2 - 6]$$
$$6[4-Y] = 3[121-6]$$
$$24 - 6Y = 3[115]$$
$$24 - 6Y = 345$$
$$-6Y = 321$$
$$Y = \frac{321}{-6}$$
$$Y = -53\frac{1}{2}$$

15. $2 = 2;\ 7 = 7;\ 8 = 2\times2\times2;$

$LCM = 2\times2\times2\times7 = 56$

$$^{7}(\cancel{56})\frac{25}{\cancel{8}} - {}^{8}(\cancel{56})\frac{11}{\cancel{7}} = {}^{28}(\cancel{56})\frac{3}{\cancel{2}}D$$
$$175 - 88 = 84D$$
$$87 = 84D$$
$$\frac{87}{84} = D$$
$$D = 1\frac{1}{28}$$

16.
$$1000(-1.203H) + 1000(.9) = 1000(-.6)$$
$$-1203H + 900 = -600$$
$$-1203H = -1500$$
$$H = \frac{-1500}{-1203}$$
$$H = 1\frac{297}{1203} \text{ or } H \approx 1.25$$

17. $.125$ or $.13$

$$\begin{array}{r} .125 \\ 8\,\overline{)\,1.000} \\ \underline{8} \\ 20 \\ \underline{16} \\ 40 \\ \underline{40} \end{array}$$

18. $.66\overline{6}$ or $.67$

$$\begin{array}{r} .66\overline{6} \\ 3\,\overline{)\,2.000} \\ \underline{1\,8} \\ 20 \\ \underline{18} \\ 20 \\ \underline{18} \end{array}$$

19. $.6$

$$\begin{array}{r} .6 \\ 5\,\overline{)\,3.0} \\ \underline{3\,0} \end{array}$$

20. $.2\overline{2}$ or $.22$

$$\begin{array}{r} .2\overline{2} \\ 9\,\overline{)\,2.00} \\ \underline{1\,8} \\ 20 \\ \underline{18} \end{array}$$

Lesson Practice 6A

1.

hours	loaves
0	2
1	5
2	8
3	11

2. on the graph

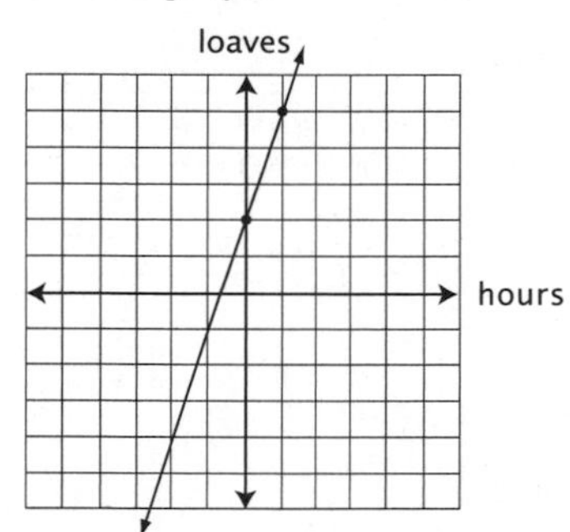

3. $L = 3H + 2$

4.

hours	rackets
0	–3
1	–1
2	1
3	3

5. on the graph

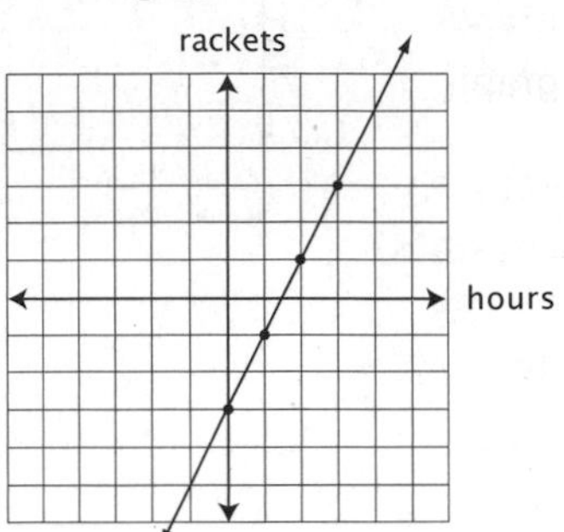

6. $R = 2H - 3$

7.

hours	steaks
0	1
1	5
2	9
3	13

8. on the graph

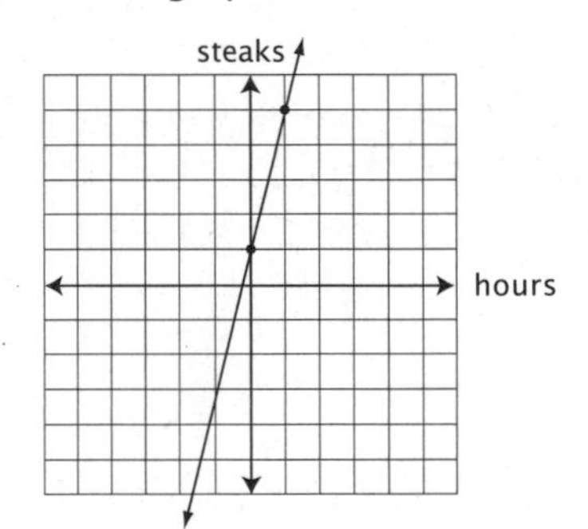

9. $S = 4H + 1$

10.

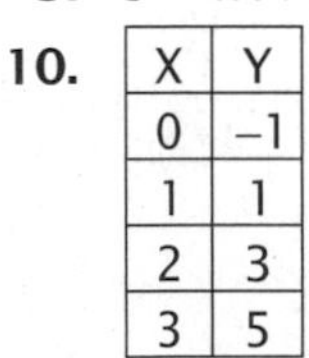

X	Y
0	–1
1	1
2	3
3	5

11. on the graph

12. Answers will vary. Your problem should start with a negative amount.

Lesson Practice 6B

1.

weeks	centimeters
0	–6
1	–4
2	–2
3	0

2. on the graph

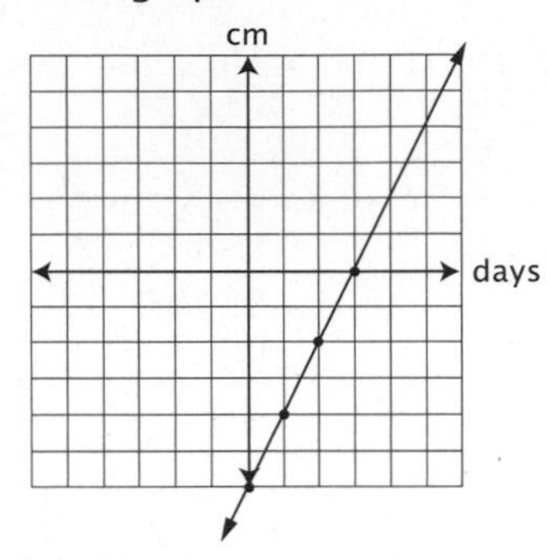

3. $C = 2W - 6$

4.

hours	fish
0	–5
1	–2
2	1
3	4

5. on the graph

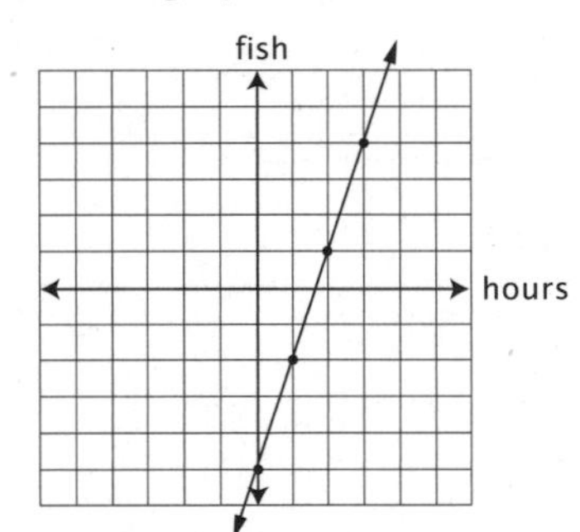

6. $F = 3H - 5$

7.

seconds	meters
0	–5
1	–3
2	–1
3	1

8. on the graph

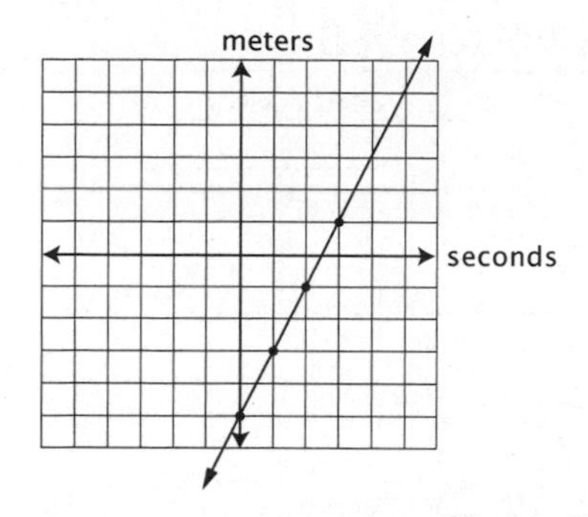

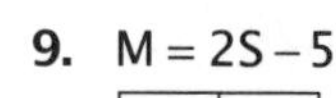

9. $M = 2S - 5$

10.

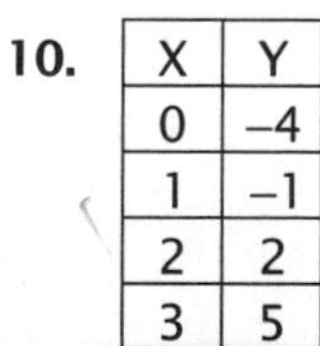

X	Y
0	–4
1	–1
2	2
3	5

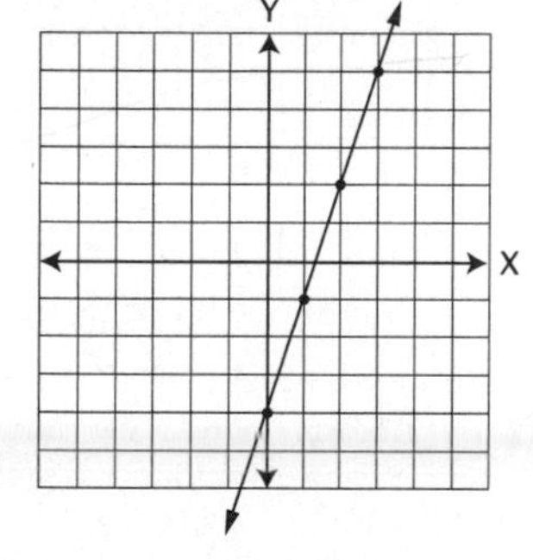

11. on the graph

12. Answers will vary. Your problem should start with a negative amount.

Systematic Review 6C

1.

days	speeches
0	1
1	3
2	5
3	7

2. on the graph

3. 2; 1; $S = 2D + 1$

4.

years	masterpieces
0	0
1	2
2	4
3	6

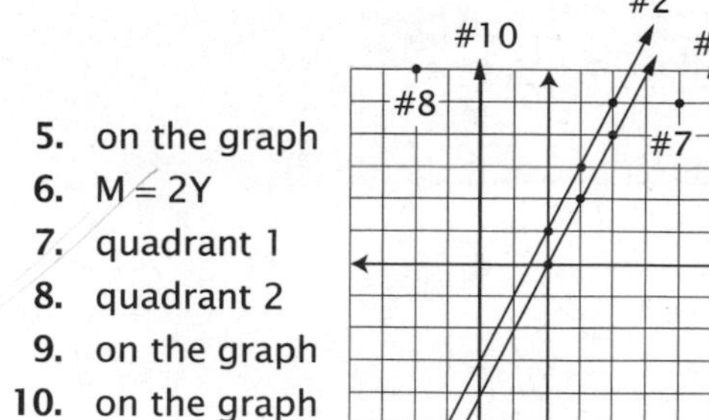

5. on the graph
6. $M = 2Y$
7. quadrant 1
8. quadrant 2
9. on the graph
10. on the graph

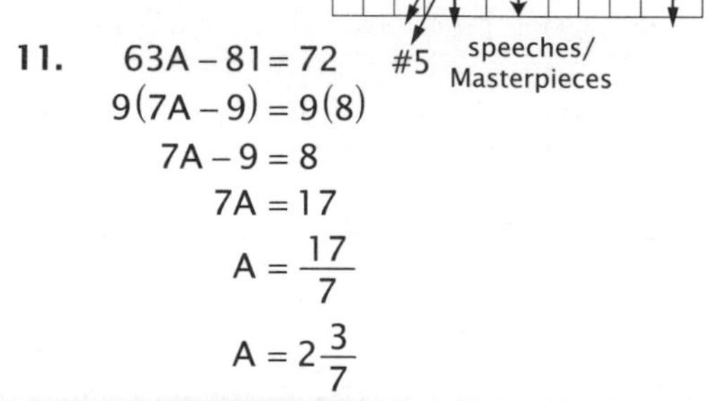

11.
$$63A - 81 = 72$$
$$9(7A - 9) = 9(8)$$
$$7A - 9 = 8$$
$$7A = 17$$
$$A = \frac{17}{7}$$
$$A = 2\frac{3}{7}$$

12.
$$48 + 54X = 36$$
$$6(8 + 9X) = 6(6)$$
$$8 + 9X = 6$$
$$9X = -2$$
$$X = \frac{-2}{9}$$

13.
$$\left[-5^2 - (-5)^2\right] \times 3 + 100 = 10X - 3X - 2X$$
$$\left[-(5 \times 5) - (-5)(-5)\right] \times 3 + 100 = 5X$$
$$[-25 - 25] \times 3 + 100 = 5X$$
$$-50 \times 3 + 100 = 5X$$
$$-150 + 100 = 5X$$
$$-50 = 5X$$
$$\frac{-50}{5} = X$$
$$X = -10$$

14.
$$100(.01) - 100(.1) + 100(.5) = 100(2Y)$$
$$1 - 10 + 50 = 200Y$$
$$41 = 200Y$$
$$Y = \frac{41}{200} \text{ or } .205$$

15.
$${}^{1}(\cancel{A})\frac{2}{\cancel{A}} - {}^{1}(\cancel{A})\frac{5}{\cancel{A}} = {}^{1}(\cancel{A})\frac{X}{\cancel{A}}$$
$$2 - 5 = X$$
$$X = -3$$

16. $^{3}\cancel{(6)}\frac{5}{\cancel{2}}X + {}^{2}\cancel{(6)}\frac{2}{\cancel{3}}X = {}^{1}\cancel{(6)}\frac{11}{\cancel{6}}$

$$15X + 4X = 11$$
$$19X = 11$$
$$X = \frac{11}{19} \text{ or } X \approx .58$$

17.
$$\begin{array}{r} .625 \\ 8\overline{)5.000} \\ \underline{4\,8} \\ 20 \\ \underline{16} \\ 40 \\ \underline{40} \end{array}$$

18. $X(X + Y + 2Q) = X^2 + XY + 2QX$

19. A

20. B

Systematic Review 6D

1.

hours	pages
0	0
1	3
2	6
3	9

2. on the graph
3. 3; 0; P = 3H
4.

customer	eggs
0	3
1	5
2	7
3	9

5. on the graph
6. E = 2C + 3
7. quadrant 2
8. quadrant 4

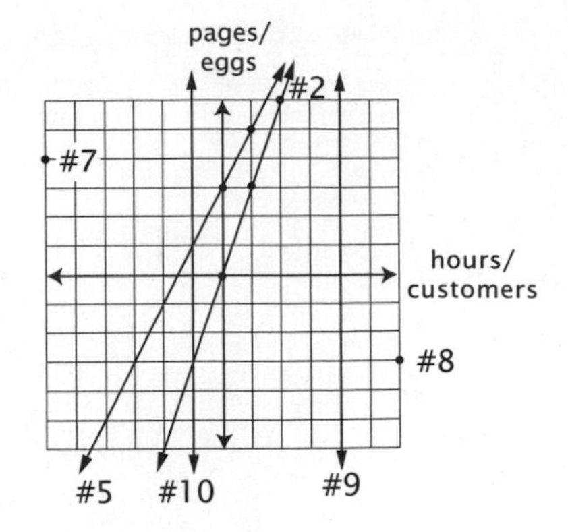

9. on the graph
10. on the graph

11.
$$-6(Y - 5 + 9) + 7(2Y + 9) = -1$$
$$-6(Y + 4) + 14Y + 63 = -1$$
$$-6Y - 24 + 14Y + 63 = -1$$
$$8Y + 39 = -1$$
$$8Y = -40$$
$$Y = \frac{-40}{8}$$
$$Y = -5$$

12.
$$3X + 3 - X - 8 + 5X + 12 = 4X - 12 - 6X + 10$$
$$7X + 7 = -2X - 2$$
$$9X = -9$$
$$X = \frac{-9}{9}$$
$$X = -1$$

13.
$$-5R + |9^2 - 3^2| + 13 = 7R + 5R$$
$$-5R + |81 - 9| + 13 = 12R$$
$$|72| + 13 = 12R + 5R$$
$$72 + 13 = 17R$$
$$85 = 17R$$
$$R = \frac{85}{17}$$
$$R = 5$$

14.
$$[8 - (-2)]^2 = 10X$$
$$[8 + 2]^2 = 10X$$
$$10^2 = 10X$$
$$100 = 10X$$
$$X = \frac{100}{10}$$
$$X = 10$$

15. $^{1}(\cancel{2A})\frac{Y}{\cancel{2A}} - {}^{2}(\cancel{2A})\frac{4}{\cancel{A}} = {}^{1}(\cancel{2A})\frac{1}{\cancel{2A}}$

$Y - 8 = 1$

$Y = 9$

16. $^{8}(\cancel{40})\frac{13}{\cancel{5}}D - {}^{5}(\cancel{40})\frac{3}{\cancel{8}}D = {}^{4}(\cancel{40})\frac{47}{\cancel{10}}$

$104D - 15D = 188$

$89D = 188$

$D = \frac{188}{89}$

$D = 2\frac{10}{89}$ or $D \approx 2.11$

17. $.9166$ or $.91\overline{6}$

$$\begin{array}{r} .9166 \\ 12\overline{)11.0000} \\ \underline{10\ 8} \\ 20 \\ \underline{12} \\ 80 \\ \underline{72} \\ 80 \end{array}$$

18. $X^2Y - 4X^2Y + BX^2Y = 0$

$X^2Y(1 - 4 + B) = 0$

$\frac{X^2Y(1-4+B)}{X^2Y} = \frac{0}{X^2Y}$

$1 - 4 + B = 0$

$-3 + B = 0$

$B = 3$

19. B

20. $A(A - B + 2AB) = A^2 - AB + 2A^2B$

Systematic Review 6E

1. 3; 1
2. on the graph
3. 1; 3 $B = M + 3$
4.

X	Y
0	2
1	3
2	4
3	5

5. on the graph
6. answers will vary
7. Y axis
8. X axis
9. on the graph
10. on the graph

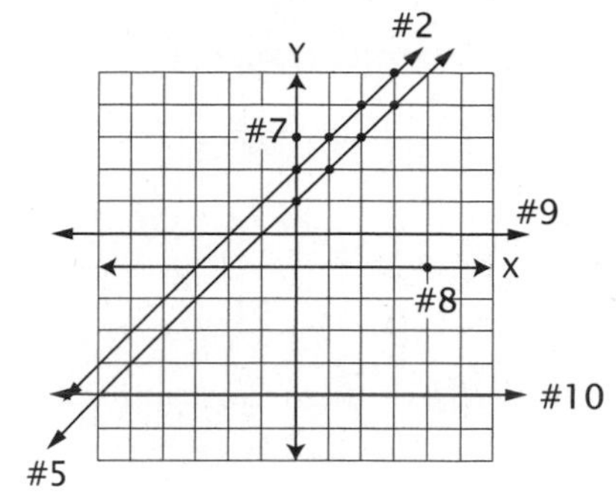

11. $4AB - 7A = 15A$

$A(4B - 7) = A(15)$

$4B - 7 = 15$

$4B = 22$

$B = \frac{22}{4} = 5\frac{1}{2}$

12. $7(B + 6 - 2B - 4) = 3^2(-4B - 8 - 9 + 2B)$

$7(-B + 2) = 9(-2B - 17)$

$-7B + 14 = -18B - 153$

$11B + 14 = -153$

$11B = -167$

$B = \frac{-167}{11}$

$B = -15\frac{2}{11}$

13. $-3(3G + 5G) + |3 - 12| = 18G + 5(-G - 4)$

$-3(8G) + |-9| = 18G - 5G - 20$

$-24G + 9 = 13G - 20$

$-37G = -29$

$G = \frac{29}{37}$

14. $100(-1.2) + 100(.07X) = 100(.3)$

$-120 + 7X = 30$

$7X = 150$

$X = \frac{150}{7}$

$X = 21\frac{3}{7}$ or $X \approx 21.43$

15. $^{4}(\cancel{40})\frac{3}{\cancel{10}} - {}^{8}(\cancel{40})\frac{8}{\cancel{5}} = {}^{5}(\cancel{40})\frac{-5}{\cancel{8}}M$

$$12 - 64 = -25M$$
$$-52 = -25M$$
$$M = \frac{-52}{-25}$$
$$M = 2\frac{2}{25} \text{ or } M = 2.08$$

16. $^{10}(\cancel{90})\frac{5}{\cancel{9}}X - {}^{15}(\cancel{90})\frac{17}{\cancel{6}} = {}^{9}(\cancel{90})\frac{7}{\cancel{10}}$

$$50X - 255 = 63$$
$$50X = 318$$
$$X = \frac{318}{50}$$
$$X = 6\frac{9}{25} \text{ or } X = 6.36$$

17. $.285$ or $\approx .29$

```
     .285
7 |2.000
   14
    60
    56
     40
     35
      5
```

18. $35\% = .35 = \frac{35}{100} = \frac{7}{20}$
19. $(-N)(-4) \div (2 \cdot 5)$
20. $3N - N + 2N + 7$

Lesson Practice 7A

1. intercept
2. up; over
3. negative
4. negative; $m = \frac{6}{-2} = -3$
5. positive; $m = \frac{8}{4} = 2$
6. positive; $m = \frac{7}{7} = 1$
7. negative; $m = \frac{6}{-3} = -2$
8. negative; $m = \frac{3}{-3} = -1$
9. positive; $m = \frac{3}{1} = 3$

Lesson Practice 7B

1. 4
2. 3
3. slope
4. negative; $m = \frac{2}{-8} = -\frac{1}{4}$
5. positive; $m = \frac{3}{5}$
6. positive; $m = \frac{4}{6} = \frac{2}{3}$
7. negative; $m = \frac{1}{-2} = -\frac{1}{2}$
8. negative; $m = \frac{2}{-6} = -\frac{1}{3}$
9. positive; $m = \frac{6}{8} = \frac{3}{4}$

Systematic Review 7C

1. -3
2. down
3. $m = \frac{6}{4} = \frac{3}{2}$; $b = -2$
4. $Y = \frac{3}{2}X - 2$
5. $m = \frac{-1}{3} = -\frac{1}{3}$; $b = 1$
6. $Y = -\frac{1}{3}X + 1$
7. $m = \frac{3}{4}$; $b = 0$
8. $Y = \frac{3}{4}X$
9. $m = \frac{2}{4} = \frac{1}{2}$; $b = -1$
10. $Y = \frac{1}{2}X - 1$
11. -5 -4 -3 -2 -1 0 1 2 3 4 5
12. -5 -4 -3 -2 -1 0 1 2 3 4 5
13. $[(7-3) \times 4^2 - 9] \div 3^3 =$

$$[4 \times 16 - 9] \div 27 =$$
$$(64 - 9) \div 27 =$$
$$55 \div 27 = 2\frac{1}{27}$$

14. $|-4-2|+8^2-7\times5+19=$
$|-6|+64-35+19=$
$6+64-35+19=54$

15. $13^2+5\div10=169+[5\div10]$
$=169+.5=169.5$

16. $5(9-2)-6(7)+2^3\cdot3=5(7)-6(7)+8\cdot3$
$=35-42+24=17$

17. $2X-5=-X+13$
$2X+X=13+5$
$3X=18$
$X=6$

18. $Y+14-3Y=0$
$Y-3Y=-14$
$-2Y=-14$
$Y=7$

19. $-3\frac{1}{2}B+\frac{2}{3}=5\frac{1}{4}+\frac{5}{6}B$
$12\left[-\frac{7}{2}B+\frac{2}{3}\right]=12\left[\frac{21}{4}+\frac{5}{6}B\right]$
$-42B+8=63+10B$
$8-63=10B+42B$
$-55=52B$
$-\frac{55}{52}=B$
$B=-\frac{55}{52}$ or $-1\frac{3}{52}$

20. $2.7T+1.09=5.3-.6T$
$100[2.7T+1.09]=100[5.3-.6T]$
$270T+109=530-60T$
$270T+60T=530-109$
$330T=421$
$T=\frac{421}{330}$ or $1\frac{91}{330}$

Systematic Review 7D

1. 4

2. 3

3. $m=\frac{3}{5};\ b=4$

4. $Y=\frac{3}{5}X+4$

5. $m=\frac{4}{-6}=-\frac{2}{3};\ b=0$

6. $Y=-\frac{2}{3}X$

7. $m=\frac{3}{-6}=-\frac{1}{2};\ b=1$

8. $Y=-\frac{1}{2}X+1$

9. $m=\frac{6}{3}=2;\ b=3$

10. $Y=2X+3$

11. -5 -4 -3 -2 -1 0 1 2 3 4 5

12. -5 -4 -3 -2 -1 0 1 2 3 4 5

13. $-|5-8|\times4-7+12=$
$-|-3|\times4-7+12=$
$(-3\times4)-7+12=$
$-12-7+12=-7$

14. $-7^2\times2-48+5=$
$(-49\times2)-48+5=$
$-98-48+5=-141$

15. $144\div9\times3-|100-121|=$
$(144\div9\times3)-|-21|=$
$(144\div9\times3)-21=$
$(16\times3)-21=27$

16. $8[17-3\times2]+6^2-(-5)^2=$
$8[17-6]+36-25=$
$8(11)+36-25=$
$88+36-25=99$

17. $4A+11=A-4$
$4A-A=-4-11$
$3A=-15$
$A=-5$

18. $-5F=-6F+8$
$-5F+6F=$
$F=8$

19. $\frac{2}{5}-\frac{1}{6}D=-\frac{3}{4}$
$60\left[\frac{2}{5}-\frac{1}{6}D\right]=60\left[-\frac{3}{4}\right]$
$24-10D=-45$
$-10D=-45-24$
$-10D=-69$
$D=\frac{69}{10}$ or $6\frac{9}{10}$

20.
$$.03M - 1.2 = -.48M$$
$$100[.03M - 1.2] = 100[-.48M]$$
$$3M - 120 = -48M$$
$$3M + 48M = 120$$
$$51M = 120$$
$$M = \frac{120}{51} \text{ or } 2\frac{6}{17}$$

Systematic Review 7E

1. up
2. slope
3. $m = \frac{2}{-5} = -\frac{2}{5};\ b = 2$
4. $Y = -\frac{2}{5}X + 2$
5. $m = \frac{2}{-8} = -\frac{1}{4};\ b = 3$
6. $Y = -\frac{1}{4}X + 3$
7. $m = \frac{3}{3} = 1;\ b = -1$
8. $Y = X - 1$
9. $m = \frac{3}{-1} = -3;\ b = -2$
10. $Y = -3X - 2$
11. 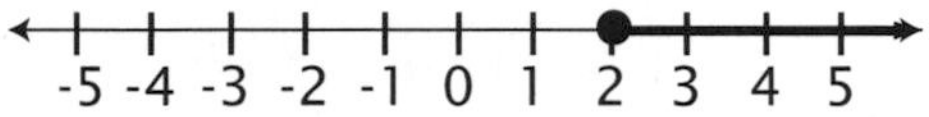

12. 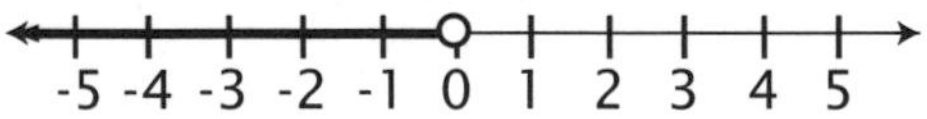

13. $11 \cdot 3^2 - 14 \times 2 =$
$(11 \cdot 9) - (14 \times 2) =$
$(99) - (28) = 71$
14. $2 \cdot 7 + 4^2 - 15 =$
$(2 \cdot 7) + 16 - 15 =$
$14 + 16 - 15 = 15$
15. $(-6)^2 + (8 - 3^2) =$
$36 + (8 - 9) =$
$36 + (-1) = 35$
16. $16 \div 8 \cdot 5 - 14 =$
$2 \cdot 5 - 14 =$
$10 - 14 = -4$
17.
$$-2B + 5 - 3 + B = B - 4B + 1 - 10$$
$$-2B + B + 5 - 3 = B - 4B + 1 - 10$$
$$-B + 2 = -3B - 9$$
$$-B + 3B = -9 - 2$$
$$2B = -11$$
$$B = -\frac{11}{2} = -5\frac{1}{2}$$
18.
$$5K + 6 - K - 9 = -2K + 6 + 3K - 3$$
$$5K - K + 6 - 9 = -2K + 3K + 6 - 3$$
$$4K - 3 = K + 3$$
$$4K - K = 3 + 3$$
$$3K = 6$$
$$K = 2$$
19.
$$4\frac{3}{10} = -\frac{2}{3} + \frac{8}{9}G$$
$$90\left(\frac{43}{10}\right) = 90\left(-\frac{2}{3} + \frac{8}{9}G\right)$$
$$387 = -60 + 80G$$
$$447 = 80G$$
$$\frac{447}{80} = G$$
$$G = \frac{447}{80} = 5\frac{47}{80}$$
20.
$$-5 - .6R = -9.8$$
$$10(-5 - .6R) = 10(-9.8)$$
$$-50 - 6R = -98$$
$$-6R = -98 + 50$$
$$-6R = -48$$
$$R = 8$$

Lesson Practice 8A

1. $Y = \frac{1}{4}X - 2,\ m = \frac{1}{4},\ b = -2$

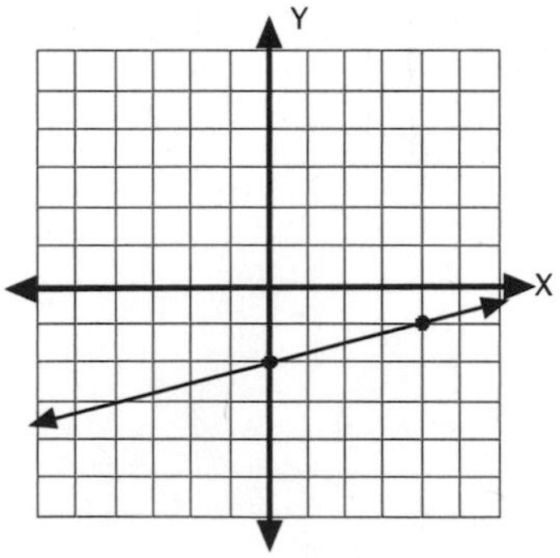

2. $Y = -X + 2$, $m = -1$, $b = 2$

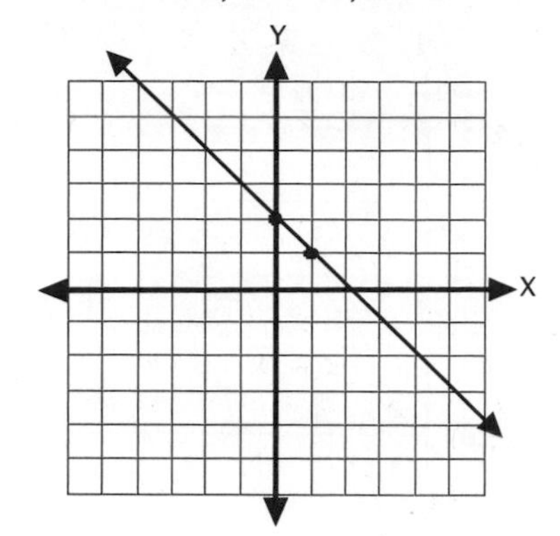

3. $Y = -2$; $m = 0$, $b = -2$

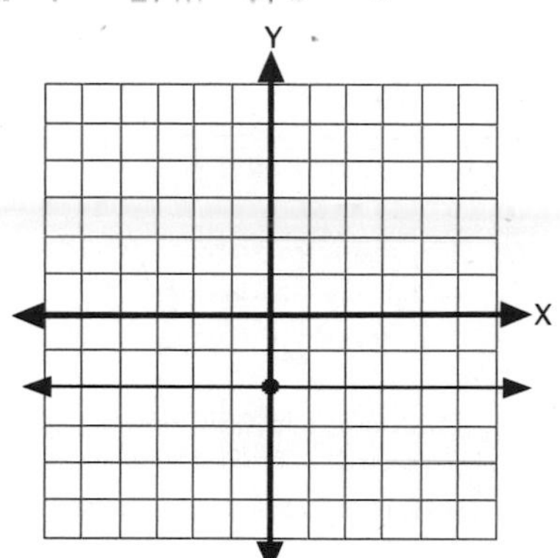

4. $Y = \frac{3}{5}X + 1$, $m = \frac{3}{5}$, $b = 1$

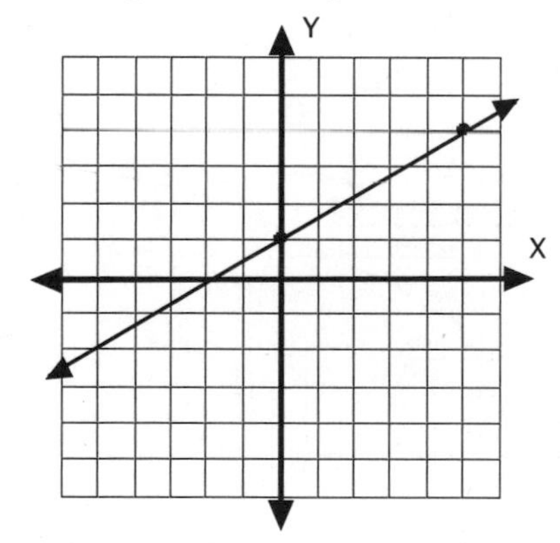

5. $Y = X$; $Y = X + 0$, $m = 1$, $b = 0$

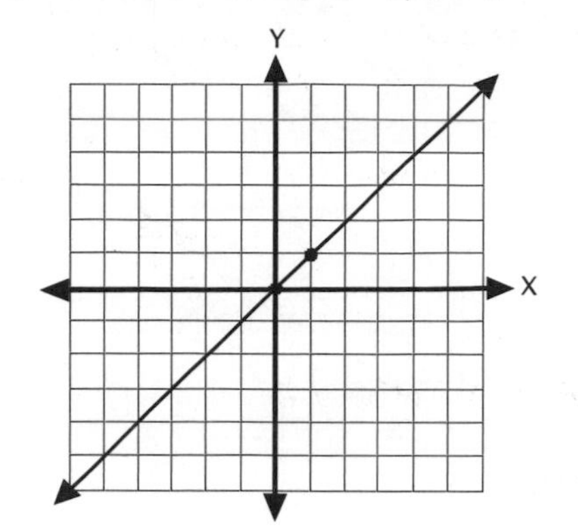

6. $X = -3$, $m =$ undefined,
because we cannot divide by zero.
$b =$ does not exist
The line never crosses the x-axis.

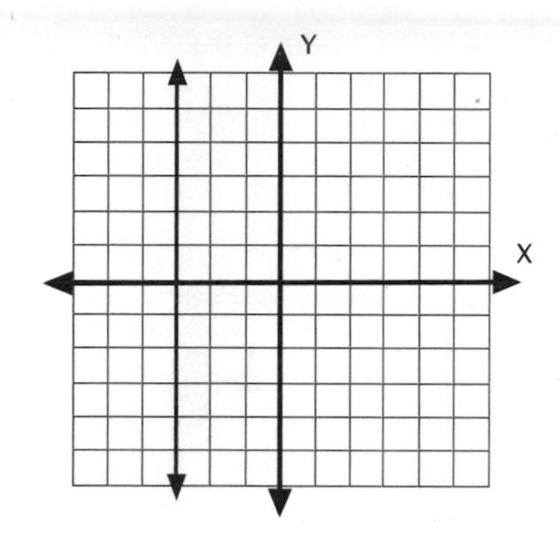

Lesson Practice 8B

1. $Y = -2X - 5$, $m = -2$, $b = -5$

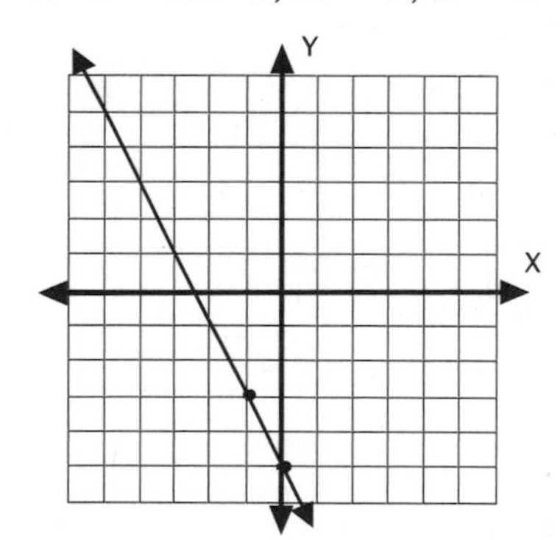

2. $Y = -\frac{3}{2}X;\ Y = -\frac{3}{2}X + 0,\ m = -\frac{3}{2},\ b = 0$

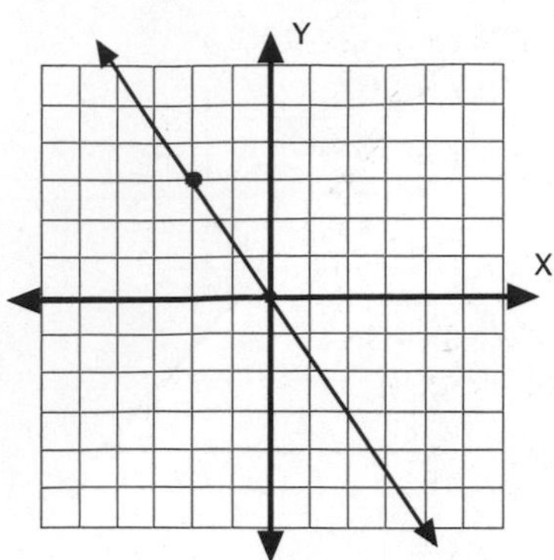

3. X = 0, m = undefined because we cannot divide by zero.
b = indeterminate, because any value of y is valid. The student may express this idea in different ways. Graph is the Y-axis.

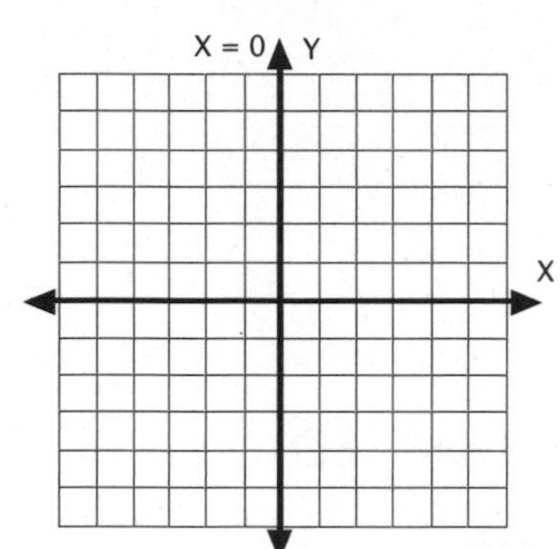

4. $Y = -3X + 2,\ m = -3,\ b = 2$

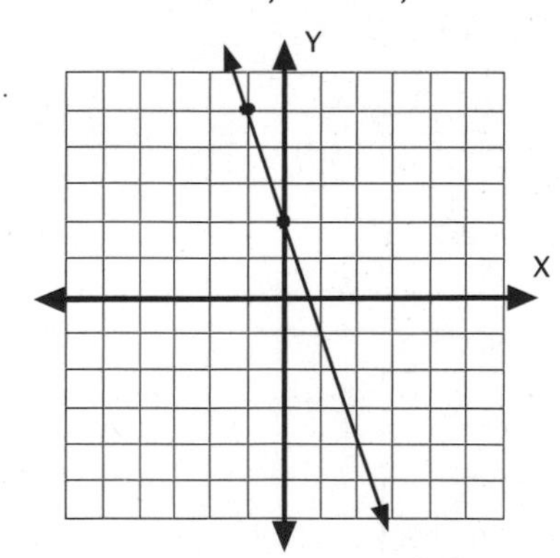

5. $Y = 2X - 1,\ m = 2,\ b = -1$

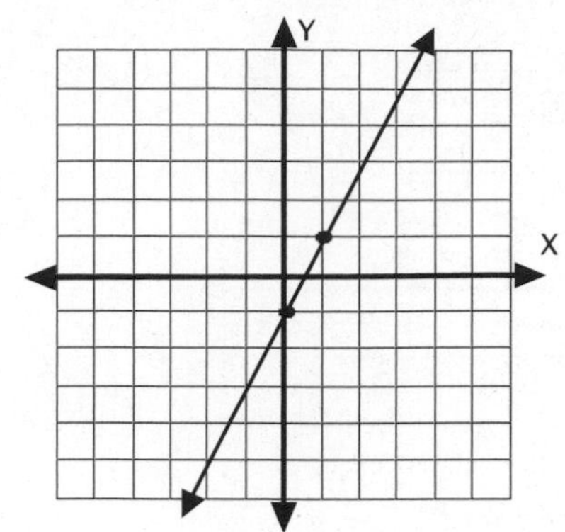

6. $Y = 4;\ Y = 0X + 4,\ m = 0,\ b = 4$

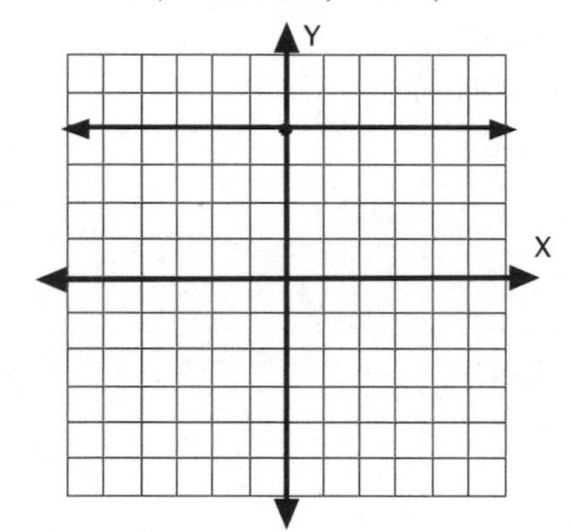

Systematic Review 8C

1.

days	dollars
0	–4
1	–5
2	–6
3	–7

2. see graph

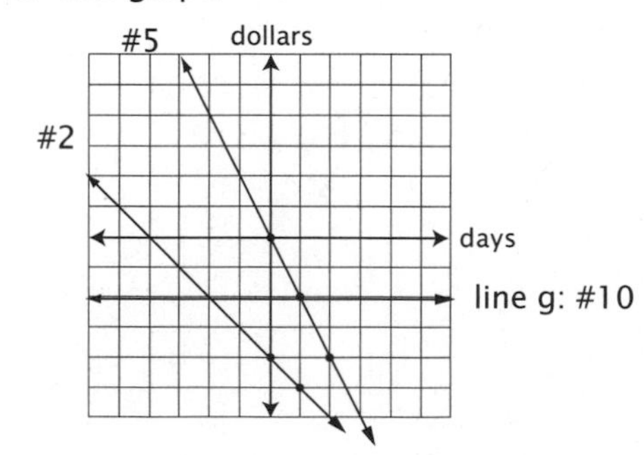

3. -1; 4; $\$ = -D - 4$

4. 0; 2

days	money
0	0
1	–2
2	–4
3	–6

5. see graph

6. -2; 0

7. slope = 4; intercept = 2

8. $Y = 4X + 2$

9. quadrants 1, 2, 3

10. see graph

11.
$$60R - 90R = 70$$
$$-30R = 70$$
$$R = -2\frac{1}{3}$$

12.
$$-18 + 54X = 27$$
$$9(-2 + 6X) = 9(3)$$
$$-2 + 6X = 3$$
$$6X = 5$$
$$X = \frac{5}{6}$$

13.
$$\left[(6+5)^2 - 1\right] \div 12 = 3X + |-2X|$$
$$(11^2 - 1) \div 12 = 3X + 2X$$
$$(121 - 1) \div 12 = 5X$$
$$120 \div 12 = 5X$$
$$10 = 5X$$
$$X = 2$$

14.
$$4B - 32B = 36B - 8BY$$
$$4B(1 - 8) = 4B(9 - 2Y)$$
$$-7 = 9 - 2Y$$
$$-16 = -2Y$$
$$Y = 8$$

15.
$$100(1.03) - 100(.8Y) = 100(5)$$
$$103 - 80Y = 500$$
$$-80Y = 397$$
$$Y = \frac{397}{-80}$$
$$Y = -4\frac{77}{80}$$

16.
$${}^{15}(\cancel{60})\frac{15}{\cancel{4}}Y = {}^{12}(\cancel{60})\frac{11}{\cancel{5}} + {}^{10}(\cancel{60})\frac{23}{\cancel{6}}$$
$$225Y = 132 + 230$$
$$225Y = 362$$
$$Y = \frac{362}{225}$$
$$Y = 1\frac{137}{225}$$

17.
$$5X - 20 = 50X + 35$$
$$-55 = 45X$$
$$X = \frac{-55}{45}$$
$$X = -1\frac{2}{9}$$

18.
$${}^{6}(\cancel{60})\frac{3}{\cancel{10}}X - {}^{10}(\cancel{60})\frac{19}{\cancel{6}}X = {}^{15}(\cancel{60})\frac{17}{\cancel{4}}$$
$$18X - 190X = 255$$
$$-172X = 255$$
$$X = \frac{255}{-172}$$
$$X = -1\frac{83}{172}$$

19. $WF \times 7 = 5; \frac{WF}{\cancel{7}} \times \cancel{7} = \frac{5}{7}; WF = \frac{5}{7}$

20. $WF \times 5 = 2; \frac{WF}{\cancel{5}} \times \cancel{5} = \frac{2}{5}; WF = \frac{2}{5}$

Systematic Review 8D

1.

days	dollars
0	–3
1	–5
2	–7
3	–9

2. see graph

3. -2; 3; $\$ = -2D - 3$

4. 2; 3

days	dollars
0	2
1	5
2	8
3	11

5. line g is the X-axis: see graph

6. 3; 2
7. slope = –1; intercept = 0
8. $Y = -X$
9. quadrants 2; 4
10. see graph

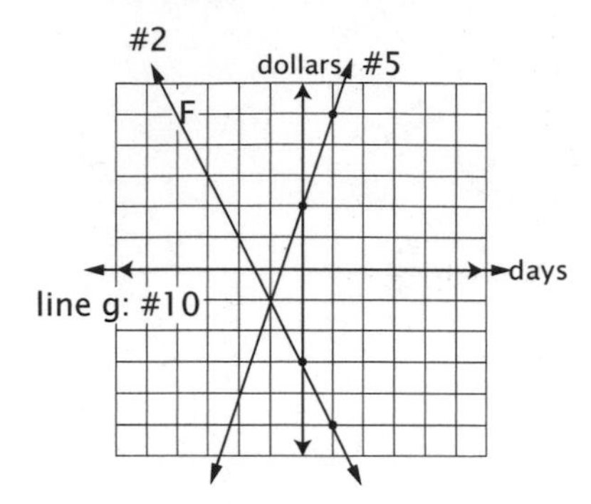

11. $$\begin{aligned} 12Y &= 6-24 \\ 12Y &= -18 \\ Y &= \frac{-18}{12} \\ Y &= -1\frac{1}{2} \end{aligned}$$

12. $$\begin{aligned} -72+60F &= 48 \\ 60F &= 120 \\ F &= 2 \end{aligned}$$

13. $$\begin{aligned} 2^2[5(4-2X+3)-3(8+9X-4X)] &= 0 \\ 4[5(7-2X)-3(8+5X)] &= 0 \\ 4[35-10X-24-15X] &= 0 \\ 4(11-25X) &= 0 \\ 44-100X &= 0 \\ 44 &= 100X \\ X &= \frac{44}{100} \\ X &= \frac{11}{25} \end{aligned}$$

14. $$\begin{aligned} -50BY+30B &= 80BY-40B \\ 10B(-5Y+3) &= 10B(8Y-4) \\ -5Y+3 &= 8Y-4 \\ 7 &= 13Y \\ Y &= \frac{7}{13} \end{aligned}$$

15. $$\begin{aligned} 1000(.018) &= 1000(.25Q)+1000(2.04) \\ 18 &= 250Q+2040 \\ -2022 &= 250Q \\ \frac{-2022}{250} &= Q \\ Q &= -8\frac{11}{125} \end{aligned}$$

16. $$\begin{aligned} {}^{3}(\cancel{24})\frac{-13}{\cancel{8}}M+{}^{8}(\cancel{24})\frac{13}{\cancel{3}} &= {}^{4}(\cancel{24})\frac{7}{\cancel{6}} \\ -39M+104 &= 28 \\ 76 &= 39M \\ \frac{76}{39} &= M \\ M &= 1\frac{37}{39} \end{aligned}$$

17. $$\begin{aligned} 10(-1.3)+10(2.6) &= 10(5.2X) \\ -13+26 &= 52X \\ 13 &= 52X \\ X &= \frac{1}{4} \end{aligned}$$

18. $$\begin{aligned} {}^{6}(\cancel{30})\frac{7}{\cancel{5}}Y &= {}^{5}(\cancel{30})\frac{25}{\cancel{6}}-{}^{10}(\cancel{30})\frac{7}{\cancel{3}} \\ 42Y &= 125-70 \\ 42Y &= 55 \\ Y &= \frac{55}{42} \\ Y &= 1\frac{13}{42} \end{aligned}$$

19. $3N-N+2N+7$
20. $WF \times 4 = 3;\ \frac{WF}{\cancel{4}} \times \cancel{4} = \frac{3}{4};\ WF = \frac{3}{4}$

Systematic Review 8E

1.

days	dollars
0	–4
1	–1
2	2
3	5

2. see graph
3. 3; 4; $\$ = 3D-4$ or $M = 3D-4$

4. $-3; 1$

days	dollars
0	−3
1	−2
2	−1
3	0

5. see graph
6. $1; -3$
7. slope $= -3$; y-intercept $= 2$
8. $Y = -3X + 2$
9. quadrants 1, 2, 4
10. see graph

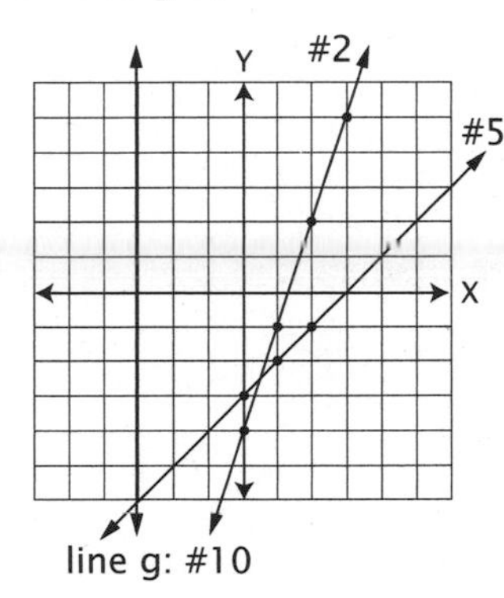

11. $$\begin{aligned} -9Q - 24Q + 15 &= 0 \\ 3(-3Q - 8Q + 5) &= 3(0) \\ -3Q - 8Q + 5 &= 0 \\ -11Q &= -5 \\ Q &= \frac{5}{11} \end{aligned}$$

12. $$\begin{aligned} 66 + 99A - 77 &= 0 \\ 11(6 + 9A - 7) &= 11(0) \\ 6 + 9A - 7 &= 0 \\ 9A - 1 &= 0 \\ 9A &= 1 \\ A &= \frac{1}{9} \end{aligned}$$

13. $$\begin{aligned} 2X(3 - 7 + 4 - 8 - 1) - 4^2 &= (-4) \\ 2X(-9) - 16 &= (-4) \\ -18X &= 12 \\ X &= \frac{-2}{3} \end{aligned}$$

14. $$\begin{aligned} 12 + 28 &= -20B \\ 40 &= -20B \\ B &= -2 \end{aligned}$$

15. $$\begin{aligned} 10(4D) - 10(.3D) &= 10(18.5) \\ 40D - 3D &= 185 \\ 37D &= 185 \\ D &= 5 \end{aligned}$$

16. $$\begin{aligned} {}^{35}(\not{7}0)\frac{13}{\not{2}} &= {}^{10}(\not{7}0)\frac{5}{\not{7}}N - {}^{14}(\not{7}0)\frac{13}{\not{5}}N \\ 455 &= 50N - 182N \\ 455 &= -132N \\ \frac{455}{-132} &= N \\ N &= -3\frac{59}{132} \end{aligned}$$

17. $$\begin{aligned} -12 &= -2A - 6 \\ -6 &= -2A \\ A &= 3 \end{aligned}$$

18. $$\begin{aligned} {}^{20}(\not{4}0)\frac{-11}{\not{2}}X + {}^{5}(\not{4}0)\frac{19}{\not{8}} &= {}^{4}(\not{4}0)\frac{9}{1\not{0}} \\ -220X + 95 &= 36 \\ -220X &= -59 \\ X &= \frac{59}{220} \end{aligned}$$

19. $(N + 1)(N - 4)$

20. $WF \times 9 = 7;\ \frac{WF}{\not{9}} \times \not{9} = \frac{7}{9};\ WF = \frac{7}{9}$

Lesson Practice 9A

1. a. $m = \frac{5}{3},\ b = 5,\ Y = \frac{5}{3}X + 5$
   b. $m = \frac{5}{3},\ b = 1,\ Y = \frac{5}{3}X + 1$
   c. $m = \frac{5}{3},\ b = -1,\ Y = \frac{5}{3}X - 1$
   d. $m = \frac{5}{3},\ b = -4,\ Y = \frac{5}{3}X - 4$
2. w. $m = -\frac{1}{2},\ b = 4,\ Y = -\frac{1}{2}X + 4$
   x. $m = -\frac{1}{2},\ b = 2,\ Y = -\frac{1}{2}X + 2$
   y. $m = -\frac{1}{2},\ b = -1,\ Y = -\frac{1}{2}X - 1$
   z. $m = -\frac{1}{2},\ b = -3,\ Y = -\frac{1}{2}X - 3$

3. A. $Y = \frac{1}{3}X - 2$

B. $Y = -3X$

C. $Y = 4 - 3X$; $Y = -3X + 4$

Lines B & C both have a slope of -3, which is the same as $Y = -3X + 2$.

Answers B & C are parallel to the given line.

4. A. $Y = \frac{1}{4}X + 5$

B. $Y = -\frac{1}{2}X + 2$

C. $Y = 4 + \frac{4}{8}X$; $Y = \frac{1}{2}X + 4$

Line C has a reduced slope of $\frac{1}{2}$, which is the same slope as $Y = \frac{1}{2}X - 5$.

Answer C is parallel to the given line.

5. A. $Y = \frac{2}{3}X + 4$

B. $Y = \frac{6}{4}X$; $Y = \frac{3}{2}X$

C. $2Y = 8 - 3X$; $2Y = -3X + 8$,

$Y = \frac{-3}{2}X + 4$

Given line: $2Y - 3X = 4$;

$2Y = 3X + 4$; $Y = \frac{3}{2}X + 2$

LIne B has a reduced slope of $\frac{3}{2}$, which is the same slope as $Y = \frac{3}{2}X + 2$.

6. A. $Y = \frac{12}{9}X - 1$; $Y = \frac{4}{3}X - 1$

B. $3Y = -4X + 0$; $Y = -\frac{4}{3}X$

C. $-2Y = 5X - 8$; $Y = -\frac{5}{2}X + 4$

Given line : $3Y + 4X = -6$; $3Y = -4X - 6$;

$Y = -\frac{4}{3}X - 2$

Line B has a slope of $-\frac{4}{3}$, which is the same slope as $Y = -\frac{4}{3}X - 2$.

Answer B is parallel to the given line.

7.
$$-Y + 2X = 4$$
$$-Y = -2x + 4$$
$$Y = 2X - 4$$

8.
$$Y - 4X = 0$$
$$Y = 4X + 0$$
$$Y = 4X$$

9.
$$-2Y - X = -2$$
$$-2Y = X - 2$$
$$Y = -\frac{1}{2}X + 1$$

10.
$$3Y - 2X = -6$$
$$3Y = 2X - 6$$
$$Y = \frac{2}{3}X - 2$$

11.
$$-4Y - 3X = -8$$
$$-4Y = 3X - 8$$
$$Y = -\frac{3}{4}X + 2$$

12.
$$Y = -\frac{5}{3}X - 2$$

$\frac{5}{3}X + Y = -2$ Adding $\frac{5}{3}X$ to both sides.

$5X + 3Y = -6$ Multiplying each term by 3.

13.
$$Y = 4X - 3$$
$$-4X + Y = -3$$

or $4X - Y = 3$ Multiplying each term by -1.

14.
$$Y = \frac{1}{4}X + 3$$
$$-\frac{1}{4}X + Y = 3$$
$$-X + 4Y = 12 \text{ or } X - 4Y = -12$$

15.
$$Y = -\frac{3}{5}X - 1$$
$$\frac{3}{5}X + Y = -1$$
$$3X + 5Y = -5$$

16.
$$Y = 3X$$
$$-3X + Y = 0 \text{ or } 3X - Y = 0$$

Lesson Practice 9B

1. $Y = \frac{6}{8}X - 3;\ Y = \frac{3}{4}X - 3$

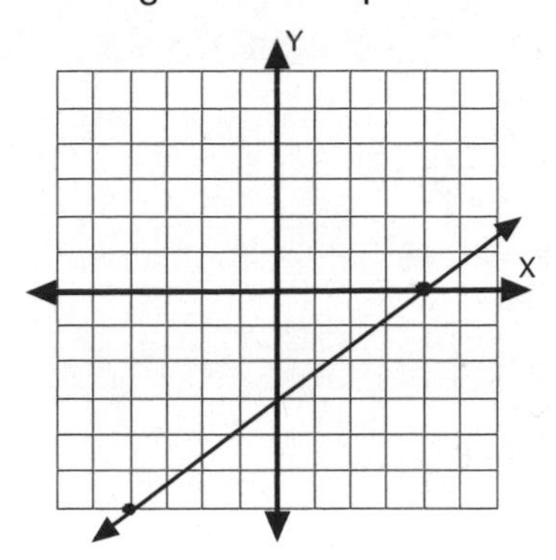

2. $Y = \frac{3}{3}X + 4;\ Y = X + 4$

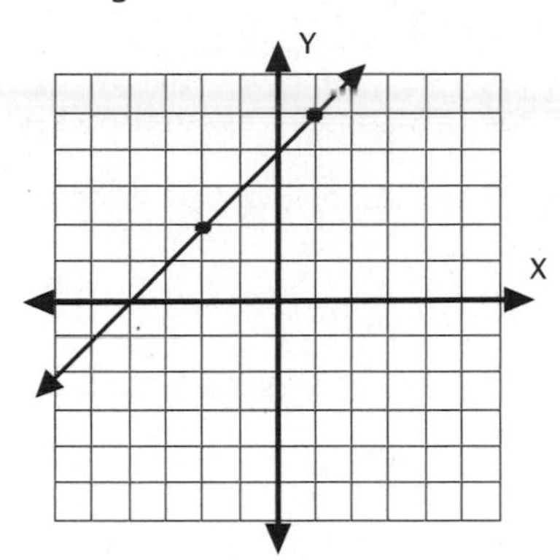

3. slope-intercept: $Y = 0X - 2$ or $Y = -2$
   standard form: $0X + Y = -2$ or $Y = -2$

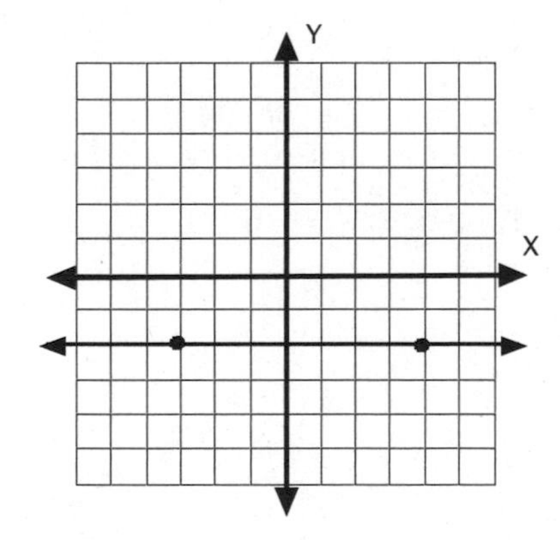

4. slope-intercept: $Y = -\frac{8}{6}X + 2;$

   $Y = -\frac{4}{3}X + 2$

   standard form: $\frac{4}{3}X + Y = 2;$

   $4X + 3Y = 6$

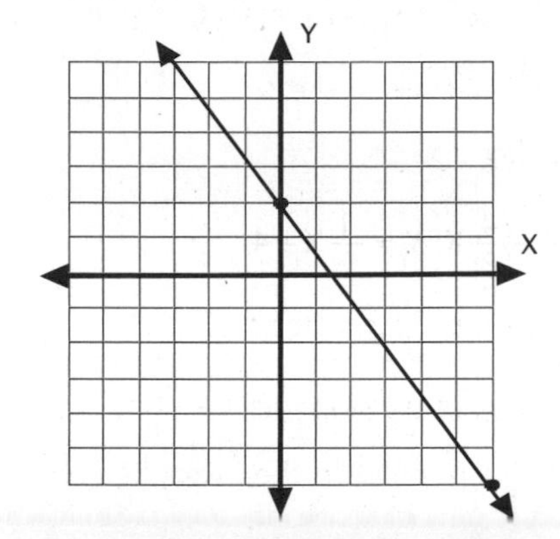

5. slope – intercept : $Y = -\frac{6}{3}X + 0;$

   $Y = -2X$

   standard form : $2X + Y = 0$

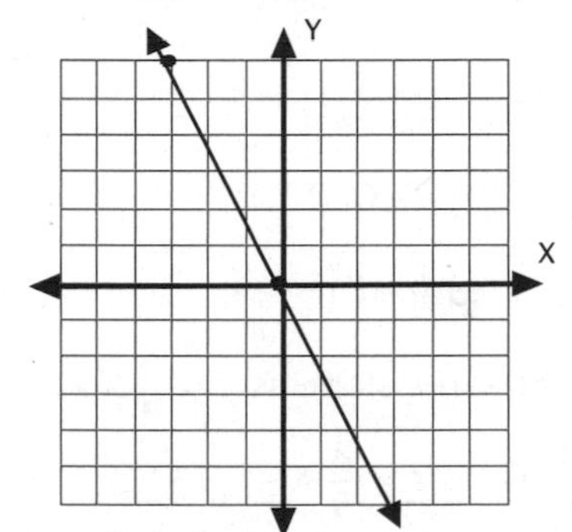

6. slope-intercept: none because slope is undefined and the Y-intercept does not exist.
   standard form: $X = 3$

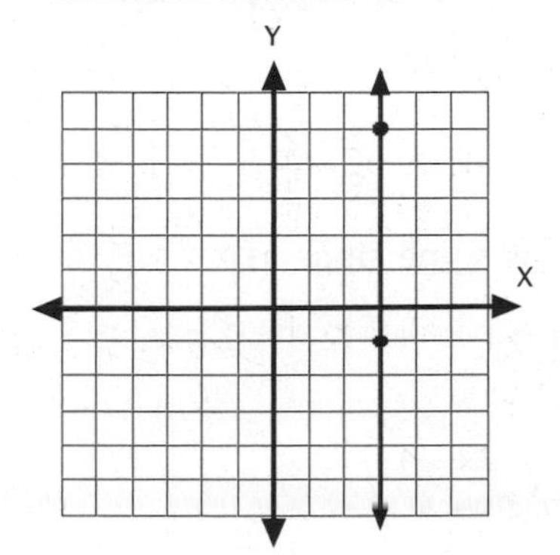

7. see graph
8. slope $= -\frac{3}{2}$

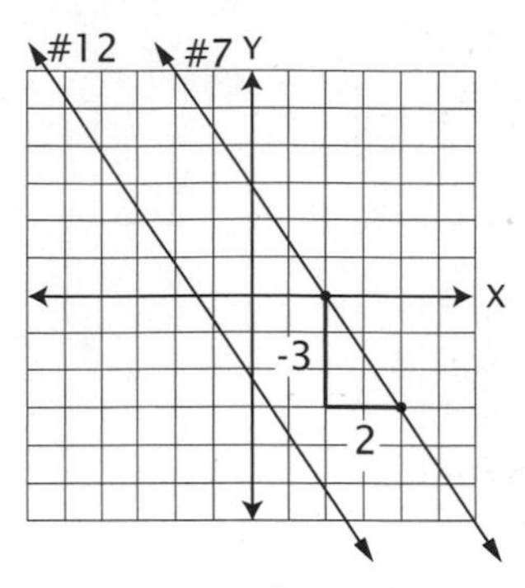

9. y-intercept = 3
10. $Y = \frac{-3}{2}X + 3$
11. C: $Y = -\frac{3}{2}X$
12. see graph on previous page
13. $Y = \frac{-3}{2}X - 2$
14. $Y + \frac{3}{2}X = -2$; $3X + 2Y = -4$
15. see graph below
16. slope $= \frac{8}{2} = 4$
17. y-intercept = −1
18. $Y = 4X - 1$
19. C
20. see graph below
21. $Y = 4X + 3$
22. $4X - Y = -3$ or $-4X + Y = 3$
    It is customary to write the standard form of the equation of a line such that the X coefficient is positive, but either form is correct.

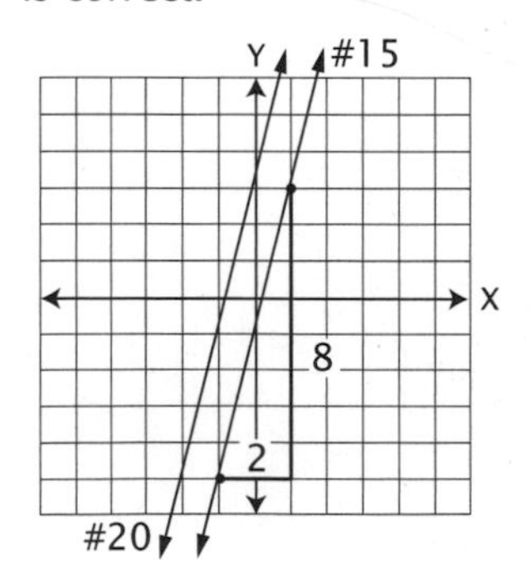

Systematic Review 9C

1. see graph
2. slope $= \frac{3}{3} = 1$
3. y-intercept = 4
4. $Y = X + 4$
   $X - Y = -4$ or $-X + Y = 4$
5. A: $Y = -X - 1$
   C: $Y = -X$
6. see graph

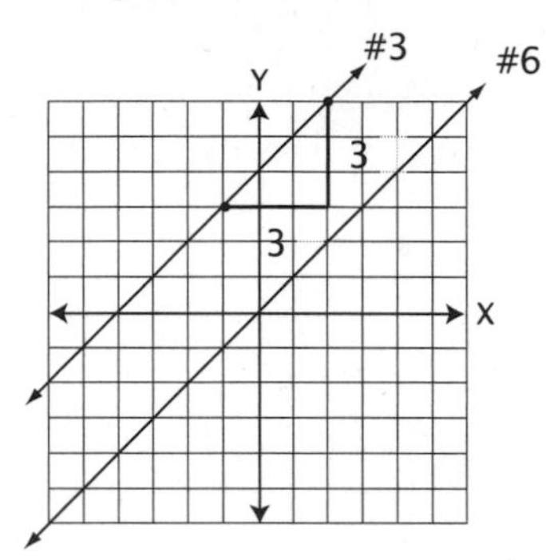

7. 2
8. $Y = -3X - 1$, so slope is −3
9. $Y - \frac{1}{3}X = 2$
   $X - 3Y = -6$ or $-X + 3Y = 6$
10. $2Y = -3X + 1$; $Y = -\frac{3}{2}X + \frac{1}{2}$
11. $$(3-11)^2 \times 2 \div 16 - 7 = 3Y - 4Y + 9$$
    $$(-8)^2 \times 2 \div 16 - 7 = -Y + 9$$
    $$64 \times 2 \div 16 - 7 = -Y + 9$$
    $$128 \div 16 = -Y + 16$$
    $$8 = -Y + 16$$
    $$-8 = -Y$$
    $$Y = 8$$
12. $$(3-5)^2 + |6-4| - X = 3X$$
    $$(-2)^2 + |2| - X = 3X$$
    $$4 + 2 = 4X$$
    $$6 = 4X$$
    $$\frac{6}{4} = X$$
    $$X = 1\frac{1}{2}$$

13. $3(A-4)-5(2A-6)=21$

$3A-12-10A+30=21$

$-7A+18=21$

$-7A=3$

$A=-\frac{3}{7}$

14. ${}^{5}(\cancel{15})\frac{4}{\cancel{3}}+{}^{3}(\cancel{15})\frac{4}{\cancel{5}}A={}^{3}(\cancel{15})\frac{11}{\cancel{5}}$

$20+12A=33$

$12A=13$

$A=\frac{13}{12}$

$A=1\frac{1}{12}$

15. $-6^2-(-6)^2=$

$-(6\times6)-(-6)(-6)=$

$-36-36=-72$

16. $5+5-(-7)=10+(+7)=17$

17. $-[-(-7)]=-[7]=-7$

18. $(-8)^2=(-8)(-8)=64$

19. $25\%=.25$

$.25\times76.98=\$19.25$

20. $45\%=.45$

$.45\times600=270$ people

Systematic Review 9D

1. see graph
2. slope $=\frac{6}{6}=1$
3. y-intercept $=-4$
4. $Y=X-4$

   $X-Y=4$ or $-X+Y=-4$
5. $C: Y=\frac{1}{4}X+2$
6. see graph

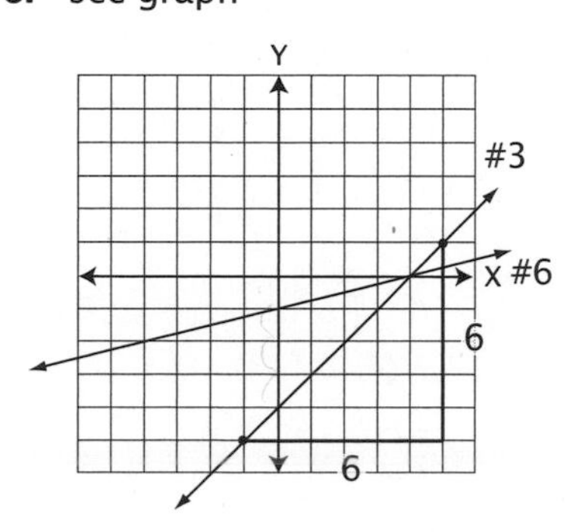

7. $Y=\frac{1}{4}X-1$

8. $Y=-2X+3$; slope $=-2$

9. $2X-Y=-5$ or $-2X+Y=5$

10. $4Y=-2X+8$; $Y=-\frac{1}{2}X+2$

11. $|-1-1-1-1|^2=(-1)^2+B(-1)\div1$

$|-4|^2=1-B\div1$

$4^2=1-B$

$16-1=-B$

$15=-B$

$B=-15$

12. $(3+5)^2+|8-11|+Z=4(Z-2)$

$8^2+|-3|+Z=4Z-8$

$64+3+8=3Z$

$75=3Z$

$Z=25$

13. $5(B-6)+4(2B+7)=102$

$5B-30+8B+28=102$

$13B-2=102$

$13B=104$

$B=8$

14. $55Q-30Q=125$

$25Q=125$

$Q=5$

15. $-\{-[-(-8)]\}=-\{-[8]\}=8$

16. $-9^2=-(9\times9)=-81$

17. $-(-4)=4$

18. $3^2+(-3)^2=9+9=18$

19. $76\%=.76$

$.76\times200=\$152$

20. $\frac{WF}{\cancel{8}}\times\cancel{8}=\frac{2}{8}$

$WF=\frac{2}{8}=\frac{1}{4}$

check: $\frac{1}{\cancel{4}}\times\frac{{}^{2}\cancel{8}}{1}=2$

Systematic Review 9E

1. see graph
2. slope $= \frac{4}{-2} = -2$
3. y-intercept $= -3$
4. $Y = -2X - 3$; $2X + Y = -3$
5. B : $Y = 3X$; C
6. see graph
   (line will have a slope of 3)

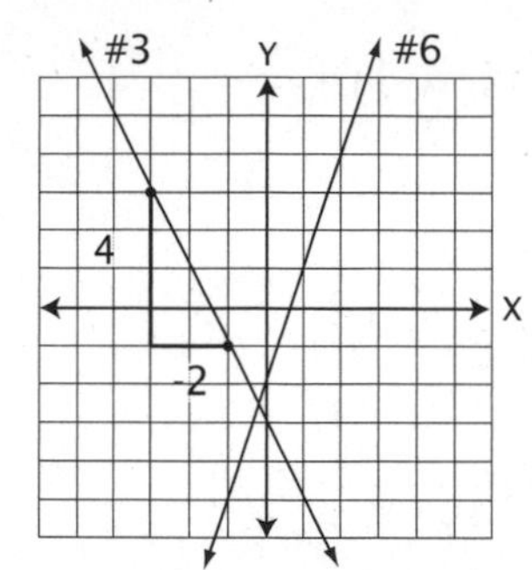

7. $Y = 3X - 2$
8. $\frac{-1}{5}$
9. $3X + Y = -6$
10. $Y + 2X = -1$; $Y = -2X - 1$
11. $$\begin{aligned} 24Y - 108Y + 96 &= 48 - 12Y \\ -84Y + 96 - 48 &= -12Y \\ 48 &= 72Y \\ Y &= \frac{2}{3} \end{aligned}$$
12. $$\begin{aligned} \{-[-(-9)] + 7^2\} \div 5 \div 2 &= Q + 4 \\ \{-9 + 49\} \div 5 \div 2 &= Q + 4 \\ 40 \div 5 \div 2 &= Q + 4 \\ 8 \div 2 &= Q + 4 \\ 4 &= Q + 4 \\ Q &= 0 \end{aligned}$$
13. $$\begin{aligned} 8(A + 3 - 9) - 4(2A + 5) &= 2A + 4 \\ 8A + 24 - 72 - 8A - 20 &= 2A + 4 \\ 24 - 72 - 20 &= 2A + 4 \\ -68 &= 2A + 4 \\ -72 &= 2A \\ A &= -36 \end{aligned}$$
14. $$\begin{aligned} (6 + 6)^2 + |100 - 1| - 14^2 &= 5 \times 9 + B \\ 12^2 + |99| - 196 &= 45 + B \\ 144 + 99 - 196 - 45 &= B \\ B &= 2 \end{aligned}$$
15. $-[-(6 - 9 + 3 - 5)] = -[-(-5)] = -5$
16. $-5^3 = -(5 \times 5 \times 5) = -125$
17. $\frac{WF}{\cancel{10}} \times \cancel{10} = \frac{3}{10}$

    check: $\frac{3}{\cancel{10}} \times \frac{\cancel{10}}{1} = 3$
18. $8.75 \div .25 = 35$ packs
19. $6\% = .06$; $.06 \times 115 = \$6.90$
20. $-N^2 - N^2$

Lesson Practice 10A

1. see graph
2. slope $= \frac{8}{2} = 4$
3. y-intercept $= 0$
4. $Y = 4X$
5. B: $Y = -\frac{1}{4}X$
6. see graph

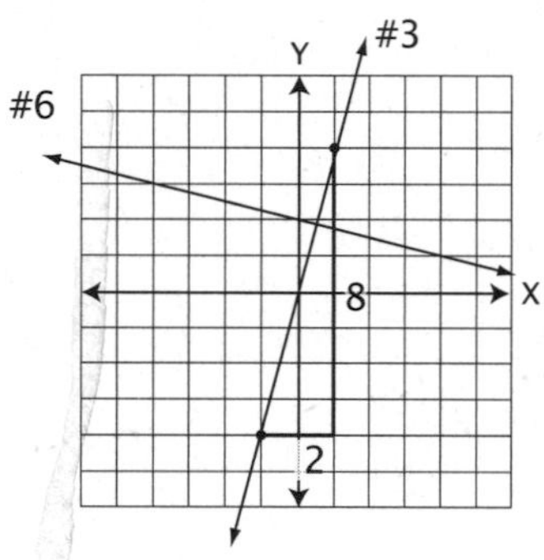

7. $Y = -\frac{1}{4}X + 2$
8. $Y + \frac{1}{4}X = 2$

   $X + 4Y = 8$
9. on the graph
10. slope $= -\frac{2}{2} = -1$
11. y-intercept $= -2$
12. $Y = -X - 2$
13. A: $Y = X - 2$
14. on the graph
15. $Y = X + 2$
16. $X - Y = -2$ or $-X + Y = 2$

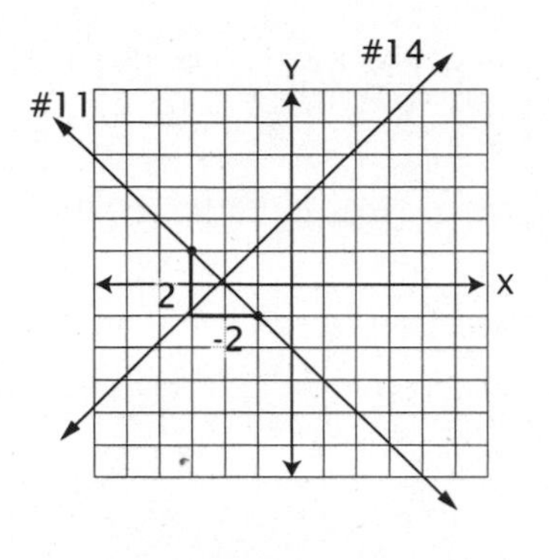

Lesson Practice 10B

1. see graph
2. slope $= \frac{-2}{8} = -\frac{1}{4}$
3. y-intercept = 2
4. $Y = -\frac{1}{4}X + 2$
5. A: $Y = 4X - 5$
6. see graph

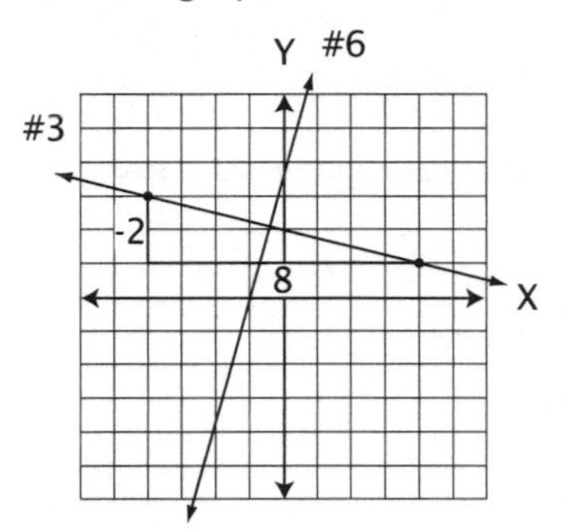

7. $Y = 4X + 4$
8. $4X - Y = -4$ or $-4X + Y = 4$
9. see graph

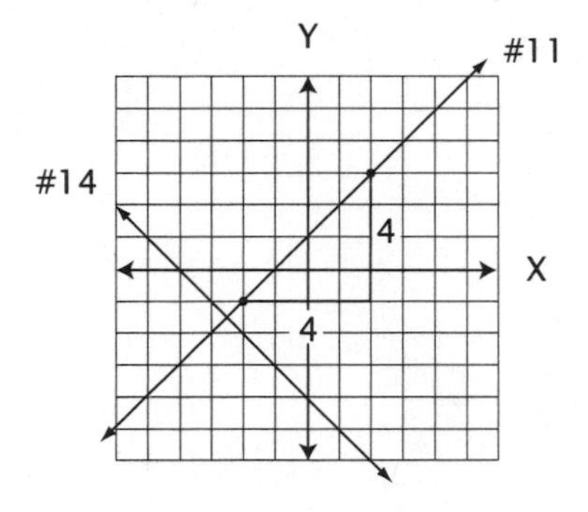

10. slope $= \frac{4}{4} = 1$
11. y-intercept = 1
12. $Y = X + 1$
13. A: $Y = -X + 1$
14. see graph
15. $Y = -X - 4$
16. $X + Y = -4$

Systematic Review 10C

1. see graph

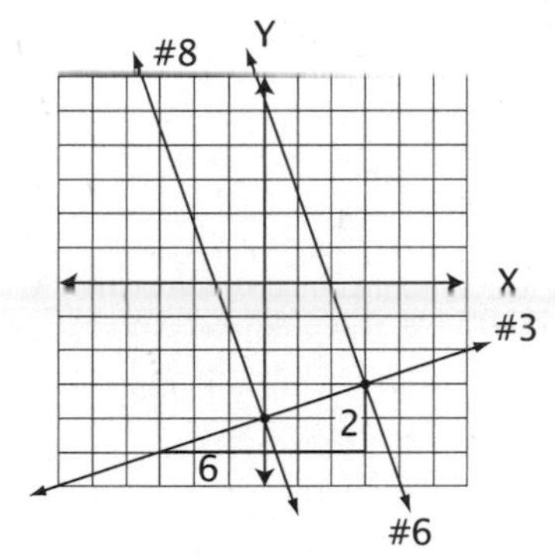

2. slope $= \frac{2}{6} = \frac{1}{3}$
3. y-intercept = −4
4. $Y = \frac{1}{3}X - 4$

   $X - 3Y = 12$ or $-X + 3Y = -12$
5. B, C: $Y = -3X - 1$
6. see graph
7. $Y = -3X + 6$

   $3X + Y = 6$
8. see graph
9. $Y = -3X - 4$

   $3X + Y = -4$
10. slopes are the same, so lines are parallel
11. $6X - X + 3 = 4X + 7$

    $5X + 3 = 4X + 7$

    $X = 4$
12. $-2X - X + 12 = X - 12$

    $-3X + 12 = X - 12$

    $24 = 4X$

    $X = 6$

13. $|-(3+7)| - 4^2 + (-4)^2 = 2R$
$$|-10| - 16 + 16 = 2R$$
$$10 - 16 + 16 = 2R$$
$$10 = 2R$$
$$R = 5$$

14. $^{9}(18)\frac{-7}{2}Y + ^{2}(18)\frac{2}{9} = ^{6}(18)\frac{-4}{3}$
$$-63Y + 4 = -24$$
$$-63Y = -28$$
$$Y = \frac{4}{9}$$

15. $100\% - 60\% = 40\%$
16. 40% of $\$12,900 = .40 \times 12,900 = \$5,160$
17. $15.3\% = .153$
$.153 \times 5160 = \$789.48$
18. $.25 \div 2 = .125$ or $12\frac{1}{2}$ cents
19. $8 \times .125 = \$1.00$ (or $4 \times .25 = \$1.00$)
20. $T = 5W + 3$; T = total and W = weeks
(different letters may be used)

Systematic Review 10D

1. see graph
2. slope $= \frac{-6}{3} = -2$
3. y-intercept $= 1$
4. $Y = -2X + 1$; $2X + Y = 1$
5. A: $Y = \frac{1}{2}X - 1$
6. see graph
7. $Y = \frac{1}{2}X - 1$
$X - 2Y = 2$ or $-X + 2Y = -2$
8. see graph
9. $Y = \frac{1}{2}X + 3$
$X - 2Y = -6$ or $-X + 2Y = 6$
10. slopes are the same,
so lines are parallel
11. $2X + 2 - X + 2X = 3X - 3 + 10 - X$
$$3X + 2 = 2X + 7$$
$$X = 5$$

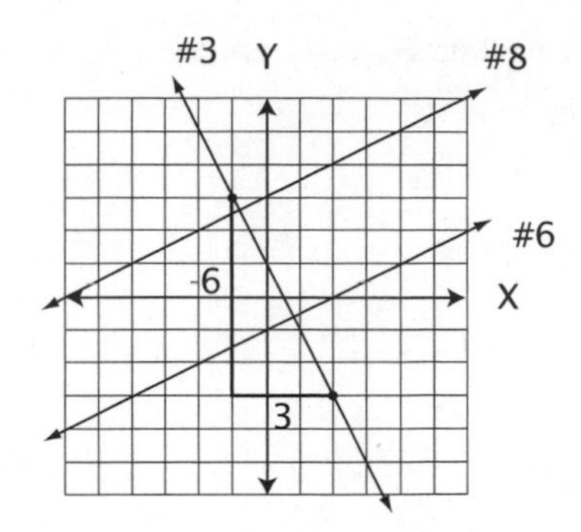

12. $3Y - 1 + 2Y - 1 - 4Y = 2Y + 3 + Y + 1$
$$Y - 2 = 3Y + 4$$
$$-6 = 2Y$$
$$Y = -3$$

13. $-(6+7)^2 + (10+5)^2 = 5M$
$$-(13)^2 + (15)^2 = 5M$$
$$-169 + 225 = 5M$$
$$56 = 5M$$
$$\frac{56}{5} = M$$
$$M = 11\frac{1}{5}$$

14. $^{20}(60)\frac{-5}{3} = ^{15}(60)\frac{-9}{4} + ^{12}(60)\frac{6}{5}A$
$$-100 = -135 + 72A$$
$$35 = 72A$$
$$A = \frac{35}{72}$$

15. $100\% - 55\% = 45\%$
16. 45% of $\$9,645 =$
$.45 \times 9,645 = \$4,340.25$
17. $15.3\% = .153$
$.153 \times 4,340.25 \approx \$664.06$
18. $2.50 \div .25 = 10$
$10 \times 2 = 20$ bits
19. $100 \div 2 = 50$
$50 \times .25 = \$12.50$
20. $L = W + 5$; L = length and W = weeks
(different letters may be used)

Systematic Review 10E

1. see graph
2. slope $= \frac{-1}{1} = -1$
3. y-intercept $= -1$
4. $Y = -X - 1$
   $X + Y = -1$
5. C: $Y = X - 2$
6. see graph

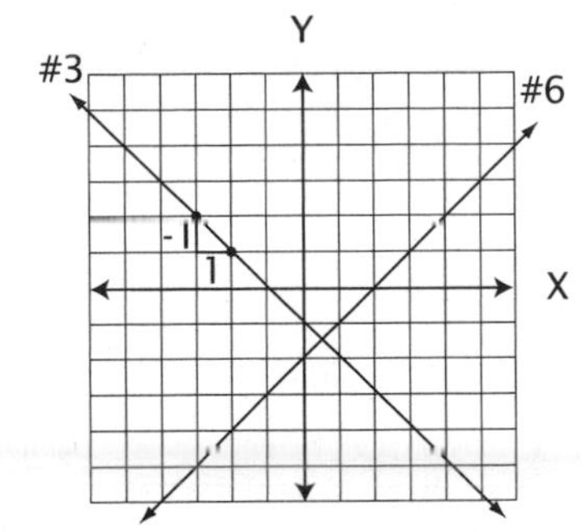

7. $Y = X - 2$
   $X - Y = 2$ or $-X + Y = -2$
8. $5Y - 3 - 2Y + 4 + 3Y = 4Y + 9 + 4Y$
   $6Y + 1 = 8Y + 9$
   $-8 = 2Y$
   $Y = -4$
9. $-M - 4 - 2M + 20 = M + 7 - 5M + 11$
   $-3M + 16 = -4M + 18$
   $M = 2$
10. $|-3 - 4 - 5 + 2| + W = 3W$
    $|-10| = 2W$
    $10 = 2W$
    $W = 5$
11. ${}^{9}(\cancel{36})\frac{13}{\cancel{4}}B = {}^{4}(\cancel{36})\frac{29}{\cancel{9}} + {}^{3}(\cancel{36})\frac{5}{\cancel{12}}$
    $117B = 116 + 15$
    $117B = 131$
    $B = \frac{131}{117}$
    $B = 1\frac{14}{117}$
12. $100\% - 48\% = 52\%$
13. 52% of \$25,813 =
    $.52 \times 25{,}813 = \$13{,}422.76$
14. $15.3\% = .153$
    $.153 \times 13{,}422.76 \approx \$2{,}053.68$
15. $20 \times 12 = 240$ pence
16. $5 \times 20 = 100$ shillings
17. $C = -20W + 1000$
    C = cash and W = weeks
18. $\sqrt{100} = 10\ (10 \times 10 = 100)$
19. $\sqrt{36} = 6\ (6 \times 6 = 36)$
20. $\sqrt{144} = 12\ (12 \times 12 = 144)$

Lesson Practice 11A

1. see graph
2. y-intercept $= -1$
3. $Y = 3X - 1$
4. $3X - Y = 1$ or $-3X + Y = -1$

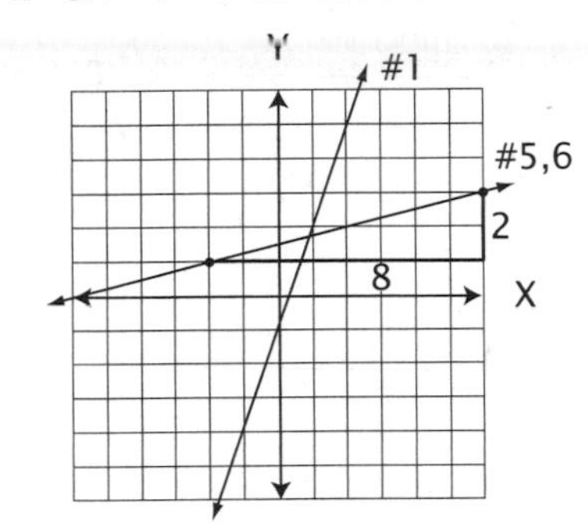

5. $\frac{3-1}{6-(-2)} = \frac{2}{8} = \frac{1}{4}$ (see graph)
6. $Y = \frac{1}{4}X + b$
   $(3) = \frac{1}{4}(6) + b$
   $3 = \frac{3}{2} + b$
   $b = 1\frac{1}{2}$ (see graph)
7. $Y = \frac{1}{4}X + 1\frac{1}{2}$
8. $Y - \frac{1}{4}X = \frac{3}{2}$
   $X - 4Y = -6$ or $-X + 4Y = 6$
9. $(2) = 5(1) + b$
   $2 - 5 = b$
   $b = -3$
   $Y = 5X - 3$

10. $(6) = 6(-3) + b$
$6 = -18 + b$
$b = 24$
$Y = 6X + 24$

11. $(1) = -4(1) + b$
$1 = -4 + b$
$b = 5$
$Y = -4X + 5$

12. $(2) = \frac{1}{2}(2) + b$
$2 = 1 + b$
$b = 1$
$Y = \frac{1}{2}X + 1$

13. $(8) = \frac{2}{3}(5) + b$
$8 = \frac{10}{3} + b$
$b = 4\frac{2}{3}$
$Y = \frac{2}{3}X + 4\frac{2}{3}$

14. $(1) = -\frac{1}{4}(2) + b$
$1 = -\frac{1}{2} + b$
$b = 1\frac{1}{2}$
$Y = -\frac{1}{4}X + 1\frac{1}{2}$

15. $\frac{5-3}{4-2} = \frac{2}{2} = 1 = m$
$(3) = 1(2) + b$
$3 = 2 + b$
$b = 1$
$Y = X + 1$

16. $\frac{1-6}{2-4} = \frac{-5}{-2} = \frac{5}{2} = m$
$(1) = \frac{5}{2}(2) + b$
$1 = 5 + b$
$b = -4$
$Y = \frac{5}{2}X - 4$

17. $\frac{0-3}{1-3} = \frac{-3}{-2} = \frac{3}{2} = m$
$(0) = \frac{3}{2}(1) + b$
$0 = \frac{3}{2} + b$
$b = -\frac{3}{2}$
$Y = \frac{3}{2}X - \frac{3}{2}$

Lesson Practice 11B

1. see graph
2. $(2) = \frac{1}{2}(3) + b;\ 2 = \frac{3}{2} + b;\ b = \frac{1}{2}$

   Estimates near $\frac{1}{2}$ are acceptable.

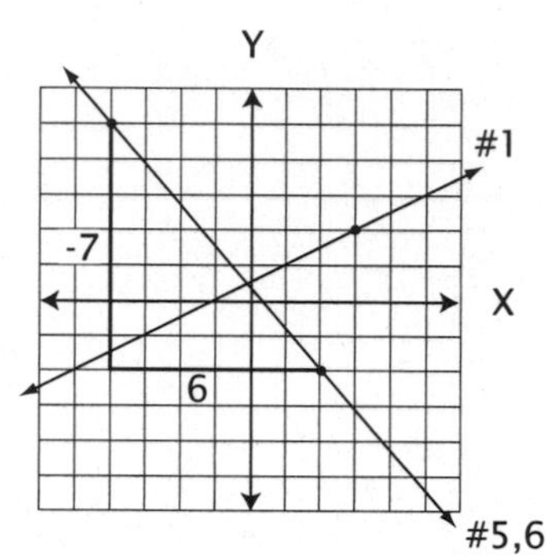

3. $Y = \frac{1}{2}X + \frac{1}{2}$
4. $X - 2Y = -1$ or $-X + 2Y = 1$
5. $\frac{-2-5}{2-(-4)} = -\frac{7}{6}$ (see graph)
6. $Y = \frac{-7}{6}X + b$

   $(-2) = \frac{-7}{6}(2) + b$

   $-2 = \frac{-14}{6} + b;\ b = \frac{1}{3}$ (see graph)
7. $Y = -\frac{7}{6}X + \frac{1}{3}$
8. $Y + \frac{7}{6}X = \frac{1}{3}$

   $7X + 6Y = 2$
9. $(2) = 8(1) + b$

   $2 = 8 + b;\ b = -6$

   $Y = 8X - 6$

10. $(2)=3(1)+b$
$2=3+b;\ b=-1$
$Y=3X-1$

11. $(0)=-2(3)+b$
$0=-6+b;\ b=6$
$Y=-2X+6$

12. $\frac{3-5}{-2-2}=\frac{-2}{-4}=\frac{1}{2}$
$(3)=\frac{1}{2}(-2)+b$
$3=\frac{-2}{2}+b;\ b=4$
$Y=\frac{1}{2}X+4$

13. $\frac{1-2}{1-5}=\frac{-1}{-4}=\frac{1}{4}$
$(1)=\frac{1}{4}(1)+b$
$1=\frac{1}{4}+b;\ b=\frac{3}{4}$
$Y=\frac{1}{4}X+\frac{3}{4}$

14. $\frac{1-(-3)}{-3-(-2)}=\frac{4}{-1}=-4$
$(1)=-4(-3)+b$
$1=12+b$
$b=-11$
$Y=-4X-11$

15. $\frac{-1-(-6)}{-2-(-5)}=\frac{5}{3}$
$(-1)=\frac{5}{3}(-2)+b$
$-1=\frac{-10}{3}+b$
$b=\frac{7}{3}$
$Y=\frac{5}{3}X+\frac{7}{3}$

16. $\frac{6-(-3)}{-1-5}=\frac{9}{-6}=\frac{-3}{2}$
$(6)=\frac{-3}{2}(-1)+b$
$6=\frac{3}{2}+b$
$b=\frac{9}{2}$
$Y=-\frac{3}{2}X+\frac{9}{2}$

17. $\frac{8-2}{-3-7}=\frac{6}{-10}=\frac{-3}{5}$
$(2)=\frac{-3}{5}(7)+b$
$2=\frac{-21}{5}+b$
$b=6\frac{1}{5}$
$Y=\frac{-3}{5}X+6\frac{1}{5}$

Systematic Review 11C

1. see graph
2. $(1)=\frac{1}{4}(-5)+b$
$1=-\frac{5}{4}+b$
$b=2\frac{1}{4}$
3. $Y=\frac{1}{4}X+2\frac{1}{4}$
$X-4Y=-9$ or $-X+4Y=9$
4. $\frac{2-2}{-3-1}=\frac{0}{-4}=0$ (see graph)
5. $(2)=0(1)+b;\ b=2$
(see graph)
6. $Y=2;\ Y=2$

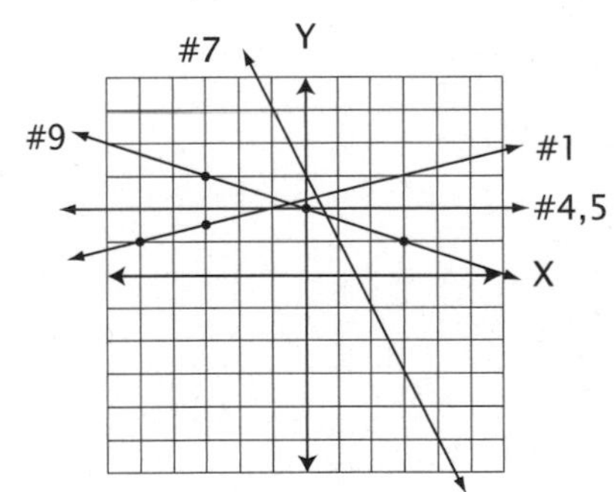

7. slope $=-2$ (see graph)
8. $(5)=-2(-1)+b$
$5=2+b$
$b=3$
$Y=-2X+3$
$2X+Y=3$
9. slope $=-\frac{1}{3}$ (see graph)

10. $(1) = -\frac{1}{3}(3) + b$

$1 = -\frac{3}{3} + b$

$b = 2$

$Y = -\frac{1}{3}X + 2$

$X + 3Y = 6$

11. distributive
12. commutative
13. commutative
14. associative
15. $\sqrt{9} = 3$
16. $45\% = .45;\ .45 \times 98 = 44.10$
17. $\frac{5 \text{ boys}}{1 \text{ girl}} = \frac{5}{1}$
18. $\frac{5 \text{ boys}}{6 \text{ total}} = \frac{5}{6}$
19. $\frac{5}{6} = 5 \div 6 \approx .83 = 83\%$
20. $\frac{5}{6} \times 48^{8} = 40$ boys

Systematic Review 11D

1. on the graph
2. $(1) = -\frac{2}{5}(1) + b$

$1 = -\frac{2}{5} + b$

$b = \frac{7}{5}$

3. $Y = -\frac{2}{5}X + 1\frac{2}{5}$

$2X + 5Y = 7$

4. $\frac{2-4}{3-(-1)} = \frac{-2}{4} = -\frac{1}{2}$ (see graph)
5. $(2) = -\frac{1}{2}(3) + b$

$2 = -\frac{3}{2} + b$

$b = 3\frac{1}{2}$

6. $Y = -\frac{1}{2}X + \frac{7}{2}$

$X + 2Y = 7$

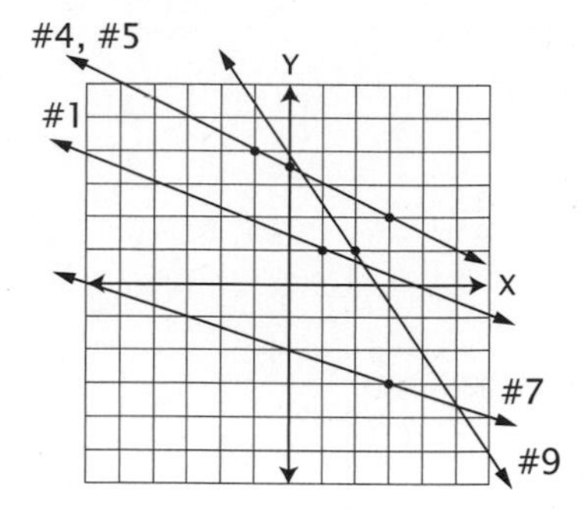

7. slope $= -\frac{1}{3}$ (see graph)
8. $(-3) = -\frac{1}{3}(3) + b$

$-3 = -\frac{3}{3} + b$

$b = -2$

$Y = -\frac{1}{3}X - 2$

$X + 3Y = -6$

9. slope $= -\frac{3}{2}$ (see graph)
10. $(1) = -\frac{3}{2}(2) + b$

$1 = -\frac{6}{2} + b$

$b = 4$

$Y = -\frac{3}{2}X + 4$

$3X + 2Y = 8$

11. true
12. false
13. false
14. true
15. $\sqrt{49} = 7$
16. $16\% = .16$

$.16 \times 32 = 5.12$

17. $\frac{5 \text{ Team S}}{8 \text{ total}} = \frac{5}{8}$

$5 \div 8 = .625 = 62.5\%$

18. $\frac{3 \text{ Team E}}{8 \text{ total}} = \frac{3}{8}$

$3 \div 8 = .375 = 37.5\%$

19. $.375 \times 640 = 240$ Team E fans
$.625 \times 640 = 400$ Team S fans
(may also be computed with fractions)

20. $Y = 20(15) + 100$
$Y = 300 + 100$
$Y = \$400$

Systematic Review 11E

1. see graph

2. $(-1) = -1(4) + b$
$-1 = -4 + b$
$b = 3$

3. $Y = -X + 3$
$X + Y = 3$

4. $\frac{5-2}{-4-1} = \frac{3}{-5} = -\frac{3}{5}$ (see graph)

5. $(2) = -\frac{3}{5}(1) + b$
$2 = -\frac{3}{5} + b$
$b = \frac{13}{5}$ (see graph)

6. $Y = -\frac{3}{5}X + 2\frac{3}{5}$
$3X + 5Y = 13$

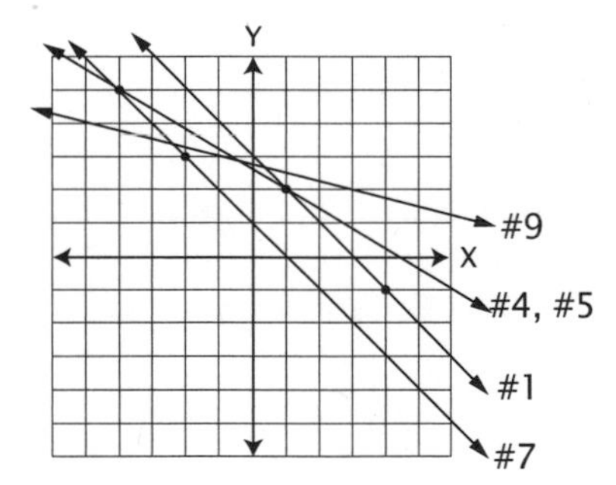

7. slope $= -1$ (see graph)

8. $(3) = -1(-2) + b$
$3 = 2 + b$
$b = 1$
$Y = -X + 1$
$X + Y = 1$

9. slope $= -\frac{1}{4}$ (see graph)

10. $(3) = -\frac{1}{4}(-1) + b$
$3 = \frac{1}{4} + b$
$b = \frac{11}{4}$
$Y = -\frac{1}{4}X + 2\frac{3}{4}$
$X + 4Y = 11$

11. $(-1)(2)(-3)(4)(-5)^2 = -\{-[-(-X)]\}$
$(-2)(-12)(25) = X$
$(24)(25) = X$
$X = 600$

12. $72A - 84A = 36AF$
$12A(6-7) = 12A(3F)$
$6 - 7 = 3F$
$-1 = 3F$
$F = \frac{-1}{3}$

13. $10(-4.2Q) - 10(1.8Q) = 10(-6)$
$-42Q - 18Q = -60$
$-60Q = -60$
$Q = 1$

14. $1000(.14) - 1000(.023) = 1000(.07C)$
$140 - 23 = 70C$
$117 = 70C$
$\frac{117}{70} = C$
$C = 1\frac{47}{70}$

15. $\frac{2}{5}$; $2 \div 5 = .4 = 40\%$

16. $\frac{3}{5}$; $3 \div 5 = .6 = 60\%$

17. $.4 \times 500 = 200$g

18. $500 - 200 = 300$g

19. $5{,}280 \times 4.5 = 23{,}760$ ft

20. 1 yd = 3 ft; $5{,}280 \div 3 = 1{,}760$ yd

Lesson Practice 12A

1. $3Y = X + 9 \Rightarrow Y = \frac{1}{3}X + 3$
2. solid; see graph

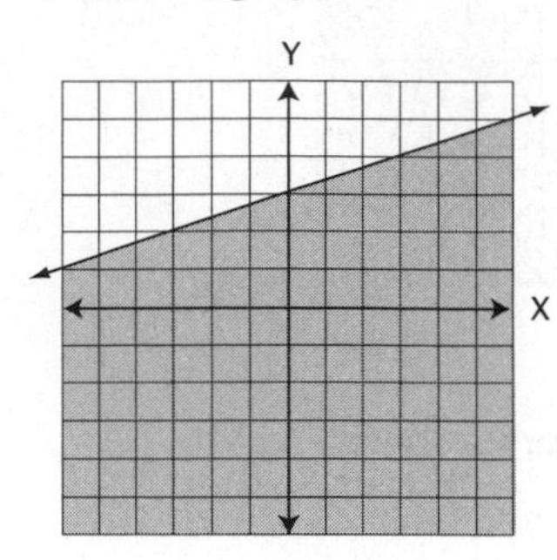

3. (0, 0): $3(0) \le (0) + 9;\ 0 \le 9$; true
   (0, 4): $3(4) \le (0) + 9;\ 12 \le 9$; false
   You may choose any points you wish, as long as they are on opposite sides of the line.
4. see graph above
5. $Y = -\frac{1}{2}X - 2$
6. dotted; see graph

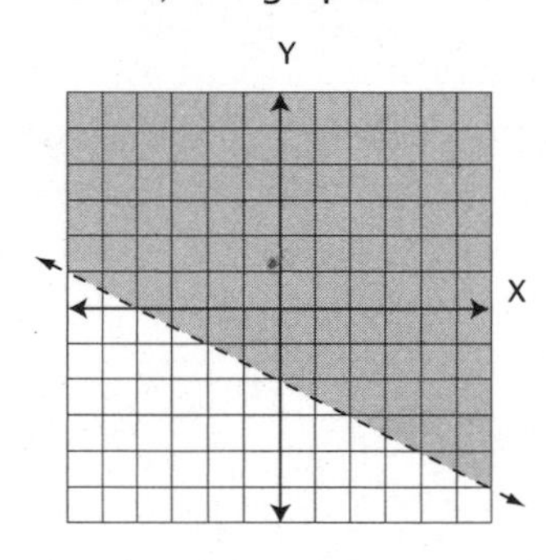

7. (0, 0): $2(0) > -(0) - 4;\ 0 > -4$; true
   (0, -3): $2(-3) > -(0) - 4;\ -6 > -4$; false
8. see graph above
9. $Y = -3X + 1$
10. solid; see graph in next column
11. $3X + Y$
    (0, 0): $3(0) + (0) \ge 1;\ 0 \ge 1$; false
    (0, 2): $3(0) + (2) \ge 1;\ 2 \ge 1$; true
12. see graph in next column
13. $Y > -X - 2$

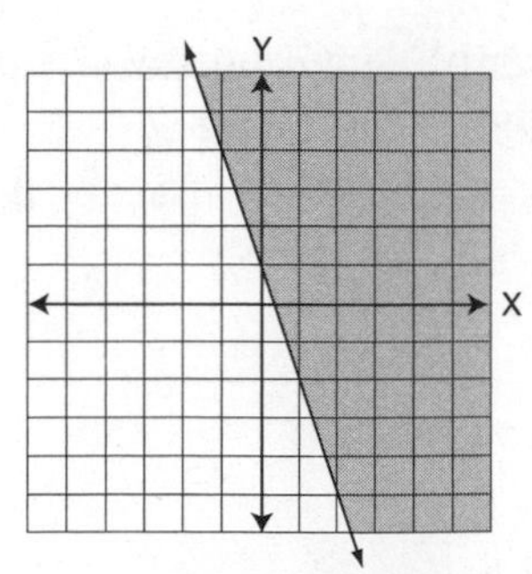

14. $-2Y < -4X + 6;\ Y > 2X - 3$
    Multiplying or dividing an inequality by a negative number reverses the direction of the sign.
15. $-4Y \ge 8X + 8;\ Y \le -2X - 2$

Lesson Practice 12B

1. $Y = 2X - 3$
2. solid; see graph

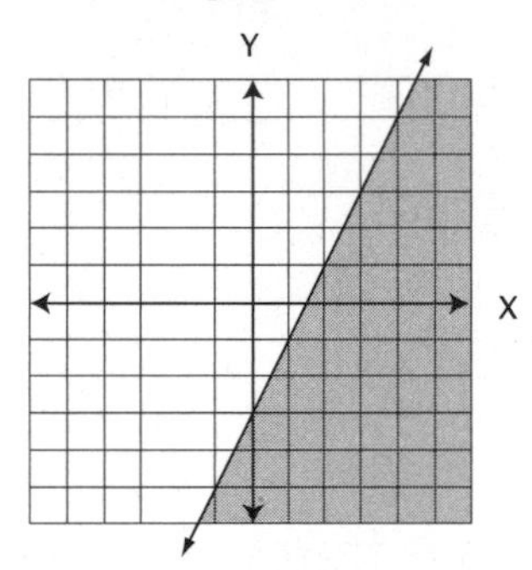

3. (0, 0): $-2(0) + (0) \le -3;\ 0 \le -3$; false
   (3, 0): $-2(3) + (0) \le -3;\ -6 \le -3$; true
   Choose any points you wish, as long as they are on opposite sides of the line.
4. see graph above
5. $Y = \frac{2}{3}X - 3$
6. solid; see graph below

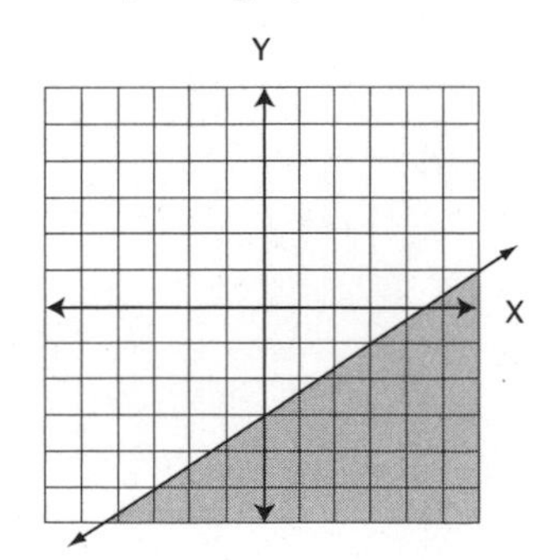

7. $(0, 0)$: $3(0) \le 2(0) - 9$; $0 \le -9$; false
   $(0,-4)$: $3(-4) \le 2(0) - 9$; $-12 \le -9$; true
8. see graph on previous page
9. $Y = \frac{1}{5}X + 1$
10. dotted; see graph below

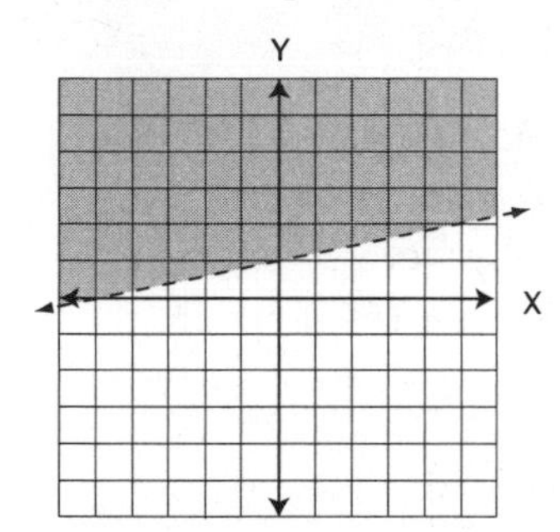

11. $(0, 0)$: $-(0) + 5(0) > 5$; $0 > 5$; false
    $(0, 2)$: $-(0) + 5(2) > 5$; $10 > 5$; true
12. see graph above
13. $Y < 3X - 5$
14. $-Y > -3X + 5$
    $Y < 3X - 5$
15. multiplying or dividing by a negative number

Systematic Review 12C

1. $Y = 2X + 1$
2. dotted; see graph in next column
3. $(0, 0)$: $-(0) > -2(0) - 1$; $0 > -1$; true
   $(0, 2)$: $-(2) > -2(0) - 1$; $-2 > -1$; false
4. see graph in next column
5. yes: $-(-2) > -2(3) - 1$; $2 > -7$; true
   Or, check visually on the graph

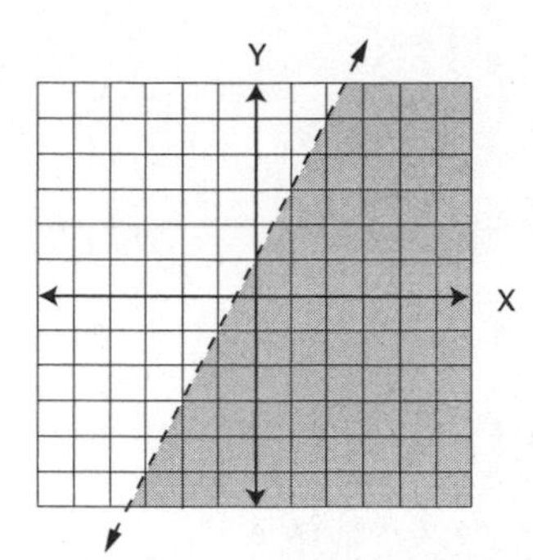

6. $Y = X - 3$
7. solid; see graph below
8. $(0, 0)$: $(0) \le (0) - 3$; $0 \le -3$; false
   $(4, 0)$: $(0) \le (4) - 3$; $0 \le 1$; true

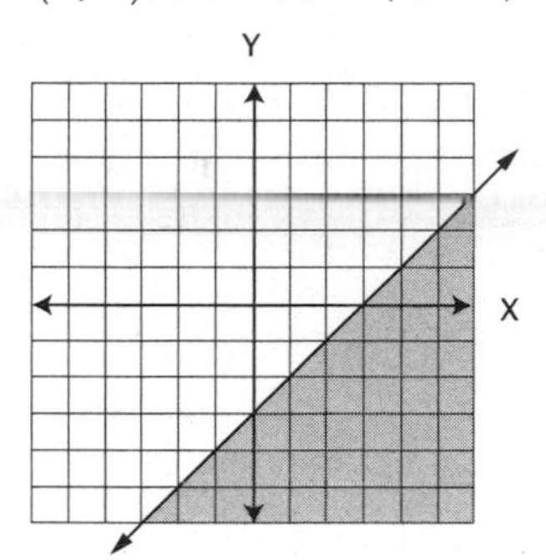

9. see graph above
10. multiplying or dividing by negative number
11. $WF \times 16 = 1$
    $WF = \frac{1}{16}$
12. $WF \times 2000 = 1$
    $WF = \frac{1}{2000}$ or .0005
13. $-2Y = -3X + 5$
    $Y = \frac{3}{2}X - \frac{5}{2}$
14. slope $= \frac{3}{2}$
15. slope $= -\frac{2}{3}$
16. y-intercept $= -2$
    $Y = 2X - 2$ or $2X - Y = 2$
17. $.16 \times 242 = 38.72$
18. quadrant 3
19. $\frac{1}{1.6} = \frac{10}{X}$
    $(X)(1) = (1.6)(10)$
    $X = 16$ km

20. $\frac{1}{1.6} = \frac{X}{10}$
$(1)(10) = (1.6)(X)$
$X = 6.25$ mi

Systematic Review 12D

1. $Y = -2$
2. dotted; see graph below
3. $(0, 0)$: $(0) + 2 < 0$; $2 < 0$; false
$(0,-3)$: $(-3) + 2 < 0$; $-1 < 0$; true
4. see graph below

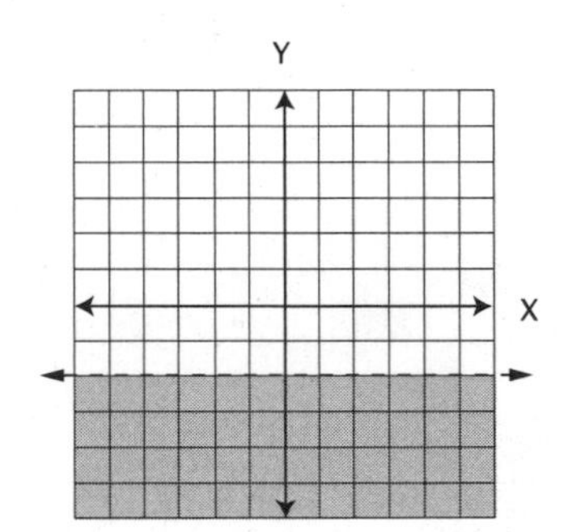

5. $4Y < -8$; $Y < -2$
6. $Y = \frac{1}{3}X + 2$
7. dotted; see graph below
8. $(0, 0)$: $(0) - 3 > \frac{1}{3}(0) - 1$; $-3 > -1$;
false
$(0, 3)$: $(3) - 3 > \frac{1}{3}(0) - 1$; $0 > -1$;
true
9. see graph below

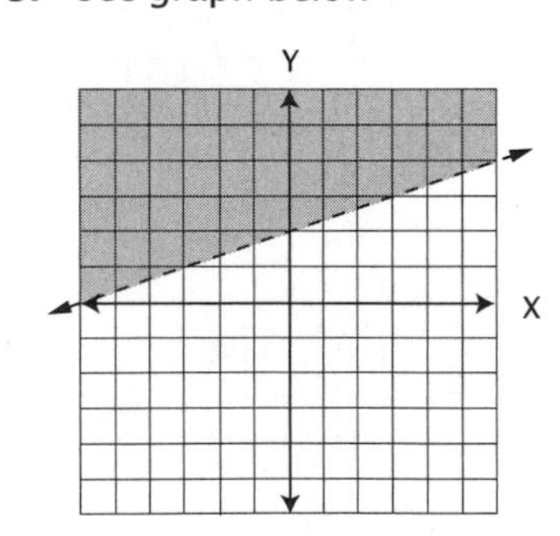

10. $Y < 2X - 1$
11. $WF \times 60 = 1$; $WF = \frac{1}{60}$
12. $WF \times 7 = 1$
$WF = \frac{1}{7} \approx .14 = 14\%$
13. $\frac{1}{.45} = \frac{10}{X}$
$(1)(X) = (.45)(10)$
$X = 4.5$ kg
14. $\frac{1}{.45} = \frac{X}{2}$
$(1)(2) = (.45)(X)$
$X = 4.44$ lb
15. $6Y - 4X - 3 = 0$
$6Y = 4X + 3$
$Y = \frac{4}{6}X + \frac{3}{6}$
$Y = \frac{2}{3}X + \frac{1}{2}$
$m = \frac{4}{6} = \frac{2}{3}$
16. slope $= -\frac{3}{2}$
17. $(1) = -\frac{1}{2}(1) + b$
$1 = -\frac{1}{2} + b$
$b = \frac{3}{2}$
$Y = -\frac{1}{2}X + \frac{3}{2}$ or $X + 2Y = 3$
18. $9 \div 25 = .36 = 36\%$
19. $6N - 5N + 8$
20. $6(10) - 5(10) + 8 = 60 - 50 + 8 = 18$

Systematic Review 12E

1. $Y = 2X + 3$
2. solid; see graph on the next page
3. $(0, 0)$: $(0) \leq 2(0) + 3$; $0 \leq 3$; true
$(-3, 0)$: $(0) \leq 2(-3) + 3$; $0 \leq -3$; false
4. see graph on the next page
5. yes: $(1) \leq 2(3) + 3$
$1 \leq 6 + 3$
$1 \leq 9$; true

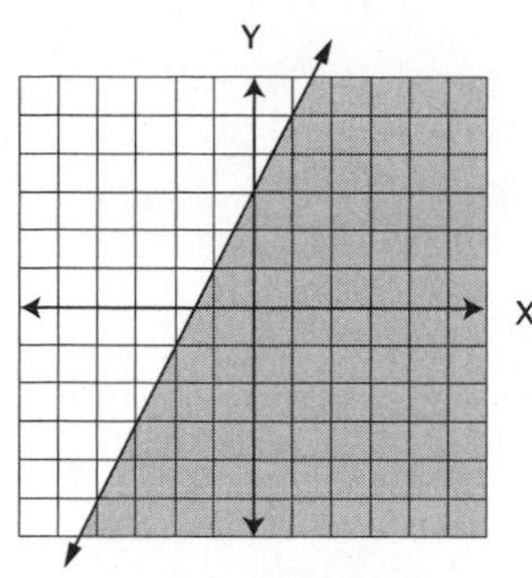

6. $X = 4$
7. solid; see graph below
8. $(0,0)$: $(0) \geq 4$; false
   $(6,0)$: $(6) \geq 4$; true
9. see graph below

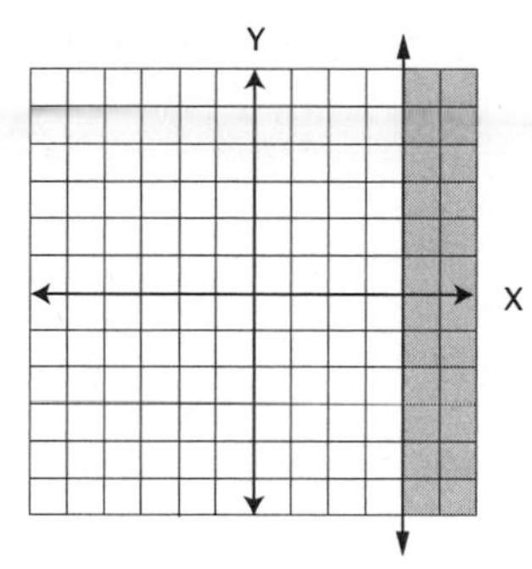

10. multiplying or dividing by a negative number
11. $WF \times 8 = 1$
    $WF = \frac{1}{8}$
12. $WF \times 4 = 1$
    $WF = \frac{1}{4} = .25 = 25\%$
13. $\frac{1}{.95} = \frac{4 \text{ qt}}{X}$
    $(1)(X) = (.95)(4)$
    $X = 3.8$ liters
14. $\frac{1}{.95} = \frac{X}{1}$
    $(1)(1) = (.95)(X)$
    $X = 1\frac{1}{19} \approx 1.05$ quarts
15. $\frac{1}{2}Y = X + 16$
    $Y = 2X + 32$
    $m = 2$
16. $m = -\frac{1}{2}$
17. $(-4) = 3(-3) + b$
    $-4 = -9 + b$
    $b = 5$
    $Y = 3X + 5$ or $3X - Y = -5$ or $-3X + Y = 5$
18. $12 \div 17 \approx .71 = 71\%$
19. $.17 \times 425 = 72.25$
20. quadrant 4

Lesson Practice 13A

1. on the graph
2. on the graph
3. $(1, 2)$
4. on the graph
5. on the graph
6. $(3, -4)$

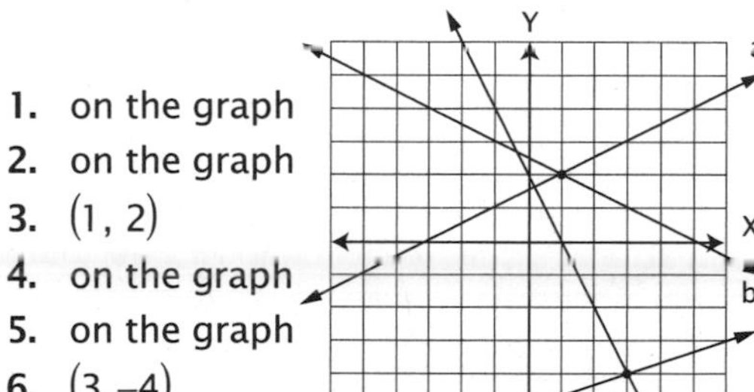

7. on the graph
8. on the graph
9. $(-3, 2)$
10. on the graph
11. on the graph
12. $(3, 1)$

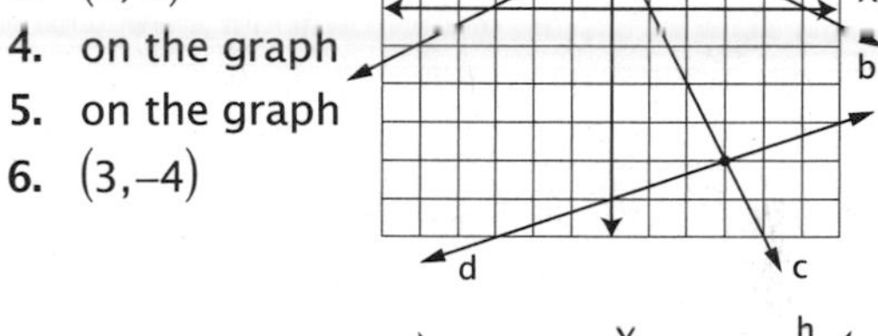

13. on the graph
14. on the graph
15. $(1, 1)$
16. on the graph
17. on the graph
18. $(-1, -3)$

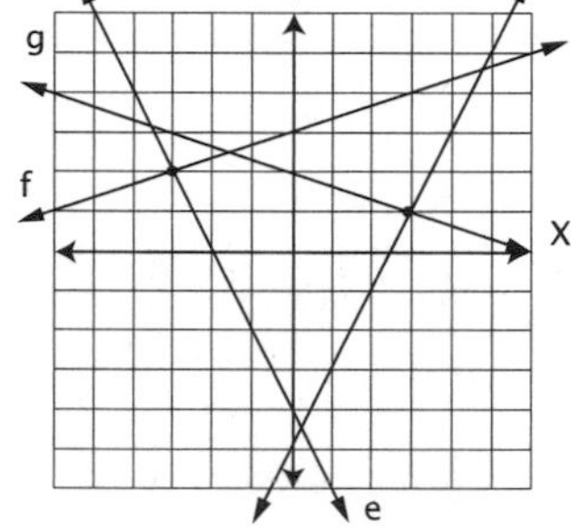

Lesson Practice 13B

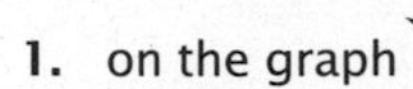

1. on the graph
2. on the graph
3. $\left(0,\ \frac{5}{2}\right)$
4. on the graph
5. on the graph
6. $(-2,-1)$

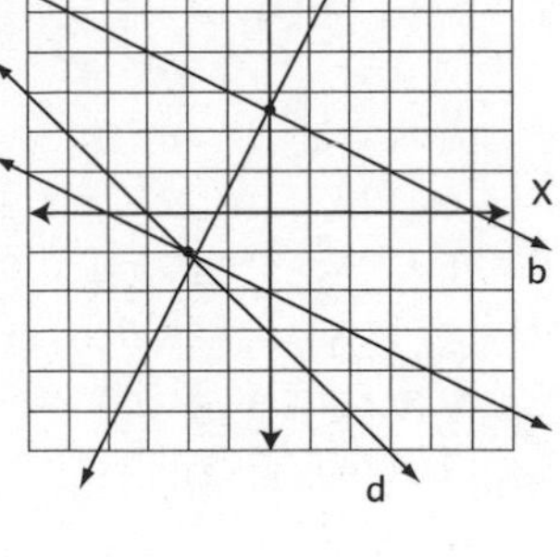

7. on the graph
   $(Y = X - 2)$
8. on the graph
   $\left(Y = -\frac{1}{3}X + 2\right)$
9. $(3,\ 1)$
10. on the graph
    $(Y = -2X - 2)$
11. on the graph
12. $(-3,\ 4)$

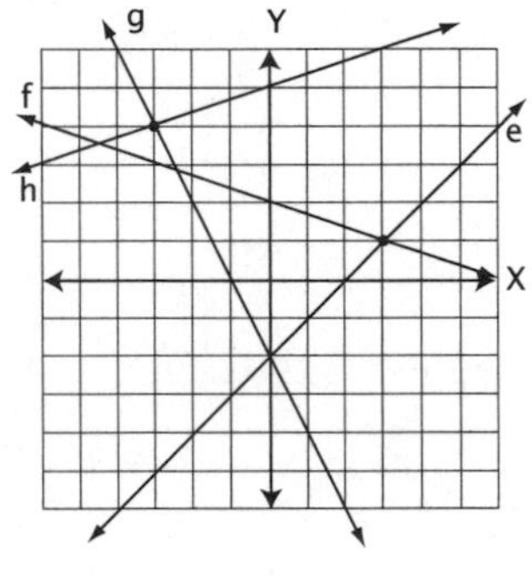

13. on the graph
    $\left(Y = -\frac{1}{4}X + 3\right)$
14. on the graph
15. $(0,\ 3)$
16. on the graph
    $\left(Y = -\frac{1}{2}X - 1\right)$
17. on the graph
18. $(2,-2)$

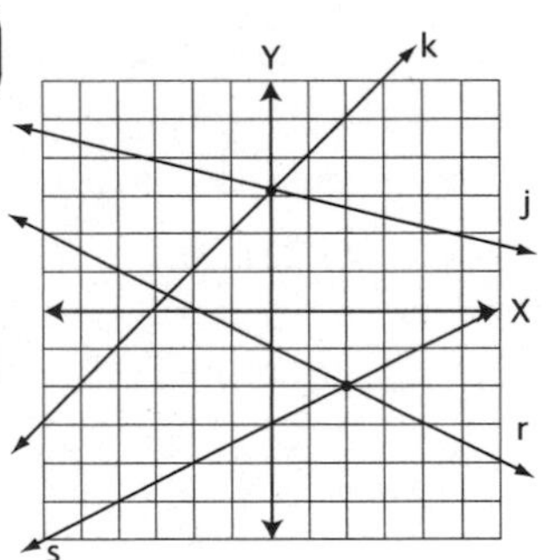

Systematic Review 13C

1. on the graph
2. on the graph

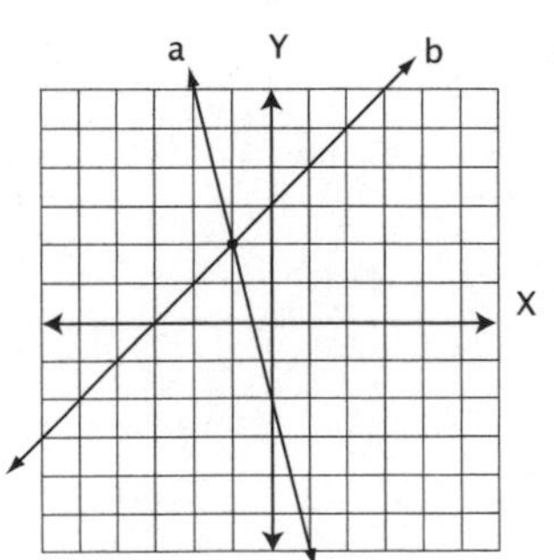

3. $(-1,\ 2)$
4. $(-3) = -4(1) + b$
   $-3 = -4 + b$
   $b = 1$
5. $Y = -4X + 1$
   $4X + Y = 1$
6. $\frac{-5-1}{-5-5} = \frac{-6}{-10} = \frac{3}{5} = m$
7. $(1) = \frac{3}{5}(5) + b$
   $1 = \frac{15}{5} + b$
   $1 = 3 + b$
   $b = -2$
8. $Y = \frac{3}{5}X - 2$
   $5Y = 3X - 10$
   $-3X + 5Y = -10$ or $3X - 5Y = 10$
9. $m = \frac{2}{3}$
   $(4) = \frac{2}{3}(4) + b$
   $4 = \frac{8}{3} + b$
   $4 - \frac{8}{3} = b$
   $\frac{4}{3} = b$
10. $Y = \frac{2}{3}X + \frac{4}{3}$
    $3Y = 2X + 4$
    $-2X + 3Y = 4$ or $2X - 3Y = -4$
11. $8X - 3X + 7 = 4X + 8$
    $5X - 4X = 8 - 7$
    $X = 1$
12. $4Q + 12 = 20$
    $4(Q + 3) = 4(5)$
    $Q + 3 = 5$
    $Q = 2$
13. $5^2 \div 5 + 3(X + 7) = 2X + 27$
    $25 \div 5 + 3X + 21 = 2X + 27$
    $5 + 3X + 21 = 2X + 27$
    $3X + 26 = 2X + 27$
    $X = 27 - 26 = 1$

14. $7^2 \times 2 - 4(Y+11) = 3Y - 2$
$49 \times 2 - 4Y - 44 = 3Y - 2$
$98 - 4Y - 44 = 3Y - 2$
$54 = 3Y + 4Y - 2$
$56 = 7Y$
$Y = 8$

15. ${}^{3}(30)\frac{6}{10} - {}^{10}(30)\frac{2}{3}X = {}^{30}(30)\frac{11}{1}$
$18 - 20X = 330$
$-20X = 312$
$X = \frac{312}{-20}$
$= -15\frac{3}{5}$ or -15.6

16. $|-8-4| - 6Y = 32 \div |-8|$
$|-12| - 6Y = 32 \div 8$
$12 - 6Y = 4$
$12 - 4 = 6Y$
$8 = 6Y$
$\frac{8}{6} = Y$
$Y = 1\frac{1}{3}$

17. 7:45 to 2:15 is $6\frac{1}{2}$ hours
$338 \div 6.5 = 52$ mph

18. $338 \div 13 = 26$ mpg

19. 32, 64, 128 (double each number)

20. 8, 13 (add the previous two numbers)

Systematic Review 13D

1. on the graph
2. on the graph

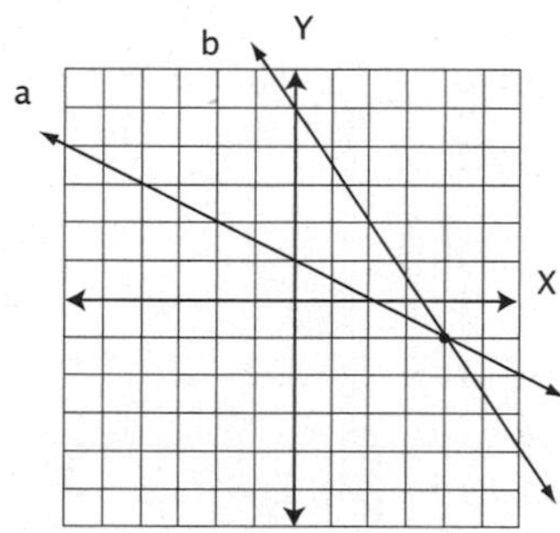

3. $(4,-1)$

4. $(1) = -\frac{3}{2}(-1) + b$
$1 = \frac{3}{2} + b$
$b = -\frac{1}{2}$

5. $Y = -\frac{3}{2}X - \frac{1}{2}$
$3X + 2Y = -1$

6. $\frac{-4-2}{1-(-4)} = \frac{-6}{5}$
$m = -\frac{6}{5}$

7. $(2) = -\frac{6}{5}(-4) + b$
$2 = \frac{24}{5} + b$
$b = -\frac{14}{5}$

8. $Y = -\frac{6}{5}X - \frac{14}{5}$
$6X + 5Y = -14$

9. $m = -\frac{4}{3}$
$(-3) = -\frac{4}{3}(2) + b$
$-3 = -\frac{8}{3} + b$
$\frac{-9}{3} + \frac{8}{3} = b$
$b = -\frac{1}{3}$

10. $Y = -\frac{4}{3}X - \frac{1}{3}$
$4X + 3Y = -1$

11. $16X - 8X = 56$
$8X = 56$
$X = \frac{56}{8} = 7$

12. $18A - 15 = 24$
$3(6A - 5) = 3(8)$
$6A - 5 = 8$
$6A = 13$
$A = \frac{13}{6} = 2\frac{1}{6}$

13. $(1-7)^2-8N+11=-3$
$$(-6)^2-8N=-3-11$$
$$36-8N=-14$$
$$-8N=-50$$
$$N=\frac{-50}{-8}=6\frac{1}{4}$$

14. $100(.78)+100(.4)=100(2X)$
$$78+40=200X$$
$$118=200X$$
$$\frac{118}{200}=X=\frac{59}{100}\text{ or }.59$$

15. $.3+\frac{1}{2}A=2A-1.8$
$$.3+.5A=2A-1.8$$
$$10(.3)+10(.5A)=10(2A)-10(1.8)$$
$$3+5A=20A-18$$
$$21=15A$$
$$\frac{21}{15}=A=1\frac{2}{5}$$

16. $(4-8)^2\times6-3\times5^2=7Y$
$$(-4)^2\times6-3\times25=7Y$$
$$16\times6-75=7Y$$
$$96-75=7Y$$
$$21=7Y$$
$$\frac{21}{7}=Y=3$$

17. 6:50 AM to 2:05 PM is 7.25 hours
$348\div7.25=48$ mph

18. $348\div14.5=24$ mpg

19. 36, 49, 64, 81 (count by 1, and square)

20. $\frac{1}{162},\frac{1}{486},\frac{1}{1458}$
$\left(\text{multiply previous number by }\frac{1}{3}\right)$

Systematic Review 13E

1. on the graph
2. on the graph

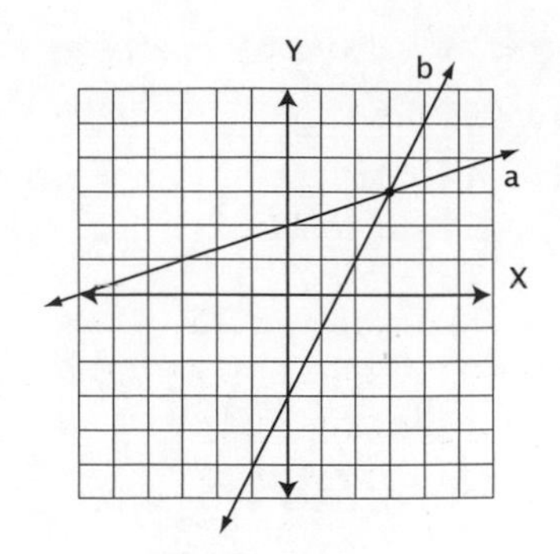

3. $(3,\ 3)$

4. $(2)=1(5)+b$
$$2=5+b$$
$$b=-3$$

5. $Y=X-3$
$X-Y=3$ or $-X+Y=-3$

6. $m=\frac{4-5}{1-3}=\frac{-1}{-2}=\frac{1}{2}$

7. $(4)=\frac{1}{2}(1)+b$
$$4=\frac{1}{2}+b$$
$$b=3\frac{1}{2}$$

8. $Y=\frac{1}{2}X+\frac{7}{2}$
$2Y=X+7$
$X-2Y=-7$ or $-X+2Y=7$

9. $m=\frac{5}{4}$
$$(-2)=\frac{5}{4}(-2)+b$$
$$-2=\frac{-10}{4}+b$$
$$\frac{-8}{4}+\frac{10}{4}=b$$
$$b=\frac{2}{4}=\frac{1}{2}$$

10. $Y=\frac{5}{4}X+\frac{1}{2}$
$4Y=5X+2$
$5X-4Y=-2$ or $-5X+4Y=2$

11. $3Q+7+2Q-5-4Q=-Q+1+Q+4$
$$Q+2=5$$
$$Q=3$$

12. $T+4+3T-6-2T=2T+5-4T-1+2T$
$$2T-2=4$$
$$2T=6$$
$$T=\frac{6}{2}=3$$

13.
$$-2.8P+.06P=5.72$$
$$100(-2.8P)+100(.06P)=100(5.72)$$
$$-280P+6P=572$$
$$-274P=572$$
$$P=\frac{572}{-274}$$
$$P=-\frac{286}{137}$$

14. $32Y-8Y=-36$
$$24Y=-36$$
$$Y=\frac{-36}{24}=-1\frac{1}{2}$$

15. $(.03)\left(\frac{3}{4}\right)(X)-.75=0$
$$(.03)(.75)(X)-.75=0$$
$$.0225X-.75=0$$
$$10{,}000(.0225)X=10{,}000(.75)$$
$$225X=7500$$
$$X=\frac{7500}{225}$$
$$X=33\frac{1}{3}\text{ or }33.3\overline{3}$$

16.
$$4\frac{2}{3}+3\frac{1}{3}X=-3$$
$$\frac{14}{3}+\frac{10}{3}X=-3$$
$$3\left(\frac{14}{3}\right)+3\left(\frac{10}{3}\right)X=3(-3)$$
$$14+10X=-9$$
$$10X=-23$$
$$X=\frac{-23}{10}$$
$$X=-2\frac{3}{10}\text{ or }-2.3$$

17. 8:20 to 2:40 is $6\frac{1}{3}$ hours

$335\div\frac{19}{3}=335\times\frac{3}{19}\approx 52.9$ mph

18. $335\div 13.4=25$ mpg

19. XD, XE, XF (X times the next letter in the alphabet)

20. 4.75, 5, 5.25, 5.5, 5.75 (add .25 each time)

Lesson Practice 14A

1. $(4,\ -2)$
2. replace X in equation 2 with its equivalent, $(Y+6)$:
$$(Y+6)+3Y=-2$$
$$4Y=-8$$
$$Y=-2$$

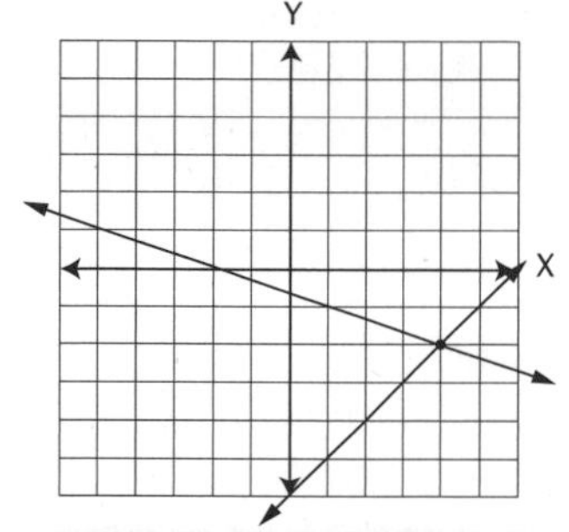

3. replace Y in equation 1 with its equivalent, (-2):
$$X=(-2)+6$$
$$X=4$$
4. $2X+3Y=0 \Rightarrow Y=-\frac{2}{3}X$

$X-2Y=7 \Rightarrow Y=\frac{1}{2}X-\frac{7}{2}$

$(3,\ -2)$

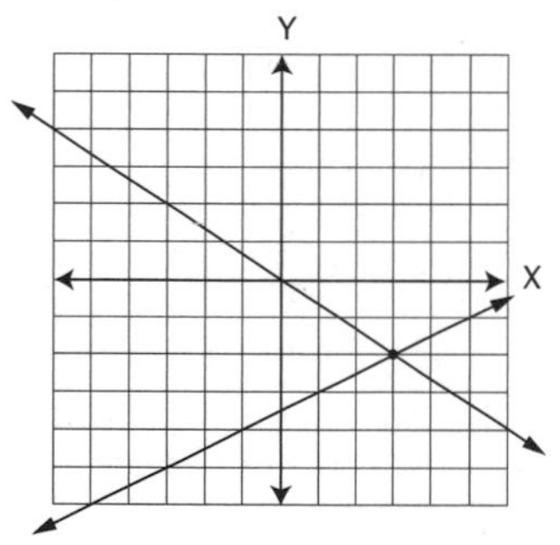

5. replace X in equation 1 with its equivalent, $(7+2Y)$:
$$2(7+2Y)+3Y=0$$
$$14+4Y+3Y=0$$
$$7Y=-14$$
$$Y=-2$$

6. replace Y in equation 1 with its equivalent, (-2):

$$2X+3(-2)=0$$
$$2X-6=0$$
$$2X=6$$
$$X=3$$

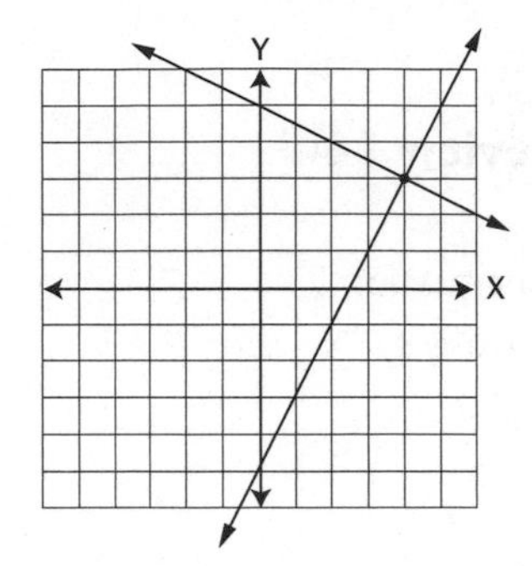

7. $X+2Y=10 \Rightarrow Y=-\frac{1}{2}X+5$

$(4, 3)$

8. replace Y in equation 2 with its equivalent, $(2X-5)$:

$$X+2(2X-5)=10$$
$$X+4X-10=10$$
$$5X=20$$
$$X=4$$

9. replace X in equation 1 with its equivalent, (4):

$$Y=2(4)-5$$
$$Y=8-5$$
$$Y=3$$

10. replace Y in equation 1 with its equivalent, $(X+3)$:

$$2X-3(X+3)=-4$$
$$2X-3X-9=-4$$
$$-X=5$$
$$X=-5$$

replace X in equation 2 with its equivalent, (-5):

$$Y=(-5)+3$$
$$Y=-2$$

$(-5, -2)$

Lesson Practice 14B

1. $(-1, 2)$

2. solve equation 1 for X:

$X+Y=1 \Rightarrow X=-Y+1$

replace X in equation 2 with its equivalent, $(-Y+1)$:

$$Y=(-Y+1)+3$$
$$Y=-Y+4$$
$$2Y=4$$
$$Y=2$$

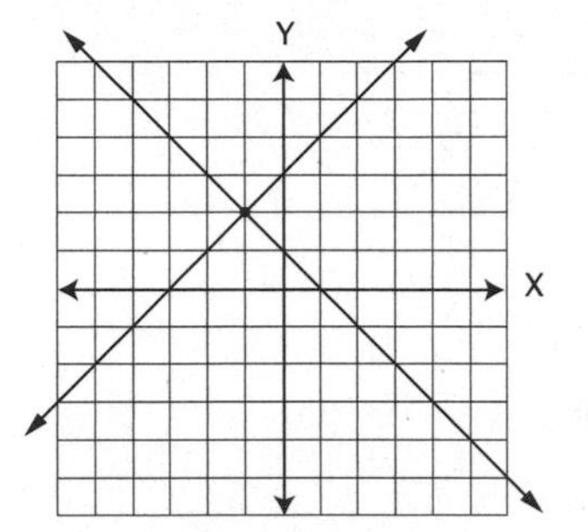

3. replace Y in equation 2 with its equivalent, (2):

$$(2)=X+3$$
$$-1=X$$

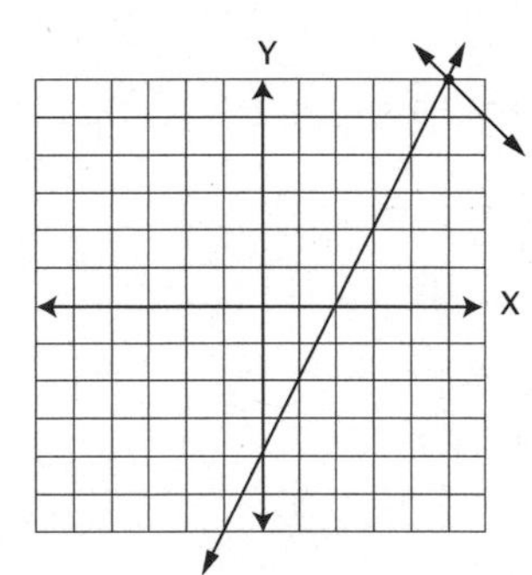

4. $2X-Y=4 \Rightarrow Y=2X-4$

$(5, 6)$

5. solve equation 2 for X:

$Y=-X+11 \Rightarrow X=-Y+11$

replace X in equation 1 with its equivalent, $(-Y+11)$:

$$2(-Y+11)-Y=4$$
$$-2Y+22-Y=4$$
$$-3Y=-18$$
$$Y=6$$

6. replace Y in equation 2 with its equivalent, (6):

$$(6) = -X + 11$$
$$-5 = -X$$
$$X = 5$$

replace X in equation 2 with its equivalent, (7):

$$5(7) - Y = 30$$
$$35 - Y = 30$$
$$-Y = -5$$
$$Y = 5$$

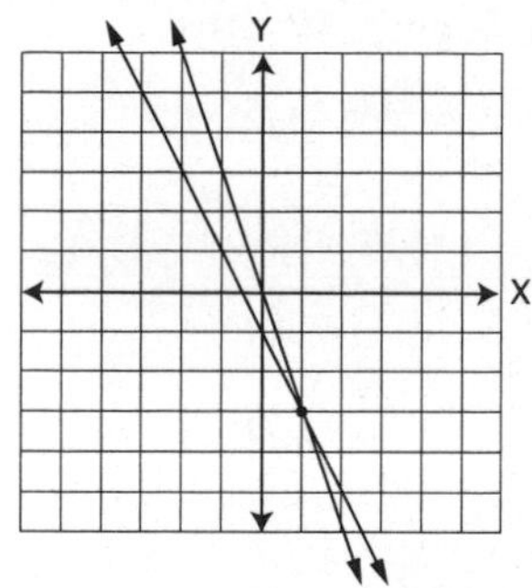

7. $2X + Y = -1 \Rightarrow Y = -2X - 1$
(1, −3)

8. solve equation 2 for X:

$$Y = -3X \Rightarrow -\frac{1}{3}Y = X$$

replace X in equation 1 with its equivalent, $\left(-\frac{1}{3}Y\right)$:

$$2\left(-\frac{1}{3}Y\right) + Y = -1$$
$$-\frac{2}{3}Y + Y = -1$$
$$\frac{1}{3}Y = -1$$
$$Y = -3$$

9. replace Y in equation 2 with its equivalent, (−3):

$$(-3) = -3X$$
$$X = 1$$

10. change equation 2 to slope-intercept form:

$$5X - Y = 30 \Rightarrow Y = 5X - 30$$

replace Y in equation 1 with its equivalent, (5X − 30):

$$2X + 3(5X - 30) = 29$$
$$2X + 15X - 90 = 29$$
$$17X = 119$$
$$X = 7$$

Systematic Review 14C

1. (3, 4)
2. replace Y in equation 2 with its equivalent, (X + 1):

$$(X + 1) = 2X - 2$$
$$1 + 2 = 2X - X$$
$$X = 3$$

3. replace X in equation 1 with its equivalent, (3):

$$Y = (3) + 1$$
$$Y = 4$$

For #1-3.

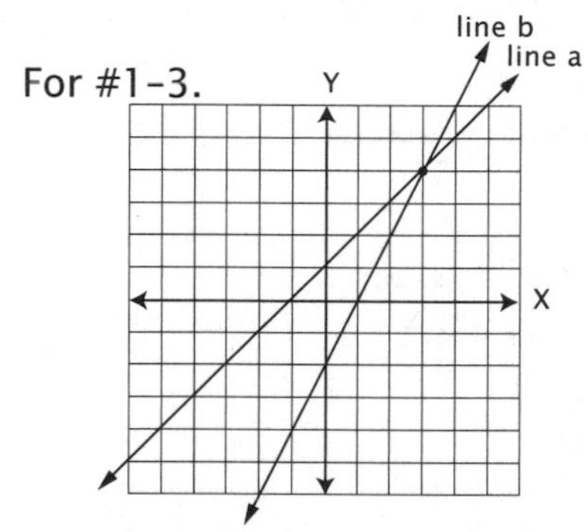

4. (−1, 3)
5. solve equation 1 for Y:

$$Y - X = 4 \Rightarrow Y = X + 4$$

replace Y in equation 2 with its equivalent, (X + 4):

$$(X + 4) + 2X = 1$$
$$3X = -3$$
$$X = -1$$

6. replace X in equation 1 with its equivalent, (−1):

$$Y - (-1) = 4$$
$$Y + 1 = 4$$
$$Y = 3$$

For #4-6.

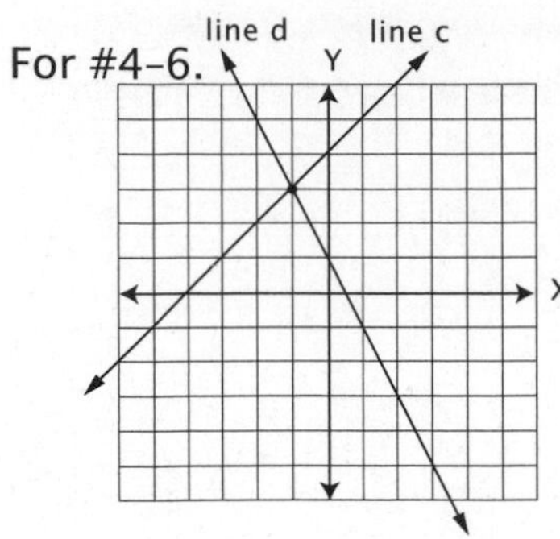

7. $m = \frac{3-5}{1-4} = \frac{-2}{-3} = \frac{2}{3}$

8. $(3) = \frac{2}{3}(1) + b$

$3 - \frac{2}{3} = b$

$\frac{9}{3} - \frac{2}{3} = b$

$b = \frac{7}{3}$

9. $Y = \frac{2}{3}X + \frac{7}{3}$

$3Y = 2X + 7$

$2X - 3Y = -7$ or $-2X + 3Y = 7$

10. $m = -\frac{4}{3}$

11. $(2) = -\frac{4}{3}(2) + b$

$2 = \frac{-8}{3} + b$

$2 + \frac{8}{3} = b$

$\frac{6}{3} + \frac{8}{3} = b$

$b = \frac{14}{3}$

12. $Y = -\frac{4}{3}X + \frac{14}{3}$

$3Y = -4X + 14$

$4X + 3Y = 14$

13. 1, 4, 9, 16, 25, 36, 49, 64, 81, 100, 121, 144, 169, 196, 225

$1^2, 2^2, 3^2, 4^2, 5^2, 6^2, 7^2, 8^2, 9^2, 10^2, 11^2, 12^2, 13^2, 14^2, 15^2$

14. 820 miles

Depending on the source, answers may vary. If a different distance is used, answers for #15 and 16 will also vary.

15. $820 \div 50 = 16.4$ hours

$16\frac{4}{10} = 16\frac{24}{60}$ or 16 hr, 24 min

$16{:}24 + 7{:}35 = 23{:}59$; 11:59 PM

16. $820 \div 25 = 32.8$ gallons

$32.8 \times 1.269 = \$41.62$

17. .923

18. $2A^2 - A^2 + 3A$

19. prime

20. $6 = 2 \times 3$; $4 = 2 \times 2$

$LCM = 2 \times 2 \times 3 = 12$

Systematic Review 14D

1. $X + Y = -6 \Rightarrow Y = -X - 6$

$(-4, -2)$

2. replace Y in equation 2 with its equivalent, $(2X + 6)$:

$X + (2X + 6) = -6$

$3X = -12$

$X = -4$

3. replace X in equation 1 with its equivalent, (-4):

$Y = 2(-4) + 6$

$Y = -8 + 6$

$Y = -2$

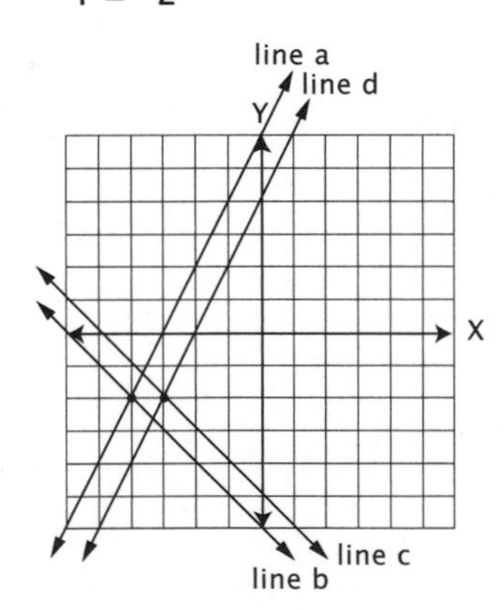

4. $(-3,-2)$
5. solve equation 2 for Y:
   $Y-2X=4 \Rightarrow Y=2X+4$
   replace Y in equation 1
   with its equivalent, $(2X+4)$:
   $$(2X+4)+X=-5$$
   $$3X=-9$$
   $$X=-3$$
6. replace X in equation 1
   with its equivalent, (-3):
   $$Y+(-3)=-5$$
   $$Y=-2$$
7. $m=\frac{4-0}{-2-0}=-2$
8. $(0)=-2(0)+b$
   $0=0+b$
   $b=0$
9. $Y=-2X$
   $2X+Y=0$
10. $m=\frac{3}{4}$
11. $(2)=\frac{3}{4}(2)+b$
    $2-\frac{6}{4}=b$
    $\frac{8}{4}-\frac{6}{4}=b$
    $b=\frac{1}{2}$
12. $Y=\frac{3}{4}X+\frac{1}{2}$
    $4Y=3X+2$
    $3X-4Y=-2$ or $-3X+4Y=2$
13. 1, 4, 9, 16, 25, 36, 49, 64, 81,
    100, 121, 144, 169, 196, 225
    $1^2, 2^2, 3^2, 4^2, 5^2, 6^2, 7^2, 8^2, 9^2,$
    $10^2, 11^2, 12^2, 13^2, 14^2, 15^2$
14. 380 miles (See note for lesson 14C.)
15. $380 \div 50 = 7.6$ hours
    $7\frac{6}{10}=7\frac{36}{60}$ or 7 hr, 36 min
    7:36 + 6:14 = 13:50; 1:50 PM
16. $380 \div 25 = 15.2$ gallons
    $15.2 \times 1.199 = \$18.22$
17. .321
18. $9A+27B-81=18C$
    $9(A+3B-9)=9(2C)$
    $A+3B-9=2C$
19. $5\times 87=5\times 3\times 29$
20. 8

Systematic Review 14E

1. $(-5, 1)$
2. solve equation 1 for Y:
   $X+Y=-4 \Rightarrow Y=-4-X$
   replace Y in equation 2
   with its equivalent, $(-4-X)$:
   $$X-(-4-X)=-6$$
   $$X+(4+X)=-6$$
   $$2X+4=-6$$
   $$2X=-10$$
   $$X=-5$$
3. replace X in equation 1
   with its equivalent, (-5):
   $$(-5)+Y=-4$$
   $$Y=1$$
4. $(-1, 0)$

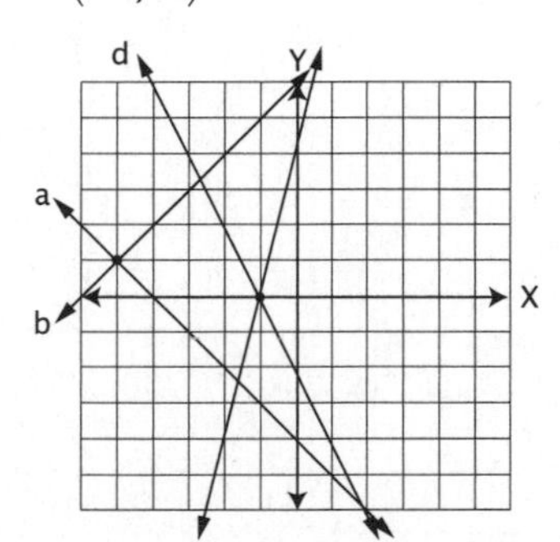

5. solve equation 1 for Y:
   $Y-4X=4 \Rightarrow Y=4X+4$
   replace Y in equation 2
   with its equivalent, $(4X+4)$:
   $$(4X+4)+2X=-2$$
   $$6X+4=-2$$
   $$6X=-6$$
   $$X=-1$$

6. replace X in equation 1 with its equivalent, (-1):

$$Y-4(-1)=4$$
$$Y+4=4$$
$$Y=0$$

7. $m=\frac{-2-1}{3-(-1)}=\frac{-3}{4}$

8.
$$(1)=-\frac{3}{4}(-1)+b$$
$$1=\frac{3}{4}+b$$
$$\frac{4}{4}-\frac{3}{4}=b$$
$$b=\frac{1}{4}$$

9. $Y=-\frac{3}{4}X+\frac{1}{4}$
$3X+4Y=1$

10. $m=\frac{3}{5}$

11.
$$(-2)=\frac{3}{5}(-3)+b$$
$$-2=\frac{-9}{5}+b$$
$$\frac{-10}{5}+\frac{9}{5}=b$$
$$b=-\frac{1}{5}$$

12. $Y=\frac{3}{5}X-\frac{1}{5}$; $3X-5Y=1$

13. 1, 4, 9, 16, 25, 36, 49, 64, 81, 100, 121, 144, 169, 196, 225
$1^2, 2^2, 3^2, 4^2, 5^2, 6^2, 7^2, 8^2, 9^2, 10^2, 11^2, 12^2, 13^2, 14^2, 15^2$

14. 804 miles; This and the following answers may vary, depending on your source of information.

15. $804\div50=16.08$ hours
$16\frac{8}{100}=16.08$ hours
$.08\times60=4.8$ min, round to 5 min
$16{:}05+4{:}42=20{:}47$; 8:47 PM

16. $804\div25=32.16$ gallons
$(32.16)(1.289)=\$41.45$

17. .368

18. $\frac{9}{10}=\frac{90}{100}=90\%$

19. no; example:
$$(1-3)-3\neq1-(3-3)$$
$$(-2)-3\neq1-(0)$$
$$-5\neq1$$

20. $.16\times24.3=3.888$

Lesson Practice 15A

1. $(-3,\ 2)$

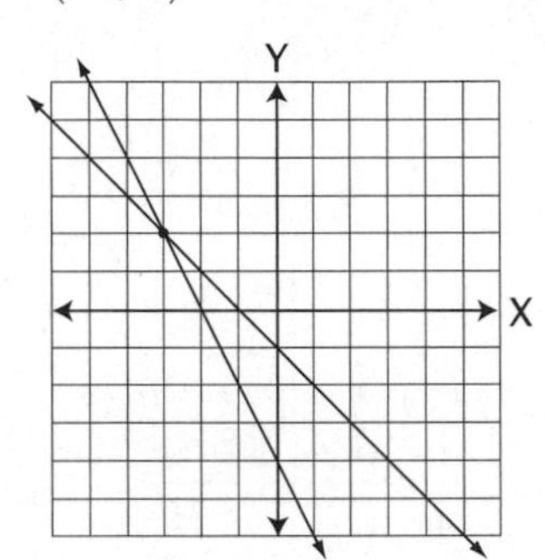

2.
$$X+Y=-1$$
$$-(2X+Y=-4)$$
$$-X\ \ =3$$
$$X=-3$$

3. $(-3)+Y=-1$; $Y=2$

4. $(-1,-1)$

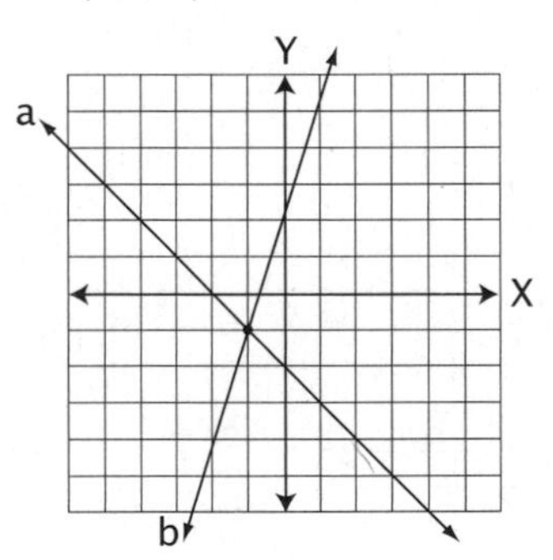

5.
$$X+Y=-2$$
$$+(3X-Y=-2)$$
$$4X\ \ =-4$$
$$X=-1$$

6. $(-1)+Y=-2$
$$Y=-1$$

7. $(4,-1)$

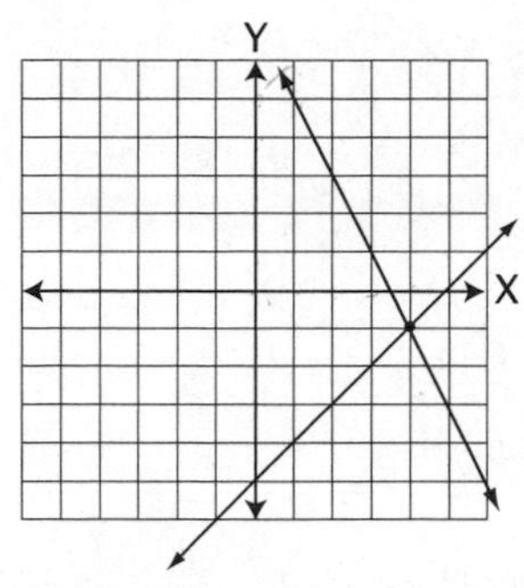

8.
$$\begin{array}{r} X - Y = 5 \\ +(2X + Y = 7) \\ \hline 3X \quad = 12 \\ X = 4 \end{array}$$

9. $(4) - Y = 5$
$4 - 5 = Y$
$-1 = Y$

10. $2(2X + 3Y = 18) \Rightarrow \begin{array}{r} 4X + 6Y = 36 \\ -(4X + Y = 6) \\ \hline 5Y = 30 \\ Y = 6 \end{array}$

$4X + (6) = 6$
$4X = 0$
$X = 0$
$(0, 6)$

Lesson Practice 15B

1. $(5, 2)$

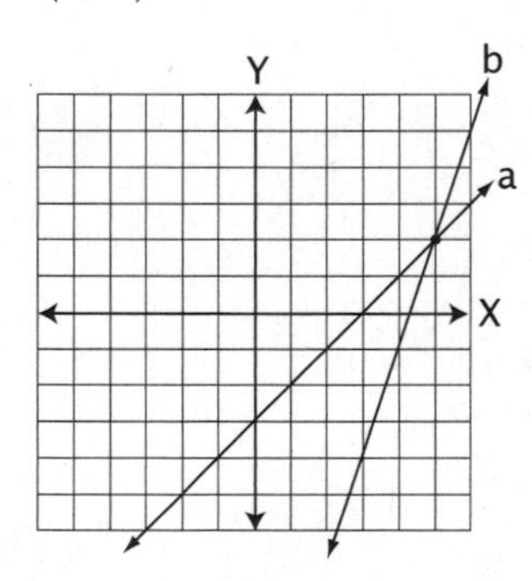

2.
$$\begin{array}{r} X - Y = 3 \\ -(3X - Y = 13) \\ \hline -2X \quad = -10 \\ X = 5 \end{array}$$

3. $(5) - Y = 3$
$Y = 2$

4. $(-2, 4)$

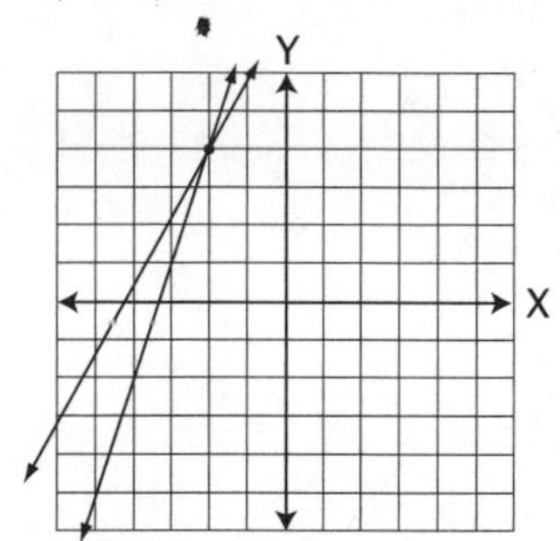

5.
$$\begin{array}{r} 3X - Y = -10 \\ -(2X - Y = -8) \\ \hline X \quad = -2 \end{array}$$

6. $2(-2) - Y = -8$
$-4 - Y = -8$
$Y = 4$

7. $(1, 1)$

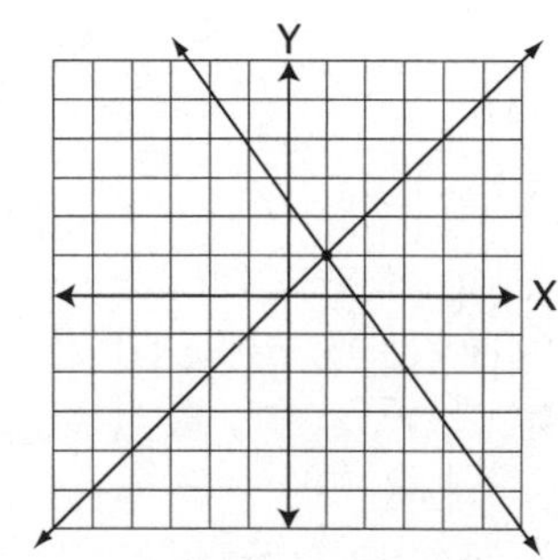

8.
$$\begin{array}{r} 3X + 2Y = 5 \\ 2(X - Y = 0) \Rightarrow +(2X - 2Y = 0) \\ \hline 5X \quad = 5 \\ X = 1 \end{array}$$

9. $(1) - Y = 0$
$Y = 1$

10.
$$\begin{array}{r} X+Y=-3 \\ +(3X-Y=-1) \\ \hline 4X \quad =-4 \\ X=-1 \end{array}$$
$(-1)+Y=-3 \Rightarrow Y=-2 \qquad (-1,-2)$

Systematic Review 15C

1. $(-2,\ 2)$ see graph
2. $$\begin{array}{r} 2Y+3X=-2 \\ +(6Y-3X=18) \\ \hline 8Y \quad =16 \\ Y=\ 2 \end{array}$$
3. $2(2)-X=6$
   $4-X=6$
   $X=-2$
4. $(1,-1)$ see graph
5. $$\begin{array}{rr} & Y+3X=2 \\ -1(Y-X=-2) \Rightarrow & -Y+\ X=2 \\ \hline & 4X=4 \\ & X=1 \end{array}$$
6. $Y-(1)=-2$
   $Y=-1$

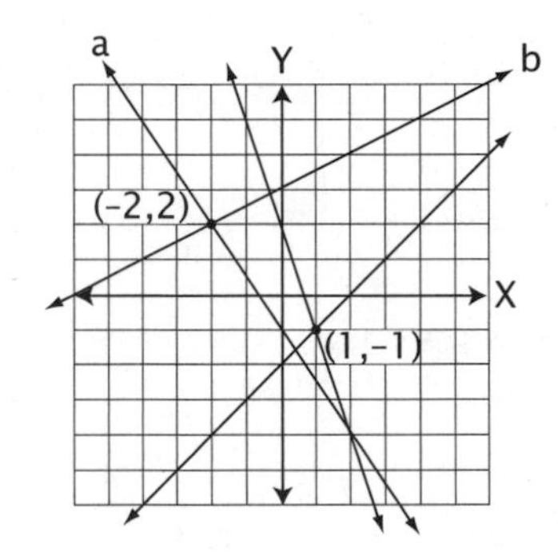

7. $-2Y>3X+6$
   $Y<-\frac{3}{2}X-3$

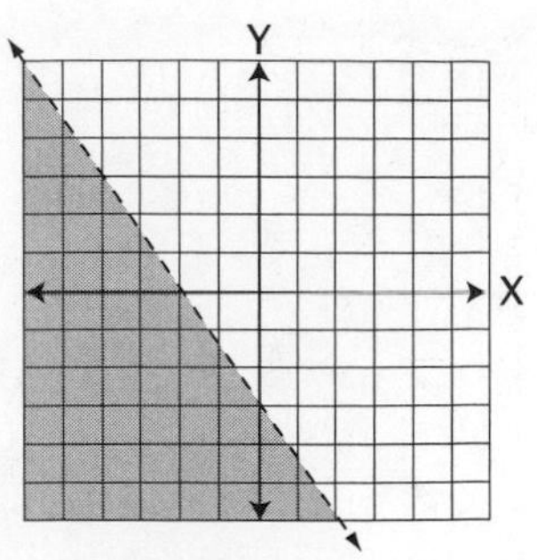

8. on the graph
9. no (see graph)
10. $m=\frac{1}{2}$
11. $(1)=\frac{1}{2}(-1)+b$
    $1=-\frac{1}{2}+b$
    $b=\frac{3}{2}$
12. $Y=\frac{1}{2}X+\frac{3}{2}$
    $2Y=X+3$
    $X-2Y=-3$ or $-X+2Y=3$
13. 1,4,9,16,25,36,49,64,81,
    100,121,144,169,196,225
    $1^2,\ 2^2,\ 3^2,\ 4^2,\ 5^2,\ 6^2,\ 7^2,\ 8^2,$
    $9^2,\ 10^2,\ 11^2,\ 12^2,\ 13^2,\ 14^2,\ 15^2$
14. $6N-4=\frac{10N}{2}$
15. $6N-4=5N$
    $N=4$
16. $(4)^2-2(4)+|3-4|=16-8+|-1|=8+1=9$
17. $.14\times 25=3.5$
18. $WF\times 16=8$
    $WF=\frac{8}{16}=\frac{1}{2}$
19. $3.14\div 2.4\approx 1.308$
20. $\frac{3}{4}\div\frac{5}{6}=\frac{3}{4}\times\frac{6}{5}=\frac{18}{20}=\frac{9}{10}$

Systematic Review 15D

1. (3, 2) see graph
2. $\begin{array}{r} 3Y+2X=12 \\ +(8Y-2X=10) \\ \hline 11Y \quad\quad =22 \\ Y=2 \end{array}$
3. $3(2)+2X=12$
   $6+2X=12$
   $2X=6$
   $X=3$
4. (2, 2) see graph

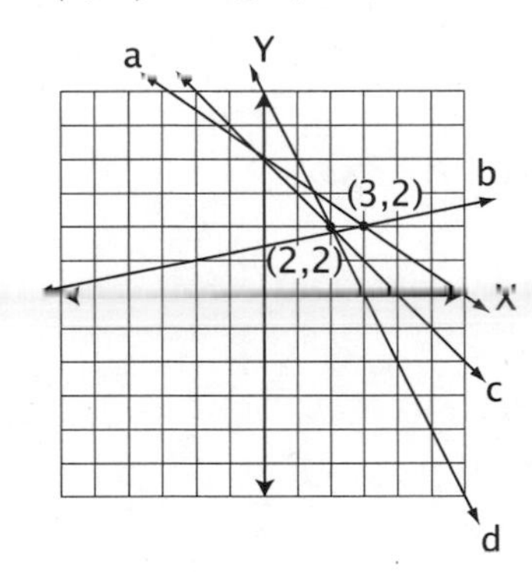

5. $\begin{array}{r} X+Y=4 \\ -2X-Y=-6 \\ \hline -X \quad =-2 \\ X=2 \end{array}$
6. $(2)+Y=4$
   $Y=2$
7. $Y \geq \frac{4}{5}X+2$
8. on the graph below
9. no (see graph)

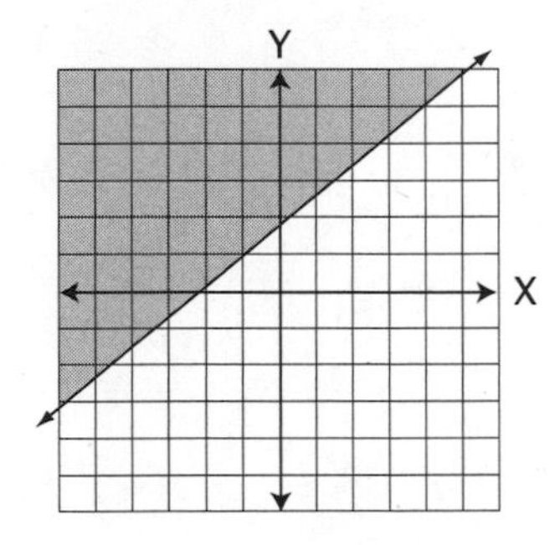

10. $m=-2$
11. $(1)=-2(-1)+b$
    $1=2+b;\ b=-1$
12. $Y=-2X-1$
    $2X+Y=-1$
13. 1, 4, 9, 16, 25, 36, 49, 64, 81, 100, 121, 144, 169, 196, 225
    $1^2, 2^2, 3^2, 4^2, 5^2, 6^2, 7^2, 8^2,$
    $9^2, 10^2, 11^2, 12^2, 13^2, 14^2, 15^2$
14. $3N-4N+8=3N$
15. $-1N+8=3N$
    $8=4N$
    $N=2$
16. $3X^2-X \div |4-3| =$
    $3(2)^2-(2) \div |4-3| =$
    $3(4)-(2) \div |1| =$
    $12-2 \div 1 =$
    $12-2=10$
17. $\frac{48}{100} \times 32 = 15\frac{36}{100} = 15\frac{9}{25}$
18. $WF \times 75 = 5$
    $WF = \frac{5}{75} = \frac{1}{15}$
19. $21.8 \div .4 = 54.5$
20. $\frac{2}{7} \div \frac{1}{2} = \frac{2}{7} \times \frac{2}{1} = \frac{4}{7}$

Systematic Review 15E

1. (1,–2) see graph
2. $\begin{array}{r} -2Y+2X=6 \\ +Y-2X=-4 \\ \hline -Y \quad\quad =2 \\ Y=-2 \end{array}$
3. $(-2)-X=-3;\ -X=-1;\ X=1$
4. (–3,1) see graph

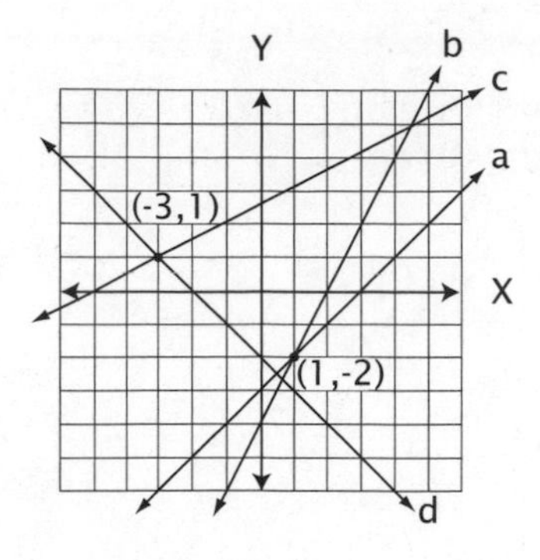

5. $$\begin{aligned} X - 2Y &= -5 \\ +\ 2X + 2Y &= -4 \\ \hline 3X \quad &= -9 \\ X &= -3 \end{aligned}$$
6. $(-3) + Y = -2;\ Y = 1$
7. $Y < 3X - 4$
8. on the graph below
9. no (see graph)

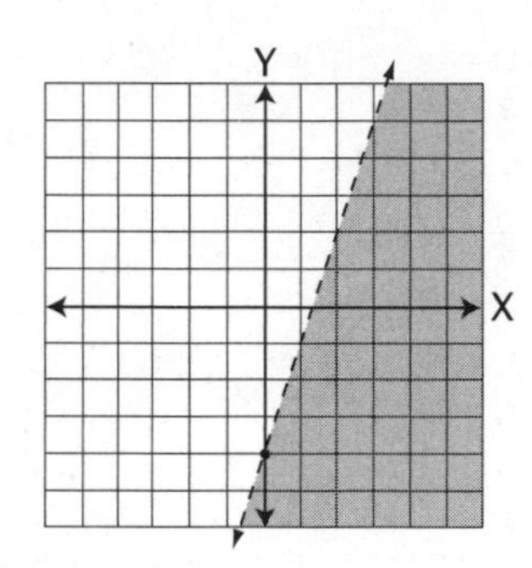

10. $m = 2$
11. $(1) = 2(1) + b$
    $1 = 2 + b;\ b = -1$
12. $Y = 2X - 1$
    $2X - Y = 1$ or $-2X + Y = -1$
13. 1, 4, 9, 16, 25, 36, 49, 64, 81, 100, 121, 144, 169, 196, 225
    $1^2, 2^2, 3^2, 4^2, 5^2, 6^2, 7^2, 8^2, 9^2, 10^2, 11^2, 12^2, 13^2, 14^2, 15^2$
14. $N - 2 + N + 5N = 6N + 1$
15. $7N - 2 = 6N + 1$
    $N = 1 + 2 = 3$
16. $$\begin{aligned} &5X - 4 + 3X \div 2 = \\ &5(3) - 4 + 3(3) \div 2 = \\ &15 - 4 + 9 \div 2 = \\ &11 + \frac{9}{2} = 15\frac{1}{2} \end{aligned}$$
17. $\frac{150}{100} \times 18 = 27$
18. $95 \div 3 = 31.67;\ 31.67 \times 2 = 63.34$
    Using a calculator without rounding the first step will give 63.33.
19. $3.14 \times 4.16 \approx 13.06$
20. $\frac{9}{16} = 9 \div 16 = .5625 \approx .56$

Lesson Practice 16A

1. $N + D = 8$
   $.05N + .10D = .65$
2. $$\begin{aligned} (N + D = 8)(-5) \Rightarrow \quad -5N - 5D &= -40 \\ (.05N + .10D = .65)(100) \Rightarrow \quad 5N + 10D &= 65 \\ \hline 5D &= 25 \\ D &= 5 \end{aligned}$$
3. $$\begin{aligned} N + D &= 8 \\ N + (5) &= 8 \\ N &= 3 \end{aligned}$$
4. $P + D = 25$
   $.01P + .10D = .88$
5. $$\begin{aligned} (P + D = 25)(-10) \Rightarrow \quad -10P - 10D &= -250 \\ (.01P + .10D = .88)(100) \Rightarrow \quad P + 10D &= 88 \\ \hline -9P \quad &= -162 \\ P &= 18 \end{aligned}$$
6. $$\begin{aligned} P + D &= 25 \\ (18) + D &= 25 \\ D &= 7 \end{aligned}$$
7. $P + N = 26$
   $.01P + .05N = .86$
8. $$\begin{aligned} (P + N = 26)(-1) \Rightarrow \quad -P - N &= -26 \\ (.01P + .05N = .86)(100) \Rightarrow \quad P + 5N &= 86 \\ \hline 4N &= 60 \\ N &= 15 \end{aligned}$$

9. $P+N=26$
$P+(15)=26$
$P=11$

10. $Q+D=13$
$.25Q+.10D=1.75$

11.
$$\begin{array}{rl} (Q+D=13)(-10) \Rightarrow & -10Q-10D=-130 \\ (.25Q+.10D=1.75)(100) \Rightarrow & \underline{25Q+10D=175} \\ & 15Q \quad = 45 \\ & Q=3 \end{array}$$

12. $Q+D=13$
$(3)+D=13$
$D=10$

Lesson Practice 16B

1. $N+D=20$
$.05N+.10D=1.75$

2.
$$\begin{array}{rl} (N+D=20)(-10) \Rightarrow & -10N-10D=-200 \\ (.05N+.10D=1.75)(100) \Rightarrow & \underline{5N+10D=\ 175} \\ & -5N \quad = -25 \\ & N=5 \end{array}$$

3. $N+D=20$
$(5)+D=20$
$D=15$

4. $P+D=39$
$.01P+.10D=1.83$

5.
$$\begin{array}{rl} (P+D=39)(-10) \Rightarrow & -10P-10D=-390 \\ (.01P+.10D=1.83)(100) \Rightarrow & \underline{P+10D=\ 183} \\ & -9P \quad =-207 \\ & P=23 \end{array}$$

6. $P+D=39$
$(23)+D=39$
$D=16$

7. $N+D=19$
$.05N+.10D=1.25$

8.
$$\begin{array}{rl} (N+D=19)(-10) \Rightarrow & -10N-10D=-190 \\ (.05N+.10D=1.25)(100) \Rightarrow & \underline{5N+10D=\ 125} \\ & -5N \quad = -65 \\ & N=13 \end{array}$$

9. $N+D=19$
$(13)+D=19$
$D=6$

10. $Q+N=40$
$.25Q+.05N=5.00$

11.
$$\begin{array}{rl} (Q+N=40)(-5) \Rightarrow & -5Q-5N=-200 \\ (.25Q+.05N=5.00)(100) \Rightarrow & \underline{25Q+5N=\ 500} \\ & 20Q \quad = 300 \\ & Q=\ 15 \end{array}$$

12. $Q+N=40$
$(15)+N=40$
$N=25$

Systematic Review 16C

1. $N+D=12$
$.05N+.10D=.85$

2.
$$\begin{array}{rl} (.05N+.10D=.85)(100) \Rightarrow & 5N+10D=\ 85 \\ (N+D=12)(-5) \Rightarrow & \underline{-5N-\ 5D=-60} \\ & 5D=\ 25 \\ & D=5 \end{array}$$

3. $N+D=12$
$N+(5)=12$
$N=7$

$.05(7)+.10(5)=.85$
$.35+.50=.85$
$.85=.85$

4. $P+N=10$
$.01P+.05N=.38$

5.
$$\begin{array}{rl} (.01P+.05N=.38)(100) \Rightarrow & P+5N=\ 38 \\ (P+N=10)(-5) \Rightarrow & \underline{-5P-5N=-50} \\ & -4P \quad =-12 \\ & P=3 \end{array}$$

6. $P+N=10$
$(3)+N=10$
$N=7$
$.01(3)+.05(7)=.38$
$.03+.35=.38$
$.38=.38$

7. 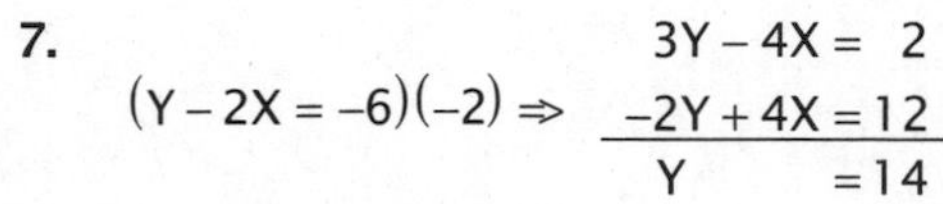
$3Y-4X=2$
$(Y-2X=-6)(-2) \Rightarrow -2Y+4X=12$
$Y=14$

8. $Y-2X=-6$
$(14)-2X=-6$
$20=2X$
$X=10;\ (10,\ 14)$

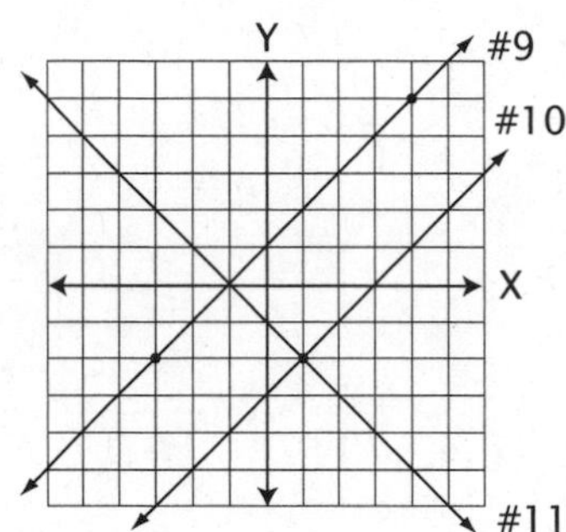

9. $m=\frac{-2-5}{-3-4}=\frac{-7}{-7}=1$
$(5)=1(4)+b \Rightarrow b=1$
$Y=X+1$

10. $(-2)=1(1)+b$
$-2=1+b$
$b=-3$
$Y=X-3$ or $X-Y=3$ or $-X+Y=-3$

11. $(-2)=-1(1)+b$
$-2=-1+b \Rightarrow b=-1$
$Y=-X-1$

12. Since the problem uses both inches and feet, we will convert the original length of the vine to feet, to make the units consistent.
24" = 2'; $Y=2X+2$ or $L=2(W)+2$

13. $Y=2(3)+2$
$Y=6+2=8'$

14. new equation: $Y=3X+2$
$Y=3(9)+2$
$Y=27+2=29'$

15. $(3+5)\times(2-7)-3-3^2=(8)\times(-5)-3-9$
$=-40-3-9=-52$

16. 4th

17. no

18. yes

19. $13^2=13\times13=169$

20. $\sqrt{64}=8$

Systematic Review 16D

1. $N+D=9$
$.05N+.10D=.60$

2. $(.05N+.10D=.60)(100) \Rightarrow 5N+10D=60$
$(N+D=9)(-5) \Rightarrow -5N-5D=-45$
$5D=15$
$D=3$

3. $N+D=9$
$N+(3)=9$
$N=6$
$.05(6)+.10(3)=.60$
$.30+.30=.60$
$.60=.60$

4. $P+N=6$
$.01P+.05N=.26$

5. $(.01P+.05N=.26)(100) \Rightarrow P+5N=26$
$(P+N=6)(-5) \Rightarrow -5P-5N=-30$
$-4P=-4$
$P=1$

6. $P+N=6$
$(1)+N=6$
$N=5$

7. $4Y+3X=-19$
$Y-3X=-1$
$5Y=-20$
$Y=-4$

8. $(-4)-3X=-1$
$-3X=3$
$X=-1$

9. $m = -\frac{5}{3}$ (from graph)

$(-1) = -\frac{5}{3}(1) + b$

$-1 = -\frac{5}{3} + b$

$b = \frac{2}{3}$

$Y = -\frac{5}{3}X + \frac{2}{3}$ or $5X + 3Y = 2$

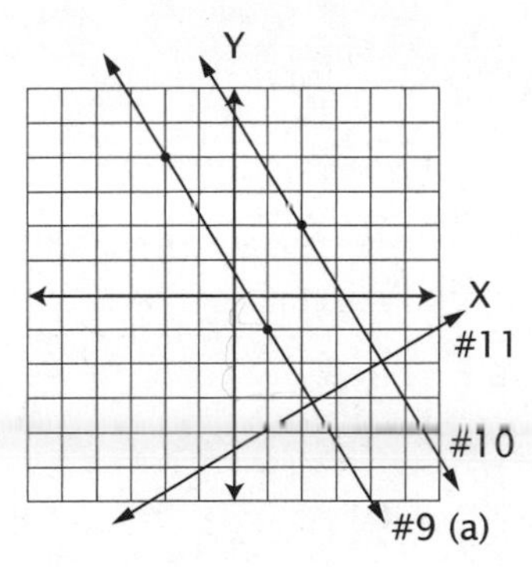

10. $(2) = -\frac{5}{3}(2) + b$

$2 = -\frac{10}{3} + b$

$b = \frac{16}{3}$

$Y = -\frac{5}{3}X + \frac{16}{3}$ or $5X + 3Y = 16$

11. $(-6) = \frac{3}{5}(-2) + b$

$-6 = \frac{-6}{5} + b$

$b = -\frac{24}{5}$

$Y = \frac{3}{5}X - \frac{24}{5}$ or $3X - 5Y = 24$ or $-3X + 5Y = -24$

12. $Y = 2X + 4$

13. $Y = X + 8$

14.

$$\begin{array}{rr} & Y = 2X + 4 \\ (Y = X + 8)(-1) \Rightarrow & -Y = -X - 8 \\ \hline & 0 = X - 4 \\ & 4 = X \end{array}$$

15. $Y = X + 8$

$Y = (4) + 8$

$Y = 12$

16. $Y = 2(12) + 4$

$Y = 24 + 4 = \$28$ for Kim

$Y = X + 8$

$Y = (12) + 8 = \$20$ for Ali

17. $-(5-9)^2 + (14-17)^2 =$

$-(-4)^2 + (-3)^2 =$

$-16 + 9 = -7$

18. 3rd

19. no

20. yes

Systematic Review 16E

1. $N + D = 14$

$.05N + .10D = 1.10$

2.

$$\begin{array}{rr} (.05N + .10D = 1.10)(100) \Rightarrow & 5N + 10D = 110 \\ (N + D = 14)(-5) \Rightarrow & -5N - 5D = -70 \\ \hline & 5D = 40 \\ & D = 8 \end{array}$$

3. $N + D = 14$

$N + (8) = 14$

$N = 6$

$.05(6) + .10(8) = 1.10$

$.30 + .80 = 1.10$

$1.10 = 1.10$

4. $P + N = 8$

$.01P + .05N = .20$

5.

$$\begin{array}{rr} (.01P + .05N = .20)(100) \Rightarrow & P + 5N = 20 \\ (P + N = 8)(-5) \Rightarrow & -5P - 5N = -40 \\ \hline & -4P = -20 \\ & P = 5 \end{array}$$

6. $P + N = 8$

$(5) + N = 8$

$N = 3$

$.01(5) + .05(3) = .20$

$.05 + .15 = .20$

$.20 = .20$

7. $Y = 2X + 2$
$Y = -4X - 4$
replace Y in equation 1
with its equivalent, $(-4X - 4)$:
$(-4X - 4) = 2X + 2$
$-6X = 6$
$X = -1$

8. replace X in equation 1
with its equivalent, (-1):
$Y = 2(-1) + 2$
$Y = -2 + 2 = 0$

9. on the graph

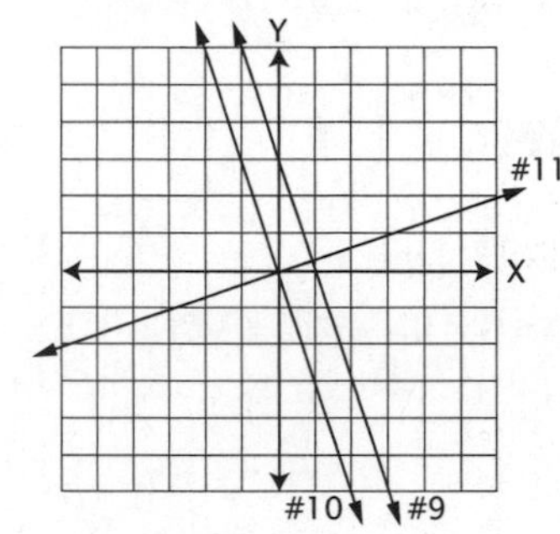

10. $(0) = -3(0) + b$
$0 = 0 + b;\ b = 0$
$Y = -3X$ or $3X + Y = 0$

11. $m = \frac{1}{3}$
$Y = \frac{1}{3}X$ or $X - 3Y = 0$ or $-X + 3Y = 0$

12. $T = 5M + 100$

13. $T = 5(15) + 100 = 175°C$

14. $T = 10M + 100$
$T = 10(15) + 100$
$T = 150 + 100 = 250°C$

15. $-[7 - (-5)]^2 + (3 - 8)^2 =$
$-[7 + 5]^2 + (-5)^2 =$
$-[12]^2 + (25) =$
$-144 + 25 = -119$

16. 2nd

17. yes

18. no

19. $25^2 = 25 \times 25 = 625$

20. $\sqrt{225} = 15$

Lesson Practice 17A

1. $N;\ N+1;\ N+2$

2. $N + (N+1) + (N+2) + 4 = 4(N+1)$

3. $N + (N+1) + (N+2) + 4 = 4(N+1)$
$3N + 7 = 4N + 4$
$7 - 4 = 4N - 3N$
$N = 3$

3; 4; 5

4. $3 + (4) + (5) + 4 = 4(4)$
$16 = 16$

5. $N;\ N+2;\ N+4$

6. $N + (N+2) = (N+4) + 4$

7. $N + (N+2) = (N+4) + 4$
$2N + 2 = N + 8$
$2N - N = 8 - 2$
$N = 6$

6; 8; 10

8. $(6) + (8) = (10) + 4$
$14 = 14$

9. $N;\ N+1;\ N+2$

10. $5(N+1) = 3[N + (N+2)] + 2$

11. $5(N+1) = 3[N + (N+2)] + 2$
$5N + 5 = 3[2N + 2] + 2$
$5N + 5 = 6N + 6 + 2$
$5N + 5 = 6N + 8$
$5 - 8 = 6N - 5N$
$-3 = N$

$-3;\ -2;\ -1$

12. $5(-2) = 3[(-3) + (-1)] + 2$
$-10 = 3[-4] + 2$
$-10 = 3[-4] + 2$
$-10 = -12 + 2$
$-10 = -10$

13. $N;\ N+2;\ N+4$

14. $N + (N+4) = 3(N+2) + 3$

15. $N + (N+4) = 3(N+2) + 3$
$2N + 4 = 3N + 6 + 3$
$2N + 4 = 3N + 9$
$4 - 9 = 3N - 2N$
$-5 = N$

$-5;\ -3;\ -1$

16. $(-5) + (-1) = 3(-3) + 3$
$-6 = -6$

Lesson Practice 17B

1. N; N+2; N+4
2. $3(N+4) = 2(N+(N+2))+2$
3. $$\begin{aligned} 3(N+4) &= 2(N+(N+2))+2 \\ 3N+12 &= 2(2N+2)+2 \\ 3N+12 &= 4N+4+2 \\ 3N+12 &= 4N+6 \\ 12-6 &= 4N-3N \\ 6 &= N \end{aligned}$$
   6; 8; 10
4. $$\begin{aligned} 3(10) &= 2((6)+(8))+2 \\ 30 &= 2(14)+2 \\ 30 &= 28+2 \\ 30 &= 30 \end{aligned}$$
5. N; N+1; N+2
6. $N+(N+2) = 20(N+1)$
7. $$\begin{aligned} N+(N+2) &= 20(N+1) \\ 2N+2 &= 20N+20 \\ 2-20 &= 20N-2N \\ -18 &= 18N \\ -1 &= N \end{aligned}$$
   −1; 0; 1
8. $$\begin{aligned} (-1)+(1) &= 20(0) \\ 0 &= 0 \end{aligned}$$
9. N; N+1; N+2
10. $5(N)+2(N+1) = 6(N+2)$
11. $$\begin{aligned} 5(N)+2(N+1) &= 6(N+2) \\ 5N+2N+2 &= 6N+12 \\ 7N+2 &= 6N+12 \\ 7N-6N &= 12-2 \\ N &= 10 \end{aligned}$$
    10; 11; 12
12. $$\begin{aligned} 5(10)+2(11) &= 6(12) \\ 50+22 &= 72 \\ 72 &= 72 \end{aligned}$$
13. N; N+2; N+4
14. $N+(N+4) = 3(N+2)+19$
15. $$\begin{aligned} N+(N+4) &= 3(N+2)+19 \\ 2N+4 &= 3N+6+19 \\ 2N-3N &= 25-4 \\ -N &= 21 \\ N &= -21 \end{aligned}$$
    −21; −19; −17
16. $$\begin{aligned} (-21)+(-17) &= 3(-19)+19 \\ -38 &= -57+19 \\ -38 &= -38 \end{aligned}$$

Systematic Review 17C

1. N; N+2; N+4
2. $5(N+4)-4(N) = 4(N+2)$
3. $$\begin{aligned} 5(N+4)-4(N) &= 4(N+2) \\ 5N+20-4N &= 4N+8 \\ 5N-4N-4N &= 8-20 \\ -3N &= -12 \\ N &= 4 \end{aligned}$$
   4; 6; 8
4. N; N+1; N+2
5. $6(N+1)+4(N) = 9(N+2)-4$
6. $$\begin{aligned} 6(N+1)+4(N) &= 9(N+2)-4 \\ 6N+6+4N &= 9N+18-4 \\ 6N+4N-9N &= 18-4-6 \\ N &= 8 \end{aligned}$$
   8; 9; 10
7. $N+D=11$
   $.05N+.10D=.80$
   $$\begin{aligned} (.05N+.10D=.80)(100) \Rightarrow&\quad 5N+10D = 80 \\ (N+D=11)(-5) \Rightarrow&\quad \underline{-5N-5D=-55} \\ &\quad 5D=25 \\ &\quad D=5 \end{aligned}$$
   $$\begin{aligned} N+D=11 \Rightarrow N+(5) &= 11 \\ N &= 6 \end{aligned}$$
   $$\begin{aligned} .05(6)+.10(5) &= .80 \\ .30+.50 &= .80 \\ .80 &= .80 \end{aligned}$$
8. $Y > 4X-4$
9. on graph

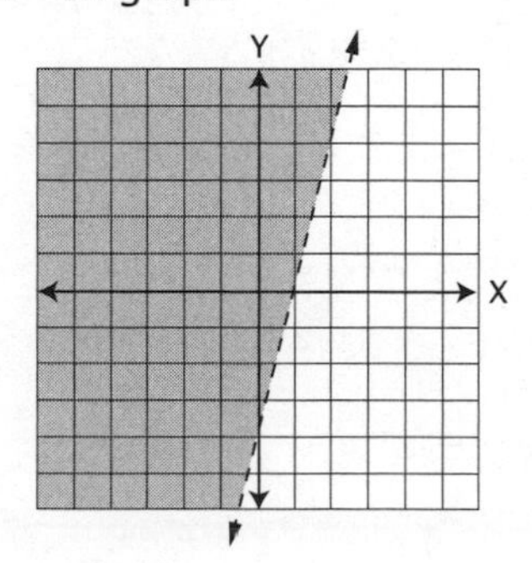

10. yes

11.
$$\begin{array}{rr} & 4Y + X = 11 \\ (2Y-3X=9)(-2) \Rightarrow & -4Y+6X=-18 \\ \hline & 7X = -7 \\ & X = -1 \end{array}$$

$$4Y+X=11 \Rightarrow 4Y+(-1)=11$$
$$4Y=12$$
$$Y=3$$

12. $4Y+X=11$
$$Y=-\frac{1}{4}X+\frac{11}{4}$$
$$(1)=-\frac{1}{4}(0)+b$$
$$1=b$$
$$Y=-\frac{1}{4}X+1 \text{ or } X+4Y=4$$

13. 21

14. $4(N+2)=23+N$

15.
$$4N+8=23+N$$
$$4N-N=23-8$$
$$3N=15$$
$$N=5$$

16.
$$[2(X-3)+1]\div X=$$
$$[2((5)-3)+1]\div(5)=$$
$$[2(2)+1]\div 5=$$
$$[4+1]\div 5=$$
$$5\div 5=1$$

17. $\frac{1}{2}+\frac{2}{3}=\frac{3}{6}+\frac{4}{6}=\frac{7}{6}=1\frac{1}{6}$

18. $.75\times 250=187.5$

19.
$$1.8-.16A=10$$
$$100(1.8)-100(.16A)=100(10)$$
$$180-16A=1000$$
$$-16A=820$$
$$A=\frac{820}{-16}=-51\frac{1}{4} \text{ or } -51.25$$

20. $2\times 2\times 2\times 2\times 2\times 3$

Systematic Review 17D

1. $N;\ N+2;\ N+4$

2. $4(N+4)+1=3(N)+2(N+2)$

3.
$$4(N+4)+1=3(N)+2(N+2)$$
$$4N+16+1=3N+2N+4$$
$$17-4=3N+2N-4N$$
$$13=N$$

13; 15; 17

4. $N;\ N+1;\ N+2$

5. $3(N)-5(N+1)=-1$

6.
$$3(N)-5(N+1)=-1$$
$$3N-5N-5=-1$$
$$-2N=-1+5$$
$$-2N=4$$
$$N=-2$$

$-2;\ -1;\ 0$

7. $N+D=15$

$.05N+.10D=1.10$

$$\begin{array}{rr} (.05N+.10D=1.10)(100) \Rightarrow & 5N+10D=110 \\ (N+D=15)(-5) \Rightarrow & -5N-\ 5D=-75 \\ \hline & 5D=\ 35 \\ & D=7 \end{array}$$

$$N+D=15 \Rightarrow N+(7)=15$$
$$N=8$$

$$.05(8)+.10(7)=1.10$$
$$.40+.70=1.10$$
$$1.10=1.10$$

8. $Y>X-2$

9. see graph below

10. yes

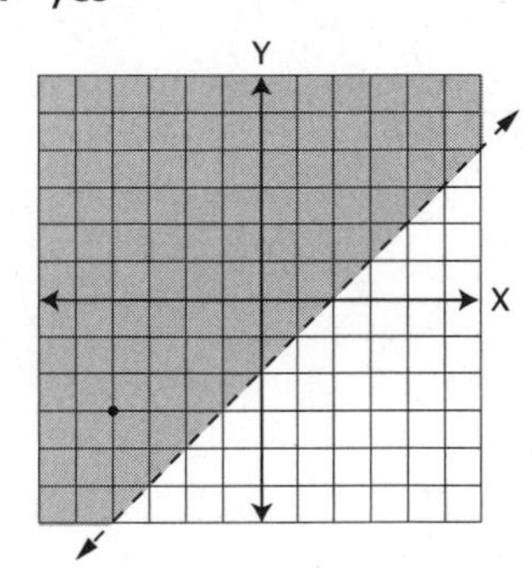

11. $$\begin{aligned} 5Y+6X&=-19\\ (5Y+3X=-22)(-1)\Rightarrow\quad -5Y-3X&=\ \ 22\\ \hline 3X&=\ \ 3\\ X&=1 \end{aligned}$$

$$\begin{aligned} 5Y+6X=-19\Rightarrow\ 5Y+6(1)&=-19\\ 5Y+6&=-19\\ 5Y&=-25\\ Y&=-5 \end{aligned}$$

$(1, -5)$

12. $$\begin{aligned} 5Y+6X=-19\Rightarrow\ Y&=-\frac{6}{5}X-\frac{19}{5}\\ (1)&=\frac{5}{6}(-4)+b\\ 1&=-\frac{20}{6}+b\\ b&=\frac{26}{6}=\frac{13}{3} \end{aligned}$$

$Y=\frac{5}{6}X+\frac{13}{3}$ or $5X-6Y=-26$ or $-5X+6Y=26$

13. $6=2\times3;\ 8=2\times2\times2;\ \text{LCM}=2\times2\times2\times3=24$
14. $2N-6+8N=4$
15. $$\begin{aligned} 2N-6+8N&=4\\ 10N&=10\\ N&=1 \end{aligned}$$
16. $$\begin{aligned} &3X-X^2+13-4X=\\ &3(1)-(1)^2+13-4(1)=\\ &3-1+13-4=11 \end{aligned}$$
17. $\frac{1}{6}+\frac{3}{4}=\frac{2}{12}+\frac{9}{12}=\frac{11}{12}$
18. $.13\times180=23.4$
19. $$\begin{aligned} 6A-16-4A&=20\\ 6A-4A&=20+16\\ 2A&=36\\ A&=18 \end{aligned}$$
20. $3\times3\times3\times5$

Systematic Review 17E

1. $N;\ N+2;\ N+4$
2. $5(N)+3(N+4)=7(N+2)+10$
3. $$\begin{aligned} 5(N)+3(N+4)&=7(N+2)+10\\ 5N+3N+12&=7N+14+10\\ 5N+3N-7N&=14+10-12\\ N&=12 \end{aligned}$$
   12; 14; 16
4. $N;\ N+1;\ N+2$
5. $7(N+2)-5(N+1)=4(N)+1$
6. $$\begin{aligned} 7(N+2)-5(N+1)&=4(N)+1\\ 7N+14-5N-5&=4N+1\\ 14-5-1&=4N-7N+5N\\ 8&=2N\\ 4&=N \end{aligned}$$
   4; 5; 6
7. $D+Q=7$
   $.10D+.25Q=1.00$
   $$\begin{aligned} (.10D+.25Q=1.00)(100)\Rightarrow\quad 10D+25Q&=100\\ (D+Q=7)(-10)\Rightarrow -10D-10Q&=-70\\ \hline 15Q&=\ \ 30\\ Q&=2 \end{aligned}$$
   $$\begin{aligned} D+Q=7\Rightarrow\quad D+(2)=7&\Rightarrow D=5\\ .10(5)+.25(2)&=1.00\\ .50+.50&=1.00\\ 1.00&=1.00 \end{aligned}$$
8. $Y\geq-2X+4$
9. on graph
10. yes

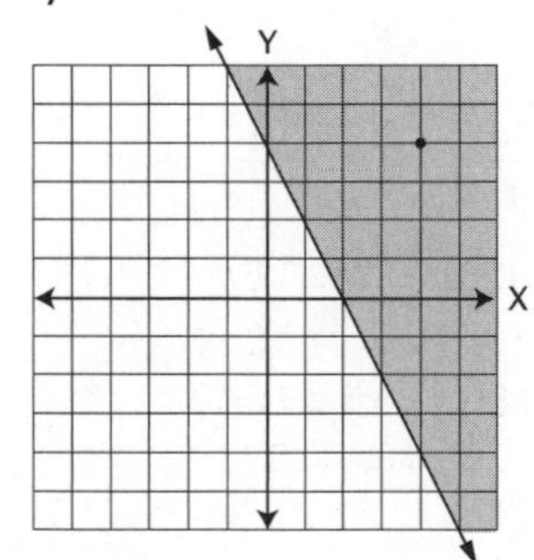

11. $Y+X=0\Rightarrow Y=-X$
    replace Y in equation 1
    with its equivalent, $(-X)$:
    $$\begin{aligned} (-X)-3X&=-4\\ -4X&=-4\\ X&=1 \quad (1,-1) \end{aligned}$$

12. $Y - 3X = -4 \Rightarrow Y = 3X - 4$
    $(-2) = 3(2) + b$
    $-2 = 6 + b$
    $b = -8$
    $Y = 3X - 8$ or $3X - Y = 8$ or $-3X + Y = -8$
13. 15's multiples: 15, 30, 45, 60, 75, 90
    25's multiples: 25, 50, 75, 100, 125
    first match, or LCM, is 75
14. $2[1+(-8)] = 18 - 2N$
15. $2[1+(-8)] = 18 - 2N$
    $2[-7] = 18 - 2N$
    $-14 = 18 - 2N$
    $-14 - 18 = -2N$
    $-32 = -2N$
    $16 = N$
16. $3Y - 5 + 6Y + (10 - 6) \Rightarrow$
    $3(16) - 5 + 6(16) + (10 - 6) =$
    $48 - 5 + 96 + 4 = 143$
17. $\frac{2}{7} - \frac{1}{14} = \frac{4}{14} - \frac{1}{14} = \frac{3}{14}$
18. $1.20 \times 100 = 120$
19. $.5X + 1.5 = 4.5$
    $10(.5X) + 10(1.5) = 10(4.5)$
    $5X + 15 = 45$
    $5X = 30$
    $X = 6$
20. $2 \times 3 \times 17$

Lesson Practice 18A

1. $15^2 = 15 \times 15 = 225$
2. $\sqrt{169} = 13$
3. $(-8)^2 = (-8)(-8) = 64$
4. $-\sqrt{100} = -10$
5. $16^2 = 16 \times 16 = 256$
6. $\sqrt{144} = 12$
7. $4^5 \cdot 4^2 = 4^{5+2} = 4^7$
8. $8^4 \cdot 8^7 = 8^{4+7} = 8^{11}$
9. $8^7 \div 8^3 = 8^{7-3} = 8^4$
10. $3^8 \cdot 3^4 = 3^{8+4} = 3^{12}$
11. $B^2B^3B^5 = B^{2+3+5} = B^{10}$
12. $C^1D^5D^4C^3D^2 = C^{1+3}D^{5+4+2} = C^4D^{11}$

13. $8^X \cdot 8^Y = 8^{X+Y}$
14. $M^{10X} \div M^{3X} = M^{10X-3X} = M^{7X}$
15. $8^9 \cdot 8^{10} \div 8^3 = 8^{9+10-3} = 8^{16}$
16. $X^{5Y} \div X^{2Y} = X^{5Y-2Y} = X^{3Y}$

Lesson Practice 18B

1. $25^2 = 25 \times 25 = 625$
2. $2^3 = 2 \times 2 \times 2 = 8$
3. $(-9)^2 = (-9)(-9) = 81$
4. $(7)^3 = 7 \times 7 \times 7 = 343$
5. $(-17)^2 = (-17)(-17) = 289$
6. $-\sqrt{81} = -9$
7. $5^3 \cdot 5^6 = 5^{3+6} = 5^9$
8. $6^4 \cdot 6^2 = 6^{4+2} = 6^6$
9. $18^{13} \div 18^9 = 18^{13-9} = 18^4$
10. $4^8 \cdot 4^5 = 4^{8+5} = 4^{13}$
11. $(4^2) = (4 \times 4) = 16$
12. $C^1C^2C^3 = C^{1+2+3} = C^6$
13. $F^3F^4E^5F^2 = E^5F^{3+4+2} = E^5F^9$
14. $B^6C^1C^3B^7 = B^{6+7}C^{1+3} = B^{13}C^4$
15. $Y^{10} \cdot Y^5 \div Y^3 = Y^{10+5-3} = Y^{12}$
16. $A^{8X} \div A^{3X} = A^{8X-3X} = A^{5X}$

Systematic Review 18C

1. $14^2 = 14 \times 14 = 196$
2. $\sqrt{121} = 11$
3. $(-9)^2 = (-9)(-9) = 81$
4. $-\sqrt{49} = -7$
5. $3^3 \cdot 3^3 = 3^{3+3} = 3^6$
6. $5^2 \cdot 5^6 = 5^{2+6} = 5^8$
7. $6^5 \div 6^2 = 6^{5-2} = 6^3$
8. $4^5 \cdot 4^2 = 4^{5+2} = 4^7$
9. $A^5A^2B^4B^1 = A^{5+2}B^{4+1} = A^7B^5$
10. $B^Y \cdot B^{2Y} = B^{Y+2Y} = B^{3Y}$
11. $A^5 \div A^1 = A^{5-1} = A^4$
12. $X^5 \cdot X^2 \div X^7 = X^{5+2-7} = X^0$ or 1

13. add
14. subtract
15. $5(N+2)-2(N)=4(N+1)-40$
16. $5(N+2)-2(N)=4(N+1)-40$
    $5N+10-2N=4N+4-40$
    $5N-2N-4N=4-40-10$
    $-N=-46$
    $N=46$

    46; 47; 48
17. $N+D=20$
    $.05N+.10D=1.60$

$$(.05N+.10D=1.60)(100) \Rightarrow 5N+10D=160$$
$$(N+D=20)(-5) \Rightarrow -5N-5D=-100$$
$$5D=60$$
$$D=12$$

$$N+D=20 \Rightarrow N+(12)=20$$
$$N=8$$

18. $6X+3Y=10$
    $3Y=-6X+10$
    $Y=-2X+\frac{10}{3}$
19. $Y=3X+2$
    $Y=X+4$
20. $Y=X+4 \Rightarrow (X+4)=3X+2$
    $4-2=3X-X$
    $2=2X$
    $1=X$

    $Y=X+4 \Rightarrow Y=(1)+4$
    $Y=5$

    In 1 year, they will have an equal height of 5 feet.

Systematic Review 18D

1. $-13^2=-(13\times13)=-169$
2. $-\sqrt{144}=-12$
3. $(-15)^2=(-15)(-15)=225$
4. $\sqrt{100}=10$
5. $7^3\cdot7^4\cdot7=7^{3+4+1}=7^8$
6. $2^8\cdot2^3\cdot2^2=2^{8+3+2}=2^{13}$
7. $X^2\cdot X^9=X^{2+9}=X^{11}$
8. $A^4A^5B^2=A^{4+5}B^2=A^9B^2$
9. $8^5\div8^3=8^{5-3}=8^2$
10. $10^5\div10=10^{5-1}=10^4$
11. $X^{10}\div X^4=X^{10-4}=X^6$
12. $X^{4Y}\cdot X^{3Y}\div X^Y=X^{4Y+3Y-Y}=X^{6Y}$
13. divide
14. multiply
15. $4(N+2)+3(N+4)=8(N)-11$
16. $4(N+2)+3(N+4)=8(N)-11$
    $4N+8+3N+12=8N-11$
    $8+12+11=8N-4N-3N$
    $31=N$

    31, 33, 35
17. $Q+D=7$
    $.25Q+.10D=1.60$

$$(.25Q+.10D=1.60)(100) \Rightarrow 25Q+10D=160$$
$$(Q+D=7)(-10) \Rightarrow -10Q-10D=-70$$
$$15Q=90$$
$$Q=6$$

$$Q+D=7 \Rightarrow (6)+D=7$$
$$D=1$$

18. $Y=37X+30$
19. $(215)=37X+30$
    $185=37X$
    $\frac{185}{37}=X=5$ weeks
20. $(326)=37X+30$
    $296=37X$
    $\frac{296}{37}=X=8$ weeks

Systematic Review 18E

1. $-11^2=-(11\times11)=-121$
2. $\sqrt{196}=14$
3. $7^2=7\times7=49$
4. $-\sqrt{225}=-15$
5. $A^2\cdot A^4=A^{2+4}=A^6$
6. $5^3\cdot5^4=5^{3+4}=5^7$
7. $A^2B^3B^6A^1C^2=A^{2+1}B^{3+6}C^2=A^3B^9C^2$
8. $X^4\div X^3=X^{4-3}=X^1=X$
9. $9^9\div9^3=9^{9-3}=9^6$

10. $11^4 \cdot 11^6 = 11^{4+6} = 11^{10}$
11. $D^{3X} \div D^{2X} = D^{3X-2X} = D^X$
12. $M^5 \cdot M^3 \div M^3 = M^{5+3-3} = M^5$
13. 10, −10
14. same base
15. $4(N+2) = 3(N) + 3(N+4)$
16. $4(N+2) = 3(N) + 3(N+4)$
$$4N + 8 = 3N + 3N + 12$$
$$8 - 12 = 3N + 3N - 4N$$
$$-4 = 2N$$
$$-2 = N \qquad -2;\ 0;\ 2$$
17. $Q + D = 10$
$$.25Q + .10D = 1.75$$
$$(.25Q + .10D = 1.75)(100) \Rightarrow 25Q + 10D = 175$$
$$(Q + D = 10)(-10) \Rightarrow -10Q - 10D = -100$$
$$15Q = 75$$
$$Q = 5$$
$$Q + D = 10 \Rightarrow (5) + D = 10$$
$$D = 5$$
18. $4X + Y = 16$
19. $Y = 25X + 50$
$$Y = 10X + 200$$
20. $Y = 10X + 200 \Rightarrow (25X + 50) = 10X + 200$
$$25X - 10X = 200 - 50$$
$$15X = 150$$
$$X = 10 \text{ hours}$$
$$Y = 10X + 200 \Rightarrow Y = 10(10) + 200$$
$$Y = 100 + 200$$
$$Y = 300 \text{ gizmos}$$

Lesson Practice 19A

1. $\frac{1}{4^2} = 4^{-2}$
2. $\frac{1}{7^2} = 7^{-2}$
3. $\frac{1}{4^{-3}} = 4^3$
4. $\frac{1}{3^{-2}} = 3^2$
5. $5^{-3} = \frac{1}{5^3}$
6. $10^{-10} = \frac{1}{10^7}$
7. $7^{-3} \cdot 7^{-8} = 7^{-3+(-8)} = 7^{-11}$
8. $6^{-2} \cdot 6^{-3} = 6^{-2+(-3)} = 6^{-5}$
9. $9^{-5} \div 9^{-2} = 9^{-5-(-2)} = 9^{-3}$
10. $3^{-8} \cdot 3^4 = 3^{-8+4} = 3^{-4}$
11. $B^{-2}B^3C^{-1}B^5C^{-5}C^1 =$
$$B^{-2+3+5}C^{-1-5+1} = B^6C^{-5}$$
12. $C^{-1}D^{-5}D^4C^3D^{-2}D^4C^1 =$
$$C^{-1+3+1}D^{-5+4-2+4} = C^3D^1$$
13. $(8^5)^4 = 8^{5\times4} = 8^{20}$
14. $(9^3)^5 = 9^{3\times5} = 9^{15}$
15. $\frac{A^{-1}B^2B^{-1}}{AB^{-3}} = A^{-1}A^{-1}B^2B^{-1}B^3$
$$= A^{-1+(-1)}B^{2+(-1)+3}$$
$$= A^{-2}B^4$$
16. $\frac{C^0B^{-3}C^3B}{C^{-3}B^4} = C^0C^3C^3B^{-3}B^1B^{-4}$
$$= C^{0+3+3}B^{-3+1+(-4)}$$
$$= C^6B^{-6}$$

Lesson Practice 19B

1. $\frac{1}{8^{-2}} = 8^2$
2. $\frac{1}{5^3} = 5^{-3}$
3. $7^{-1} = \frac{1}{7^1} = \frac{1}{7}$
4. $X^{-6} = \frac{1}{X^6}$
5. $4^{-8} \cdot 4^5 = 4^{-8+5} = 4^{-3}$
6. $6^{-4} \cdot 6^{-2} = 6^{-4+(-2)} = 6^{-6}$
7. $(3^{-3})^2 = 3^{-3\times2} = 3^{-6}$
8. $(A^4)^{-5} = A^{4\times-5} = A^{-20}$
9. $(4^{-2})^3 = 4^{-2\times3} = 4^{-6}$

10. $C^0D^{-5}D^6C^1C^2C^3 =$
$C^{0+1+2+3}D^{-5+6} = C^6D^1$

11. $E^0F^3F^4E^{-5}F^{-2}E^{-6} = E^{0+(-6)+(-5)}F^{3+4+(-2)}$
$= E^{-11}F^5$

12. $B^{-6}C^1C^2C^3C^{-4}B^7 = B^{-6+7}C^{1+2+3+(-4)} = B^1C^2 = BC^2$

13. $Y^{-10} \cdot Y^5 \div Y^3 = Y^{-10+5-3} = Y^{-8}$

14. $A^{8X} \div A^{3X} = A^{8X-3X} = A^{5X}$

15. $\dfrac{X^{-5}Y^2X^3Y^2}{Y^{-3}Y^4X^2} = X^{-5}Y^2X^3Y^2Y^3Y^{-4}X^{-2}$
$= X^{-5+3+(-2)}Y^{2+2+3+(-4)}$
$= X^{-4}Y^3$

16. $\dfrac{A^{-3}B^2A^5B^3}{B^4A^{-3}A^5} = A^{-3}B^2A^5B^3B^{-4}A^3A^{-5}$
$= A^{-3+5+3+(-5)}B^{2+3+(-4)}$
$= A^0B^1 = 1B = B$

13. $(-8)^2 = (-8)(-8) = 64$

14. $\sqrt{25} = 5$

15. $\dfrac{E^{-1}F^2F^3E^4}{F^{-2}E^{-3}E^5} = E^{-1}F^2F^3E^4F^2E^3E^{-5}$
$= E^{-1+4+3+(-5)}F^{2+3+2}$
$= E^1F^7 = EF^7$

16. $1\times10^3 + 3\times10^2 + 7\times10^0 + 8\times10^{-2} =$
$1000 + 300 + 7 + .08 = 1,307.08$

17. $3(N) + 4(N+2) = -13(N+4)$

18. $3(N) + 4(N+2) = -13(N+4)$
$3N + 4N + 8 = -13N - 52$
$3N + 4N + 13N = -52 - 8$
$20N = -60$
$N = -3$ $\quad$ $-3;\ -1;\ 1$

19. $N + D = 7$
$.05N + .10D = .45$
$(.05N + .10D = .45)(100) \Rightarrow \quad 5N + 10D = 45$
$(N + D = 7)(-5) \Rightarrow \quad -5N - 5D = -35$
$5D = 10$
$D = 2$

$N + D = 7 \Rightarrow N + (2) = 7$
$N = 5$

20. $5X + 10Y - 20 = 0$
$10Y = -5X + 20$
$Y = -\frac{1}{2}X + 2$

Systematic Review 19C

1. $\dfrac{1}{3^2} = 3^{-2}$
2. $2^{-4} = \dfrac{1}{2^4}$
3. $\dfrac{1}{7^{-2}} = 7^2$
4. $Y^{-5} = \dfrac{1}{Y^5}$
5. $4^5 \cdot 4^{-2} = 4^{5+(-2)} = 4^3$
6. $5^{-2} \cdot 5^{-6} = 5^{-2+(-6)} = 5^{-8}$
7. $A^{-8}B^{-2}A^3A^4B^5 =$
$A^{-8+3+4}B^{-2+5} = A^{-1}B^3$
8. $D^{-2}C^3C^4D^4C^{-2}D^4 =$
$D^{-2+4+4}C^{3+4+(-2)} = C^5D^6$
9. $4^{-10} \cdot 4^6 = 4^{-10+6} = 4^{-4}$
10. $X^5 \div X^4 = X^{5-4} = X^1 = X$
11. $(3^3)^2 = 3^{3\times2} = 3^6$
12. $(2^5)^7 = 2^{5\times7} = 2^{35}$

Systematic Review 19D

1. $\dfrac{1}{4^{-5}} = 4^5$
2. $5^{-8} = \dfrac{1}{5^8}$
3. $\dfrac{1}{X^5} = X^{-5}$
4. $A^{-1} = \dfrac{1}{A^1} = \dfrac{1}{A}$
5. $X^A \cdot X^B = X^{A+B}$
6. $3^{-2} \cdot 3^8 = 3^{-2+8} = 3^6$
7. $E^0F^5E^{-1}F^{-2}E^3F^3 = E^{0+(-1)+3}F^{5+(-2)+3} = E^2F^6$

8. $C^{-8}B^{5}C^{1}C^{2}B^{-6}C^{4} = B^{5+(-6)}C^{-8+1+2+4}$
$= C^{-1}B^{-1}$

9. $7^{-3} \div 7^{-6} = 7^{-3-(-6)} = 7^{3}$

10. $X^{10Y} \div X^{5Y} = X^{10Y-5Y} = X^{5Y}$

11. $\left(10^{3}\right)^{4} = 10^{3\times4} = 10^{12}$

12. $(1{,}000)^{5} = \left(10^{3}\right)^{5} = 10^{15}$

13. $-5^{2} = -(5\times5) = -25$

14. $-\sqrt{36} = -6$

15. $\dfrac{C^{5}D^{4}D^{-3}}{D^{-2}C^{1}C^{-3}D^{4}} = C^{5}D^{4}D^{-3}D^{2}C^{-1}C^{3}D^{-4}$
$= C^{5+(-1)+3}D^{4+(-3)+2+(-4)}$
$= C^{7}D^{-1}$

16. $2\times10^{4} + 5\times10^{1} + 6\times10^{-1} + 9\times10^{-2} =$
$20{,}000 + 50 + .6 + .09 = 20{,}050.69$

17. $3(N) + 6(N+2) = 8(N+4) - 14$

18. $3(N) + 6(N+2) = 8(N+4) - 14$
$3N + 6N + 12 = 8N + 32 - 14$
$3N + 6N - 8N = 32 - 14 - 12$
$N = 6$

6; 8; 10

19. $Q + D = 11$
$.25Q + .10D = 2.15$

$(.25Q + .10D = 2.15)(100) \Rightarrow 25Q + 10D = 215$
$(Q + D = 11)(-10) \Rightarrow -10Q - 10D = -110$
$15Q = 105$
$Q = 7$

$Q + D = 11 \Rightarrow (7) + D = 11$
$D = 4$

20. $Y - X = 0 \Rightarrow Y = X$
$Y - 3X = -4 \Rightarrow (X) - 3X = -4$
$-2X = -4$
$X = 2$

$Y - X = 0 \Rightarrow Y - (2) = 0$
$Y = 2$

Systematic Review 19E

1. $\dfrac{1}{7^{-3}} = 7^{3}$

2. $10^{-7} = \dfrac{1}{10^{7}}$

3. $\dfrac{1}{A^{X}} = A^{-X}$

4. $8^{-X} = \dfrac{1}{8^{X}}$

5. $A^{2}\cdot A^{-4} = A^{2+(-4)} = A^{-2}$

6. $5^{6} \div 5^{4} = 5^{6-4} = 5^{2}$

7. $10^{11}\cdot10^{-3} \div 10^{5} = 10^{11+(-3)-5} = 10^{3}$

8. $D^{2}C^{-3}C^{-4}D^{8}C^{2}D^{-4} =$
$C^{-3+(-4)+2}D^{2+8+(-4)} = C^{-5}D^{6}$

9. $M^{-X}\cdot M^{X} = M^{-X+X} = M^{0} = 1$

10. $X^{2Y} \div X^{4Y} = X^{2Y-4Y} = X^{-2Y}$

11. $\left[\left(11^{2}\right)^{5}\right]^{3} = 11^{2\times5\times3} = 11^{30}$

12. $(49)^{3} = \left(7^{2}\right)^{3} = 7^{6}$

13. $(15)^{2} = 15\times15 = 225$

14. $\sqrt{81} = 9$

15. $\dfrac{X^{1}Y^{2}X^{4}Y^{-1}}{X^{-3}Y^{4}} = X^{1}Y^{2}X^{4}Y^{-1}X^{3}Y^{-4}$
$= X^{1+4+3}Y^{2+(-1)+(-4)}$
$= X^{8}Y^{-3}$

16. $4.093 = 4\times10^{0} + 9\times10^{-2} + 3\times10^{-3}$

17. $2(N) + 3(N+1) - (N+2) = 21$

18. $2(N) + 3(N+1) - (N+2) = 21$
$2N + 3N + 3 - N - 2 = 21$
$2N + 3N - N = 21 - 3 + 2$
$4N = 20$
$N = 5$

5; 6; 7

19. $Q + N = 30$

$.25Q + .05N = 4.30$

$$\begin{aligned}(.25Q + .05N = 4.30)(100) &\Rightarrow 25Q + 5N = 430\\ (Q + N = 30)(-5) &\Rightarrow \underline{-5Q - 5N = -150}\\ &\qquad 20Q = 280\\ &\qquad Q = 14\end{aligned}$$

$$\begin{aligned}Q + N = 30 \Rightarrow (14) + N &= 30\\ N &= 16\end{aligned}$$

20. $Y = -2X + 9$

$2X + Y = 9$

Lesson Practice 20A

1. $X^2 + 11X + 2$

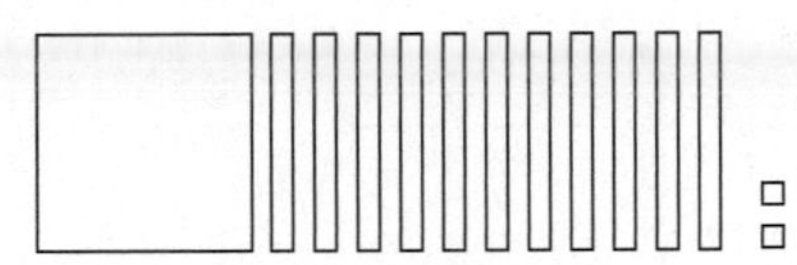

2. $X^2 + 6X + 8$

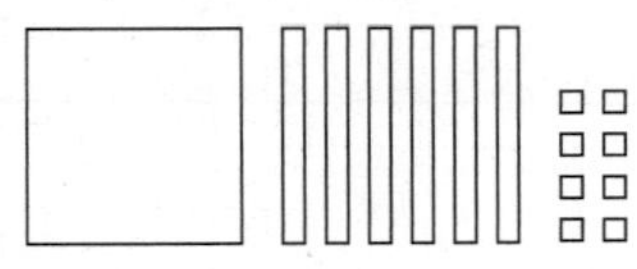

3. $X^2 - 8$

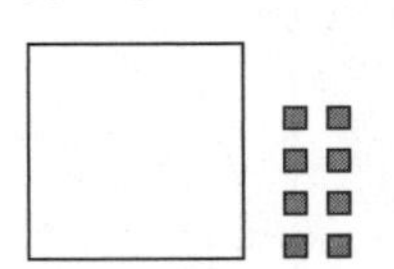

4.

$$\begin{array}{r} X^2 - 6X + 3\\ \underline{3X^2 + 7X - 9}\\ 4X^2 + X - 6\end{array}$$

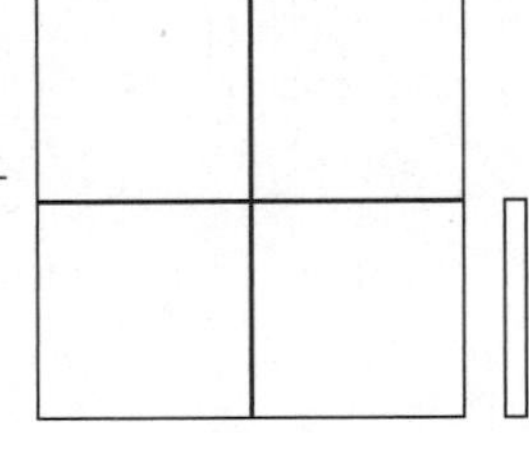

5.

$$\begin{array}{r} X^2 \qquad - 8\\ \underline{X^2 + 6X - 7}\\ 2X^2 + 6X - 15\end{array}$$

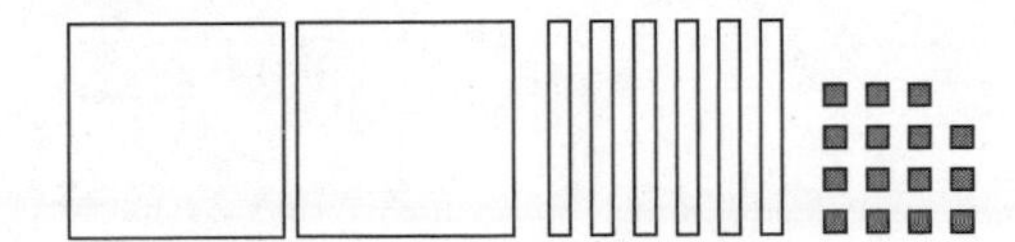

6.

$$\begin{array}{r} 2X^2 + 10X + 7\\ \underline{2X^2 - 8X - 9}\\ 4X^2 + 2X - 2\end{array}$$

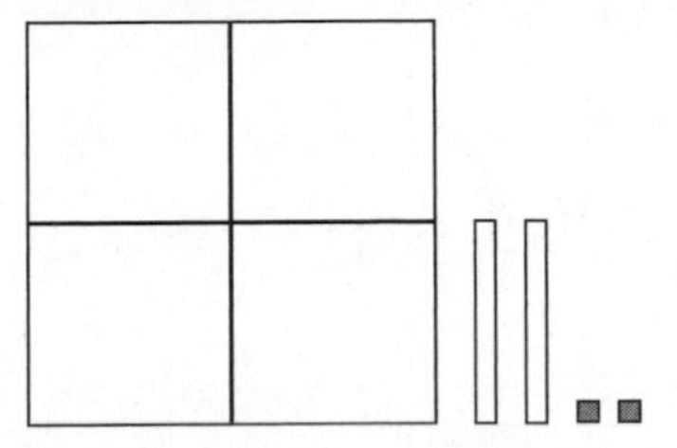

7. $(X + 1)(X + 2) = X^2 + 3X + 2$

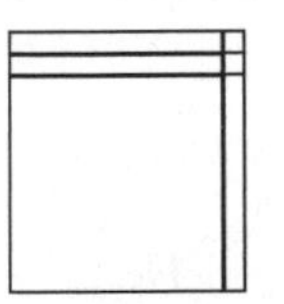

8. $(X + 4)(X + 3) = X^2 + 7X + 12$

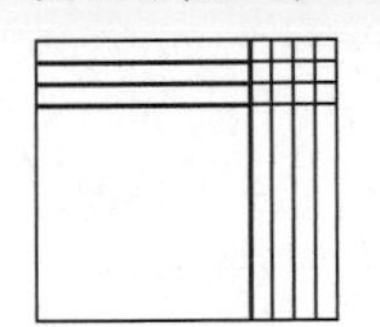

9. $(X + 1)(X + 5) = X^2 + 6X + 5$

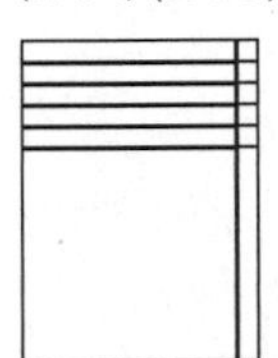

10.

$$\begin{array}{r} 3X + 2\\ \underline{\times X + 1}\\ 3X + 2\\ \underline{3X^2 + 2X \quad}\\ 3X^2 + 5X + 2\end{array}$$

11.

$$\begin{array}{r} 5X + 5\\ \underline{\times X + 2}\\ 10X + 10\\ \underline{5X^2 + 5X \quad}\\ 5X^2 + 15X + 10\end{array}$$

12.

$$\begin{array}{r} 2X + 1\\ \underline{\times X + 5}\\ 10X + 5\\ \underline{2X^2 + X \quad}\\ 2X^2 + 11X + 5\end{array}$$

13.

$$\begin{array}{r} X + 8\\ \underline{\times 3X + 5}\\ 5X + 40\\ \underline{3X^2 + 24X \quad}\\ 3X^2 + 29X + 40\end{array}$$

14.
$$\begin{array}{r} X+3 \\ \times\ 2X+1 \\ \hline X+3 \\ 2X^2+6X \quad \\ \hline 2X^2+7X+3 \end{array}$$

15.
$$\begin{array}{r} 3X+2 \\ \times\ 2X+1 \\ \hline 3X+2 \\ 6X^2+4X \quad \\ \hline 6X^2+7X+2 \end{array}$$

16.
$$\begin{array}{r} 4X+2 \\ \times\ X+3 \\ \hline 12X+6 \\ 4X^2+\ 2X \quad \\ \hline 4X^2+14X+6 \end{array}$$

17.
$$\begin{array}{r} 2X-5 \\ \times\ X+2 \\ \hline 4X-10 \\ 2X^2-5X \quad \\ \hline 2X^2-\ X-10 \end{array}$$

18.
$$\begin{array}{r} 3X+5 \\ \times\ 3X-1 \\ \hline -3X-5 \\ 9X^2-15X \quad \\ \hline 9X^2+12X-5 \end{array}$$

Lesson Practice 20B

1. X^2-3X-7

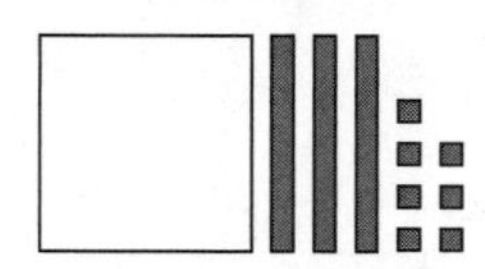

2. $2X^2-7X-3$

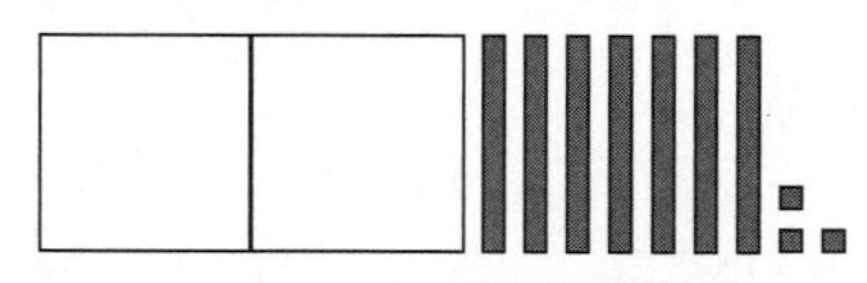

3. X^2+5X+9

4.
$$\begin{array}{r} X^2+3X+\ 2 \\ X^2+7X+12 \\ \hline 2X^2+10X+14 \end{array}$$

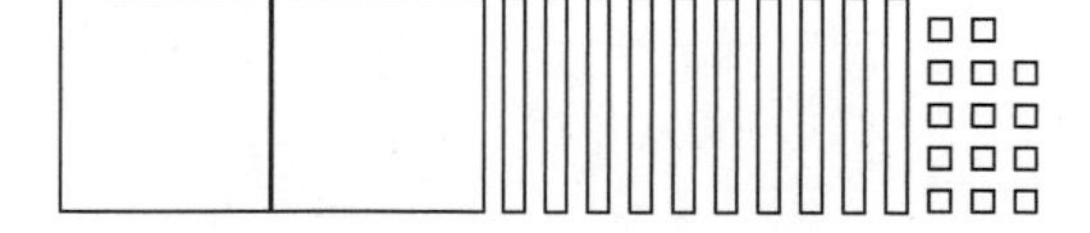

5.
$$\begin{array}{r} X^2+6X+5 \\ 3X^2-\ X-2 \\ \hline 4X^2+5X+3 \end{array}$$

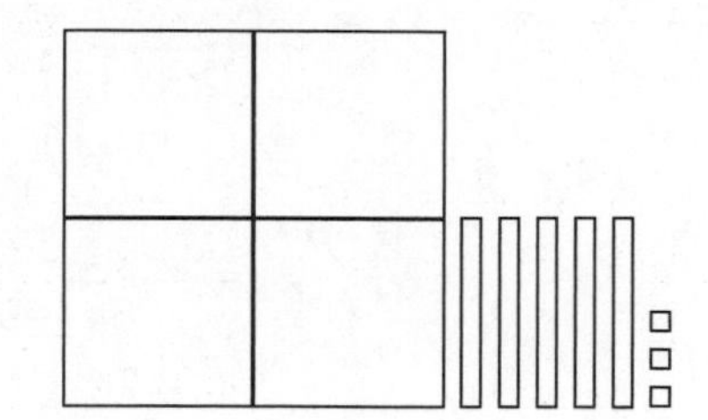

6.
$$\begin{array}{r} 5X^2-\ 5X-10 \\ 2X^2+11X+\ 5 \\ \hline 7X^2+\ 6X-\ 5 \end{array}$$

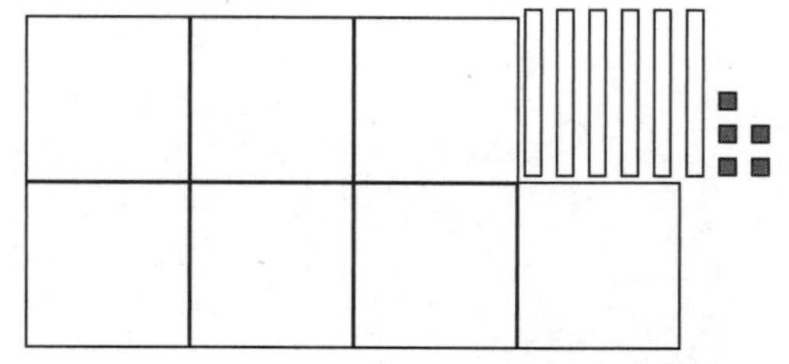

7. $(X+4)(X+5)=X^2+9X+20$

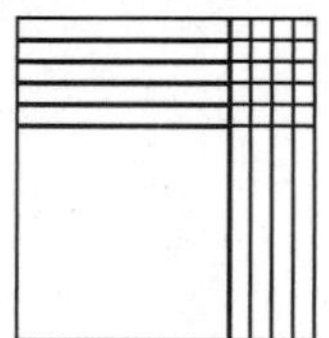

8. $(X+7)(X+3)=X^2+10X+21$

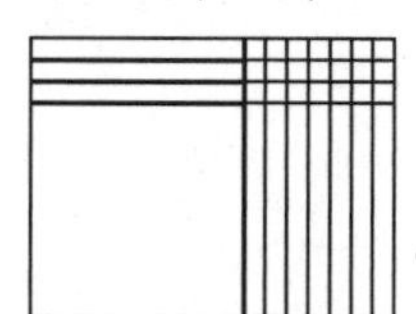

9. $(X+4)(X+8)=X^2+12X+32$

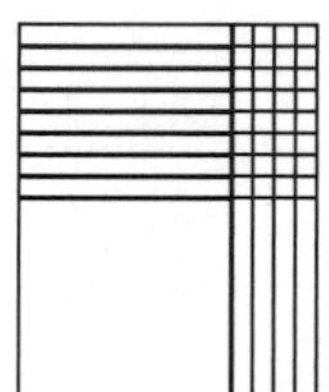

10.
$$\begin{array}{r} 7X+1 \\ \times\ X+2 \\ \hline 14X+2 \\ 7X^2+\quad X \quad \\ \hline 7X^2+15X+2 \end{array}$$

11.
$$\begin{array}{r} 3X+\ 7 \\ \times\ X+\ 6 \\ \hline 18X+42 \\ 3X^2+\ 7X \quad \\ \hline 3X^2+25X+42 \end{array}$$

12.
$$\begin{array}{r} 2X+8 \\ \times\ 3X+1 \\ \hline 2X+8 \\ 6X^2+24X \\ \hline 6X^2+26X+8 \end{array}$$

13.
$$\begin{array}{r} X+8 \\ \times\ X-3 \\ \hline -3X-24 \\ X^2+8X \\ \hline X^2+5X-24 \end{array}$$

14.
$$\begin{array}{r} 2X-1 \\ \times\ X+9 \\ \hline 18X-9 \\ 2X^2+\ \ X \\ \hline 2X^2+17X-9 \end{array}$$

15.
$$\begin{array}{r} 3X+5 \\ \times\ X+2 \\ \hline 6X+10 \\ 3X^2+\ 5X \\ \hline 3X^2+11X+10 \end{array}$$

16.
$$\begin{array}{r} 4X-2 \\ \times\ X-3 \\ \hline -12X+6 \\ 4X^2-\ 2X \\ \hline 4X^2-14X+6 \end{array}$$

17.
$$\begin{array}{r} 5X+2 \\ \times\ 3X-3 \\ \hline -15X-6 \\ 15X^2-6X \\ \hline 15X^2-9X-6 \end{array}$$

18.
$$\begin{array}{r} 3X+7 \\ \times\ 4X+2 \\ \hline 6X+14 \\ 12X^2+28X \\ \hline 12X^2+34X+14 \end{array}$$

Systematic Review 20C

1.
$$\begin{array}{r} 3X^2+7X+6 \\ X^2+2X+3 \\ \hline 4X^2+9X+9 \end{array}$$

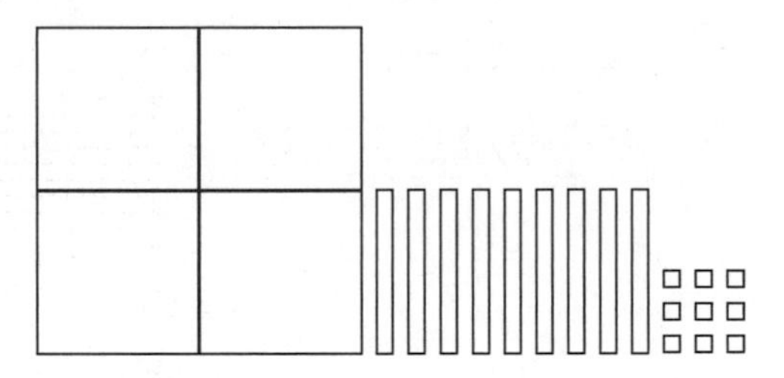

2.
$$\begin{array}{r} 2X^2+5X+1 \\ X^2+3X+4 \\ \hline 3X^2+8X+5 \end{array}$$

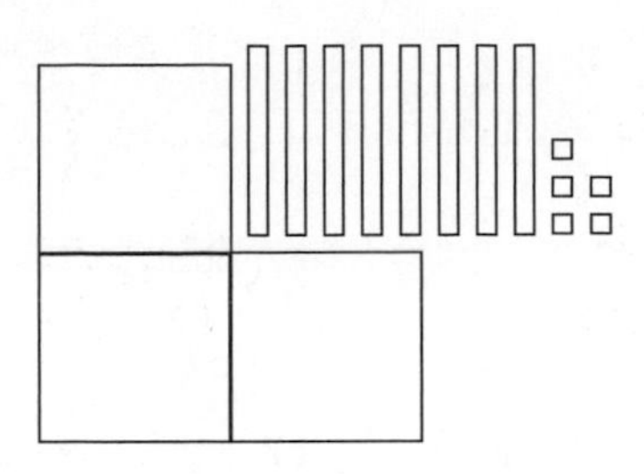

3.
$$\begin{array}{r} 4X^2+8X+2 \\ -X^2+3X-1 \\ \hline 3X^2+11X+1 \end{array}$$

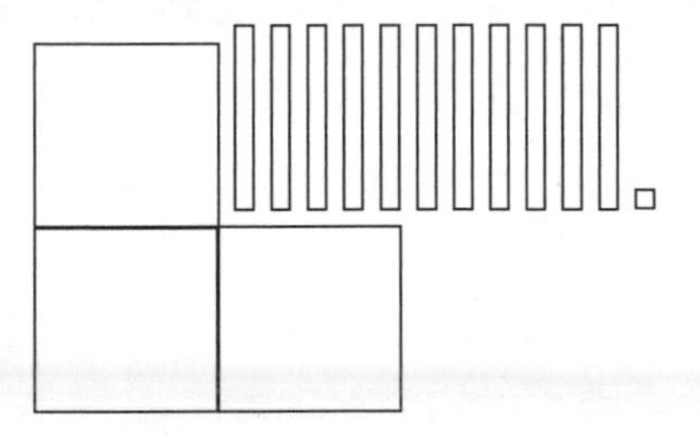

4. $(X+4)(X+8)=X^2+12X+32$

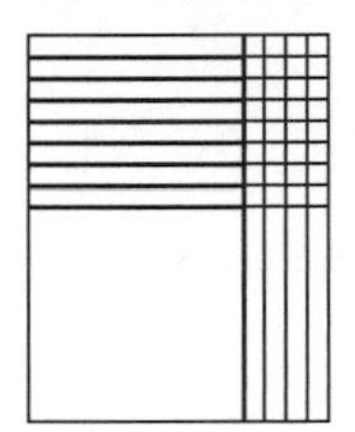

5. $(X+5)(X+2)=X^2+7X+10$

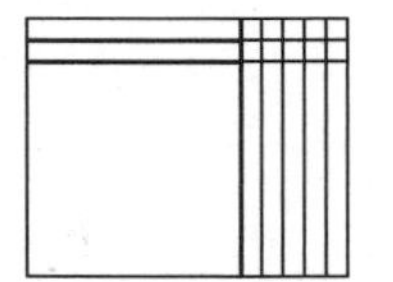

6. $(X+2)(X+6)=X^2+8X+12$

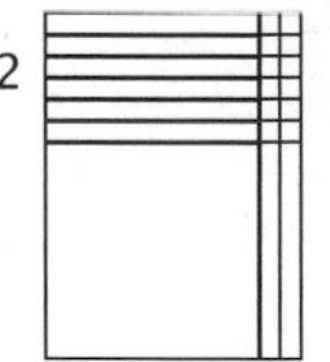

7.
$$\begin{array}{r} 3X+6 \\ \times\ X+2 \\ \hline 6X+12 \\ 3X^2+\ 6X \\ \hline 3X^2+12X+12 \end{array}$$

8.
$$\begin{array}{r} 2X+5 \\ \times\, X+3 \\ \hline 6X+15 \\ 2X^2+\ 5X \\ \hline 2X^2+11X+15 \end{array}$$

9.
$$\begin{array}{r} 4X-5 \\ \times\, X+1 \\ \hline 4X-5 \\ 4X^2-5X \\ \hline 4X^2-\ X-5 \end{array}$$

10. $\frac{1}{X^{-4}} = X^4$

11. $X^{-3} = \frac{1}{X^3}$

12. $5^2 \times 3^0 \times 5^{-4} = 5^{2+(-4)} \times 1 = 5^{-2}$

13. $A^4 \div A^7 = A^{4-7} = A^{-3}$

14. $(5^2)^5 = 5^{2\times5} = 5^{10}$

15. $(5)^{12} = (5)^{3\times4} = (5^3)^4$

16. $\sqrt{196} = 14$

17. $C^{-5} \times C^2 = C^{-5+2} = C^{-3}$

18.
$$\begin{array}{r} X+4 \\ \times\, X+5 \\ \hline 5X+20 \\ X^2+4X \\ \hline X^2+9X+20 \end{array}$$

19. $A = X^2+9X+20 = (6)^2+9(6)+20$
$= 36+54+20 = 110$ square units

20. $(X+4)(2) \Rightarrow$ $2X+8$
$(X+5)(2) \Rightarrow$ $\times\, 2X+10$
$$\begin{array}{r} 2X\ +8 \\ \times\, 2X+10 \\ \hline 20X+80 \\ 4X^2+16X \\ \hline 4X^2+36X+80 \end{array}$$

Systematic Review 20D

1.
$$\begin{array}{r} X^2-3X-7 \\ 2X^2+4X-4 \\ \hline 3X^2+\ X-11 \end{array}$$

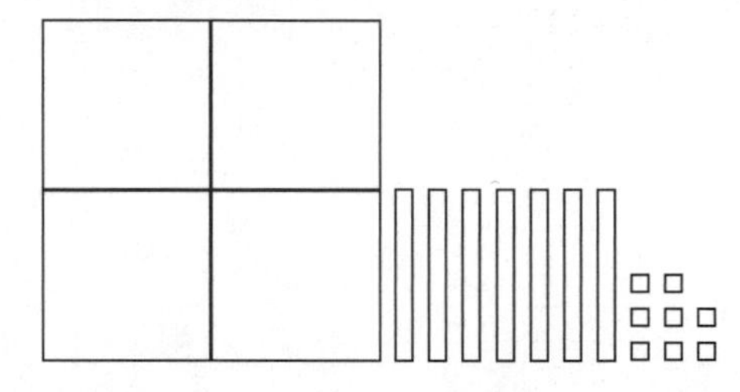

2.
$$\begin{array}{r} X^2+11X+2 \\ 3X^2-\ 4X+6 \\ \hline 4X^2+\ 7X+8 \end{array}$$

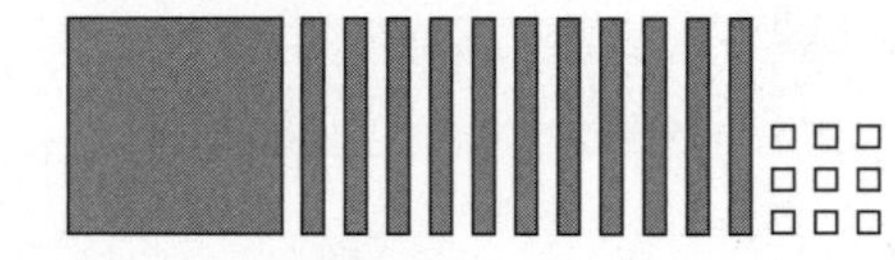

3.
$$\begin{array}{r} X^2-10X-\ 5 \\ -2X^2-\ \ X+14 \\ \hline -X^2-11X+\ 9 \end{array}$$

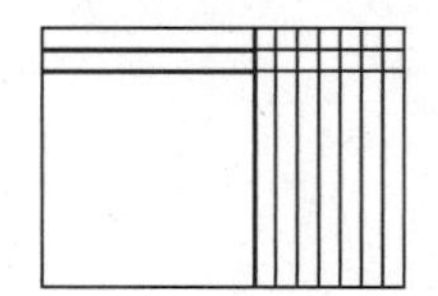

4. $(X+2)(X+7) = X^2+9X+14$

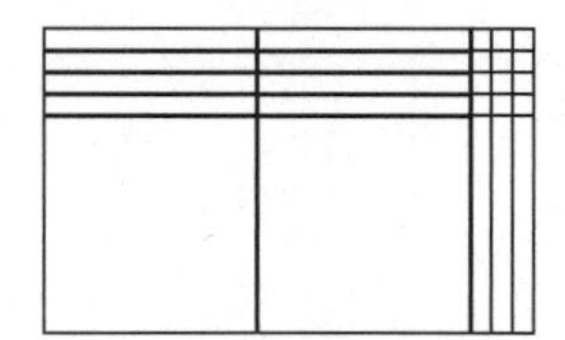

5. $(2X+3)(X+4) = 2X^2+11X+12$

6. $(X+1)(X+9) = X^2+10X+9$

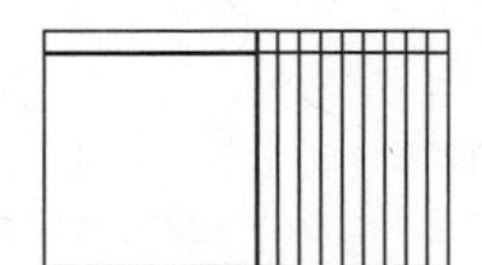

7.
$$\begin{array}{r} 2X+4 \\ \times X+3 \\ \hline 6X+12 \\ 2X^2+\ 4X \quad \\ \hline 2X^2+10X+12 \end{array}$$

8.
$$\begin{array}{r} 3X-1 \\ \times X+4 \\ \hline 12X-4 \\ 3X^2-\ X \quad \\ \hline 3X^2+11X-4 \end{array}$$

9.
$$\begin{array}{r} 2X-3 \\ \times X-4 \\ \hline -8X+12 \\ 2X^2-3X \quad \\ \hline 2X^2-11X+12 \end{array}$$

10. $\frac{1}{X^4}=X^{-4}$

11. $\frac{1}{Y^{-5}}=Y^5$

12. $3^7\times4^3\times4^{-2}=3^74^{3+(-2)}=3^74^1$ or $3^7\times4$

13. $B^5\div B^1=B^{5-1}=B^4$

14. $(8^3)^6=8^{3\times6}=8^{18}$

15. $(2)^{15}=(2)^{3\times5}=(2^3)^5$

16. $\sqrt{225}=15$

17. $D^{-3}\times D^8\times D^{-7}=D^{-3+8+(-7)}=D^{-2}$

18.
$$\begin{array}{r} 2X+4 \\ \times X+4 \\ \hline 8X+16 \\ 2X^2+\ 4X \quad \\ \hline 2X^2+12X+16 \end{array}$$

19. $A=2X^2+12X+16=$

$2(10)^2+12(10)+16=$

$2(100)+120+16=$

$200+120+16=336$ square units

20.
$$\begin{array}{r} 2X^2+12X+16 \\ X^2+\ 3X+\ 1 \\ \hline 3X^2+15X+17 \end{array}$$

Systematic Review 20E

1.
$$\begin{array}{r} X^2+3X-2 \\ X^2+4X+3 \\ \hline 2X^2+7X+1 \end{array}$$

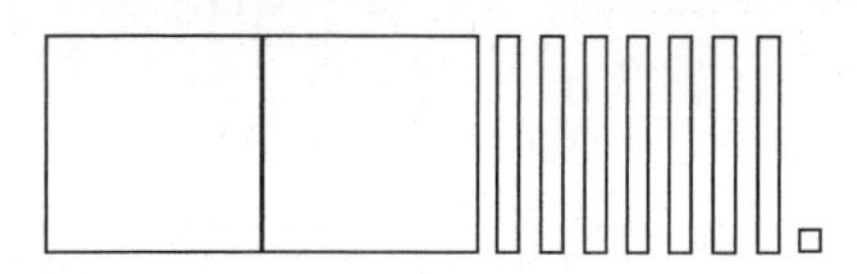

2.
$$\begin{array}{r} 3X^2+2X-1 \\ 2X^2-2X+8 \\ \hline 5X^2 \qquad +7 \end{array}$$

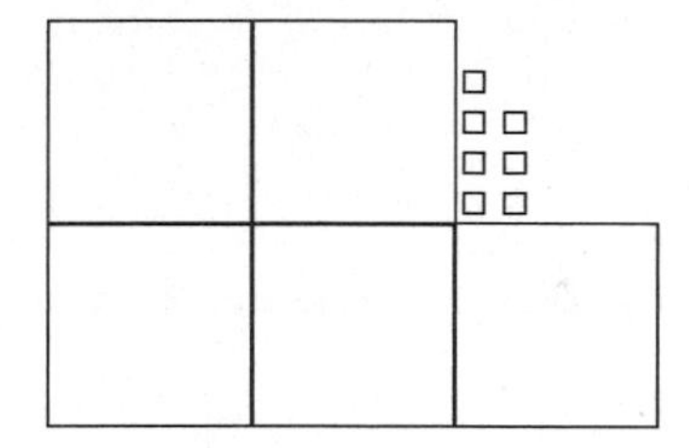

3.
$$\begin{array}{r} 5X^2+4X+\ 7 \\ -X^2+3X+\ 7 \\ \hline 4X^2+7X+14 \end{array}$$

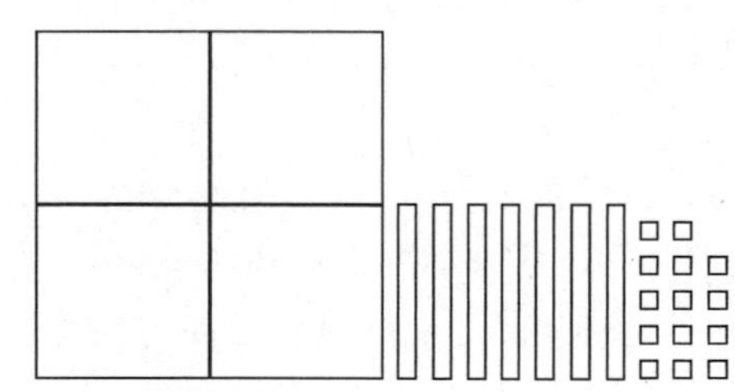

4. $(X+3)(X+3)=X^2+6X+9$

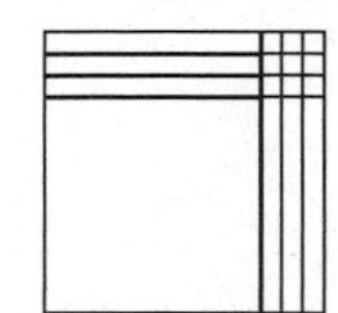

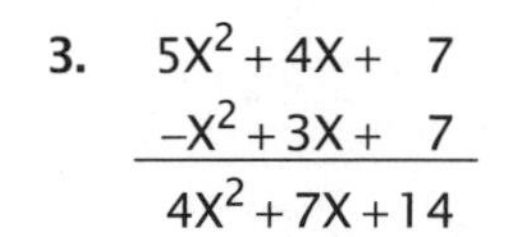

5. $(2X+4)(X+2)=2X^2+8X+8$

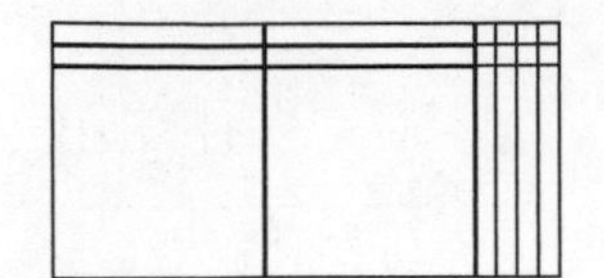

6. $(3X)(X+2)=3X^2+6X$

7.
$$\begin{array}{r} 2X-3 \\ \times\ X-2 \\ \hline -4X+6 \\ 2X^2-3X\quad \\ \hline 2X^2-7X+6 \end{array}$$

8.
$$\begin{array}{r} X-1 \\ \times\ X-6 \\ \hline -6X+6 \\ X^2-\ X\quad \\ \hline X^2-7X+6 \end{array}$$

9.
$$\begin{array}{r} 2X+2 \\ \times\ X-3 \\ \hline -6X-6 \\ 2X^2+2X\quad \\ \hline 2X^2-4X-6 \end{array}$$

10. $\frac{1}{X^5}=X^{-5}$

11. $Y^{-2}=\frac{1}{Y^2}$

12. $7^{-2}\times7^5\div7^{-2}=7^{-2+5-(-2)}=7^5$

13. $A^7\div B^3=A^7B^{-3}$

14. $(5^2)^5=5^{2\times5}=5^{10}$

15. $(5)^{12}=(5)^{3\times4}=(5^3)^4$

16. $-\sqrt{169}=-13$

17. $C^0C^{-4}D^8D^{-7}D^{-3}C^3=C^{0+(-4)+3}D^{8+(-7)+(-3)}$
$=C^{-1}D^{-2}$

18.
$$\begin{array}{r} 3N+4 \\ +\ 2N+5 \\ \hline 5N+9 \end{array}$$

19. $5N+9=5(10)+9=50+9=59$

20.
$$\begin{array}{r} 2Y+\ 7 \\ \times\ 7Y+\ 5 \\ \hline 10Y+35 \\ 14Y^2+49Y\quad\quad \\ \hline 14Y^2+59Y+35 \end{array}$$

Lesson Practice 21A

1.
$$\begin{array}{r} X+2 \\ \times\ X+2 \\ \hline 2X+4 \\ X^2+2X\quad \\ \hline X^2+4X+4 \end{array}$$

(X + 2) (X + 2)

2.
$$\begin{array}{r} X+3 \\ \times\ X+2 \\ \hline 2X+6 \\ X^2+3X\quad \\ \hline X^2+5X+6 \end{array}$$

(X + 2) (X + 3)

3.
$$\begin{array}{r} X+10 \\ \times\ X+\ 1 \\ \hline X+10 \\ X^2+10X\quad \\ \hline X^2+11X+10 \end{array}$$

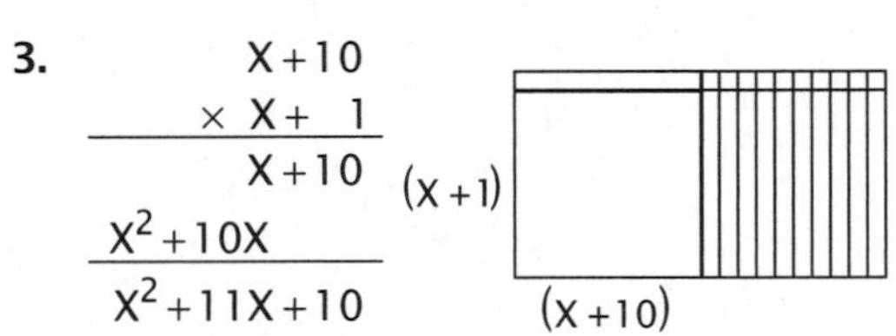

4.
$$\begin{array}{r} X+4 \\ \times\ X+2 \\ \hline 2X+8 \\ X^2+4X\quad \\ \hline X^2+6X+8 \end{array}$$

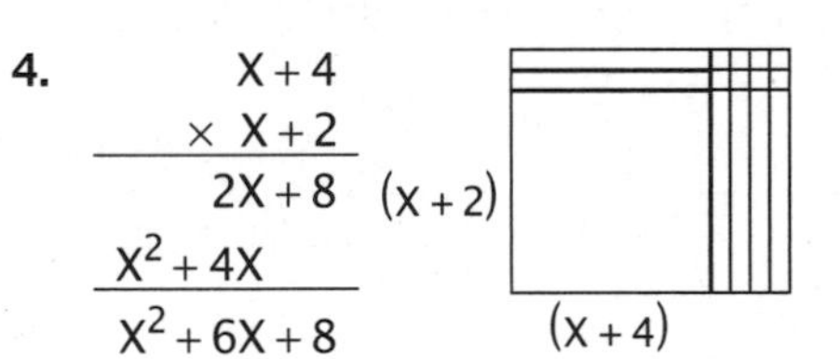

5.
$$\begin{array}{r} X+7 \\ \times\ X+1 \\ \hline X+7 \\ X^2+7X\quad \\ \hline X^2+8X+7 \end{array}$$

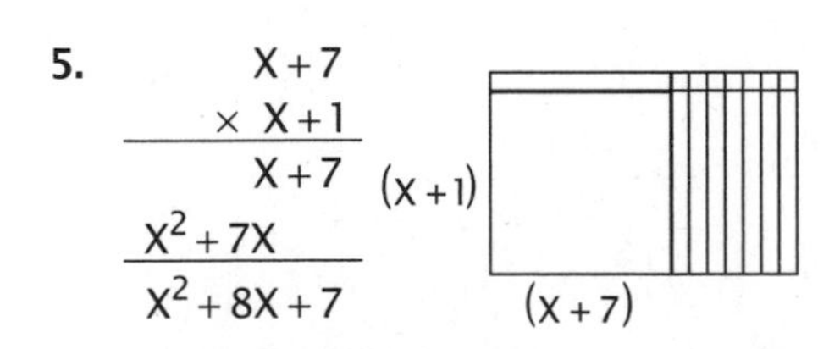

6.
$$\begin{array}{r} X+6 \\ \times\ X+2 \\ \hline 2X+12 \\ X^2+6X \\ \hline X^2+8X+12 \end{array}$$

(X+2) (X+6)

7.
$$\begin{array}{r} X+11 \\ \times\ X+1 \\ \hline X+11 \\ X^2+11X \\ \hline X^2+12X+11 \end{array}$$

(X+1) (X+11)

8.
$$\begin{array}{r} X+6 \\ \times\ X+1 \\ \hline X+6 \\ X^2+6X \\ \hline X^2+7X+6 \end{array}$$

(X+1) (X+6)

9.
$$\begin{array}{r} X+7 \\ \times\ X+2 \\ \hline 2X+14 \\ X^2+7X \\ \hline X^2+9X+14 \end{array}$$

(X+2) (X+7)

10.
$$\begin{array}{r} X+15 \\ \times\ X+1 \\ \hline X+15 \\ X^2+15X \\ \hline X^2+16X+15 \end{array}$$

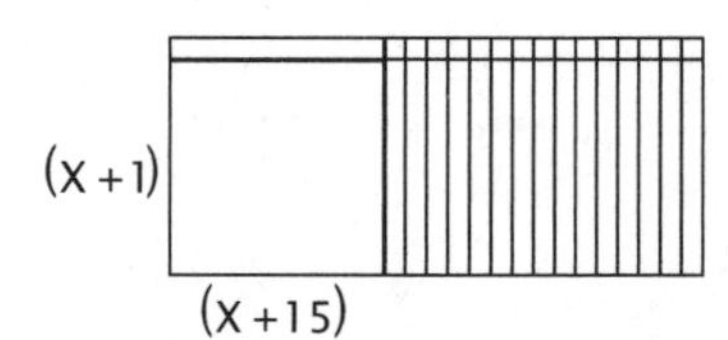

11.
$$\begin{array}{r} X+2 \\ \times\ X+1 \\ \hline X+2 \\ X^2+2X \\ \hline X^2+3X+2 \end{array}$$

(X+1) (X+2)

12.
$$\begin{array}{r} X+3 \\ \times\ X+1 \\ \hline X+3 \\ X^2+3X \\ \hline X^2+4X+3 \end{array}$$

(X+1) (X+3)

13.
$$\begin{array}{r} X+8 \\ \times\ X+1 \\ \hline X+8 \\ X^2+8X \\ \hline X^2+9X+8 \end{array}$$

(X+1) (X+8)

14.
$$\begin{array}{r} X+18 \\ \times\ X+1 \\ \hline X+18 \\ X^2+18X \\ \hline X^2+19X+18 \end{array}$$

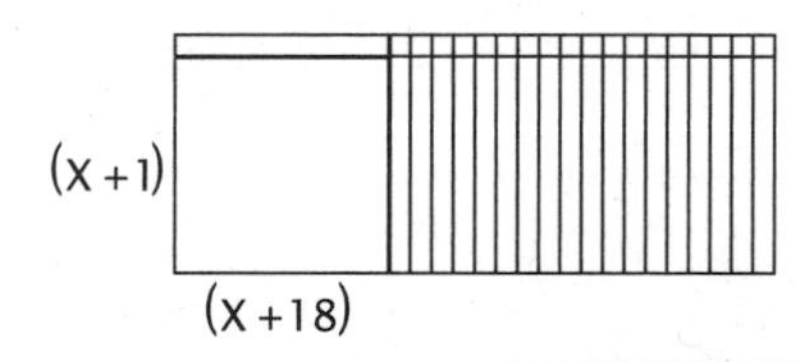

15.
$$\begin{array}{r} X+5 \\ \times\ X+4 \\ \hline 4X+20 \\ X^2+5X \\ \hline X^2+9X+20 \end{array}$$

(X+4) (X+5)

16.
$$\begin{array}{r} X+7 \\ \times\ X+3 \\ \hline 3X+21 \\ X^2+7X \\ \hline X^2+10X+21 \end{array}$$

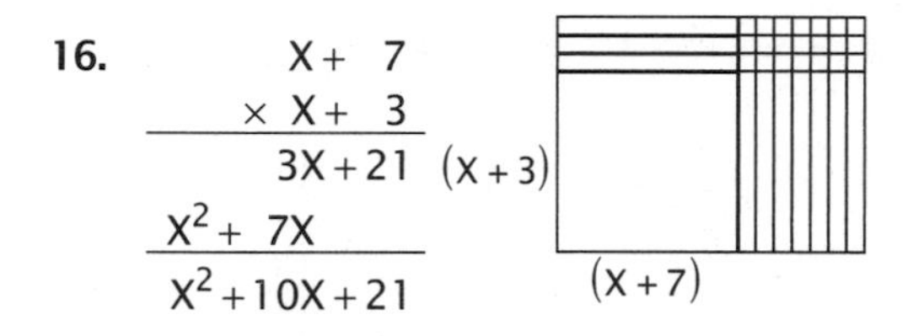

Lesson Practice 21B

1.
$$\begin{array}{r} X+8 \\ \times\ X+2 \\ \hline 2X+16 \\ X^2+8X \\ \hline X^2+10X+16 \end{array}$$

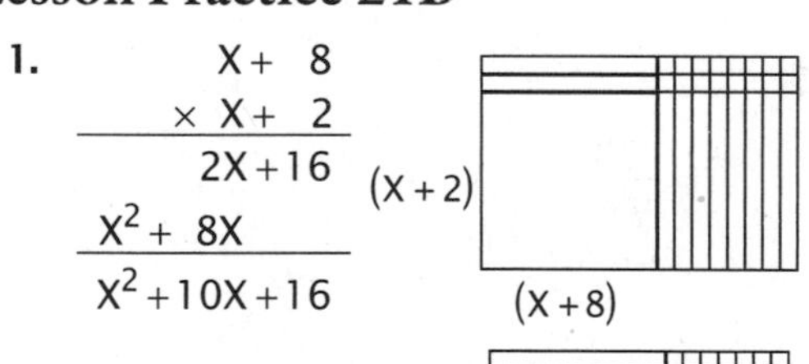

2.
$$\begin{array}{r} X+7 \\ \times\ X+4 \\ \hline 4X+28 \\ X^2+7X \\ \hline X^2+11X+28 \end{array}$$

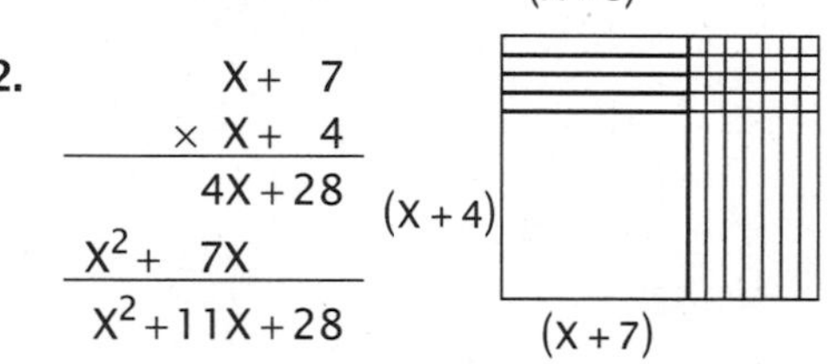

3.

$$\begin{array}{r} X+11 \\ \times\ X+2 \\ \hline 3X+22 \\ X^2+11X \quad \\ \hline X^2+13X+22 \end{array}$$

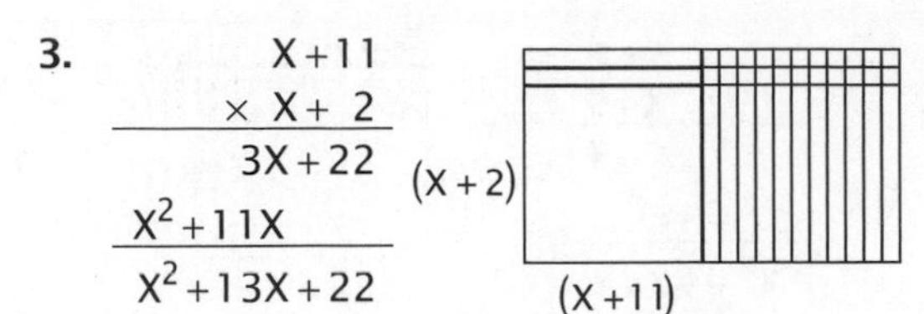

4.

$$\begin{array}{r} X+4 \\ \times\ X+3 \\ \hline 3X+12 \\ X^2+4X \quad \\ \hline X^2+7X+12 \end{array}$$

(X+3) (X+4)

5.

$$\begin{array}{r} X+5 \\ \times\ X+3 \\ \hline 3X+15 \\ X^2+5X \quad \\ \hline X^2+8X+15 \end{array}$$

(X+3) (X+5)

6.

$$\begin{array}{r} X+6 \\ \times\ X+5 \\ \hline 5X+30 \\ X^2+6X \quad \\ \hline X^2+11X+30 \end{array}$$

(X+5) (X+6)

7.

$$\begin{array}{r} X+4 \\ \times\ X+1 \\ \hline X+4 \\ X^2+4X \quad \\ \hline X^2+5X+4 \end{array}$$

(X+1) (X+4)

8.

$$\begin{array}{r} X+5 \\ \times\ X+1 \\ \hline X+5 \\ X^2+5X \quad \\ \hline X^2+6X+5 \end{array}$$

(X+1) (X+5)

9.

$$\begin{array}{r} X+4 \\ \times\ X+4 \\ \hline 4X+16 \\ X^2+4X \quad \\ \hline X^2+8X+16 \end{array}$$

(X+4) (X+4)

10.

$$\begin{array}{r} X+10 \\ \times\ X+2 \\ \hline 2X+20 \\ X^2+10X \quad \\ \hline X^2+12X+20 \end{array}$$

(X+2) (X+10)

11.

$$\begin{array}{r} X+9 \\ \times\ X+2 \\ \hline 2X+18 \\ X^2+9X \quad \\ \hline X^2+11X+18 \end{array}$$

(X+2) (X+9)

12.

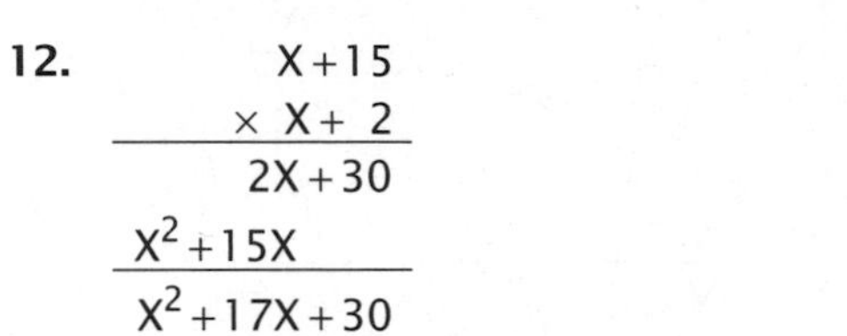

$$\begin{array}{r} X+15 \\ \times\ X+2 \\ \hline 2X+30 \\ X^2+15X \quad \\ \hline X^2+17X+30 \end{array}$$

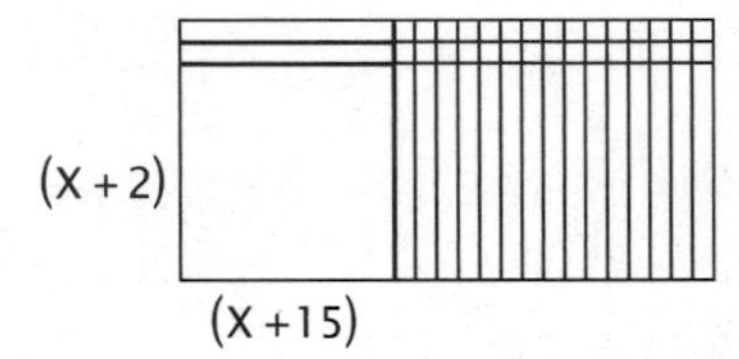

13.

$$\begin{array}{r} X+5 \\ \times\ X+2 \\ \hline 2X+10 \\ X^2+5X \quad \\ \hline X^2+7X+10 \end{array}$$

(X+2) (X+5)

14.

$$\begin{array}{r} X+1 \\ \times\ X+1 \\ \hline X+1 \\ X^2+X \quad \\ \hline X^2+2X+1 \end{array}$$

(X+1) (X+1)

15.

$$\begin{array}{r} X+5 \\ \times\ X+5 \\ \hline 5X+25 \\ X^2+5X \quad \\ \hline X^2+10X+25 \end{array}$$

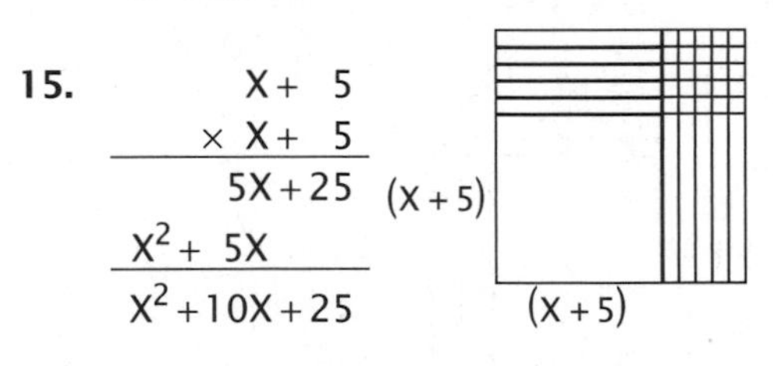

16.
$$\begin{array}{r} X+1 \\ \times\ X+25 \\ \hline 25X+25 \\ X^2+X \\ \hline X^2+26X+25 \end{array}$$

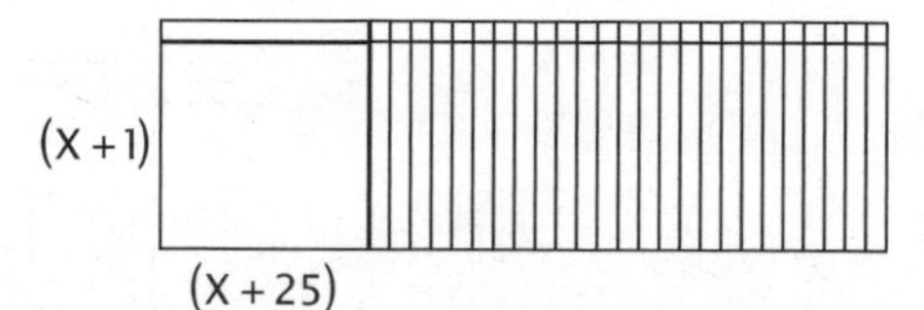

Systematic Review 21C

1. $X^2+7X+12=(X+4)(X+3)$

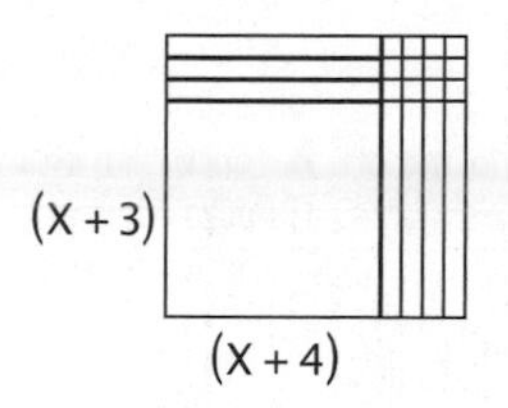

2. $X^2+10X+16=(X+8)(X+2)$

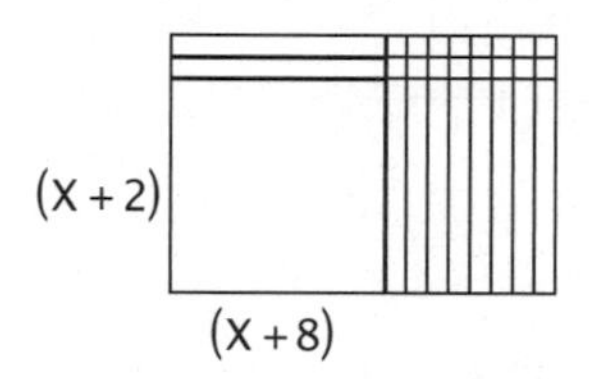

3. $X^2+11+24=(X+8)(X+3)$

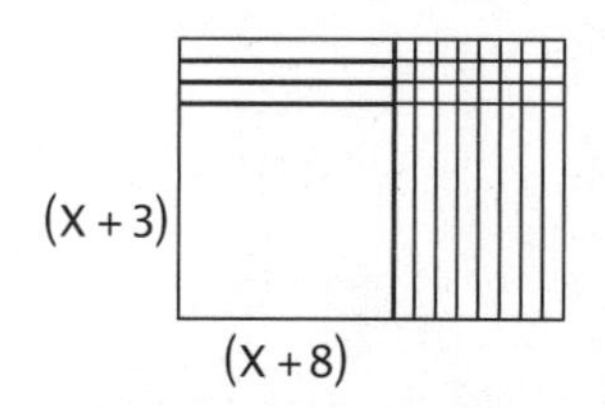

4. $X^2+8X+12=(X+6)(X+2)$

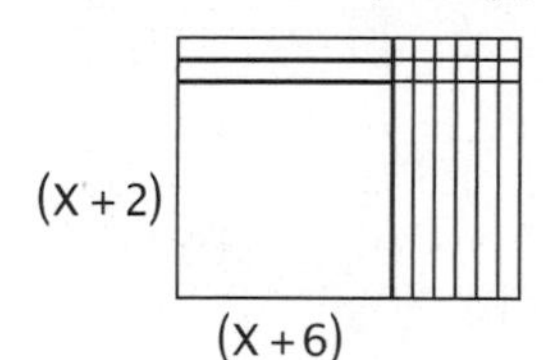

5.
$$\begin{array}{r} X+4 \\ \times\ X+2 \\ \hline 2X+8 \\ X^2+4X \\ \hline X^2+6X+8 \end{array}$$

(X+2) (X+4)

6.
$$\begin{array}{r} X+5 \\ \times\ X+3 \\ \hline 3X+15 \\ X^2+5X \\ \hline X^2+8X+15 \end{array}$$

(X+3) (X+5)

7. $X^2+7X+6=(X+6)(X+1)$

8.
$$\begin{array}{r} X+6 \\ \times\ X+1 \\ \hline X+6 \\ X^2+6X \\ \hline X^2+7X+6 \end{array}$$

9. $X^2+2X+1=(X+1)(X+1)$

10.
$$\begin{array}{r} X+1 \\ \times\ X+1 \\ \hline X+1 \\ X^2+X \\ \hline X^2+2X+1 \end{array}$$

11.
$$\begin{array}{r} 2X^2-7X-3 \\ X^2+5X+9 \\ \hline 3X^2-2X+6 \end{array}$$

12.
$$\begin{array}{r} 6X^2+2X+1 \\ X^2-4X+3 \\ \hline 7X^2-2X+4 \end{array}$$

13.
$$\begin{aligned} (P^{-4})^2P^3P^1 &= P^{-4\times2}\times P^{3+1} \\ &= P^{-8}\times P^4 \\ &= P^{-8+4}=P^{-4} \end{aligned}$$

14.
$$\begin{aligned} (R^{-2}S^3)^{-3} &= R^{(-2)(-3)}S^{(3)(-3)} \\ &= R^6S^{-9} \end{aligned}$$

15. $15^2=15\times15=225$

16. $\sqrt{16}=4$

17.
$$\begin{aligned} 11N+2(N+2) &= 6(N+4)+1 \\ 11N+2N+4 &= 6N+24+1 \\ 11N+2N-6N &= 24+1-4 \\ 7N &= 21 \\ N &= 3 \end{aligned}$$

3; 5; 7

18.
$$\begin{aligned} (.10D+.05N=.60)(100) &\Rightarrow 10D+5N = 60 \\ (D+N=9)(-5) &\Rightarrow -5D-5N = -45 \\ & \quad\; 5D = 15 \\ & \quad\; D = 3 \end{aligned}$$

$$\begin{aligned} D+N=9 \Rightarrow (3)+N &= 9 \\ N &= 9-3 \\ N &= 6 \end{aligned}$$

19. $7X-Y=-3$ or $-7X+Y=3$

20. $4Y<3X-5 \Rightarrow Y<\frac{3}{4}X-\frac{5}{4}$

test point $(0,0)$:

$4(0)<3(0)-5$

$0<0-5$

$0<-5$; false

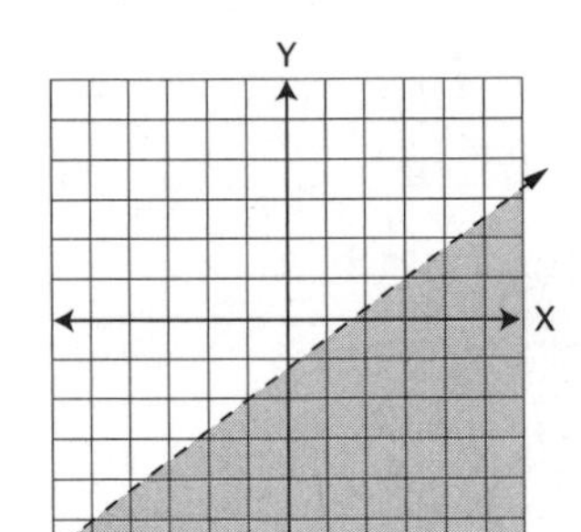

Systematic Review 21D

1. $X^2+11X+28=(X+7)(X+4)$

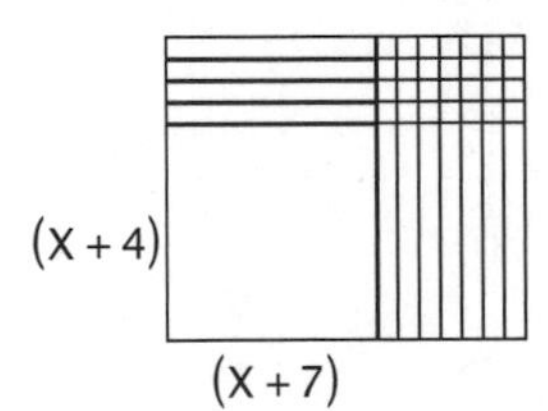

(X + 4)

(X + 7)

2. $X^2+4X+4=(X+2)(X+2)$

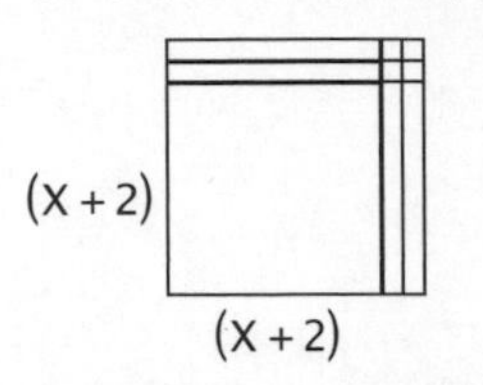

(X + 2)

(X + 2)

3. $X^2+6X+8=(X+4)(X+2)$

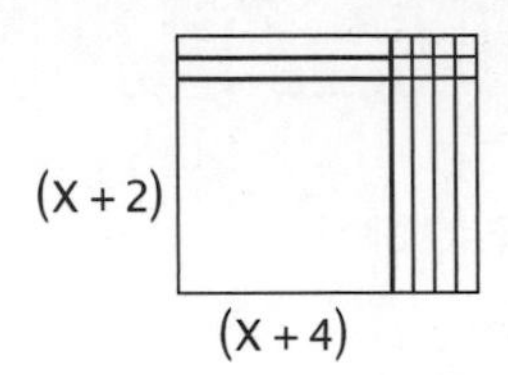

(X + 2)

(X + 4)

4. $X^2+8X+16=(X+4)(X+4)$

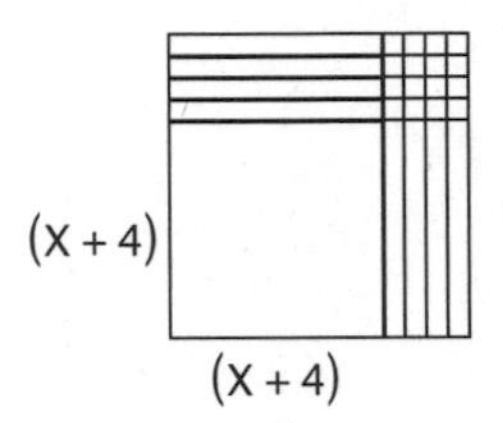

(X + 4)

(X + 4)

5.
$$\begin{array}{r} X+5 \\ \times\ X+1 \\ \hline X+5 \\ X^2+5X \quad \\ \hline X^2+6X+5 \end{array}$$

6.
$$\begin{array}{r} X+3 \\ \times\ X+3 \\ \hline 3X+9 \\ X^2+3X \quad \\ \hline X^2+6X+9 \end{array}$$

7. $X^2+12X+32=(X+8)(X+4)$

8.
$$\begin{array}{r} X+\ 8 \\ \times\ X+\ 4 \\ \hline 4X+32 \\ X^2+8X \quad \\ \hline X^2+12X+32 \end{array}$$

9. $X^2+20X+100=(X+10)(X+10)$

10.
$$\begin{array}{r} X+\ 10 \\ \times\ X+\ 10 \\ \hline 10X+100 \\ X^2+10X \quad \\ \hline X^2+20X+100 \end{array}$$

11.
$$\begin{array}{r} X^2+X-4 \\ \underline{X^2+3X+3} \\ 2X^2+4X-1 \end{array}$$

12.
$$\begin{array}{r} 2X^2+7X+6 \\ \underline{5X^2-4X+10} \\ 7X^2+3X+16 \end{array}$$

13. $\left[\left(P^5\right)^3\right]^{-2} = P^{5\times3\times-2} = P^{-30}$

14. $\left(S^6R^{-3}S^2\right)^0 = 1$

Anything to the 0 power $=1$.

15. $11^2 = 11\times11 = 121$

16. $\sqrt{144} = 12$

17. $14(N+2)+4(N) = 12(N+4)-2$

$14N+28+4N = 12N+48-2$

$14N+4N-12N = 48-2-28$

$6N = 18$

$N = 3$

3, 5, 7

18.

$(.10D+.05N=1.80)(100) \Rightarrow 10D+5N = 180$

$(D+N=27)(-5) \Rightarrow -5D-5N = -135$

$5D = 45$

$D = 9$

$D+N = 27 \Rightarrow (9)+N = 27$

$N = 27-9$

$N = 18$

19. on the graph

20. $m = -\frac{2}{3}$

$(-3) = -\frac{2}{3}(3)+b$

$-3 = -\frac{6}{3}+b$

$b = -1$

$Y = -\frac{2}{3}X-1$

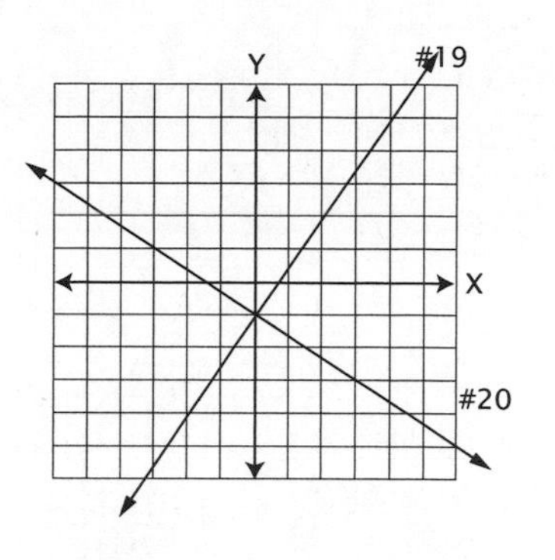

Systematic Review 21E

1. $X^2+8X+7 = (X+7)(X+1)$

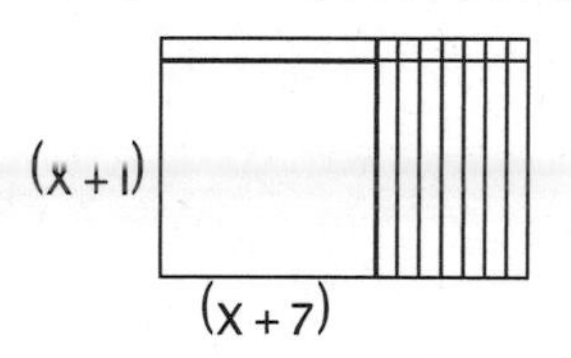

2. $X^2+5X+6 = (X+3)(X+2)$

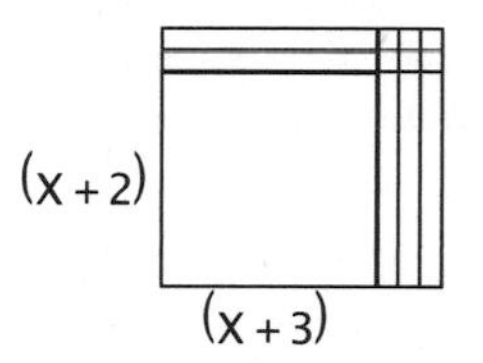

3. $X^2+9X+20 = (X+5)(X+4)$

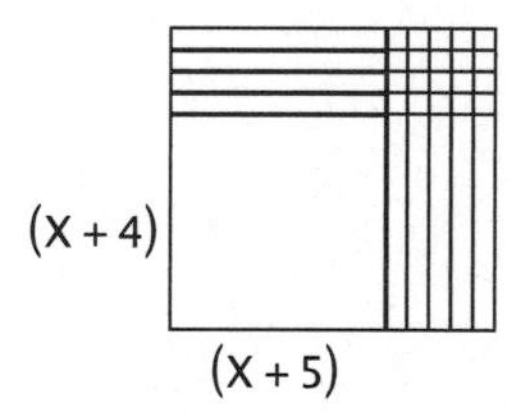

4. $X^2+8X+15 = (X+5)(X+3)$

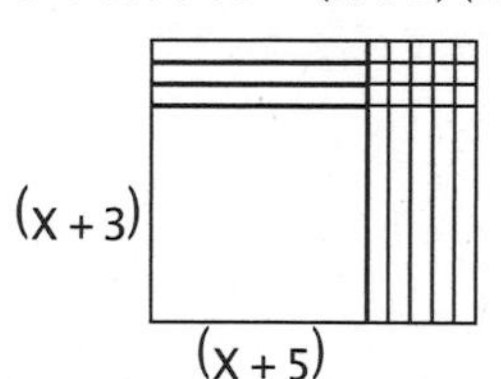

5.
$$\begin{array}{r} X+1 \\ \times\ X+9 \\ \hline 9X+9 \\ X^2+\ \ X\quad \\ \hline X^2+10X+9 \end{array}$$

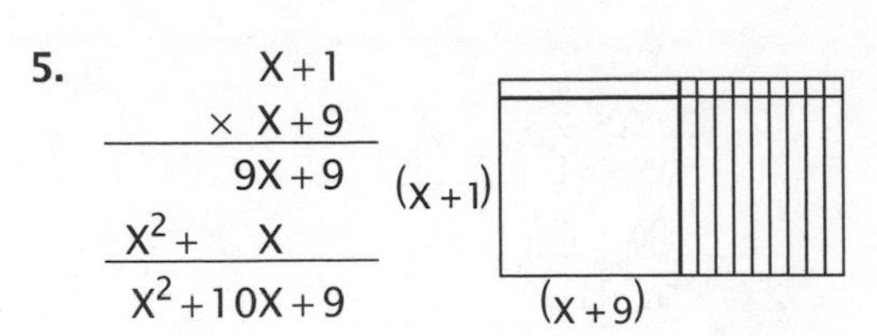

6.
$$\begin{array}{r} X+7 \\ \times\ X+2 \\ \hline 2X+14 \\ X^2+7X\quad \\ \hline X^2+9X+14 \end{array}$$

(X + 2) (X + 7)

7. $X^2+7X+12=(X+3)(X+4)$

8.
$$\begin{array}{r} X+3 \\ \times\ X+4 \\ \hline 4X+12 \\ X^2+3X\quad \\ \hline X^2+7X+12 \end{array}$$

9. $X^2+10X+21=(X+3)(X+7)$

10.
$$\begin{array}{r} X+3 \\ \times\ X+7 \\ \hline 7X+21 \\ X^2+\ 3X\quad \\ \hline X^2+10X+21 \end{array}$$

11.
$$\begin{array}{r} 4X^2-4X+1 \\ X^2+2X-1 \\ \hline 5X^2-2X\quad \end{array}$$

12.
$$\begin{array}{r} 2X^2+3X+3 \\ X^2+7X-2 \\ \hline 3X^2+10X+1 \end{array}$$

13. $(P^3)^0 P^4P^{-1}=P^{3\times0}P^{4+(-1)}=P^0P^3=P^3$

14. $(S^2R^0S^0)^{-2}R^5=(S^2\times1\times1)^{-2}R^5$
$=(S^2)^{-2}R^5$
$=S^{2\times-2}R^5=S^{-4}R^5$

15. $13^2=13\times13=169$

16. $\sqrt{25}=5$

17. $(N+1)+7(N+2)=5(N)$
$N+1+7N+14=5N$
$N+7N-5N=-1-14$
$3N=-15$
$N=-5$

$-5;\ -4;\ -3$

18.
$(.01P+.05N=.76)(100)\Rightarrow\ P+5N=76$
$(P+N=20)(-1)\Rightarrow\ -P-N=-20$
$4N=56$
$N=14$

$P+N=20\Rightarrow P+(14)=20$
$P=20-14$
$P=6$

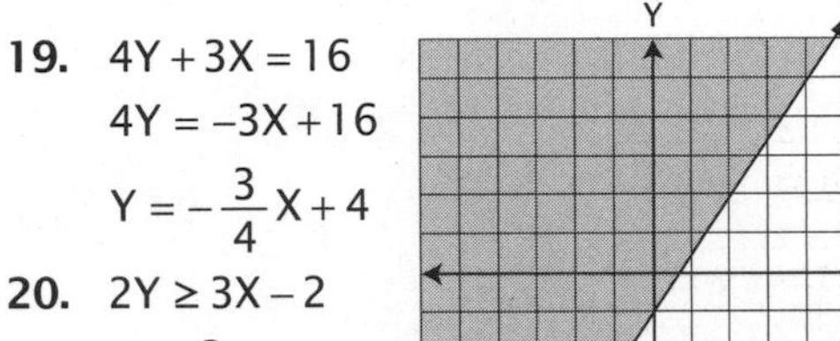

19. $4Y+3X=16$
$4Y=-3X+16$
$Y=-\frac{3}{4}X+4$

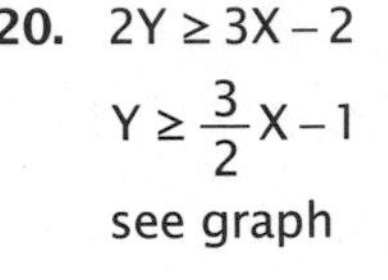

20. $2Y\geq3X-2$
$Y\geq\frac{3}{2}X-1$
see graph

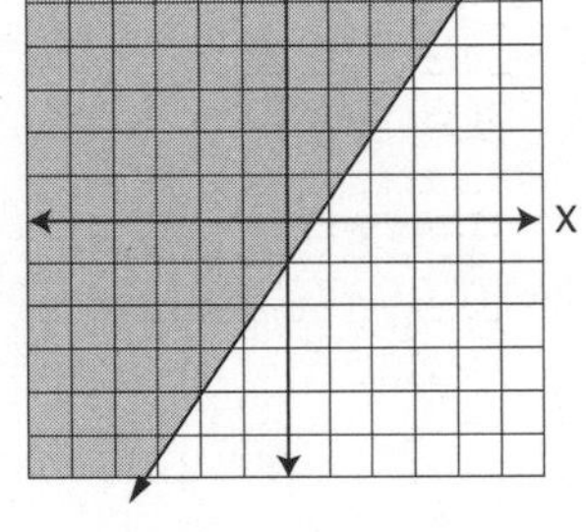

Lesson Practice 22A

1.
$$\begin{array}{r} 2X+1 \\ \times\ X+1 \\ \hline 2X+1 \\ 2X^2+\ \ X\quad \\ \hline 2X^2+3X+1 \end{array}$$

(X + 1) (2X + 1)

2.
$$\begin{array}{r} 3X+1 \\ \times\ X+4 \\ \hline 12X+4 \\ 3X^2+\ \ \ X\quad \\ \hline 3X^2+13X+4 \end{array}$$

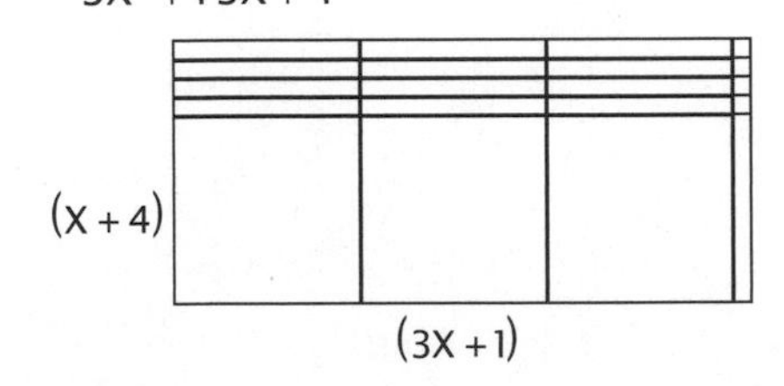

3. $4X^2+8X+4=4(X^2+2X+1)=4(X+1)(X+1)$

$$\begin{array}{r} X+1 \\ \times\ X+1 \\ \hline X+1 \\ X^2+\ X \\ \hline X^2+2X+1 \end{array}$$

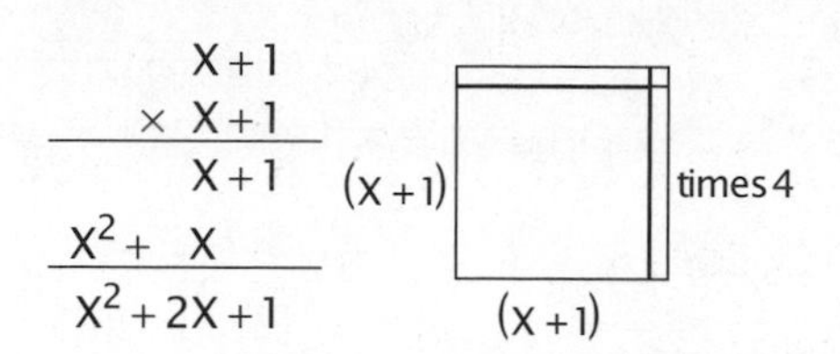

4.

$$\begin{array}{r} 2X+1 \\ \times\ X+5 \\ \hline 10X+5 \\ 2X^2+\ \ \ X \\ \hline 2X^2+11X+5 \end{array}$$

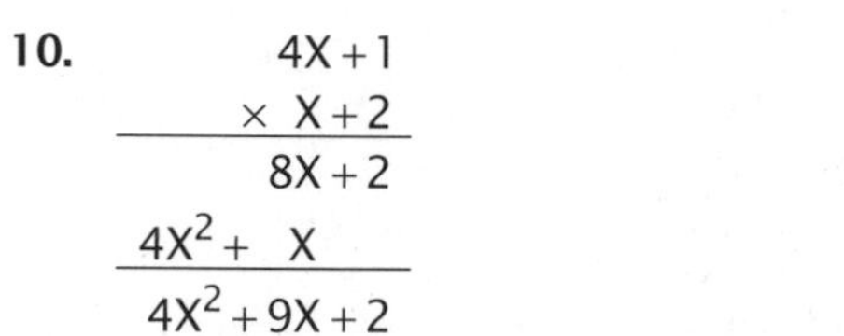

5.

$$\begin{array}{r} 2X+3 \\ \times\ X+6 \\ \hline 12X+18 \\ 2X^2+\ 3X \\ \hline 2X^2+15X+18 \end{array}$$

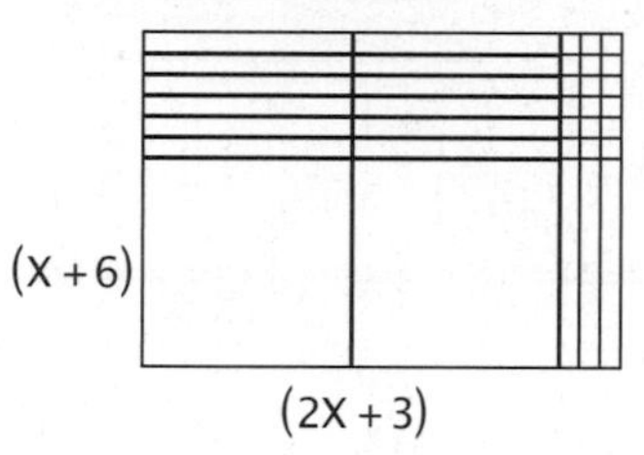

6.

$$\begin{array}{r} 3X+1 \\ \times\ X+2 \\ \hline 6X+2 \\ 3X^2+\ X \\ \hline 3X^2+7X+2 \end{array}$$

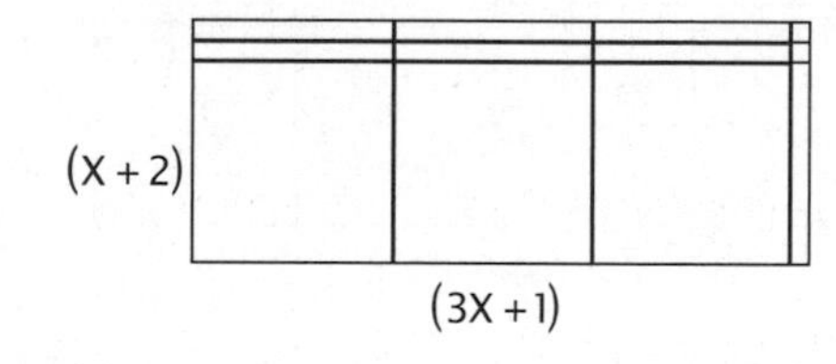

7.

$$\begin{array}{r} 2X+5 \\ \times\ X+2 \\ \hline 4X+10 \\ 2X^2+5X \\ \hline 2X^2+9X+10 \end{array}$$

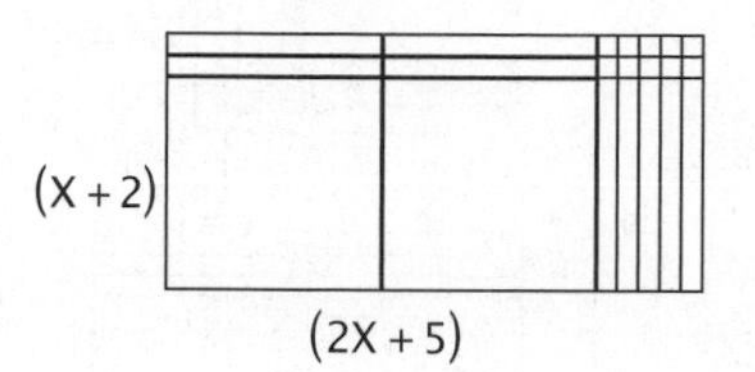

8. $4X^2+10X+4=2(2X^2+5X+2)=2(2X+1)(X+2)$

$$\begin{array}{r} 2X+1 \\ \times\ X+2 \\ \hline 4X+2 \\ 2X^2+\ X \\ \hline 2X^2+5X+2 \end{array}$$

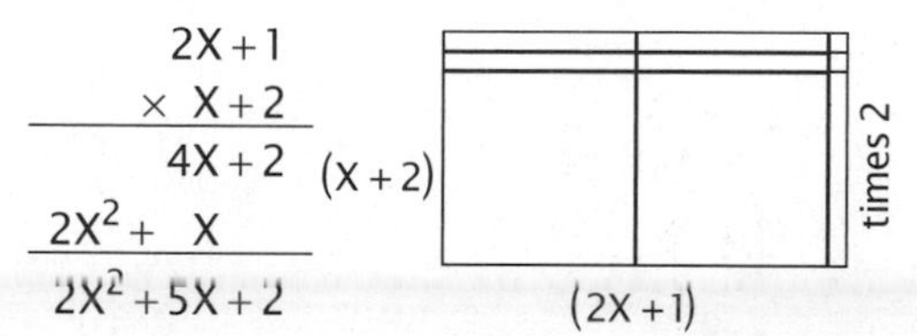

9.

$$\begin{array}{r} 2X+3 \\ \times\ X+3 \\ \hline 6X+9 \\ 2X^2+3X \\ \hline 2X^2+9X+9 \end{array}$$

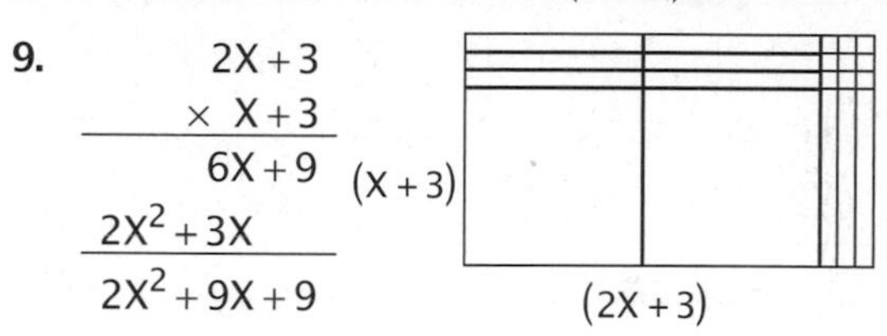

10.

$$\begin{array}{r} 4X+1 \\ \times\ X+2 \\ \hline 8X+2 \\ 4X^2+\ \ X \\ \hline 4X^2+9X+2 \end{array}$$

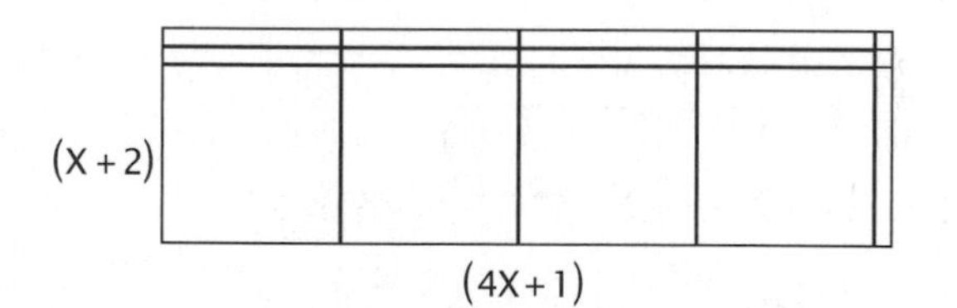

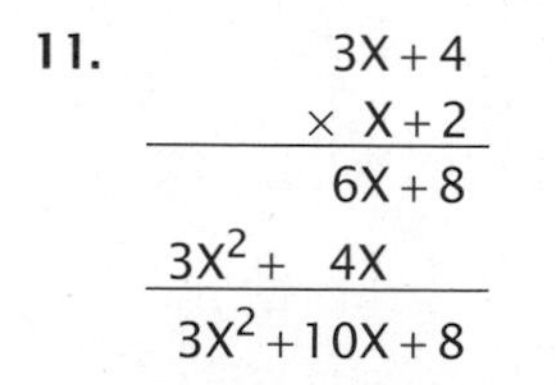

11.

$$\begin{array}{r} 3X+4 \\ \times\ X+2 \\ \hline 6X+8 \\ 3X^2+\ \ 4X \\ \hline 3X^2+10X+8 \end{array}$$

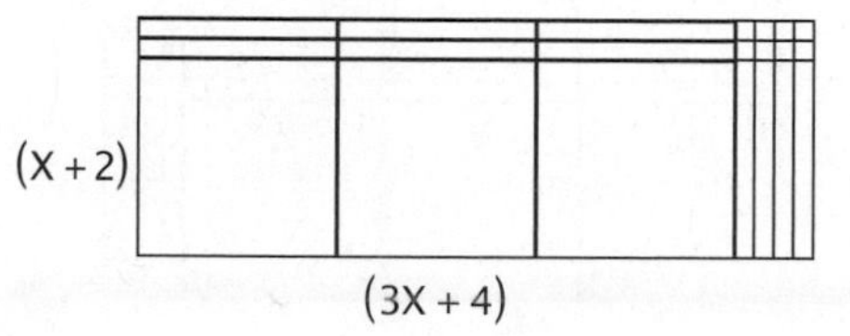

12. 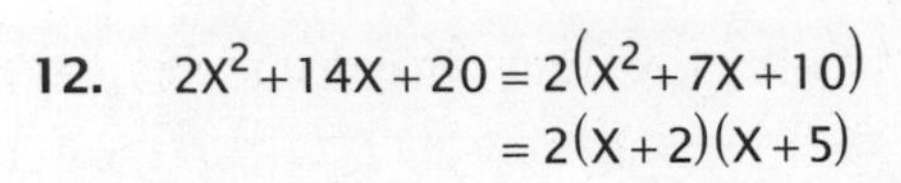

$2X^2+14X+20=2(X^2+7X+10)$
$=2(X+2)(X+5)$

$$\begin{array}{r} X+\ 2 \\ \times\ X+\ 5 \\ \hline 5X+10 \\ X^2+2X \\ \hline X^2+7X+10 \end{array}$$

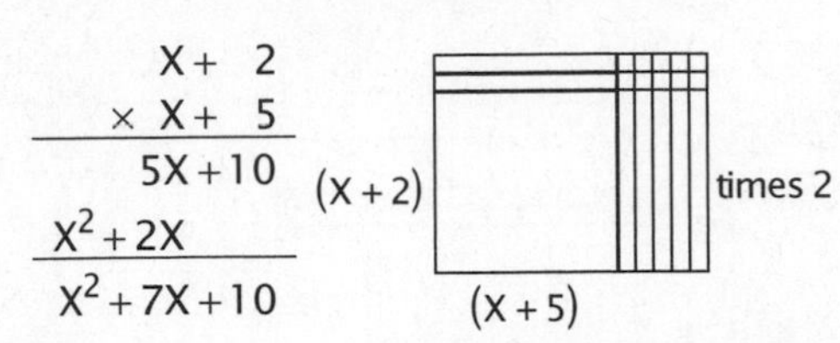

13.

$$\begin{array}{r} 2X+1 \\ \times\ X+3 \\ \hline 6X+3 \\ 2X^2+\ X \\ \hline 2X^2+7X+3 \end{array}$$

(X+3) (2X+1)

14.

$$\begin{array}{r} 4X+3 \\ \times\ X+1 \\ \hline 4X+3 \\ 4X^2+3X \\ \hline 4X^2+7X+3 \end{array}$$

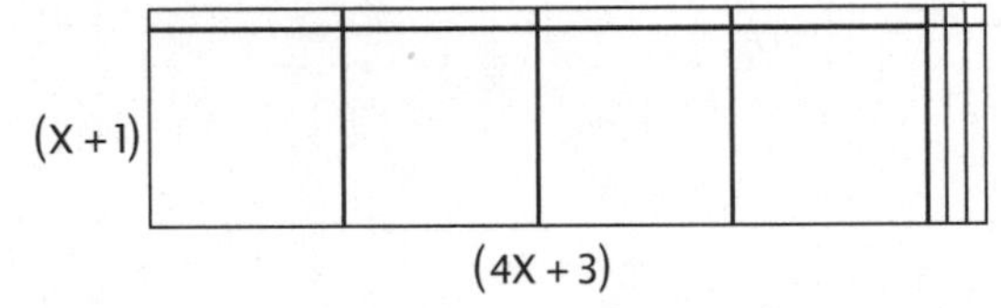

15.

$$\begin{array}{r} 2X+9 \\ \times\ X+2 \\ \hline 4X+18 \\ 2X^2+9X \\ \hline 2X^2+13X+18 \end{array}$$

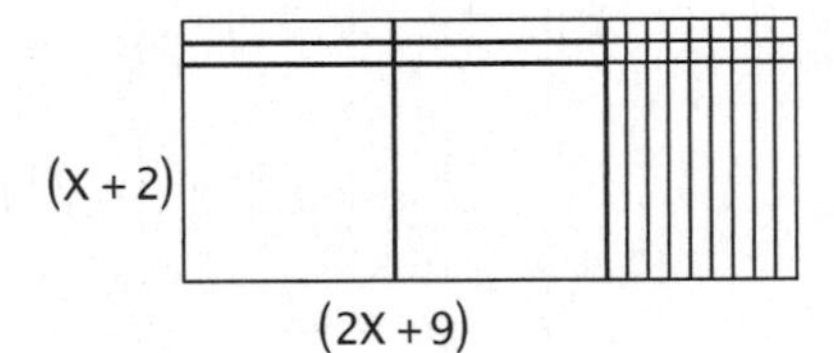

16.

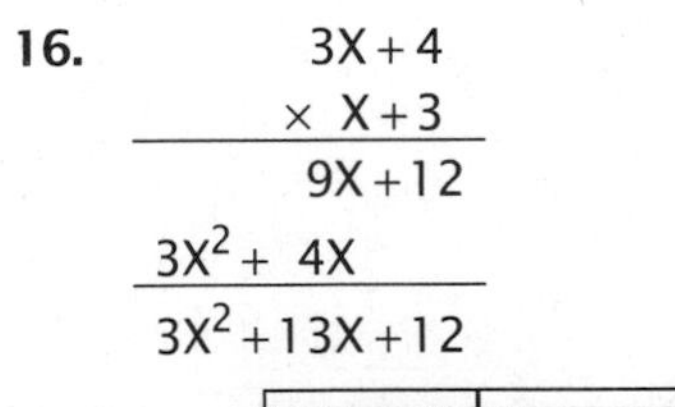

$$\begin{array}{r} 3X+4 \\ \times\ X+3 \\ \hline 9X+12 \\ 3X^2+\ 4X \\ \hline 3X^2+13X+12 \end{array}$$

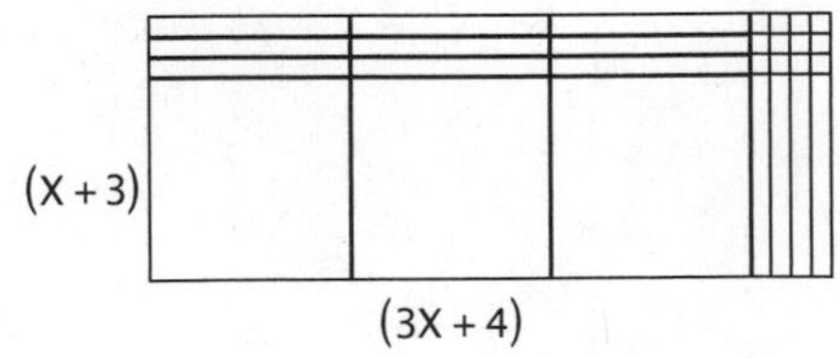

Lesson Practice 22B

1.

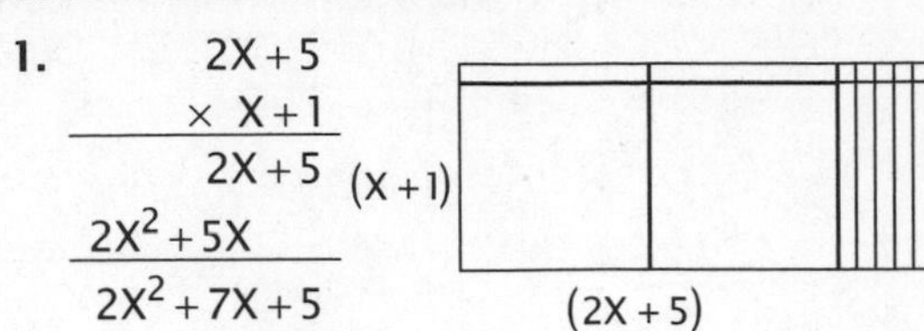

$$\begin{array}{r} 2X+5 \\ \times\ X+1 \\ \hline 2X+5 \\ 2X^2+5X \\ \hline 2X^2+7X+5 \end{array}$$

2.

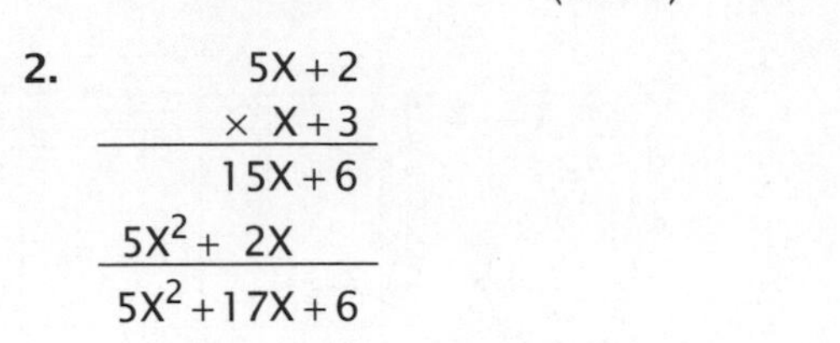

$$\begin{array}{r} 5X+2 \\ \times\ X+3 \\ \hline 15X+6 \\ 5X^2+\ 2X \\ \hline 5X^2+17X+6 \end{array}$$

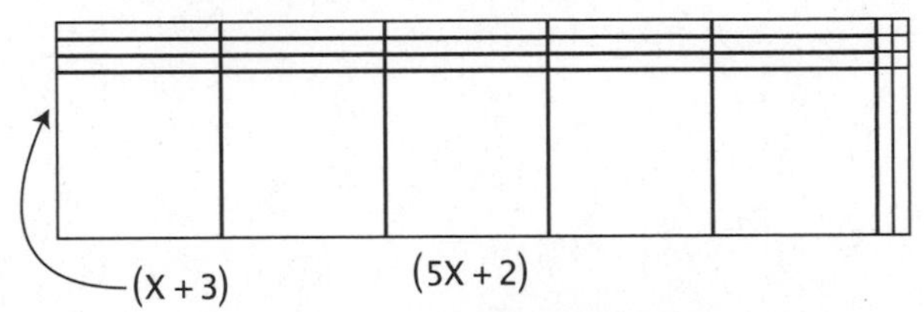

3.

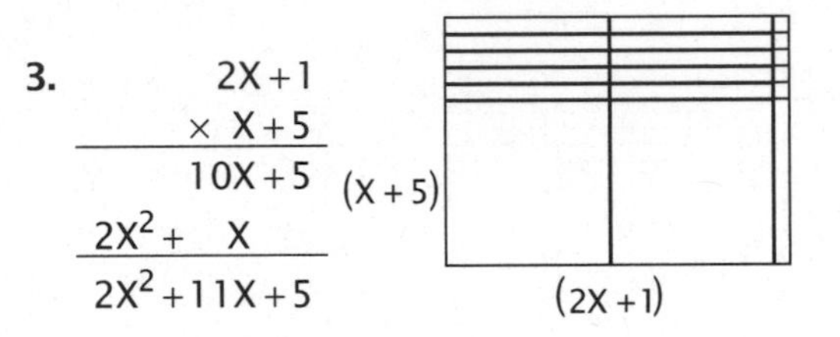

$$\begin{array}{r} 2X+1 \\ \times\ X+5 \\ \hline 10X+5 \\ 2X^2+\ X \\ \hline 2X^2+11X+5 \end{array}$$

4.

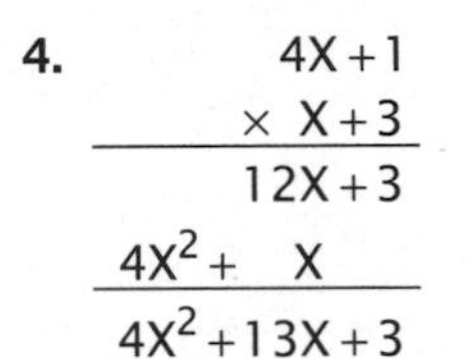

$$\begin{array}{r} 4X+1 \\ \times\ X+3 \\ \hline 12X+3 \\ 4X^2+\ X \\ \hline 4X^2+13X+3 \end{array}$$

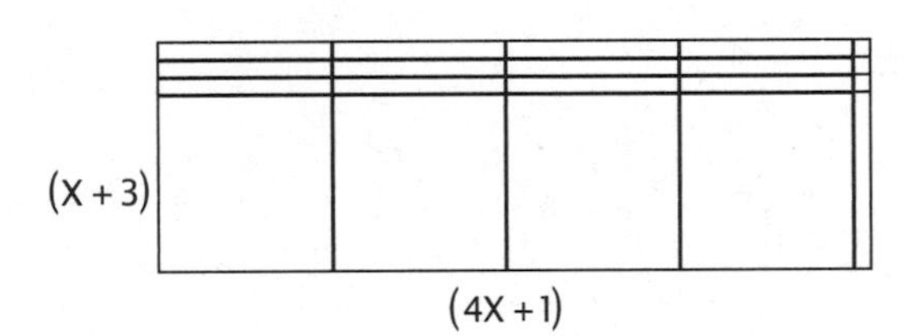

5. $2X^2+16X+30=2(X^2+8X+15)=2(X+5)(X+3)$

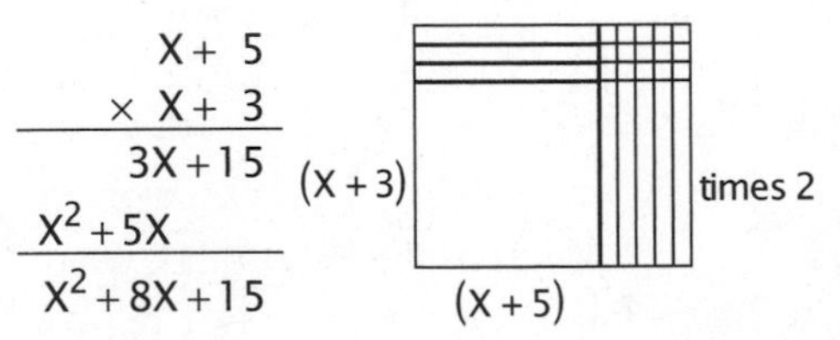

$$\begin{array}{r} X+\ 5 \\ \times\ X+\ 3 \\ \hline 3X+15 \\ X^2+5X \\ \hline X^2+8X+15 \end{array}$$

6. $3X^2 + 9X + 6 = 3(X^2 + 3X + 2) = 3(X + 1)(X + 2)$

$$\begin{array}{r} X+1 \\ \times\ X+2 \\ \hline 2X+2 \\ X^2+\ X \\ \hline X^2+3X+2 \end{array}$$

(X + 1) times 3

(X + 2)

7.

$$\begin{array}{r} 2X+9 \\ \times\ X+1 \\ \hline 2X+9 \\ 2X^2+\ 9X \\ \hline 2X^2+11X+9 \end{array}$$

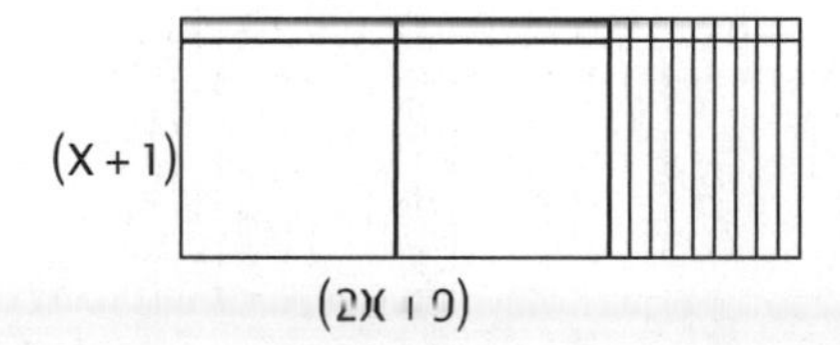

8.

$$\begin{array}{r} 3X+\ 2 \\ \times\ X+\ 7 \\ \hline 21X+14 \\ 3X^2+\ 2X \\ \hline 3X^2+23X+14 \end{array}$$

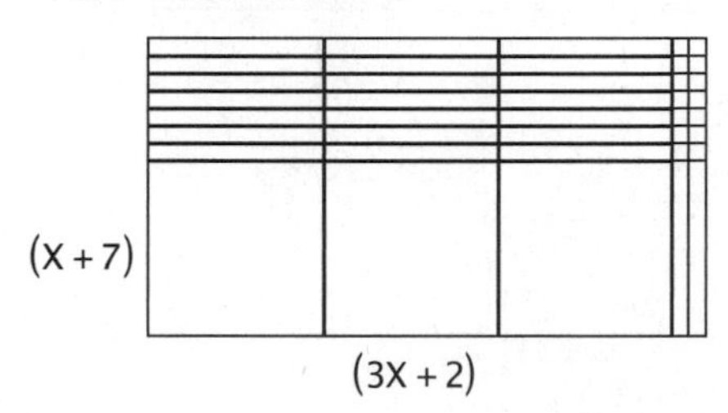

9.

$$\begin{array}{r} 2X+\ 3 \\ \times\ X+\ 5 \\ \hline 10X+15 \\ 2X^2+\ 3X \\ \hline 2X^2+13X+15 \end{array}$$

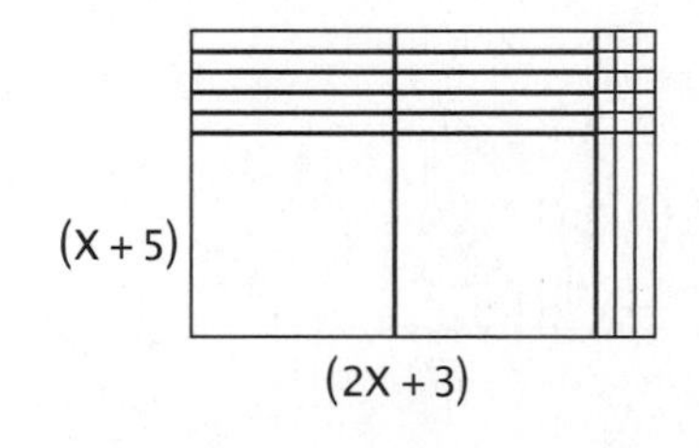

10. $(5)(X^2 + 10X + 21) =$

$$\begin{array}{r} X+\ 7 \\ \times\ X+\ 3 \\ \hline 3X+21 \\ X^2+\ 7X \\ \hline X^2+10X+21 \end{array}$$

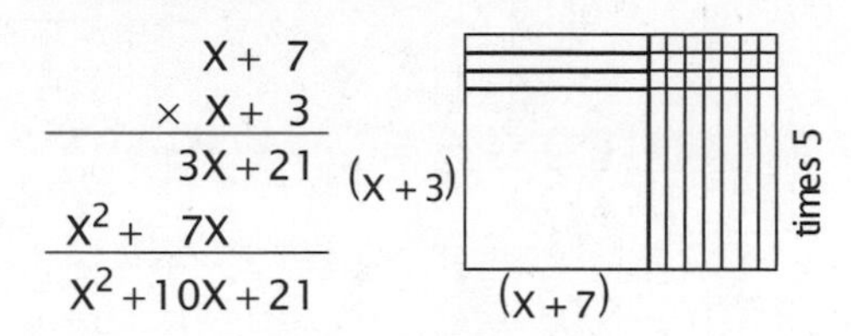

11. $6X^2 + 36X + 48 = (6)(X^2 + 6X + 8)$
$= 6(X + 4)(X + 2)$

$$\begin{array}{r} X+4 \\ \times\ X+2 \\ \hline 2X+8 \\ X^2+4X \\ \hline X^2+6X+8 \end{array}$$

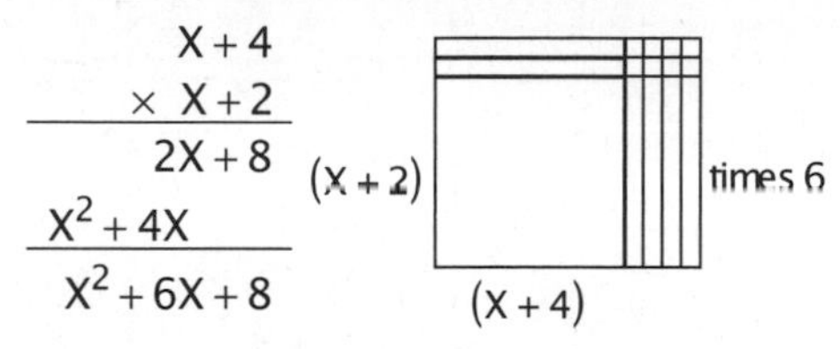

12.

$$\begin{array}{r} 3X+\ 8 \\ \times\ X+\ 2 \\ \hline 6X+16 \\ 3X^2+\ 8X \\ \hline 3X^2+14X+16 \end{array}$$

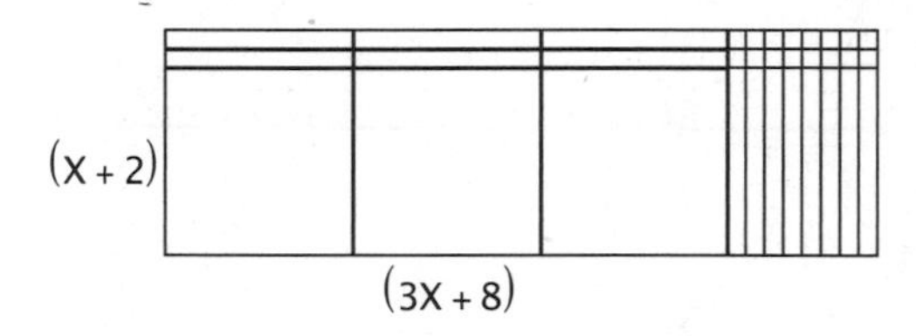

13. $4X^2 + 14X + 6 = (2)(2X^2 + 7X + 3)$
$= 2(2X + 1)(X + 3)$

$$\begin{array}{r} 2X+1 \\ \times\ X+3 \\ \hline 6X+3 \\ 2X^2+\ X \\ \hline 2X^2+7X+3 \end{array}$$

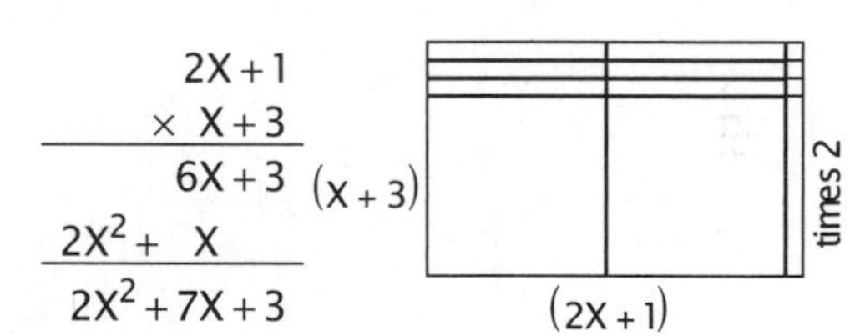

14.

$$\begin{array}{r} 5X+2 \\ \times\ X+1 \\ \hline 5X+2 \\ 5X^2+2X \\ \hline 5X^2+7X+2 \end{array}$$

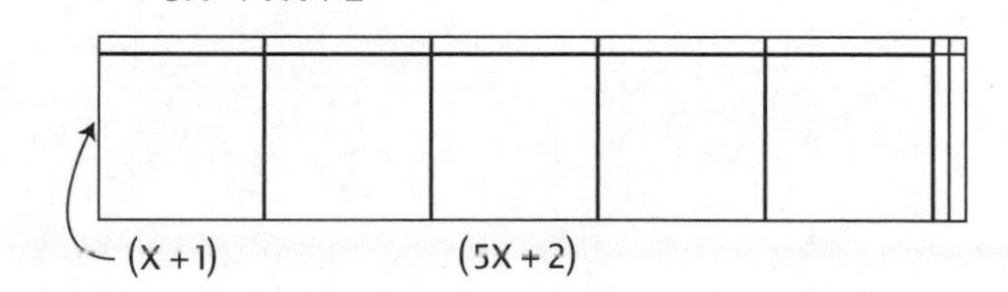

15.
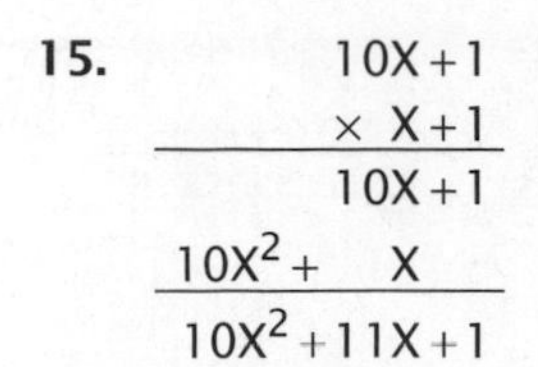

$$\begin{array}{r} 10X+1 \\ \times\ X+1 \\ \hline 10X+1 \\ 10X^2+\ \ X\quad\ \ \\ \hline 10X^2+11X+1 \end{array}$$

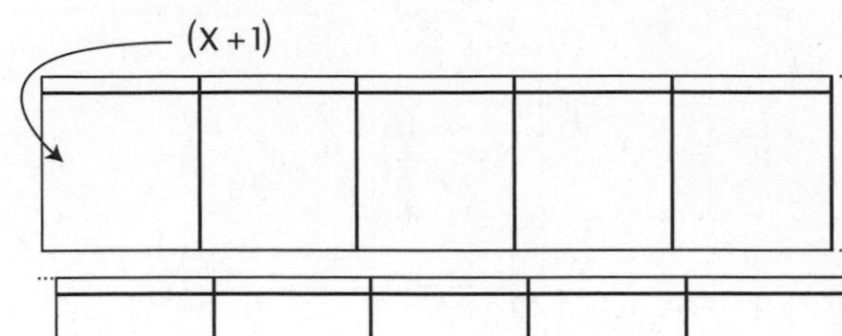

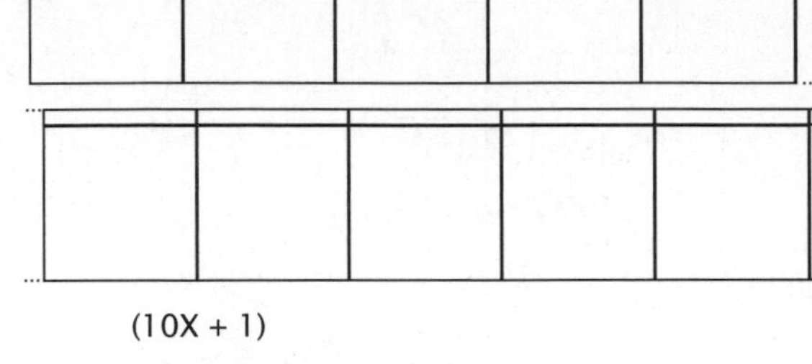

16.
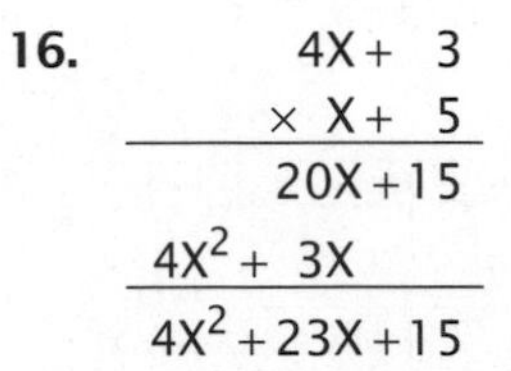

$$\begin{array}{r} 4X+\ 3 \\ \times\ X+\ 5 \\ \hline 20X+15 \\ 4X^2+\ 3X\qquad \\ \hline 4X^2+23X+15 \end{array}$$

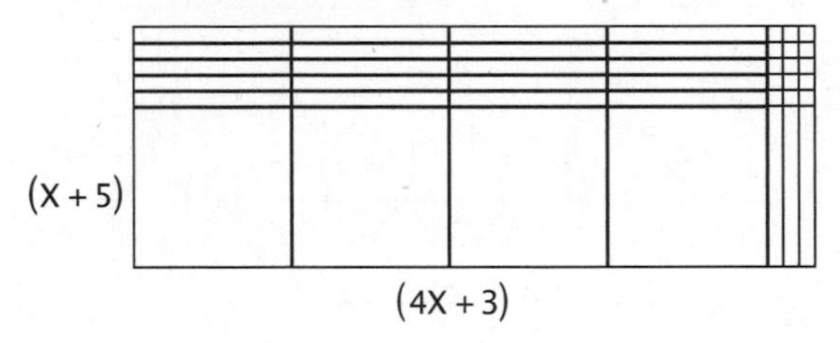

Systematic Review 22C

1. $(3X+4)(X+1)$

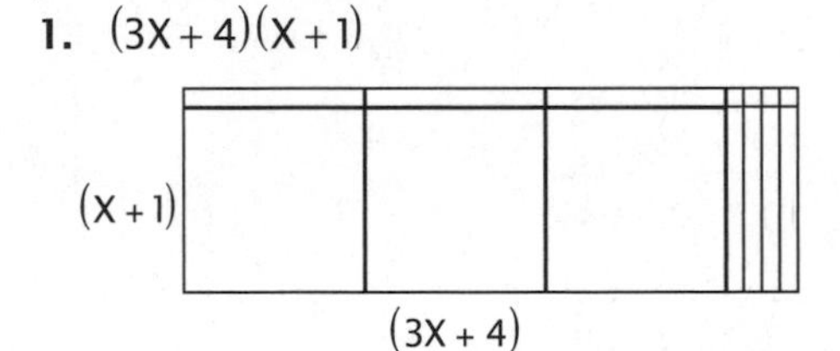

2. $(2X+3)(X+2)$

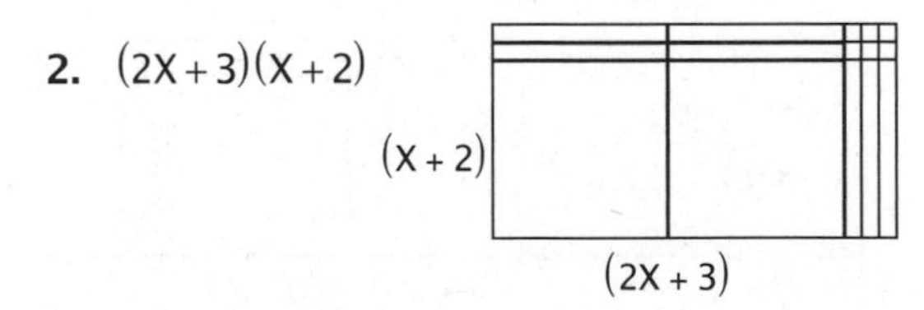

3.
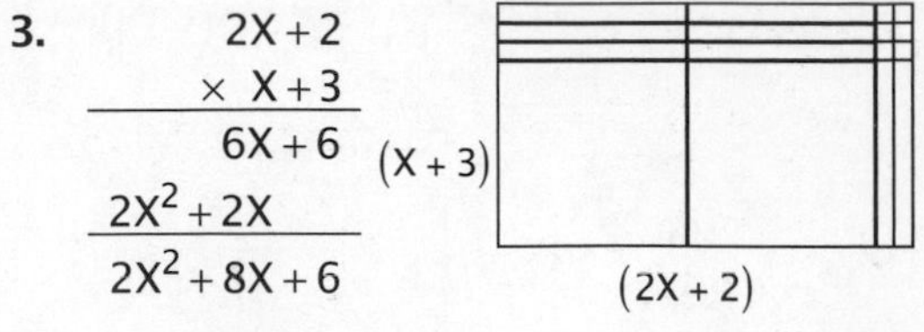

$$\begin{array}{r} 2X+2 \\ \times\ X+3 \\ \hline 6X+6 \\ 2X^2+2X\qquad \\ \hline 2X^2+8X+6 \end{array}$$

4.
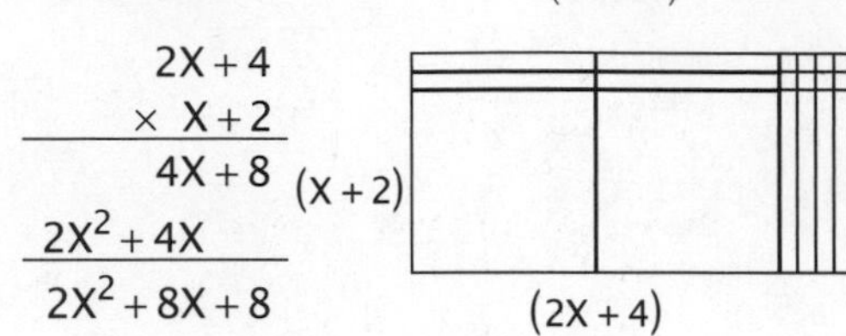

$$\begin{array}{r} 2X+4 \\ \times\ X+2 \\ \hline 4X+8 \\ 2X^2+4X\qquad \\ \hline 2X^2+8X+8 \end{array}$$

5. $(3X+4)(X+3)$

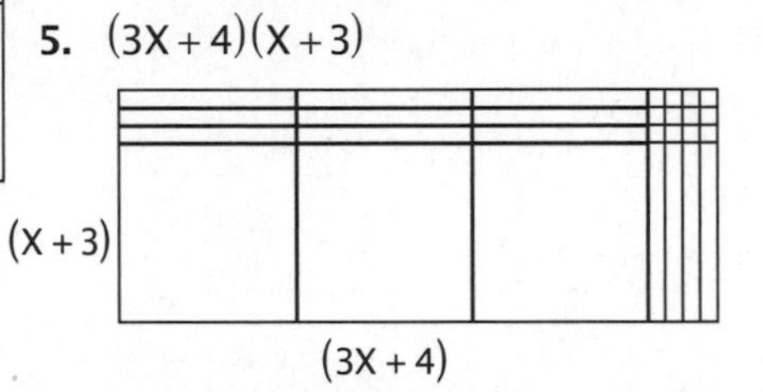

6.
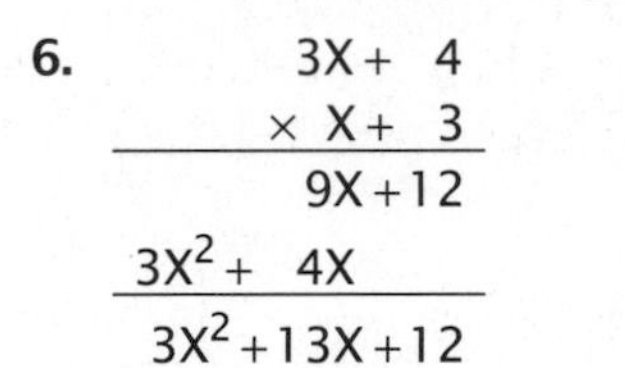

$$\begin{array}{r} 3X+\ 4 \\ \times\ X+\ 3 \\ \hline 9X+12 \\ 3X^2+\ 4X\qquad \\ \hline 3X^2+13X+12 \end{array}$$

7.
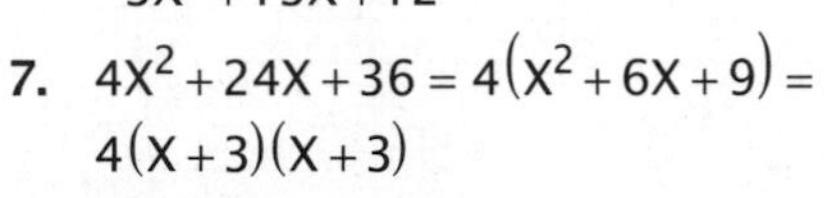

$4X^2+24X+36=4(X^2+6X+9)=$
$4(X+3)(X+3)$

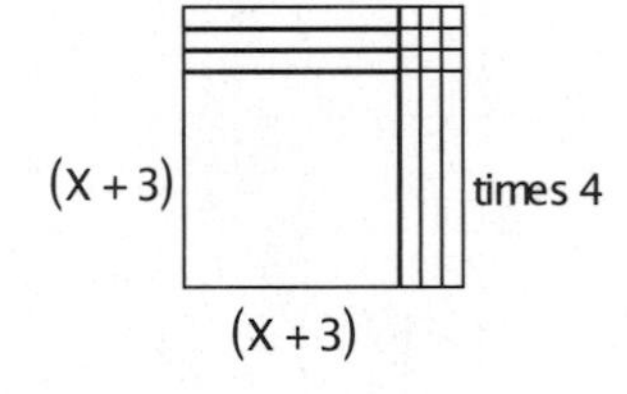

8.
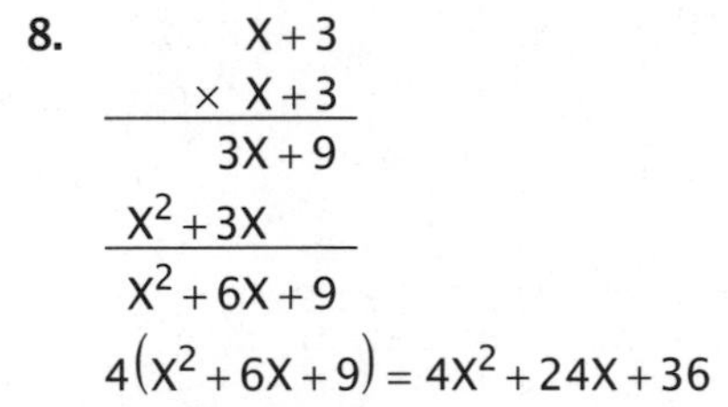

$$\begin{array}{r} X+3 \\ \times\ X+3 \\ \hline 3X+9 \\ X^2+3X\qquad \\ \hline X^2+6X+9 \end{array}$$

$4(X^2+6X+9)=4X^2+24X+36$

9. $(2X+1)(2X+3)$

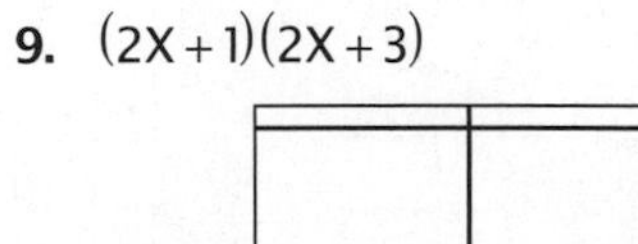

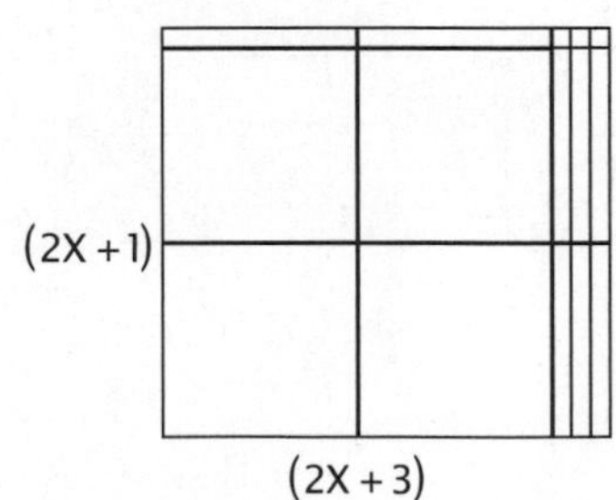

10.
$$\begin{array}{r} 2X+1 \\ \times\ 2X+3 \\ \hline 6X+3 \\ 4X^2+2X \quad \\ \hline 4X^2+8X+3 \end{array}$$

11. $B^2 \times B^6 \times B^{-5} = B^{2+6+(-5)} = B^3$

12. $A^B \cdot A^C = A^{B+C}$

13. $\frac{X^{-3}Y^2X^{-1}}{Y^{-3}X^{-5}} = X^{-3}Y^2X^{-1}Y^3X^5$
$= X^{-3+(-1)+5}Y^{2+3}$
$= X^1Y^5$ or XY^5

14. $\frac{A^3A^{-2}B^1}{B^{-2}A^4} = A^3A^{-2}B^1B^2A^{-4}$
$= A^{3+(-2)+(-4)}B^{1+2}$
$= A^{-3}B^3$

15. $6\times10^6 + 8\times10^4 + 2\times10^3 + 7\times10^{-2} =$
$6{,}000{,}000 + 80{,}000 + 2{,}000 + .07 =$
$6{,}082{,}000.07$

16. $2Y = 3X - 2$
$Y = \frac{3}{2}X - 1$; see graph

17. $m = \frac{3}{2}$
$(4) = \frac{3}{2}(0) + b$
$4 = 0 + b$
$b = 4$
$Y = \frac{3}{2}X + 4$ or $3X - 2Y = -8$
or $-3X + 2Y = 8$
see graph

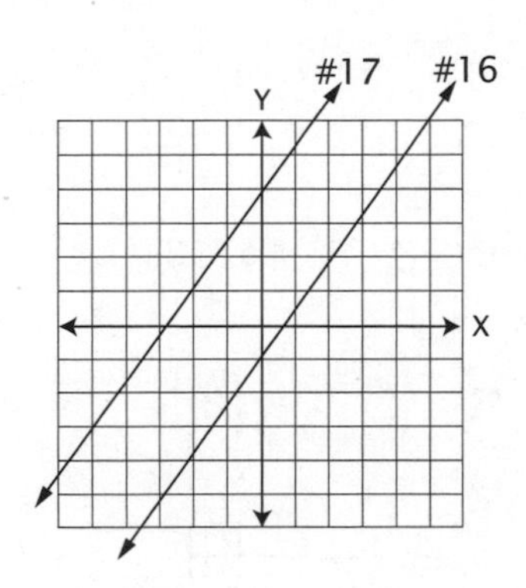

18.

hours	amoeba
1	2
2	4
3	8
4	16

19.

hours	amoeba
1	2^1
2	2^2
3	2^3
4	2^4

20. 2^6 after 6 hours
2^X after X hours

Systematic Review 22D

1. $(3X+5)(X+2)$

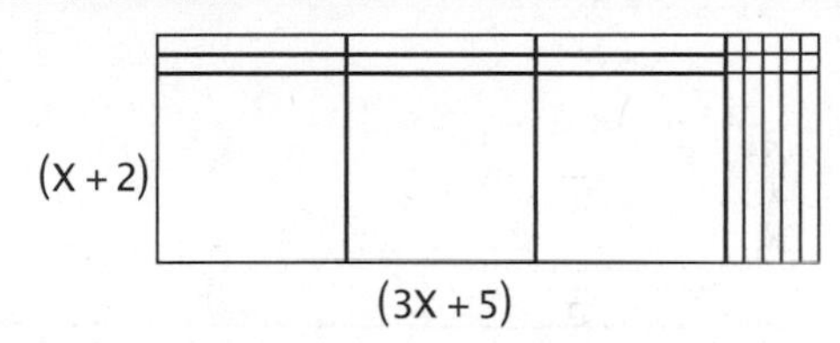

2. $4X^2 + 10X + 4 = 2(2X^2 + 5X + 2)$
$= 2(2X+1)(X+2)$

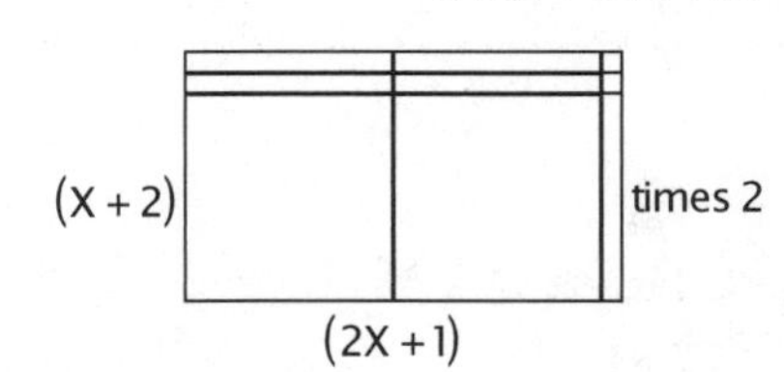

3.
$$\begin{array}{r} 3X+3 \\ \times\ X+2 \\ \hline 6X+6 \\ 3X^2+3X \quad \\ \hline 3X^2+9X+6 \end{array}$$

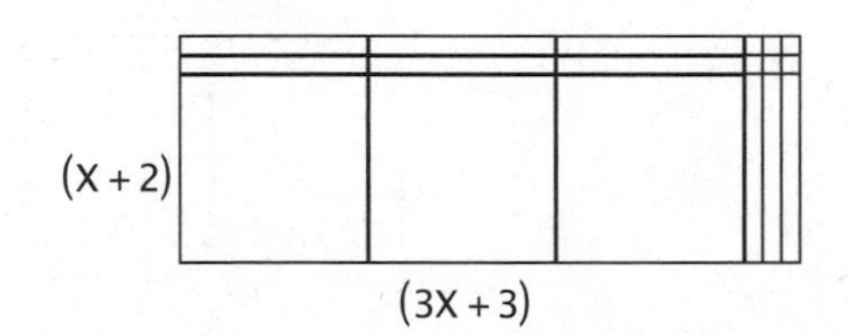

4. $\begin{array}{r} 2X+1 \\ \times\ 3X \\ \hline 6X^2+3X \end{array}$

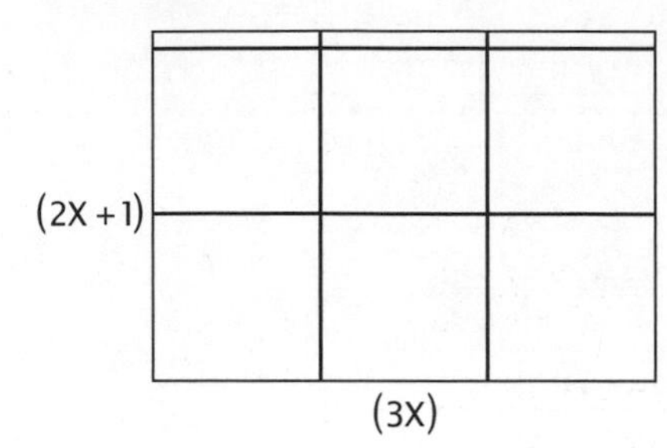

5. $3X^2+8X+5$

$(3X+5)(X+1)$

6. $\begin{array}{r} 3X+5 \\ \times\ X+1 \\ \hline 3X+5 \\ 3X^2+5X \\ \hline 3X^2+8X+5 \end{array}$

7. $(4X+7)(X+1)$

8. $\begin{array}{r} 4X+7 \\ \times\ X+1 \\ \hline 4X+7 \\ 4X^2+\ 7X \\ \hline 4X^2+11X+7 \end{array}$

9. $(X+3)(X+2)$

10. $\begin{array}{r} X+3 \\ \times\ X+2 \\ \hline 2X+6 \\ X^2+3X \\ \hline X^2+5X+6 \end{array}$

11. $C^{-4}\times C^3\times C^0 = C^{-4+3+0} = C^{-1}$ or $\frac{1}{C}$

12. $8^5 \div 8^3 = 8^{5-3} = 8^2$

13. $\frac{B^5B^2C^{-5}}{B^{-4}C^{-3}} = B^5B^2C^{-5}B^4C^3$

$= B^{5+2+4}C^{-5+3}$

$= B^{11}C^{-2}$ or $\frac{B^{11}}{C^2}$

14. $\frac{D^6C^{-4}D^2}{D^{-4}C^0C^2} = D^6C^{-4}D^2D^4C^{-0}C^{-2}$

$= C^{-4+(-0)+(-2)}D^{6+2+4}$

$= C^{-6}D^{12}$ or $\frac{D^{12}}{C^6}$

15. $86{,}900.4 =$

$8\times10^4+6\times10^3+9\times10^2+4\times10^{-1}$

16. $3Y = 2X+6$

$Y = \frac{2}{3}X+2$

see graph

17. $m = \frac{2}{3}$

$(-3) = \frac{2}{3}(-3)+b$

$-3 = \frac{-6}{3}+b$

$-3 = -2+b$

$b = -1$

$Y = \frac{2}{3}X-1$ or $2X-3Y = 3$

or $-2X+3Y = -3$

see graph

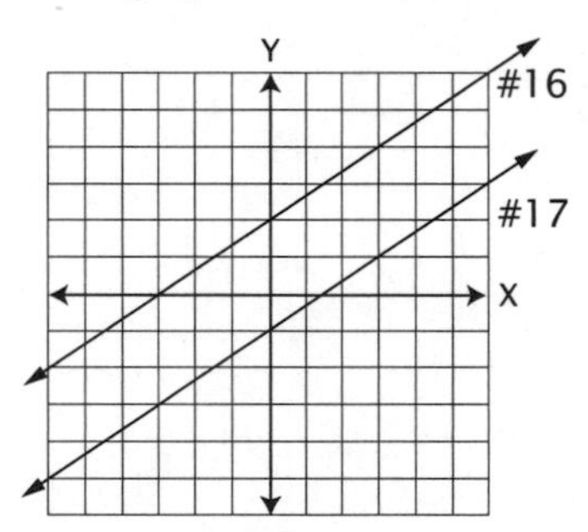

18.

weeks	dollars
2	9
3	27
4	81
5	243

19.

weeks	dollars
1	3^1
2	3^2
3	3^3
4	3^4
5	3^5

20. 20 weeks $= 3^{20} \approx \$3,486,800,000$ (rounc
May be shown on your
calculator as 3.4868×10^9

Systematic Review 22E

1. $(2X+3)(2X+3)$

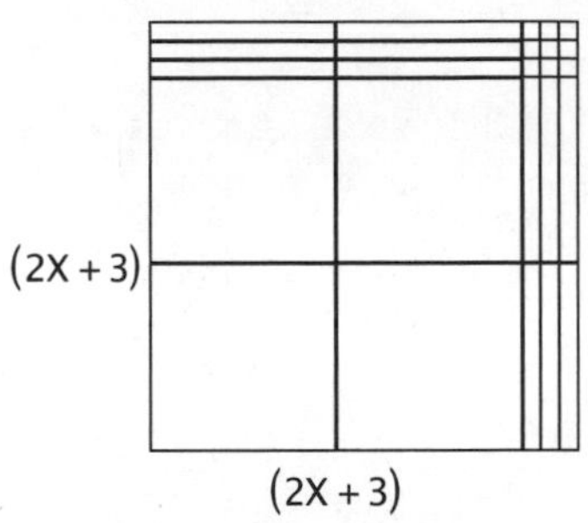

2. $2X^2 + 12X + 16 = (2)(X^2 + 6X + 8) =$
$(2)(X+4)(X+2)$

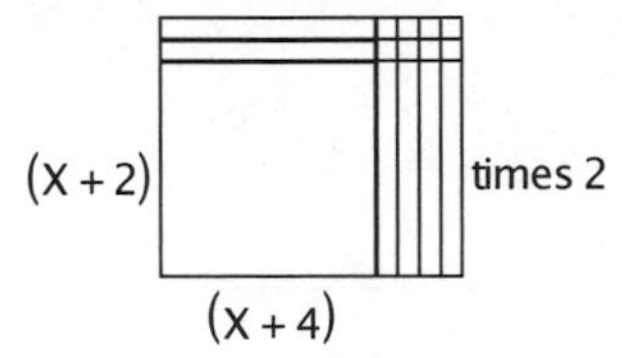

3.
$$\begin{array}{r} 2X+2 \\ \times\ X+1 \\ \hline 2X+2 \\ 2X^2+2X \\ \hline 2X^2+4X+2 \end{array}$$

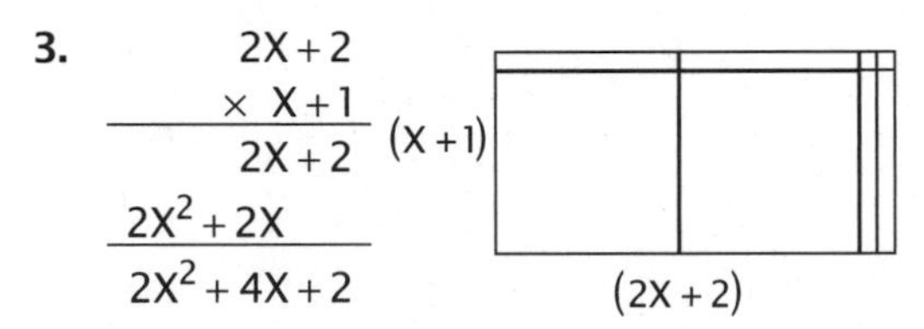

4.
$$\begin{array}{r} 2X+\ \ 4 \\ \times\ X+\ \ 5 \\ \hline 10X+20 \\ 2X^2+\ \ 4X \\ \hline 2X^2+14X+20 \end{array}$$

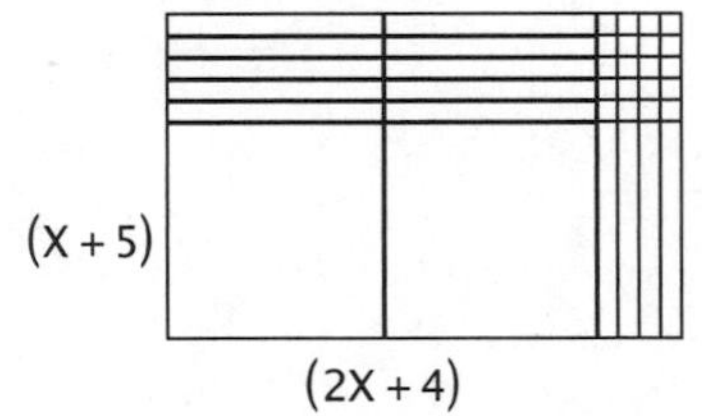

5. $(4X+3)(X+2)$

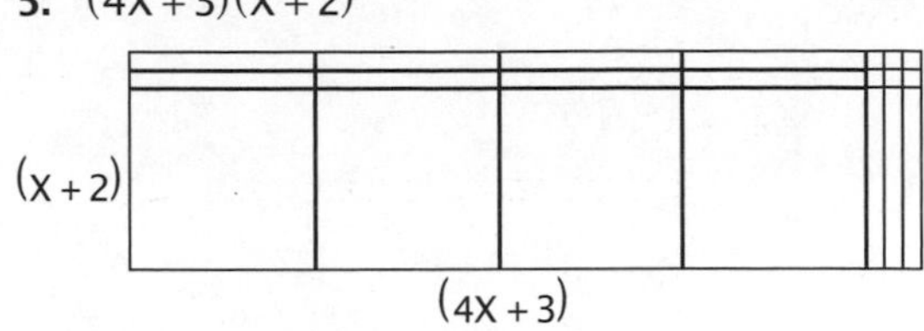

6.
$$\begin{array}{r} 4X+3 \\ \times\ X+2 \\ \hline 8X+6 \\ 4X^2+3X \\ \hline 4X^2+11X+6 \end{array}$$

7. $(2X+1)(X+5)$

8.
$$\begin{array}{r} 2X+1 \\ \times\ X+5 \\ \hline 10X+5 \\ 2X^2+\ \ X \\ \hline 2X^2+11X+5 \end{array}$$

(X + 5)
(2X + 1)

9. $(X+3)(X+1)$

10.
$$\begin{array}{r} X+3 \\ \times\ X+1 \\ \hline X+3 \\ X^2+3X \\ \hline X^2+4X+3 \end{array}$$

(X + 1)
(X + 3)

11. $B^2B^6C^2B^{-5}C^{-5} = B^{2+6+(-5)}C^{2+(-5)}$
$= B^3C^{-3} \text{ or } \dfrac{B^3}{C^3}$

12. $Y^5 \cdot Y^A = Y^{5+A}$

13. $\dfrac{D^8C^{-3}A^{-2}}{A^0D^{-7}C^2} = D^8C^{-3}A^{-2}A^{-0}D^7C^{-2}$
$= A^{-2+(-0)}C^{-3+(-2)}D^{8+7}$
$= A^{-2}C^{-5}D^{15} \text{ or } \dfrac{D^{15}}{A^2C^5}$

14. $\dfrac{A^5D^{-6}A^{-7}}{C^{-3}D^{-8}} = A^5D^{-6}A^{-7}C^3D^8$
$= A^{5+(-7)}C^3D^{(-6)+8}$
$= A^{-2}C^3D^2 \text{ or } \dfrac{C^3D^2}{A^2}$

15. $3 \times 10^5 + 5 \times 10^0 + 2 \times 10^{-2} + 8 \times 10^{-3} =$
$300,000 + 5 + .02 + .008 = 300,005.028$

16. $5Y + 4X = 10$

$5Y = -4X + 10$

$Y = -\frac{4}{5}X + 2$

see graph

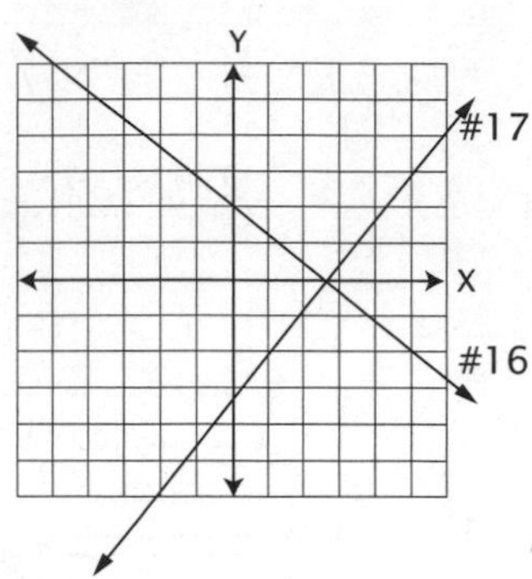

17. $m = \frac{5}{4}$

$(-2) = \frac{5}{4}(1) + b$

$-2 = \frac{5}{4} + b$

$b = -\frac{13}{4}$

$Y = \frac{5}{4}X - \frac{13}{4}$ or $5X - 4Y = 13$

or $-5X + 4Y = -13$ (see graph)

18.

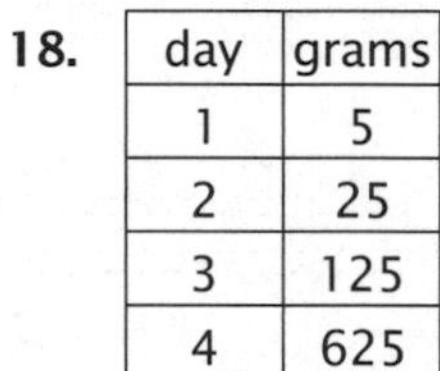

day	grams
1	5
2	25
3	125
4	625

19.

day	grams
1	5^1
2	5^2
3	5^3
4	5^4

20. 8 days $= 5^8$

Y days $= 5^Y$

Lesson Practice 23A

1. $(X-5)(X-2)$

$$\begin{array}{r} X-5 \\ \times\ X-5 \\ \hline -2X+10 \\ X^2-5X \quad \\ \hline X^2-7X+10 \end{array}$$

(X - 2) (X - 5)

2. $(X-6)(X-1)$

$$\begin{array}{r} X-6 \\ \times\ X-1 \\ \hline -X+6 \\ X^2-6X \quad \\ \hline X^2-7X+6 \end{array}$$

(X - 1) (X - 6)

3. $(X-7)(X-2)$

$$\begin{array}{r} X-7 \\ \times\ X-2 \\ \hline -2X+14 \\ X^2-7X \quad \\ \hline X^2-9X+14 \end{array}$$

(X - 2) (X - 7)

4. $(X-4)(X-3)$

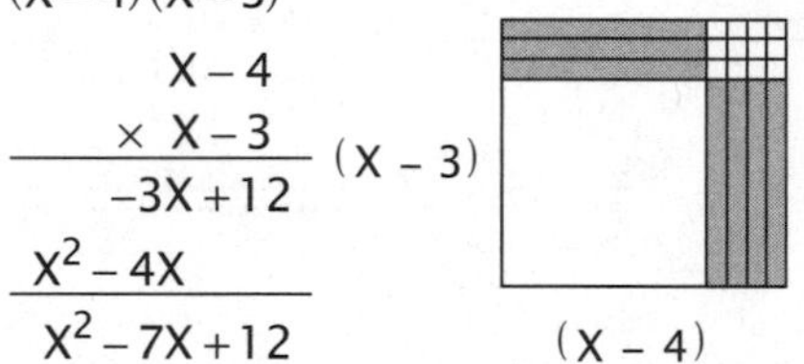

$$\begin{array}{r} X-4 \\ \times\ X-3 \\ \hline -3X+12 \\ X^2-4X \quad \\ \hline X^2-7X+12 \end{array}$$

(X - 3) (X - 4)

5. $(X-8)(X-1)$

$$\begin{array}{r} X-8 \\ \times\ X-1 \\ \hline -X+8 \\ X^2-8X \quad \\ \hline X^2-9X+8 \end{array}$$

(X - 1) (X - 8)

6. $(X-7)(X-3)$

$$\begin{array}{r} X-7 \\ \times\ X-3 \\ \hline -3X+21 \\ X^2-7X \quad \\ \hline X^2-10X+21 \end{array}$$

(X - 3) (X - 7)

7. $(X-9)(X-3)$

$$\begin{array}{r} X-9 \\ \times\ X-3 \\ \hline -3X+27 \\ X^2-\ 9X \\ \hline X^2-12X+27 \end{array}$$

(X - 3)

(X - 9)

8. $(X-5)(X-6)$

$$\begin{array}{r} X-5 \\ \times\ X-6 \\ \hline -6X+30 \\ X^2-\ 5X \\ \hline X^2-11X+30 \end{array}$$

(X - 5)

(X - 6)

9. $(X-9)(X-10)$

$$\begin{array}{r} X-9 \\ \times\ X-10 \\ \hline -10X+90 \\ X^2-\ 9X \\ \hline X^2-19X+90 \end{array}$$

(X - 9)

(X - 10)

10. $(X-11)(X-3)$

$$\begin{array}{r} X-11 \\ \times\ X-\ 3 \\ \hline -3X+33 \\ X^2-11X \\ \hline X^2-14X+33 \end{array}$$

(X - 3)

(X - 11)

11. $(X+7)(X-3)$

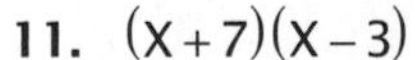

$$\begin{array}{r} X+\ 7 \\ \times\ X-\ 3 \\ \hline -3X-21 \\ X^2+7X \\ \hline X^2+4X-21 \end{array}$$

(X - 3)

(X + 7)

12. $(X+7)(X-5)$

$$\begin{array}{r} X+\ 7 \\ \times\ X-\ 5 \\ \hline -5X-35 \\ X^2+7X \\ \hline X^2+2X-35 \end{array}$$

(X - 5)

(X + 7)

13. $(X+6)(X-3)$

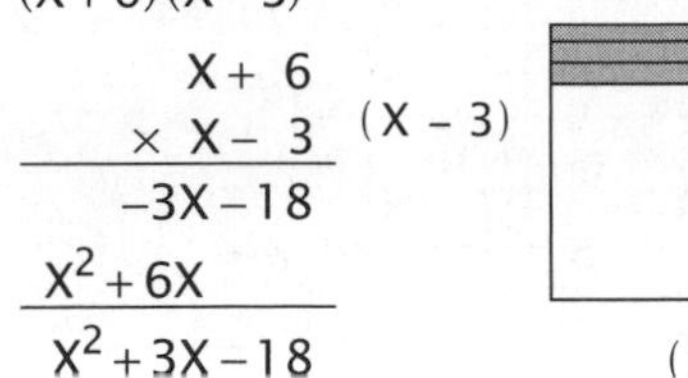

$$\begin{array}{r} X+\ 6 \\ \times\ X-\ 3 \\ \hline -3X-18 \\ X^2+6X \\ \hline X^2+3X-18 \end{array}$$

(X - 3)

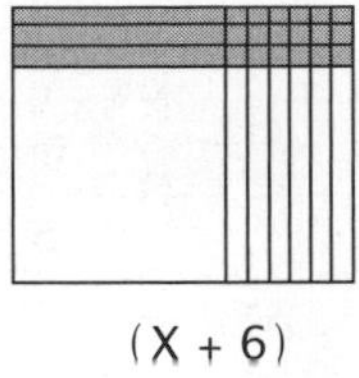

(X + 6)

14. $(X-9)(X+4)$

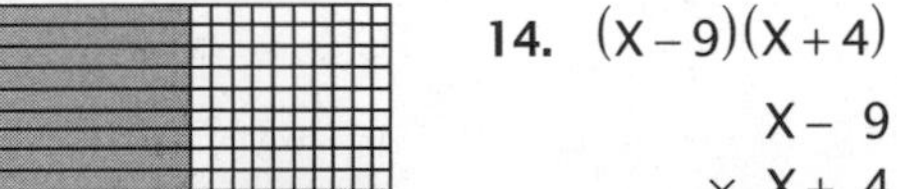

$$\begin{array}{r} X-\ 9 \\ \times\ X+\ 4 \\ \hline 4X-36 \\ X^2-9X \\ \hline X^2-5X-36 \end{array}$$

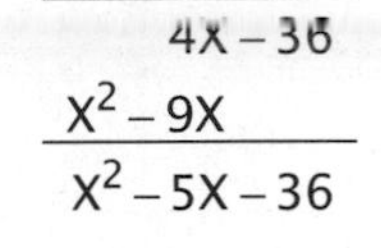

(X + 4)

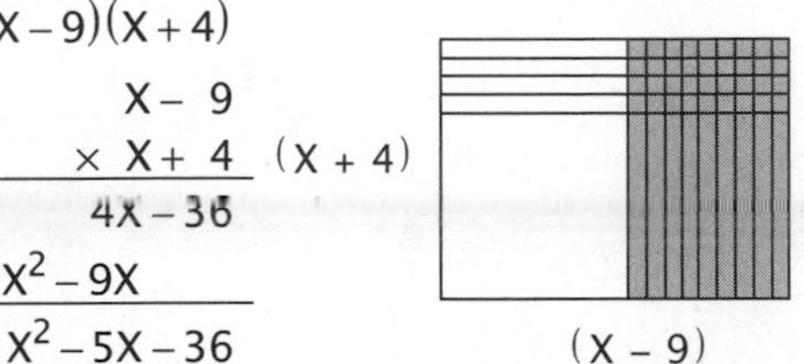

(X - 9)

15. $(2X+1)(X-5)$

$$\begin{array}{r} 2X+1 \\ \times\ X-5 \\ \hline -10X-5 \\ 2X^2+\ X \\ \hline 2X^2-9X-5 \end{array}$$

(X - 5)

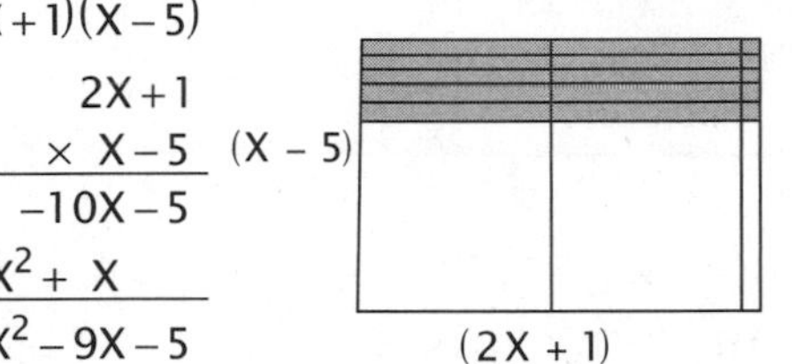

(2X + 1)

16. $(2X-3)(X+4)$

$$\begin{array}{r} 2X-\ 3 \\ \times\ X+\ 4 \\ \hline 8X-12 \\ 2X^2-3X \\ \hline 2X^2+5X-12 \end{array}$$

(X + 4)

(2X - 3)

Lesson Practice 23B

1. $(X-4)(X-2)$

$$\begin{array}{r} X-4 \\ \times\ X-2 \\ \hline -2X+8 \\ X^2-4X \\ \hline X^2-6X+8 \end{array}$$

(X - 2)

(X - 4)

2. $(X-10)(X-8)$

$$\begin{array}{r} X-10 \\ \times\ X-\ 8 \\ \hline -8X+80 \\ X^2-10X \\ \hline X^2-18X+80 \end{array}$$

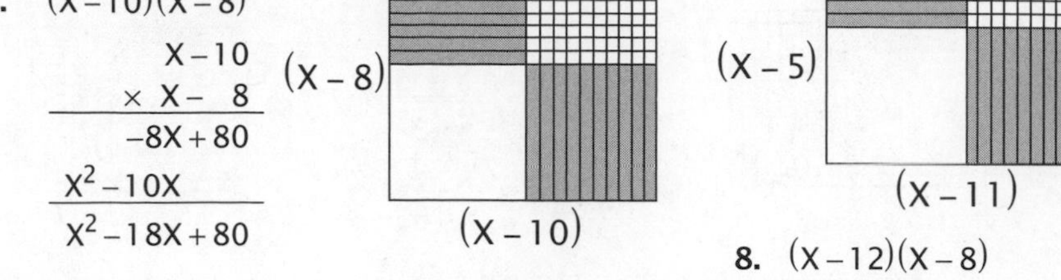

3. $(X-5)(X-3)$

$$\begin{array}{r} X-\ 5 \\ \times\ X-\ 3 \\ \hline -3X+15 \\ X^2-5X \\ \hline X^2-8X+15 \end{array}$$

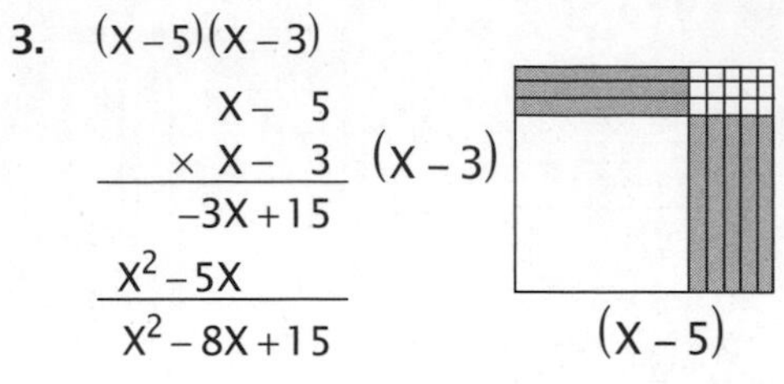

4. $(X-5)(X-4)$

$$\begin{array}{r} X-\ 5 \\ \times\ X-\ 4 \\ \hline -4X+20 \\ X^2-5X \\ \hline X^2-9X+20 \end{array}$$

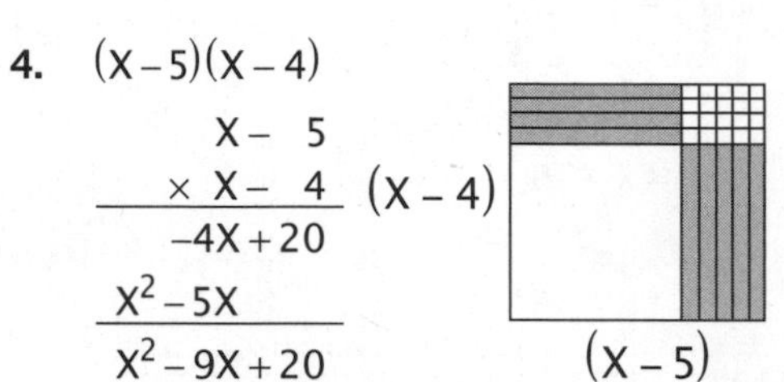

5. $(X-9)(X-1)$

$$\begin{array}{r} X-9 \\ \times\ X-1 \\ \hline -X+9 \\ X^2-9X \\ \hline X^2-10X+9 \end{array}$$

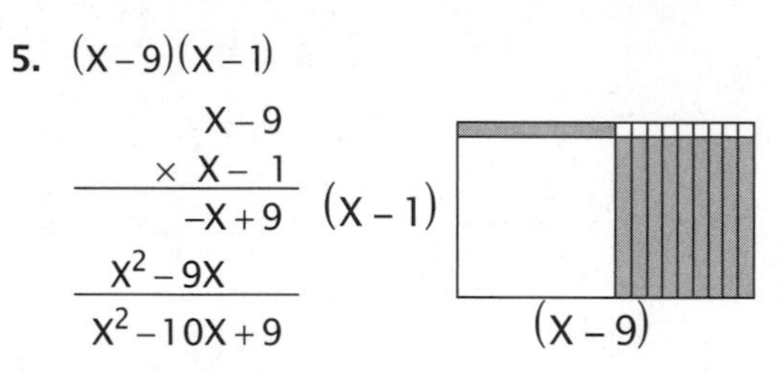

6. $(X-1)(X-3)$

$$\begin{array}{r} X-1 \\ \times\ X-3 \\ \hline -3X+3 \\ X^2-\ X \\ \hline X^2-4X+3 \end{array}$$

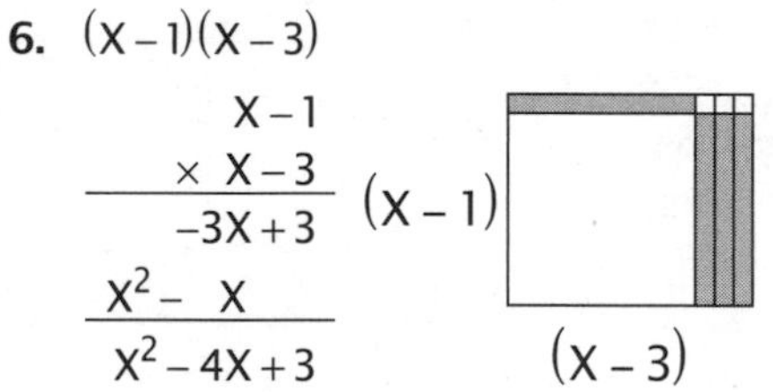

7. $(X-11)(X-5)$

$$\begin{array}{r} X-\ 11 \\ \times\ X-\ 5 \\ \hline -5X+55 \\ X^2-11X \\ \hline X^2-16X+55 \end{array}$$

8. $(X-12)(X-8)$

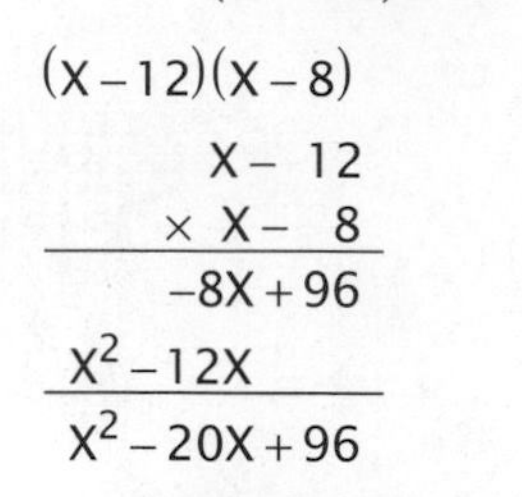

$$\begin{array}{r} X-\ 12 \\ \times\ X-\ 8 \\ \hline -8X+96 \\ X^2-12X \\ \hline X^2-20X+96 \end{array}$$

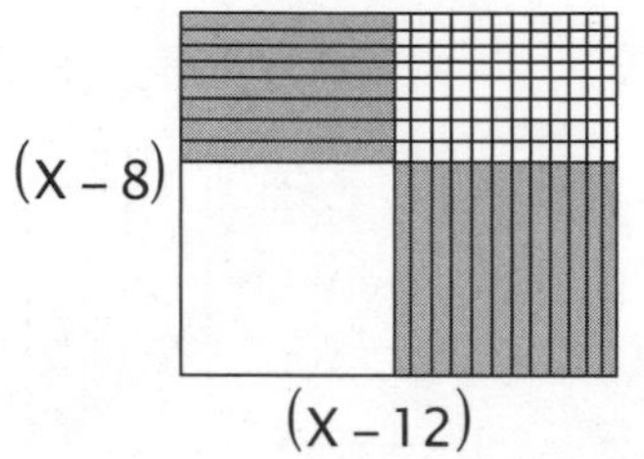

9. $(X-7)(X-6)$

$$\begin{array}{r} X-\ 7 \\ \times\ X-\ 6 \\ \hline -6X+42 \\ X^2-\ 7X \\ \hline X^2-13X+42 \end{array}$$

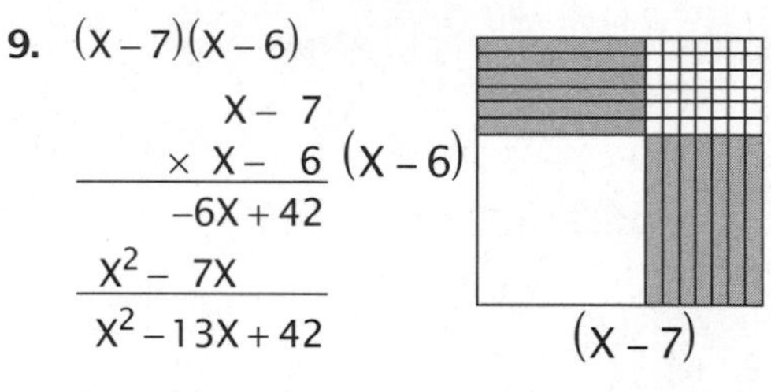

10. $(X-8)(X-3)$

$$\begin{array}{r} X-\ 8 \\ \times\ X-\ 3 \\ \hline -3X+24 \\ X^2-\ 8X \\ \hline X^2-11X+24 \end{array}$$

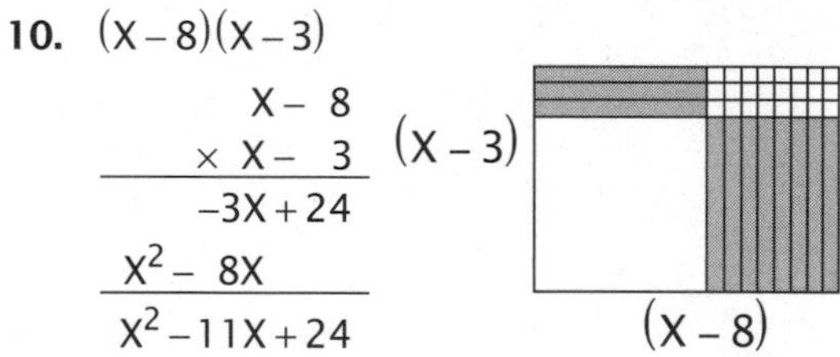

11. $(X+3)(X-1)$

$$\begin{array}{r} X+3 \\ \times\ X-1 \\ \hline -X-3 \\ X^2+3X \\ \hline X^2+2X-3 \end{array}$$

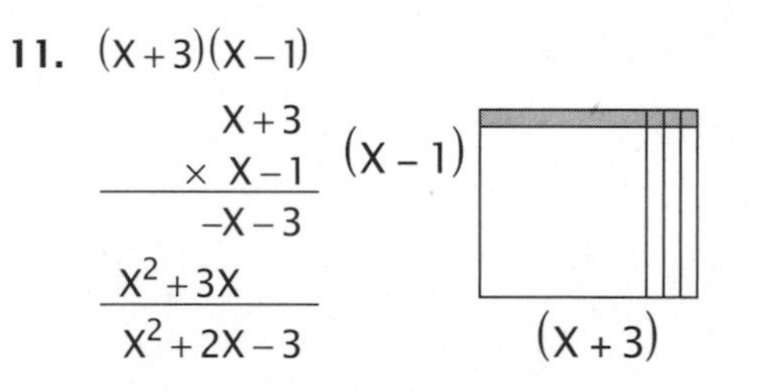

12. $(X+6)(X-3)$

$$\begin{array}{r} X+\ 6 \\ \times\ X-\ 3 \\ \hline -3X-18 \\ X^2+6X \\ \hline X^2+3X-18 \end{array}$$

$(X-3)$ $(X+6)$

13. $(X-5)(X+4)$

$$\begin{array}{r} X-\ 5 \\ \times\ X+\ 4 \\ \hline 4X-20 \\ X^2-5X \\ \hline X^2-\ X-20 \end{array}$$

$(X+4)$ $(X-5)$

14. $(X+5)(X-3)$

$$\begin{array}{r} X+\ 5 \\ \times\ X-\ 3 \\ \hline -3X-15 \\ X^2+5X \\ \hline X^2+2X-15 \end{array}$$

15. $(5X-1)(X+2)$

$$\begin{array}{r} 5X-\ 1 \\ \times\ X+\ 2 \\ \hline 10X-\ 2 \\ 5X^2+X \\ \hline 5X^2+9X-2 \end{array}$$

16. $(4X-1)(X+2)$

$$\begin{array}{r} 4X-1 \\ \times\ X+2 \\ \hline 8X-2 \\ 4X^2-\ X \\ \hline 4X^2+7X-2 \end{array}$$

Systematic Review 23C

1. $(X-5)(X+2)$

$$\begin{array}{r} X-5 \\ \times\ X+2 \\ \hline 2X-10 \\ X^2-5X \\ \hline X^2-3X-10 \end{array}$$

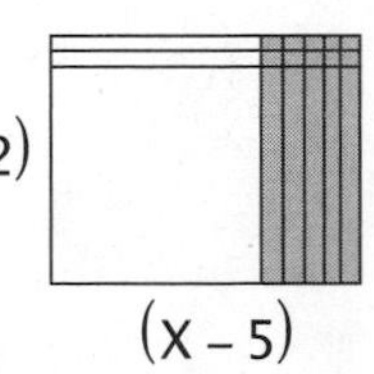

$(X+2)$ $(X-5)$

2. $(X+4)(X-1)$

$$\begin{array}{r} X+4 \\ \times\ X-1 \\ \hline -X-4 \\ X^2+4X \\ \hline X^2+3X-4 \end{array}$$

$(X-1)$ $(X+4)$

3.

$$\begin{array}{r} X-3 \\ \times\ X-9 \\ \hline -9X+27 \\ X^2-3X \\ \hline X^2-12X+27 \end{array}$$

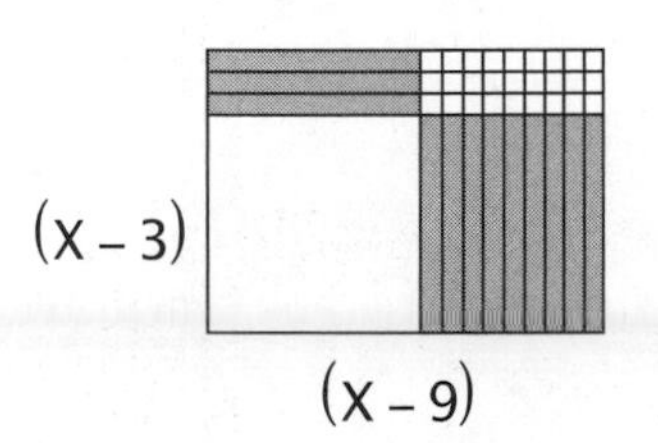

$(X-3)$ $(X-9)$

4.

$$\begin{array}{r} X-3 \\ \times\ X-3 \\ \hline -3X+9 \\ X^2-3X \\ \hline X^2-6X+9 \end{array}$$

$(X-3)$ $(X-3)$

5. $(X+2)(X-1)$

6.

$$\begin{array}{r} X+2 \\ \times\ X-1 \\ \hline -X-2 \\ X^2+2X \\ \hline X^2+\ X-2 \end{array}$$

7. $(X+5)(X-2)$

8.

$$\begin{array}{r} X+5 \\ \times\ X-2 \\ \hline -2X-10 \\ X^2+5X \\ \hline X^2+3X-10 \end{array}$$

9. $(2X+1)(X+3)$

10.

$$\begin{array}{r} 2X+1 \\ \times\ X+3 \\ \hline 6X+3 \\ 2X^2+\ X \\ \hline 2X^2+7X+3 \end{array}$$

11. $3^4 \times 3^{-2} \div 3^3 = 3^{4+(-2)-3} = 3^{-1}$

12. $\frac{7^{-10}}{7^5} = 7^{-10}7^{-5} = 7^{-10+(-5)} = 7^{-15}$

13. $\frac{A^5B^2A^{-4}}{A^3B^7} = A^5B^2A^{-4}A^{-3}B^{-7} =$

$A^{5+(-4)+(-3)}B^{2+(-7)} = A^{-2}B^{-5}$

14. $2AB^{-2} + \frac{4B^{-1}}{B^{-1}A^{-1}} + \frac{3A^2}{B^2A^1} =$

$2AB^{-2} + 4B^{-1}B^1A^1 + 3A^2B^{-2}A^{-1} =$

$2AB^{-2} + 4B^{-1+1}A + 3A^{2+(-1)}B^{-2} =$

$2AB^{-2} + 4A + 3AB^{-2} =$

$5AB^{-2} + 4A$

15. $Y = -4X$

$3Y = 2X + 7 \Rightarrow 3(-4X) = 2X + 7$

$-12X = 2X + 7$

$-14X = 7$

$X = -\frac{1}{2}$

$Y = -4X \Rightarrow Y = -4\left(-\frac{1}{2}\right)$

$Y = 2$

$\left(-\frac{1}{2}, 2\right)$

16. $7(N+2) + 2(N) - 6(N+4) = -1$

$7N + 14 + 2N - 6N - 24 = -1$

$7N + 2N - 6N = -1 - 14 + 24$

$3N = 9$

$N = 3$

3, 5, 7

17.

$(.10D + .05N = .95)(100) \Rightarrow \quad 10D + 5N = 95$

$(D + N = 12)(-5) \Rightarrow \quad -5D - 5N = -60$

$5D = 35$

$D = 7$

$D + N = 12 \Rightarrow (7) + N = 12$

$N = 5$

18. $\frac{2}{3} \div \frac{5}{6} \times \frac{1}{2} = \frac{2}{3} \times \frac{6}{5} \times \frac{1}{2} = \frac{2}{5}$

19. $(100)(.2X - .02X + 1.4 = 2.09)$

$20X - 2X + 140 = 209$

$18X = 209 - 140$

$18X = 69$

$X = \frac{69}{18} = \frac{23}{6} = 3\frac{5}{6}$

20. $5\frac{1}{2}\% = 5.5\% = .055$

$.055 \times 400 = 22$

Systematic Review 23D

1. $(X-2)(X+1)$

$$\begin{array}{r} X-2 \\ \times X+1 \\ \hline X-2 \\ X^2-2X \\ \hline X^2-X-2 \end{array}$$

(X + 1) (X − 2)

2. $(X-1)(X+3)$

$$\begin{array}{r} X-1 \\ \times X+3 \\ \hline 3X-3 \\ X^2-X \\ \hline X^2+2X-3 \end{array}$$

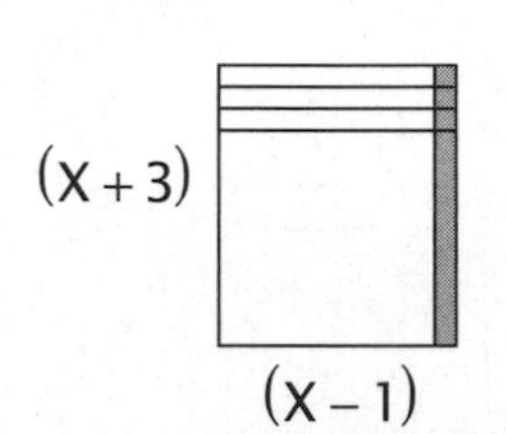

3.

$$\begin{array}{r} X-3 \\ \times X+9 \\ \hline 9X-27 \\ X^2-3X \\ \hline X^2+6X-27 \end{array}$$

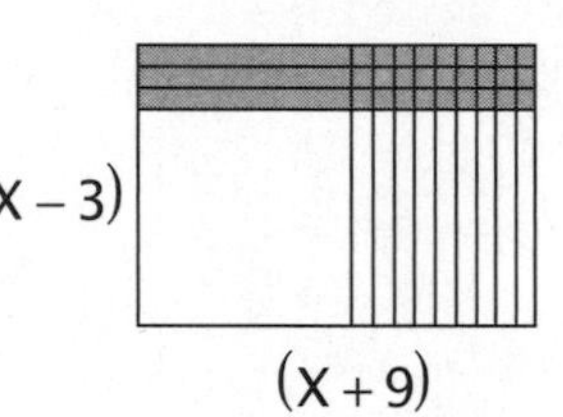

4.

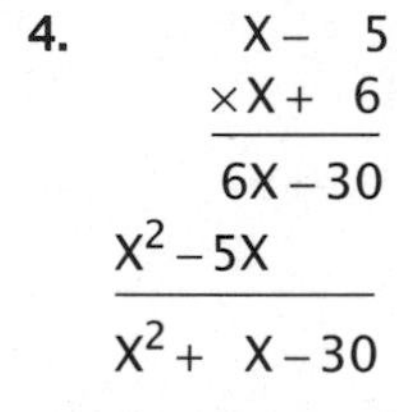

$$\begin{array}{r} X-5 \\ \times X+6 \\ \hline 6X-30 \\ X^2-5X \\ \hline X^2+X-30 \end{array}$$

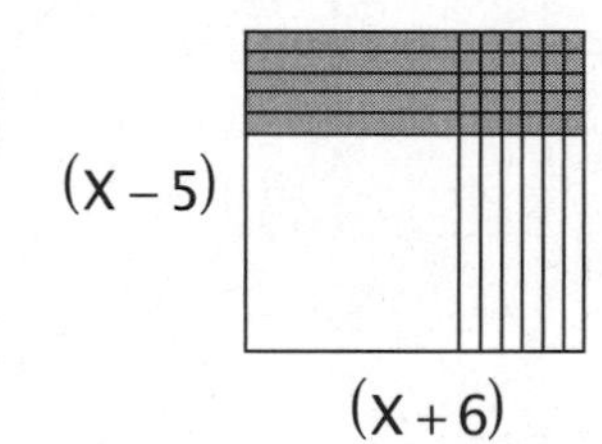

5. $(X-4)(X+1)$

6.

$$\begin{array}{r} X-4 \\ \times X+1 \\ \hline X-4 \\ X^2-4X \\ \hline X^2-3X-4 \end{array}$$

7. $(X-3)(X+1)$

8.
$$\begin{array}{r} X-3 \\ \times X+1 \\ \hline X-3 \\ X^2-3X \\ \hline X^2-2X-3 \end{array}$$

9. $(X-3)(X+2)$

10.
$$\begin{array}{r} X-3 \\ \times X+2 \\ \hline 2X-6 \\ X^2-3X \\ \hline X^2-X-6 \end{array}$$

11. $(10^2)^7 = 10^{2\times7} = 10^{14}$

12. $\left[(5^2)^4\right]^3 = 5^{2\times4\times3} = 5^{24}$

13. $\dfrac{D^{-4}D^{3}D^{-2}}{D^{4}D^{-5}} = D^{-4}D^{3}D^{-2}D^{-4}D^{5}$

$= D^{-4+3+(-2)+(-4)+5} = D^{-2}$

14. $BB^2 + \dfrac{3B^{-1}}{B^{-4}} + \dfrac{5B^4}{B^{-1}} =$

$B^1B^2 + 3B^{-1}B^4 + 5B^4B^1 =$

$B^{1+2} + 3B^{-1+4} + 5B^{4+1} =$

$B^3 + 3B^3 + 5B^5 = 4B^3 + 5B^5$

15. $Y = -4X + 5$

$2Y = 4X - 3 \Rightarrow 2(-4X+5) = 4X-3$

$-8X + 10 = 4X - 3$

$10 + 3 = 4X + 8X$

$13 = 12X$

$\dfrac{13}{12} = X = 1\dfrac{1}{12}$

$Y = -4X + 5 \Rightarrow Y = -4\left(\dfrac{13}{12}\right) + 5$

$Y = \dfrac{-52}{12} + 5$

$Y = -\dfrac{13}{3} + \dfrac{15}{3} = \dfrac{2}{3}$

$\left(\dfrac{13}{12}, \dfrac{2}{3}\right)$

16. $4(N+1) + 3(N+2) - 8(N) + 11 = 0$

$4N + 4 + 3N + 6 - 8N + 11 = 0$

$4N + 3N - 8N = -4 - 6 - 11$

$-N = -21$

$N = 21$

21; 22; 23

17. $(.10D + .05N = 3.30)(100)$

$\Rightarrow 10D + 5N = 330$

$(D + N = 45)(-5) \Rightarrow -5D - 5N = -225$

$5D = 105$

$D = 21$

$(D + N = 45) \Rightarrow (21) + N = 45$

$N = 24$

18. $\dfrac{1}{2} \div \dfrac{1}{2} \times \dfrac{3}{4} = \dfrac{1}{2} \times \dfrac{2}{1} \times \dfrac{3}{4} = \dfrac{3}{4}$

19. $(100)(1.03X + .2X - .73X = .45)$

$103X + 20X - 73X = 45$

$50X = 45$

$X = \dfrac{45}{50} = \dfrac{9}{10}$ or .9

20. $5\dfrac{2}{5}\% = 5.4\% = .054$

$.054 \times 250 = 13.5$

Systematic Review 23E

1. $(X-3)(X+1)$
$$\begin{array}{r} X-3 \\ \times X+1 \\ \hline X-3 \\ X^2-3X \\ \hline X^2-2X-3 \end{array}$$

(X + 1)

(X – 3)

2. $(X+4)(X-1)$
$$\begin{array}{r} X+4 \\ \times X-1 \\ \hline -X-4 \\ X^2+4X \\ \hline X^2+3X-4 \end{array}$$

(X – 1)

(X + 4)

3. $$\begin{array}{r} X-4 \\ \times X+2 \\ \hline 2X-8 \\ X^2-4X \\ \hline X^2-2X-8 \end{array}$$

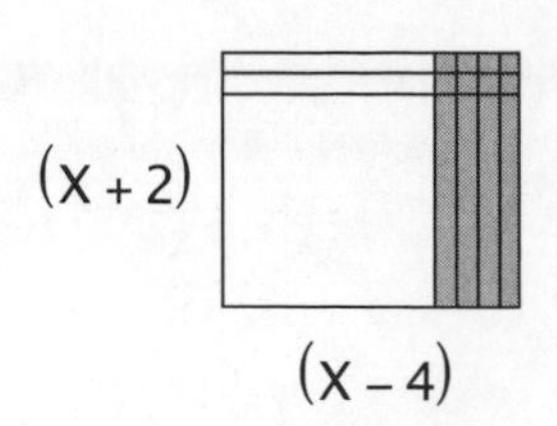

4. $$\begin{array}{r} X-3 \\ \times X+5 \\ \hline 5X-15 \\ X^2-3X \\ \hline X^2+2X-15 \end{array}$$

(X – 3)

(X + 5)

5. $(X-5)(X-2)$

6. $$\begin{array}{r} X-5 \\ \times X-2 \\ \hline -2X+10 \\ X^2-5X \\ \hline X^2-7X+10 \end{array}$$

7. $(3X-7)(X-1)$

8. $$\begin{array}{r} 3X-7 \\ \times X-1 \\ \hline -3X+7 \\ 3X^2-7X \\ \hline 3X^2-10X+7 \end{array}$$

9. $3X^2+15X-18=3(X^2+5X-6)=3(X+6)(X-1)$

10. $$\begin{array}{r} X+6 \\ \times X-1 \\ \hline -X-6 \\ X^2+6X \\ \hline X^2+5X-6 \end{array}$$

$(X^2+5X-6)(3)=3X^2+15X-18$

11. $5^4\times5^{-6}\div5^2=5^{4+(-6)-2}=5^{-4}$

12. $\frac{1}{6^{-1}}=6^1=6$

13. $4Q^{-1}Y^{-2}+\frac{5QY^{-3}}{Q^{-1}Y^{-2}}=$

$4Q^{-1}Y^{-2}+5QY^{-3}Q^1Y^2=$

$4Q^{-1}Y^{-2}+5Q^{1+1}Y^{-3+2}=$

$4Q^{-1}Y^{-2}+5Q^2Y^{-1}$

14. $5M^4N^2M^{-1}+\frac{2NM^4}{N^{-3}M}=$

$5M^{4+(-1)}N^2+2N^1M^4N^3M^{-1}=$

$5M^3N^2+2M^3N^4$

15. $$\begin{array}{l} X-Y=-2 \\ 3X+Y=18 \\ \hline 4X=16 \\ X=4 \end{array}$$

$X-Y=-2 \Rightarrow (4)-Y=-2$

$4+2=Y$

$6=Y$

16. $11(N)+2(N+2)=6(N+4)+1$

$11N+2N+4=6N+24+1$

$11N+2N-6N=24+1-4$

$7N=21$

$N=3$

3; 5; 7

17.

$$\begin{array}{rr} (.25Q+.10D=2.00)(100)\Rightarrow & 25Q+10D=200 \\ (Q+D=14)(-25)\Rightarrow & -25Q-25D=-350 \\ \hline & -15D=-150 \\ & D=10 \end{array}$$

$Q+D=14\Rightarrow Q+(10)=14$

$Q=4$

18. $\frac{3}{7}\times\frac{14}{15}\div\frac{1}{2}=\frac{3}{7}\times\frac{14}{15}\times\frac{2}{1}=\frac{4}{5}$

19. $36-8F=20F+12$

$36-12=20F+8F$

$24=28F$

$\frac{24}{28}=F=\frac{6}{7}$

20. $6.8\%=.068$

$.068\times95=6.46$

Lesson Practice 24A

1. $\sqrt{X^2+4X+4}=X+2$

check: $$\begin{array}{r} X+2 \\ \times X+2 \\ \hline 2X+4 \\ X^2+2X \\ \hline X^2+4X+4 \end{array}$$

2. $\sqrt{X^2+6X+9} = X+3$

check:
$$\begin{array}{r} X+3 \\ \times\ X+3 \\ \hline 3X+9 \\ X^2+3X \\ \hline X^2+6X+9 \end{array}$$

3. $\sqrt{X^2+10X+25} = X+5$

check:
$$\begin{array}{r} X+\ 5 \\ \times\ X+\ 5 \\ \hline 5X+25 \\ X^2+\ 5X \\ \hline X^2+10X+25 \end{array}$$

4.
$$\begin{array}{r} X+2 \\ X+3\overline{\smash{)}\,X^2+5X+6} \\ -(X^2+3X) \\ \hline 2X+6 \\ -(2X+6) \\ \hline 0 \end{array}$$

check:
$$\begin{array}{r} X+2 \\ \times\ X+3 \\ \hline 3X+6 \\ X^2+2X \\ \hline X^2+5X+6 \end{array}$$

5.
$$\begin{array}{r} X+\ 6\ R\ 6 \\ X+5\overline{\smash{)}\,X^2+11X+36} \\ -(X^2+\ 5X) \\ \hline 6X+36 \\ -(6X+30) \\ \hline 6 \end{array}$$

check:
$$\begin{array}{r} X+\ 6 \\ \times X+\ 5 \\ \hline 5X+30 \\ X^2+6X \\ \hline X^2+11X+30 \\ +\ 6 \\ \hline X^2+11X+36 \end{array}$$

6.
$$\begin{array}{r} X+\ 4 \\ X+3\overline{\smash{)}\,X^2+7X+12} \\ -(X^2+3X) \\ \hline 4X+12 \\ -(4X+12) \\ \hline 0 \end{array}$$

check:
$$\begin{array}{r} X+\ 4 \\ \times X+\ 3 \\ \hline 3X+12 \\ X^2+4X \\ \hline X^2+7X+12 \end{array}$$

7.
$$\begin{array}{r} X+\ 2 \\ X+8\overline{\smash{)}\,X^2+10X+16} \\ -(X^2+\ 8X) \\ \hline 2X+16 \\ -(2X+16) \\ \hline 0 \end{array}$$

check:
$$\begin{array}{r} X+\ 2 \\ \times X+\ 8 \\ \hline 8X+16 \\ X^2+\ 2X \\ \hline X^2+10X+16 \end{array}$$

8.
$$\begin{array}{r} X+\ 7 \\ X+3\overline{\smash{)}\,X^2+10X+21} \\ -(X^2+\ 3X) \\ \hline 7X+21 \\ -(7X+21) \\ \hline 0 \end{array}$$

check:
$$\begin{array}{r} X+\ 7 \\ \times X+\ 3 \\ \hline 3X+21 \\ X^2+\ 7X \\ \hline X^2+10X+21 \end{array}$$

9.

$$\begin{array}{r} 2X+1 \\ X+3\,\overline{)\,2X^2+7X+3} \\ \underline{-(2X^2+6X)} \\ X+3 \\ \underline{-(X+3)} \\ 0 \end{array}$$

check:

$$\begin{array}{r} 2X+1 \\ \underline{\times X+3} \\ 6X+3 \\ \underline{2X^2+X} \\ 2X^2+7X+3 \end{array}$$

10.

$$\begin{array}{r} X^2+5X+7 \\ X+4\,\overline{)\,X^3+9X^2+27X+28} \\ \underline{-(X^3+4X^2)} \\ 5X^2+27X \\ \underline{-(5X^2+20X)} \\ 7X+28 \\ \underline{-(7X+28)} \\ 0 \end{array}$$

check:

$$\begin{array}{r} X^2+5X+7 \\ \underline{\times X+4} \\ 4X^2+20X+28 \\ \underline{X^3+5X^2+7X} \\ X^3+9X^2+27X+28 \end{array}$$

11.

$$\begin{array}{r} X^2+3X+9 \\ X+1\,\overline{)\,X^3+4X^2+12X+9} \\ \underline{-(X^3+X^2)} \\ 3X^2+12X \\ \underline{-(3X^2+3X)} \\ 9X+9 \\ \underline{-(9X+9)} \\ 0 \end{array}$$

check:

$$\begin{array}{r} X^2+3X+9 \\ \underline{\times X+1} \\ X^2+3X+9 \\ \underline{X^3+3X^2+9X} \\ X^3+4X^2+12X+9 \end{array}$$

Lesson Practice 24B

1. $\sqrt{X^2+12X+36}=X+6$

check:

$$\begin{array}{r} X+6 \\ \underline{\times X+6} \\ 6X+36 \\ \underline{X^2+6X} \\ X^2+12X+36 \end{array}$$

2. $\sqrt{X^2+14X+49}=X+7$

check:

$$\begin{array}{r} X+7 \\ \underline{\times X+7} \\ 7X+49 \\ \underline{X^2+7X} \\ X^2+14X+49 \end{array}$$

3. $\sqrt{4X^2+4X+1}=2X+1$

check:

$$\begin{array}{r} 2X+1 \\ \underline{\times 2X+1} \\ 2X+1 \\ \underline{4X^2+2X} \\ 4X^2+4X+1 \end{array}$$

4.

$$\begin{array}{r} X+7 \\ X+3\overline{)X^2+10X+21} \\ \underline{-(X^2+3X)} \\ 7X+21 \\ \underline{-(7X+21)} \\ 0 \end{array}$$

check:

$$\begin{array}{r} X+7 \\ \underline{\times X+3} \\ 3X+21 \\ \underline{X^2+7X} \\ X^2+10X+21 \end{array}$$

5.

$$\begin{array}{r} X+5 \\ X+2\overline{)X^2+7X+10} \\ \underline{-(X^2+2X)} \\ 5X+10 \\ \underline{-(5X+10)} \\ 0 \end{array}$$

check:

$$\begin{array}{r} X+5 \\ \underline{\times X+2} \\ 2X+10 \\ \underline{X^2+5X} \\ X^2+7X+10 \end{array}$$

6.

$$\begin{array}{r} X+6 \\ X+1\overline{)X^2+7X+6} \\ \underline{-(X^2+X)} \\ 6X+6 \\ \underline{-(6X+6)} \\ 0 \end{array}$$

check:

$$\begin{array}{r} X+6 \\ \underline{\times X+1} \\ X+6 \\ \underline{X^2+6X} \\ X^2+7X+6 \end{array}$$

7.

$$\begin{array}{r} X+5 \\ X+3\overline{)X^2+8X+15} \\ \underline{-(X^2+3X)} \\ 5X+15 \\ \underline{-(5X+15)} \\ 0 \end{array}$$

check:

$$\begin{array}{r} X+5 \\ \underline{\times X+3} \\ 3X+15 \\ \underline{X^2+5X} \\ X^2+8X+15 \end{array}$$

8.

$$\begin{array}{r} X+5 \\ X+4\overline{)X^2+9X+20} \\ \underline{-(X^2+4X)} \\ 5X+20 \\ \underline{-(5X+20)} \\ 0 \end{array}$$

check:

$$\begin{array}{r} X+5 \\ \underline{\times X+4} \\ 4X+20 \\ \underline{X^2+5X} \\ X^2+9X+20 \end{array}$$

9.

$$\begin{array}{r} X+3 \\ X-2\overline{)X^2+X-6} \\ \underline{-(X^2-2X)} \\ 3X-6 \\ \underline{-(3X-6)} \\ 0 \end{array}$$

check:

$$\begin{array}{r} X+3 \\ \underline{\times X-2} \\ -2X-6 \\ \underline{X^2+3X} \\ X^2+X-6 \end{array}$$

10.

$$
\begin{array}{r}
X^2 - 3X + 5 \\
X-2 \overline{)\, X^3 - 5X^2 + 11X - 10} \\
\underline{-(X^3 - 2X^2)} \\
-3X^2 + 11X \\
\underline{-(-3X^2 + 6X)} \\
5X - 10 \\
\underline{-(5X - 10)} \\
0
\end{array}
$$

check:

$$
\begin{array}{r}
X^2 - 3X + 5 \\
\underline{\times X - 2} \\
-2X^2 + 6X - 10 \\
\underline{X^3 - 3X^2 + 5X \quad\quad} \\
X^3 - 5X^2 + 11X - 10
\end{array}
$$

11.

$$
\begin{array}{r}
X^2 + 4X - 7 \text{ R } 5 \\
X-3 \overline{)\, X^3 + X^2 - 19X + 26} \\
\underline{-(X^3 - 3X^2)} \\
4X^2 - 19X \\
\underline{-(4X^2 - 12X)} \\
-7X + 26 \\
\underline{-(-7X + 21)} \\
5
\end{array}
$$

check:

$$
\begin{array}{r}
X^2 + 4X - 7 \\
\underline{\times X - 3} \\
-3X^2 - 12X + 21 \\
\underline{X^3 + 4X^2 - 7X \quad\quad} \\
X^3 + X^2 - 19X + 21 \\
\underline{+ 5} \\
X^3 + X^2 - 19X + 26
\end{array}
$$

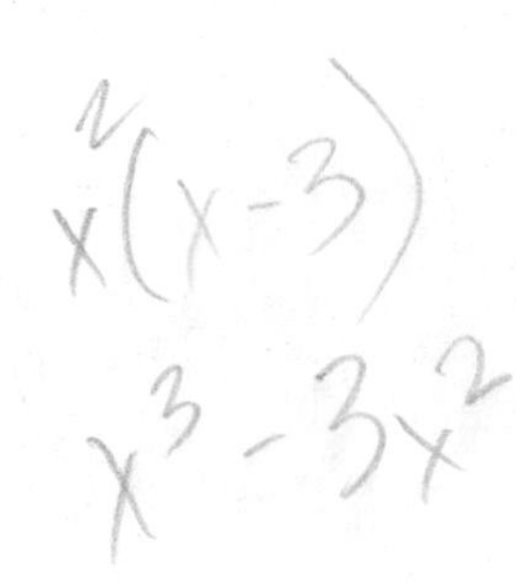

Systematic Review 24C

1.

$$
\begin{array}{r}
4X + 6 \text{ R } -5 \\
X+1 \overline{)\, 4X^2 + 10X + 1} \\
\underline{-(4X^2 + 4X)} \\
6X + 1 \\
\underline{-(6X + 6)} \\
-5
\end{array}
$$

2.

$$
\begin{array}{r}
4X + 6 \\
\underline{\times X + 1} \\
4X + 6 \\
\underline{4X^2 + 6X \quad\quad} \\
4X^2 + 10X + 6 \\
\underline{+ \quad (-5)} \\
4X^2 + 10X + 1
\end{array}
$$

3.

$$
\begin{array}{r}
2X + 2 \text{ R } 3 \\
2X+1 \overline{)\, 4X^2 + 6X + 5} \\
\underline{-(4X^2 + 2X)} \\
4X + 5 \\
\underline{-(4X + 2)} \\
3
\end{array}
$$

4.

$$
\begin{array}{r}
2X + 2 \\
\underline{\times 2X + 1} \\
2X + 2 \\
\underline{4X^2 + 4X \quad\quad} \\
4X^2 + 6X + 2 \\
\underline{+ \quad 3} \\
4X^2 + 6X + 5
\end{array}
$$

5.

$$
\begin{array}{r}
X + 5 \\
X+4 \overline{)\, X^2 + 9X + 20} \\
\underline{-(X^2 + 4X)} \\
5X + 20 \\
\underline{-(5X + 20)} \\
0
\end{array}
$$

6.

$$
\begin{array}{r}
X + 4 \\
\underline{\times X + 5} \\
5X + 20 \\
\underline{X^2 + 4X \quad\quad} \\
X^2 + 9X + 20
\end{array}
$$

7. $\sqrt{X^2 + 2X + 1} = X + 1$

8.
$$\begin{array}{r} X+1 \\ \times X+1 \\ \hline X+1 \\ X^2+X \quad \\ \hline X^2+2X+1 \end{array}$$

9. $(X^4)^3(Y^2)^6(Y^2)(Y^0) = X^{4\times3}Y^{2\times6}Y^{2+0} =$
$X^{12}Y^{12}Y^2 = X^{12}Y^{12+2} = X^{12}Y^{14}$

10. $\frac{A^5}{A^{-3}} = A^5A^3 = A^{5+3} = A^8$

11. $X^5X^{-2} \div X^{-4} = X^{5+(-2)-(-4)} = X^7$

12. $2XY^{-1} - \frac{3YY^{-2}}{X^{-1}} + 4X^{-1}Y^{-1} =$

$\frac{2X}{Y} - \frac{3Y^{1+(-2)}X}{1} + \frac{4}{XY} =$

$\frac{2X}{Y} - \frac{3Y^{-1}X}{1} + \frac{4}{XY} = \frac{2X}{Y} - \frac{3X}{Y} + \frac{4}{XY} =$

$\frac{-X}{Y} + \frac{4}{XY}$ or, using common denominators to add:

$\frac{-X^2}{XY} + \frac{4}{XY} = \frac{4-X^2}{XY}$

13. $.234 \times .21 = .04914$

14. $540 \div .15 = 3600$

15. $(-7)(-9) = 63$

16. $|4-8+1| = |-4+1| = |-3| = 3$

17.
$$\begin{array}{r} 6X^2-3X+2 \\ +X^2+5X-1 \\ \hline 7X^2+2X+1 \end{array}$$

18.
$$\begin{array}{r} X^2+4X-8 \\ +X^2-4X-9 \\ \hline 2X^2 \quad\quad -17 \end{array}$$

19. 97 is prime, so 1 and 97

20. addition and multiplication

Systematic Review 24D

1.
$$\begin{array}{r} 2X-3 \text{ R } 13 \\ X+1 \overline{\big) 2X^2-X+10} \\ -(2X^2+2X) \quad \\ \hline -3X+10 \\ -(-3X-3) \\ \hline 13 \end{array}$$

2.
$$\begin{array}{r} 2X-3 \\ \times X+1 \\ \hline 2X-3 \\ 2X^2-3X \quad\quad \\ \hline 2X^2-X-3 \\ + \quad\quad 13 \\ \hline 2X^2-X+10 \end{array}$$

3.
$$\begin{array}{r} 3X+2 \\ X+3 \overline{\big) 3X^2+11X+6} \\ -(3X^2+9X) \quad \\ \hline 2X+6 \\ -(2X+6) \\ \hline 0 \end{array}$$

4.
$$\begin{array}{r} 3X+2 \\ \times X+3 \\ \hline 9X+6 \\ 3X^2+2X \quad\quad \\ \hline 3X^2+11X+6 \end{array}$$

5.
$$\begin{array}{r} 3X-2 \text{ R } -1 \\ X+4 \overline{\big) 3X^2+10X-9} \\ -(3X^2+12X) \quad \\ \hline -2X-9 \\ -(-2X-8) \\ \hline -1 \end{array}$$

6.
$$\begin{array}{r} 3X-2 \\ \times X+4 \\ \hline 12X-8 \\ 3X^2-2X \quad\quad \\ \hline 3X^2+10X-8 \\ -1 \\ \hline 3X^2+10X-9 \end{array}$$

7. $\sqrt{X^2+8X+16} = X+4$

8.
$$\begin{array}{r} X+4 \\ \times X+4 \\ \hline 4X+16 \\ X^2+4X \quad\quad \\ \hline X^2+8X+16 \end{array}$$

9. $\left(A^5B^7B^3\right)^{-2}\left(A^4\right)=\left(A^5B^{7+3}\right)^{-2}A^4=$
$\left(A^5B^{10}\right)^{-2}A^4=A^{5\times-2}B^{10\times-2}A^4=$
$A^{-10}B^{-20}A^4=A^{-10+4}B^{-20}=$
$A^{-6}B^{-20}$

10. $\dfrac{B^4}{AB^{-2}}=\dfrac{B^4B^2}{A}=\dfrac{B^{4+2}}{A}=\dfrac{B^6}{A}=B^6A^{-1}$

11. $.586\times1.5=.879$

12. $125\div2.5=50$

13. $(-7)-9=-16$

14. $|10\div2-8|=|5-8|=|-3|=3$

15.
$$\begin{array}{r} 7X^2+4X-1 \\ +\ -2X^2+3X+6 \\ \hline 5X^2+7X+5 \end{array}$$

16.
$$\begin{array}{r} X^2+11X+5 \\ +\ X^2-\ 8X-6 \\ \hline 2X^2+\ 3X-1 \end{array}$$

17. $216=2\times2\times2\times3\times3\times3$

18. addition and multiplication

19. $24\div6=4$ hours

20. $24\div3=8$ hours

Systematic Review 24E

1.
$$\begin{array}{r} X+4 \\ 2X+2\,\overline{\big)\,2X^2+10X+8} \\ -\left(2X^2+2X\right) \\ \hline 8X+8 \\ -(8X+8) \\ \hline 0 \end{array}$$

2.
$$\begin{array}{r} 2X+2 \\ \times X+4 \\ \hline 8X+8 \\ 2X^2+\ 2X\quad \\ \hline 2X^2+10X+8 \end{array}$$

3.
$$\begin{array}{r} 3X-2 \\ X+4\,\overline{\big)\,3X^2+10X-8} \\ -\left(3X^2+12X\right) \\ \hline -2X-8 \\ -(-2X-8) \\ \hline 0 \end{array}$$

4.
$$\begin{array}{r} 3X-2 \\ \times X+4 \\ \hline 12X-8 \\ 3X^2-\ 2X\quad \\ \hline 3X^2+10X-8 \end{array}$$

5.
$$\begin{array}{r} 2X+\ 4\ R\ 3 \\ 2X-5\,\overline{\big)\,4X^2-\ 2X-17} \\ -\left(4X^2-10X\right) \\ \hline 8X-17 \\ -(8X-20) \\ \hline 3 \end{array}$$

6.
$$\begin{array}{r} 2X-\ 5 \\ \times 2X+\ 4 \\ \hline 8X-20 \\ 4X^2-10X\quad \\ \hline 4X^2-\ 2X-20 \\ +\qquad\quad 3 \\ \hline 4X^2-\ 2X-17 \end{array}$$

7. $\sqrt{X^2+6X+9}=X+3$

8.
$$\begin{array}{r} X+3 \\ \times X+3 \\ \hline 3X+9 \\ X^2+3X\quad \\ \hline X^2+6X+9 \end{array}$$

9. $(4)^3=2^?$
$\left(2^2\right)^3=2^{2\times3}=2^6$

10. $\dfrac{\left(X^4Y^{-2}\right)^3}{X^3Y^5X^{-1}}=\dfrac{X^{4\times3}Y^{-2\times3}}{X^{3+(-1)}Y^5}=\dfrac{X^{12}Y^{-6}}{X^2Y^5}=$
$X^{12}Y^{-6}X^{-2}Y^{-5}=X^{12+(-2)}Y^{-6+(-5)}=$
$X^{10}Y^{-11}$

11. $(10)^4=\left(10^1\right)^?$
$10^{4\times1}=10^4=\left(10^1\right)^4$

12. $3A^2B^3A+\dfrac{6A^3B^3}{A^{-1}}-7B^3A^3=$
$3A^{2+1}B^3+6A^3A^1B^3-7B^3A^3=$
$3A^3B^3+6A^{3+1}B^3-7A^3B^3=$
$6A^4B^3-4A^3B^3$

13. $1.68+.045=1.725$

14. $49\div.007=7{,}000$

15.
$$\begin{array}{r} 2X^2+4X-\ \ 6 \\ \underline{+X^2+\ X-10} \\ 3X^2+5X-16 \end{array}$$

16.
$$\begin{array}{r} 5X^2+11X-3 \\ \underline{+-4X^2-\ 5X+7} \\ X^2+\ \ 6X+4 \end{array}$$

17. $132=2\times2\times3\times11$

18. $2X$

19. $18\div9=2$ hours

20. $18\div3=6$ hours

Lesson Practice 25A

1. $X^2-4=(X-2)(X+2)$
$$\begin{array}{r} X+2 \\ \underline{\times X-2} \\ -2X-4 \\ \underline{X^2+2X} \\ X^2\qquad-4 \end{array}$$

2. $X^2-16=(X-4)(X+4)$
$$\begin{array}{r} X+\ 4 \\ \underline{\times X-\ 4} \\ -4X-16 \\ \underline{X^2+4X} \\ X^2\qquad-16 \end{array}$$

3. $X^2-25=(X-5)(X+5)$
$$\begin{array}{r} X+\ 5 \\ \underline{\times X-\ 5} \\ -5X-25 \\ \underline{X^2+5X} \\ X^2\qquad-25 \end{array}$$

4. $Y^2-144=(Y-12)(Y+12)$
$$\begin{array}{r} Y+\ 12 \\ \underline{\times Y-\ 12} \\ -12Y-144 \\ \underline{Y^2+12Y} \\ Y^2\qquad-144 \end{array}$$

5. $X^2-100=(X-10)(X+10)$
$$\begin{array}{r} X+\ 10 \\ \underline{\times X-\ 10} \\ -10X-100 \\ \underline{X^2+10X} \\ X^2\qquad-100 \end{array}$$

6. $X^2-81=(X-9)(X+9)$
$$\begin{array}{r} X+\ 9 \\ \underline{\times X-\ 9} \\ -9X-81 \\ \underline{X^2+9X} \\ X^2\qquad-81 \end{array}$$

7. $X^2-49=(X-7)(X+7)$
$$\begin{array}{r} X+\ 7 \\ \underline{\times X-\ 7} \\ -7X-49 \\ \underline{X^2+7X} \\ X^2\qquad-49 \end{array}$$

8. $X^2-64=(X-8)(X+8)$
$$\begin{array}{r} X+\ 8 \\ \underline{\times X-\ 8} \\ -8X-64 \\ \underline{X^2+8X} \\ X^2\qquad-64 \end{array}$$

9. $A^2-121=(A-11)(A+11)$
$$\begin{array}{r} A+\ 11 \\ \underline{\times A-\ 11} \\ -11A-121 \\ \underline{A^2\_11A} \\ A^2\qquad-121 \end{array}$$

10. $X^2-Y^2=(X-Y)(X+Y)$
$$\begin{array}{r} X+Y \\ \underline{\times X-Y} \\ -XY-Y^2 \\ \underline{X^2+XY} \\ X^2\qquad-Y^2 \end{array}$$

11. $B^2 - 4 = (B-2)(B+2)$

$$\begin{array}{r} B+2 \\ \times \quad B-2 \\ \hline -2B-4 \\ B^2+2B \quad \\ \hline B^2 \qquad -4 \end{array}$$

12. $X^2 - 9 = (X-3)(X+3)$

$$\begin{array}{r} X+3 \\ \times X-3 \\ \hline -3X-9 \\ X^2+3X \quad \\ \hline X^2 \qquad -9 \end{array}$$

13. $\begin{array}{r} 65 \\ \times\ 65 \\ \hline 4225 \end{array}$

14. $\begin{array}{r} 35 \\ \times\ 35 \\ \hline 1225 \end{array}$

15. $\begin{array}{r} 48 \\ \times\ 42 \\ \hline 2016 \end{array}$

16. $\begin{array}{r} 85 \\ \times\ 85 \\ \hline 7225 \end{array}$

Lesson Practice 25B

1. $X^2 - 1 = (X-1)(X+1)$

$$\begin{array}{r} X+1 \\ \times X-1 \\ \hline -X-1 \\ X^2+X \quad \\ \hline X^2 \qquad -1 \end{array}$$

2. $X^2 - 36 = (X-6)(X+6)$

$$\begin{array}{r} X+\ 6 \\ \times X-\ 6 \\ \hline -6X-36 \\ X^2+6X \quad \\ \hline X^2 \qquad -36 \end{array}$$

3. $Y^2 - 16 = (Y-4)(Y+4)$

$$\begin{array}{r} Y+4 \\ \times Y-4 \\ \hline -4Y-16 \\ Y^2+4Y \quad \\ \hline Y^2 \qquad -16 \end{array}$$

4. $A^2 - B^2 = (A-B)(A+B)$

$$\begin{array}{r} A+B \\ \times A-B \\ \hline -AB-B^2 \\ A^2+AB \quad \\ \hline A^2 \qquad -B^2 \end{array}$$

5. $A^2 - 49 = (A-7)(A+7)$

$$\begin{array}{r} A+\ 7 \\ \times A-\ 7 \\ \hline -7A-49 \\ A^2+7A \quad \\ \hline A^2 \qquad -49 \end{array}$$

6. $B^2 - 25 = (B-5)(B+5)$

$$\begin{array}{r} B+\ 5 \\ \times B-\ 5 \\ \hline -5B-25 \\ B^2+5B \quad \\ \hline B^2 \qquad -25 \end{array}$$

7. $Y^2 - X^2 = (Y-X)(Y+X)$

$$\begin{array}{r} Y+X \\ \times Y-X \\ \hline -XY-X^2 \\ Y^2+XY \quad \\ \hline Y^2 \qquad -X^2 \end{array}$$

8. $X^2 - 4 = (X-2)(X+2)$

$$\begin{array}{r} X+2 \\ \times X-2 \\ \hline -2X-4 \\ X^2+2X \quad \\ \hline X^2 \qquad -4 \end{array}$$

9. $A^2 - 144 = (A-12)(A+12)$

$$\begin{array}{r} A+\ 12 \\ \times A-\ 12 \\ \hline -12A-144 \\ A^2+12A \quad \\ \hline A^2 \qquad -144 \end{array}$$

10. $4X^2 - 4Y^2 = (4)(X^2 - Y^2) =$
$(4)(X - Y)(X + Y)$

$$\begin{array}{r} X + Y \\ \times \quad X - Y \\ \hline -XY - Y^2 \\ X^2 + XY \quad\quad \\ \hline X^2 \quad\quad - Y^2 \end{array}$$

11. $(B - 8)(B + 8)$
12. $(X - 9)(X + 9)$
13. $$\begin{array}{r} 57 \\ 53 \\ \hline 3021 \end{array}$$
14. $$\begin{array}{r} 75 \\ 75 \\ \hline 5625 \end{array}$$
15. $$\begin{array}{r} 35 \\ 35 \\ \hline 1225 \end{array}$$
16. $$\begin{array}{r} 96 \\ 94 \\ \hline 9024 \end{array}$$

Systematic Review 25C

1. $X^2 - 16 = (X - 4)(X + 4)$
2. $$\begin{array}{r} X + 4 \\ \times X - 4 \\ \hline -4X - 16 \\ X^2 + 4X \quad\quad \\ \hline X^2 \quad\quad -16 \end{array}$$
3. $X^2 - 36 = (X - 6)(X + 6)$
4. $$\begin{array}{r} X + 6 \\ \times X - 6 \\ \hline -6X - 36 \\ X^2 + 6X \quad\quad \\ \hline X^2 \quad\quad -36 \end{array}$$
5. $$\begin{array}{r} 2X + 5 \text{ R } 10 \\ X - 1 \overline{\smash{)}\, 2X^2 + 3X + 5} \\ -(2X^2 - 2X) \quad\quad \\ \hline 5X + 5 \\ -(5X - 5) \\ \hline 10 \end{array}$$
6. $$\begin{array}{r} 2X + 5 \\ \times X - 1 \\ \hline -2X - 5 \\ 2X^2 + 5X \quad\quad \\ \hline 2X^2 + 3X - 5 \\ + 10 \\ \hline 2X^2 + 3X + 5 \end{array}$$
7. $\sqrt{4X^2} = 2X$
8. $\sqrt{4(10)^2} = \sqrt{400} = 20$
9. $$\begin{array}{r} 45 \\ \times 45 \\ \hline 2025 \end{array}$$
10. $$\begin{array}{r} 37 \\ \times 33 \\ \hline 1221 \end{array}$$
11. $(X - 7)(X - 11)$
12. $$\begin{array}{r} X - 7 \\ \times X - 11 \\ \hline -11X + 77 \\ X^2 - 7X \quad\quad \\ \hline X^2 - 18X + 77 \end{array}$$
13. $(2^5)^5 = 2^{5 \times 5} = 2^{25}$
14. $2Y - 3X + 6 = 0$
$2Y = 3X - 6$
$Y = \frac{3}{2}X - 3$; slope $= \frac{3}{2}$
15. origin
16. $(D + 2)(X + 3) = D(X + 3) + 2(X + 3) =$
$DX + 3D + 2X + 6$
17. $$\begin{array}{r} 300{,}000{,}000 \\ \times \quad 1{,}000 \\ \hline \$300{,}000{,}000{,}000 \end{array}$$ (not enough)

18.

$$\begin{aligned} 5(24Y+12X=36) &\Rightarrow 120Y+60X=180 \\ 12(5Y-5X=10) &\Rightarrow \underline{60Y-60X=120} \\ & \quad 180Y = 300 \\ & \quad Y=\frac{300}{180} \\ & \quad Y=\frac{5}{3} \end{aligned}$$

$$\begin{aligned} 5Y-5X=10 \Rightarrow 5\left(\frac{5}{3}\right)-5X&=10 \\ \frac{25}{3}-5X&=10 \\ \frac{25}{3}-10&=5X \\ \frac{25}{3}-\frac{30}{3}&=5X \\ \frac{-5}{3}&=5X \\ \frac{-5}{3}\div 5&=X \\ X&=-\frac{1}{3} \end{aligned}$$

19. $3Y \le 2X+6$

$Y \le \frac{2}{3}X+2$

see graph

20. $Y \le \frac{2}{3}X+2$

$(-4) \le \frac{2}{3}(-3)+2$

$-4 \le \frac{-6}{3}+2$

$-4 \le -2+2$

$-4 \le 0$ true

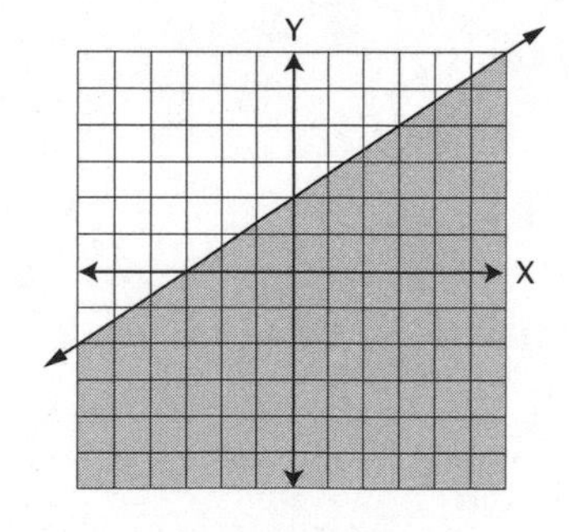

Systematic Review 25D

1. $X^2-4=(X-2)(X+2)$

2.

$$\begin{array}{r} X+2 \\ \times X-2 \\ \hline -2X-4 \\ X^2+2X \quad \\ \hline X^2 \quad -4 \end{array}$$

3. $X^2-25=(X-5)(X+5)$

4.

$$\begin{array}{r} X+5 \\ \times X-5 \\ \hline -5X-25 \\ X^2+5X \quad \\ \hline X^2 \quad -25 \end{array}$$

5.

$$\begin{array}{r} 2X+3 \\ X+2 \overline{\smash{)}\, 2X^2+7X+6} \\ \underline{-(2X^2+4X)} \\ 3X+6 \\ \underline{-(3X+6)} \\ 0 \end{array}$$

6.

$$\begin{array}{r} 2X+3 \\ \times X+2 \\ \hline 4X+6 \\ 2X^2+3X \quad \\ \hline 2X^2+7X+6 \end{array}$$

7. $\sqrt{X^2+10X+25}=X+5$

8.

$$\begin{aligned} \sqrt{(10)^2+10(10)+25} &= (10)+5 \\ \sqrt{100+100+25} &= 15 \\ \sqrt{225} &= 15 \\ 15 &= 15 \end{aligned}$$

9.

$$\begin{array}{r} 65 \\ \times\ 65 \\ \hline 4225 \end{array}$$

10.

$$\begin{array}{r} 78 \\ \times\ 72 \\ \hline 5616 \end{array}$$

11. $X^2+3X-4=(X+4)(X-1)$

12.
$$\begin{array}{r} X+4 \\ \times X-1 \\ \hline -X-4 \\ X^2+4X \quad \\ \hline X^2+3X-4 \end{array}$$

13. $(49)^3 = 7^?$

$(7^2)^3 = 7^{2\times3} = 7^6$

14. $4Y + 8X + 2 = 0$

$4Y = -8X - 2$

$Y = \frac{-8}{4}X - \frac{2}{4}$

$Y = -2X - \frac{1}{2}$

slope $= -2$

15. $(A+B)(C+D+E) =$

$A(C+D+E) + B(C+D+E) =$

$AC + AD + AE + BC + BD + BE$

16.
$$\begin{array}{r} 300{,}000{,}000 \\ \times \quad 10{,}000 \\ \hline \$3{,}000{,}000{,}000{,}000 \end{array}$$
(not enough)

17-18.

Rate	Time	Distance
20mph	1 hr	20 mi
10mph	2 hr	20 mi
5mph	4 hr	20 mi
4mph	5 hr	20 mi
1mph	20 hr	20 mi

19-20.

Rate	Time	Distance
12mph	1 hr	12 mi
6mph	2 hr	12 mi
4mph	3 hr	12 mi
3mph	4 hr	12 mi
2mph	6 hr	12 mi
1 mph	12 hr	12 mi

Systematic Review 25E

1. $X^2 - 9 = (X-3)(X+3)$

2.
$$\begin{array}{r} X+3 \\ \times X-3 \\ \hline -3X-9 \\ \times X^2+3X \quad \\ \hline X^2 \qquad -9 \end{array}$$

3. $X^2 - Y^2 = (X-Y)(X+Y)$

4.
$$\begin{array}{r} X+Y \\ \times X-Y \\ \hline -XY-Y^2 \\ X^2+XY \quad \\ \hline X^2 \qquad -Y^2 \end{array}$$

5.
$$\begin{array}{r} 2X^2 + \ X \ R-8 \\ X+4 \overline{)2X^3+9X^2+4X-8} \\ -(2X^3+8X^2) \qquad \\ \hline X^2+4X \quad \\ -(X^2+4X) \quad \\ \hline -8 \end{array}$$

6.
$$\begin{array}{r} 2X^2+X \\ \times X+4 \\ \hline 8X^2+4X \\ 2X^3+\ X^2 \qquad \\ \hline 2X^3+9X^2+4X \\ -8 \\ \hline 2X^3+9X^2+4X-8 \end{array}$$

7. $\sqrt{4X^2+4X+1} = 2X+1$

8. $\sqrt{4(10)^2+4(10)+1} = 2(10)+1$

$\sqrt{4(100)+40+1} = 20+1$

$\sqrt{441} = 21$

$21 = 21$

9.
$$\begin{array}{r} 85 \\ \times \ 85 \\ \hline 7225 \end{array}$$

10.
$$\begin{array}{r} 59 \\ \times \ 51 \\ \hline 3009 \end{array}$$

11. $X^2 - 10X + 24 = (X-6)(X-4)$

12.
$$\begin{array}{r} X-\ 6 \\ \times X-\ 4 \\ \hline -\ 4X+24 \\ X^2-\ 6X\quad\ \ \\ \hline X^2-10X+24 \end{array}$$

13. $(Q+R)(X+Y)=Q(X+Y)+R(X+Y)=$
$QX+QY+RX+RY$

14. $\dfrac{\$5{,}000{,}000{,}000{,}000}{300{,}000{,}000}=\dfrac{\$50{,}000}{3}\approx$
$\$16{,}666.67$

15.
$$\begin{array}{r} \$5{,}000{,}000{,}000{,}000 \\ \times\qquad\qquad\quad .08 \\ \hline \$400{,}000{,}000{,}000.00 \end{array}$$
$400 billion in interest each year

16. $300\div 50=6$ hours

17. $300\div 60=5$ hours

18. $6.5\times 46=299$ miles

19. $46+8=54$ mph
$299\div 54\approx 5.54$ hours

20.
$$\begin{aligned} 4R-32R&=36R+8XR \\ -28R-36R&=8XR \\ -64R&=8XR \\ \frac{-64R}{8R}&=X=-8 \end{aligned}$$

Lesson Practice 26A

1. $X^4-9=(X^2-3)(X^2+3)$
2. $X^4-Y^4=(X^2-Y^2)(X^2+Y^2)=$
$(X-Y)(X+Y)(X^2+Y^2)$
3. $2X^3-16X=2X(X^2-8)$
4. $X^8-Y^4=(X^4-Y^2)(X^4+Y^2)=$
$(X^2-Y)(X^2+Y)(X^4+Y^2)$
5. $2X^3+10X^2+12X=2X(X^2+5X+6)=$
$2X(X+2)(X+3)$
6. $5X^3+5X^2-30X=5X(X^2+X-6)=$
$5X(X+3)(X-2)$
7. $2X^3+11X^2+5X=X(2X^2+11X+5)=$
$X(2X+1)(X+5)$
8. $3X^2-12X=3X(X-4)$
9. $2X^3-18X=2X(X^2-9)=$
$2X(X-3)(X+3)$
10. $5X^4-20X^3-25X^2=$
$5X^2(X^2-4X-5)=5X^2(X-5)(X+1)$
11. $4X^3+16X^2-48X=$
$4X(X^2+4X-12)=4X(X+6)(X-2)$
12. $2X^4-32=2(X^4-16)=$
$2(X^2-4)(X^2+4)=2(X-2)(X+2)(X^2+4)$
13. $X^3+5X^2+4X=X(X^2+5X+4)=$
$X(X+4)(X+1)$
14. $3X^3+6X^2-9X=3X(X^2+2X-3)=$
$3X(X+3)(X-1)$
15. $2X^3+7X^2-4X=X(2X^2+7X-4)=$
$X(2X-1)(X+4)$
16. $4X^3-16X=4X(X^2-4)=$
$4X(X-2)(X+2)$

Lesson Practice 26B

1. $X^4-9X^2=X^2(X^2-9)=$
$X^2(X-3)(X+3)$
2. $3X^3-75X=3X(X^2-25)=$
$3X(X-5)(X+5)$
3. $4X^4-4X^2=4X^2(X^2-1)=$
$4X^2(X-1)(X+1)$
4. $5X^5-5X=5X(X^4-1)=$
$5X(X^2-1)(X^2+1)=$
$5X(X-1)(X+1)(X^2+1)$
5. $-2X^2-16X-30=-2(X^2+8X+15)=$
$-2(X+5)(X+3)$
6. $3X^3+9X^2-30X=3X(X^2+3X-10)=$
$3X(X+5)(X-2)$
7. $5X^3-5X^2-30X=5X(X^2-X-6)=$
$5X(X-3)(X+2)$
8. $X^3+11X^2+30X=X(X^2+11X+30)=$
$X(X+6)(X+5)$
9. $-4X^2-28X-40=-4(X^2+7X+10)=$
$-4(X+2)(X+5)$

10. $-3X^3 - 24X^2 - 36X =$
$-3X(X^2 + 8X + 12) = -3X(X+2)(X+6)$

11. $2X^3 - 8X^2 - 10X = 2X(X^2 - 4X - 5) =$
$2X(X-5)(X+1)$

12. $5X^5 - X^4 - 6X^3 = X^3(5X^2 - X - 6) =$
$X^3(5X-6)(X+1)$

13. $-3X^3 - 12X^2 + 36X =$
$-3X(X^2 + 4X - 12) = -3X(X+6)(X-2)$

14. $X^4 + 3X^3 - 4X^2 = X^2(X^2 + 3X - 4) =$
$X^2(X+4)(X-1)$

15. $4X^3 - 36X = 4X(X^2 - 9) =$
$4X(X-3)(X+3)$

16. $2X^4 - 32X^2 = 2X^2(X^2 - 16) =$
$2X^2(X-4)(X+4)$

Systematic Review 26C

1. $X^4 - 16 = (X^2 - 4)(X^2 + 4) =$
$(X-2)(X+2)(X^2+4)$

2.
$$\begin{aligned} (10)^4 - 16 &= ((10)-2)((10)+2)((10)^2+4) \\ 10{,}000 - 16 &= ((8)(12)(100+4)) \\ 9{,}984 &= (96)(104) \\ 9{,}984 &= 9{,}984 \end{aligned}$$

3. $16X^2 - 9 = (4X-3)(4X+3)$

4.
$$\begin{aligned} 16(10)^2 - 9 &= (4(10)-3)(4(10)+3) \\ 16(100) - 9 &= (40-3)(40+3) \\ 1{,}600 - 9 &= (37)(43) \\ 1{,}591 &= 1{,}591 \end{aligned}$$

5.
$$\begin{array}{r} 3X - 8 \text{ R } 7 \\ X+2 \overline{\big)\, 3X^2 - 2X - 9} \\ -(3X^2 + 6X) \\ \hline -8X - 9 \\ -(-8X - 16) \\ \hline 7 \end{array}$$

6.
$$\begin{array}{r} 3X - 8 \\ \times X + 2 \\ \hline 6X - 16 \\ 3X^2 - 8X \quad \\ \hline 3X^2 - 2X - 16 \\ + \quad 7 \\ \hline 3X^2 - 2X - 9 \end{array}$$

7. $(X-3)(X-4) = X^2 - 7X + 12$

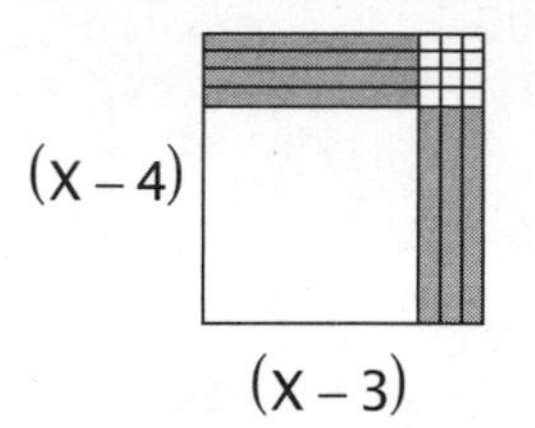

8.
$$\begin{array}{r} X - 3 \\ \times X - 4 \\ \hline -4X + 12 \\ X^2 - 3X \quad \\ \hline X^2 - 7X + 12 \end{array}$$

9.
$$\begin{array}{r} 75 \\ \times\ 75 \\ \hline 5625 \end{array}$$

10.
$$\begin{array}{r} 41 \\ \times\ 49 \\ \hline 2009 \end{array}$$

11. $2X^2 + 4X + 2 = 2(X^2 + 2X + 1) =$
$2(X+1)(X+1)$
$$\begin{array}{r} X + 1 \\ \times X + 1 \\ \hline X + 1 \\ X^2 + X \quad \\ \hline X^2 + 2X + 1 \end{array}$$
$2(X^2 + 2X + 1) = 2X^2 + 4X + 2$

12. $6X^2 - 600 = 6(X^2 - 100) =$
$6(X-10)(X+10)$
$$\begin{array}{r} X + 10 \\ \times X - 10 \\ \hline -10X - 100 \\ X^2 + 10X \quad \\ \hline X^2 \quad\quad -100 \end{array}$$
$6(X^2 - 100) = 6X^2 - 600$

13. $\frac{3}{7} = \frac{6}{Q}$

$\frac{3}{7} \times \frac{Q}{1} = \frac{6}{Q} \times \frac{Q}{1}$

$\frac{3Q}{7} = \frac{6}{1}$

$\frac{3Q}{7} \times \frac{7}{1} = \frac{6}{1} \times \frac{7}{1}$

$3Q = 42$

$Q = \frac{42}{3} = 14$

14. $\frac{2}{9} = \frac{X}{36}$

$\frac{2}{9} \times \frac{36}{1} = \frac{X}{36} \times \frac{36}{1}$

$\frac{72}{9} = X = 8$

15. $.015 = .25Q - .44$

$1000(.015) = 1000(.25Q - .44)$

$15 = 250Q - 440$

$15 + 440 = 250Q$

$455 = 250Q$

$\frac{455}{250} = Q = 1.82$

16. $-4X - 16 = -5X + 43$

$-4X + 5X = 43 + 16$

$X = 59$

17. $49{,}703 = 4 \times 10^4 + 9 \times 10^3 + 7 \times 10^2 + 3 \times 10^0$

18. $1 \times 10^{-2} + 5 \times 10^{-4} =$

$.01 + .0005 = .0105$

19. $12(N+1) + 4(N) = 9(N+2) + 8$

$12N + 12 + 4N = 9N + 18 + 8$

$12N + 4N - 9N = 18 + 8 - 12$

$7N = 14$

$N = \frac{14}{7} = 2$

2; 3; 4

20. $(2X+3)(A+4) =$

$(2X)(A+4) + 3(A+4) = 2XA + 8X + 3A + 12$

Systematic Review 26D

1. $X^3 - 9X = X(X^2 - 9) = X(X-3)(X+3)$

2. $(10)^3 - 9(10) = (10)((10)-3)((10)+3)$

$1000 - 90 = (10)(7)(13)$

$910 = 910$

3. $X^4 - 81 = (X^2 - 9)(X^2 + 9) =$

$(X-3)(X+3)(X^2+9)$

4.

$(10)^4 - 81 = ((10)-3)((10)+3)((10)^2+9)$

$10{,}000 - 81 = (7)(13)(100+9)$

$9{,}919 = (7)(13)(109)$

$9{,}919 = 9{,}919$

5.

$$\begin{array}{r|l} & 2X - 1 \text{ R } -11 \\ X - 3 & 2X^2 - 7X - 8 \\ & -(2X^2 - 6X) \\ & \quad -X - 8 \\ & \quad -(-X + 3) \\ & \quad\quad -11 \end{array}$$

6.

$$\begin{array}{r} 2X - 1 \\ \times X - 3 \\ \hline -6X + 3 \\ 2X^2 - X \quad \\ \hline -11 \\ \hline 2X^2 - 7X - 8 \end{array}$$

7. $(X-2)(X-1) = X^2 - 3X + 2$

8.

$$\begin{array}{r} X - 2 \\ \times X - 1 \\ \hline -X + 2 \\ X^2 - 2X \quad \\ \hline X^2 - 3X + 2 \end{array}$$

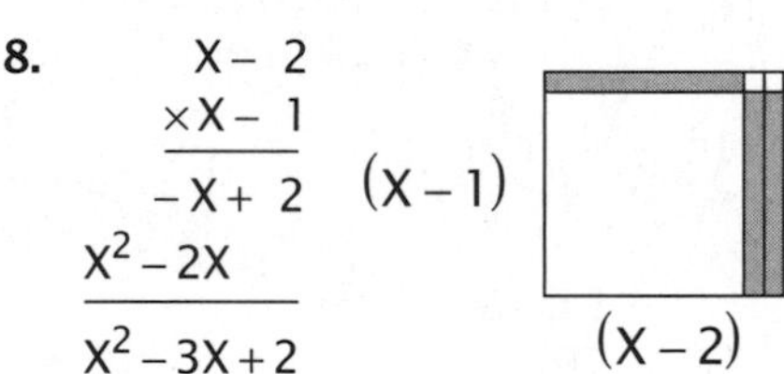

9.

$$\begin{array}{r} 95 \\ \times 95 \\ \hline 9025 \end{array}$$

10.

$$\begin{array}{r} 24 \\ \times 26 \\ \hline 624 \end{array}$$

11. $5X^2 - 45 = 5(X^2 - 9) =$

$5(X-3)(X+3)$

12. $4X^2 - 324$
$4(X^2 - 81)$
$4(X+9)(X-9)$

13.
$$\frac{4}{11} = \frac{P}{110}$$
$$4 \times 110 = 11P$$
$$440 = 11P$$
$$\frac{440}{11} = P = 40$$

14.
$$\frac{5}{8} = \frac{C}{15}$$
$$5 \times 15 = 8C$$
$$75 = 8C$$
$$\frac{75}{8} = C = 9\frac{3}{8}$$

15. $-50BY + 30B = 80BY - 40B$
divide all terms by 10B:
$$-5Y + 3 = 8Y - 4$$
$$3 + 4 = 8Y + 5Y$$
$$7 = 13Y$$
$$\frac{7}{13} = Y$$

16.
$$2.07 - .9X = 5X + .83$$
$$100(2.07 - .9X) = 100(5X + .83)$$
$$207 - 90X = 500X + 83$$
$$207 - 83 = 500X + 90X$$
$$124 = 590X$$
$$\frac{124}{590} = X = \frac{62}{295}$$

17.
$$(.25Q + .10D = 2.30)(100) \Rightarrow \quad 25Q + 10D = 230$$
$$(Q + D = 14)(-10) \Rightarrow \quad -10Q - 10D = -140$$
$$15Q = 90$$
$$Q = \frac{90}{15}$$
$$Q = 6$$

$$Q + D = 14 \Rightarrow (6) + D = 14$$
$$D = 14 - 6$$
$$D = 8$$

18. $4.2 \times 180 = 756$ miles

19. $180 - 30 = 150$ mph
$756 \div 150 = 5.04$ hours

20. $(X + A)(C + B) =$
$(X)(C + B) + (A)(C + B)$

Systematic Review 26E

1. $X^4 - 25X^2 = X^2(X^2 - 25) =$
$X^2(X-5)(X+5)$

2.
$$(10)^4 - 25(10)^2 = (10)^2((10)-5)((10)+5)$$
$$10{,}000 - 25(100) = 100(5)(15)$$
$$10{,}000 - 2{,}500 = 7{,}500$$
$$7{,}500 = 7{,}500$$

3. $5X^3 - 45X = 5X(X^2 - 9) =$
$5X(X-3)(X+3)$

4.
$$5(10)^3 - 45(10) = 5(10)((10)-3)((10)+3)$$
$$5(1000) - 450 = 50(7)(13)$$
$$5{,}000 - 450 = 4{,}550$$
$$4{,}550 = 4{,}550$$

5.
$$\begin{array}{r} 2X - 7 \text{ R } 29 \\ X+4 \overline{\smash{)}\,2X^2 + X + 1} \\ \underline{-(2X^2 + 8X)} \\ -7X + 1 \\ \underline{-(-7X - 28)} \\ 29 \end{array}$$

6.
$$\begin{array}{r} 2X - 7 \\ \underline{\times X + 4} \\ 8X - 28 \\ \underline{2X^2 - 7X } \\ 2X^2 + X - 28 \\ \underline{+ 29} \\ 2X^2 + X + 1 \end{array}$$

7. $(2X-3)(X-2) = 2X^2 - 7X + 6$

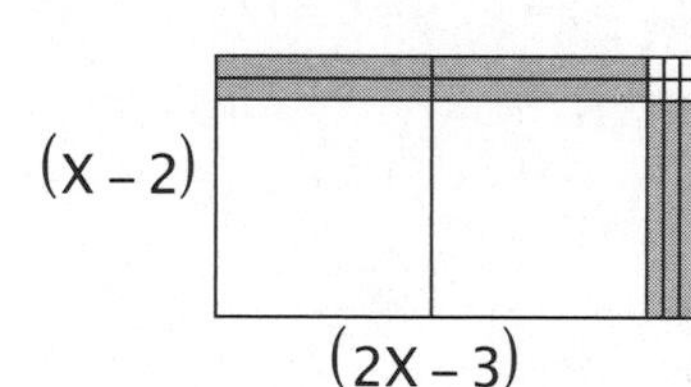

8.
$$\begin{array}{r} 2X - 3 \\ \underline{\times X - 2} \\ -4X + 6 \\ \underline{2X^2 - 3X } \\ 2X^2 - 7X + 6 \end{array}$$

9. $\begin{array}{r} 25 \\ \times 25 \\ \hline 625 \end{array}$

10. $\begin{array}{r} 32 \\ \times\ 38 \\ \hline 1216 \end{array}$

11. $\frac{12}{72} = \frac{A}{8}$
$12 \times 8 = 72A$
$96 = 72A$
$\frac{96}{72} = A = 1\frac{1}{3}$

12. $\frac{5}{12} = \frac{20}{Y}$
$5Y = 12 \times 20$
$5Y = 240$
$Y = \frac{240}{5} = 48$

13. $-.35Y + .55Y = 2.2$
$100(-.35Y + .55Y) = 100(2.2)$
$-35Y + 55Y = 220$
$20Y = 220$
$Y = \frac{220}{20} = 11$

14. $WF \times 100 = 1$
$WF = \frac{1}{100}$

15. $.0378 = 3 \times 10^{-2} + 7 \times 10^{-3} + 8 \times 10^{-4}$

16. $2 \times 10^6 + 6 \times 10^4 + 1 \times 10^3 =$
$2{,}000{,}000 + 60{,}000 + 1{,}000 =$
$2{,}061{,}000$

17. $2(N) + 2(N+2) - 5 = 7 + (N+4)$
$2N + 2N + 4 - 5 = 7 + N + 4$
$2N + 2N - N = 7 + 4 - 4 + 5$
$3N = 12$
$N = \frac{12}{3} = 4$

4; 6; 8

18. $442 \div 52 = 8.5$ hours

19. $212 \times 1 = 212$ miles

20. $(3X+2)(X+3) = 3X(X+3) + 2(X+3) =$
$(3X^2 + 9X) + (2X + 6)$

Lesson Practice 27A

1. $X^2 - 2X - 15 = 0$
$(X-5)(X+3) = 0$

2. $X - 5 = 0 \qquad X + 3 = 0$
$X = 5 \qquad X = -3$

3. $(5)^2 - 2(5) - 15 = 0$
$25 - 10 - 15 = 0$
$0 = 0$

$(-3)^2 - 2(-3) - 15 = 0$
$9 + 6 - 15 = 0$
$0 = 0$

4. $X^3 - 3X^2 + 2X = 0$
$X(X^2 - 3X + 2) = 0$
$X(X-2)(X-1) = 0$

5. $X = 0 \quad X - 2 = 0 \quad X - 1 = 0$
$X = 2 \qquad X = 1$

6. $(0)^3 - 3(0)^2 + 2(0) = 0$
$0 + 0 + 0 = 0$
$0 = 0$

$(2)^3 - 3(2)^2 + 2(2) = 0$
$8 - 3(4) + 2(2) = 0$
$8 - 12 + 4 = 0$
$0 = 0$

$(1)^3 - 3(1)^2 + 2(1) = 0$
$1 - 3 + 2 = 0$
$0 = 0$

7. $X^3 - X = 0$
$X(X^2 - 1) = 0$
$X(X-1)(X+1) = 0$

8. $X = 0 \quad X - 1 = 0 \quad X + 1 = 0$
$X = 1 \qquad X = -1$

9. $(0)^3 - (0) = 0$
$0 = 0$

$(1)^3 - (1) = 0$
$1 - 1 = 0$
$0 = 0$

$(-1)^3 - (-1) = 0$
$-1 - (-1) = 0$
$0 = 0$

10. $2X^2 - 7X + 3 = 0$
$(2X-1)(X-3) = 0$

11. $2X-1=0 \quad X-3=0$
$2X=1 \quad X=3$
$X=\frac{1}{2}$

12. $2\left(\frac{1}{2}\right)^2-7\left(\frac{1}{2}\right)+3=0$
$2\left(\frac{1}{4}\right)-\frac{7}{2}+3=0$
$\frac{1}{2}-\frac{7}{2}+3=0$
$-\frac{6}{2}+3=0$
$-3+3=0$
$0=0$

$2(3)^2-7(3)+3=0$
$2(9)-21+3=0$
$18-21+3=0$
$0=0$

Lesson Practice 27B

1. $X^2+X=56$
$X^2+X-56=0$
$(X-7)(X+8)=0$

2. $X+8=0 \quad X-7=0$
$X=-8 \quad X=7$

3. $(-8)^2+(-8)=56 \quad (7)^2+(7)=56$
$64-8=56 \quad 49+7=56$
$56=56 \quad 56=56$

4. $X^2-11X+30=0$
$(X-5)(X-6)=0$

5. $X-5=0 \quad X-6=0$
$X=5 \quad X=6$

6. $(5)^2-11(5)+30=0$
$25-55+30=0$
$0=0$

$(6)^2-11(6)+30=0$
$36-66+30=0$
$0=0$

7. $X^2-15X+56=0$
$(X-7)(X-8)=0$

8. $X-7=0 \quad X-8=0$
$X=7 \quad X=8$

9. $(7)^2-15(7)+56=0$
$49-105+56=0$
$0=0$

$(8)^2-15(8)+56=0$
$64-120+56=0$
$0=0$

10. $X^2-13X+40=0$
$(X-5)(X-8)=0$

11. $X-5=0 \quad X-8=0$
$X=5 \quad X=8$

12. $(5)^2-13(5)+40=0$
$25-65+40=0$
$0=0$

$(8)^2-13(8)+40=0$
$64-104+40=0$
$0=0$

Systematic Review 27C

1. $2X^2+7X+6=0$
$(2X+3)(X+2)=0$

$2X+3=0 \quad X+2=0$
$2X=-3 \quad X=-2$
$X=\frac{-3}{2}$

2. $2\left(-\frac{3}{2}\right)^2+7\left(-\frac{3}{2}\right)+6=0$
$2\left(\frac{9}{4}\right)+\left(-\frac{21}{2}\right)+6=0$
$\frac{18}{4}-\frac{21}{2}+\frac{12}{2}=0$
$0=0$

$2(-2)^2+7(-2)+6=0$
$2(4)-14+6=0$
$8-14+6=0$
$0=0$

3. $X^2+6X+8=0$
$(X+2)(X+4)=0$
$X+2=0 \quad X+4=0$
$X=-2 \quad X=-4$

4. $(-2)^2+6(-2)+8=0$
$4-12+8=0$
$0=0$

$(-4)^2+6(-4)+8=0$
$16-24+8=0$

5. $X^2+3X+4=14$
$X^2+3X+4-14=0$
$X^2+3X-10=0$
$(X+5)(X-2)=0$
$X+5=0 \quad X-2=0$
$X=-5 \quad X=2$

6. $(-5)^2+3(-5)+4=14$
$25-15+4=14$
$14=14$

$(2)^2+3(2)+4=14$
$4+6+4=14$
$14=14$

7. $(X-6)(X-6)=X^2-12X+36$

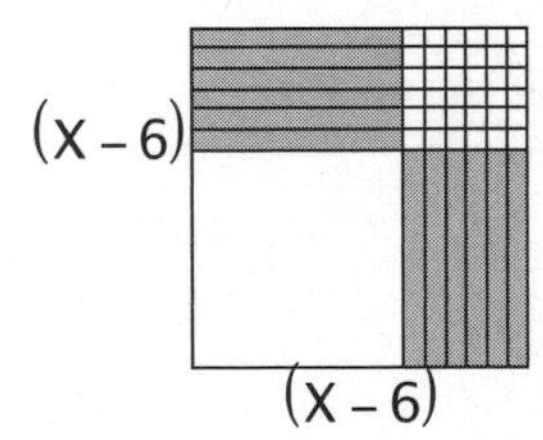

8.
$$\begin{array}{r} X-\ 6 \\ \times X-\ 6 \\ \hline -6X+36 \\ X^2-\ 6X\quad\ \ \\ \hline X^2-12X+36 \end{array}$$

9. $X^2-16=(X-4)(X+4)$

10. $X^2-49=(X-7)(X+7)$

11. $-4^2+(-2)^2=-(4\times4)+(-2)(-2)=$
$-16+4=-12$

12. $3^{-1}\times3^1=3^{-1+1}=3^0=1$

13. $(X^2)^2(X^{-3})^{-1}=X^{2\times2}X^{-3\times-1}=$
$X^4X^3=X^{4+3}=X^7$

14. $\dfrac{2X^2X^{-1}Y}{Y^3}-\dfrac{3X^0Y^3}{X^2}+\dfrac{5Y^{-2}}{X^{-1}}$
$=2X^2X^{-1}Y^1Y^{-3}-3X^0Y^3X^{-2}+5Y^{-2}X^1$
$=2X^{2+(-1)}Y^{1+(-3)}-3X^{0+(-2)}Y^3+5XY^{-2}$
$=2XY^{-2}-3X^{-2}Y^3+5XY^{-2}$
$=7XY^{-2}-3X^{-2}Y^3$ or $\dfrac{7X}{Y^2}-\dfrac{3Y^3}{X^2}$

15. $2X+4Y-8=0$
$4Y=-2X+8$
$Y=\dfrac{-2}{4}X+2$
$Y=-\dfrac{1}{2}X+2$

16. $m=2$ (negative reciprocal)

17. 11

18. $100=2\times2\times5\times5$

19.
$$\begin{array}{rr} & Y=\ \ X-3 \\ (Y=2X-4)(-1)\Rightarrow & -Y=-2X+4 \\ \hline & 0=\ -X+1 \\ & X=1 \end{array}$$
$Y=X-3\Rightarrow \ Y=(1)-3$
$Y=-2$
$(1,-2)$

20. $(2X+3)(2X+1)=$
$(2X)(2X+1)+(3)(2X+1)=$
$(4X^2+2X)+(6X+3)$

Systematic Review 27D

1. $2X^2+9X+4=0$
$(2X+1)(X+4)=0$
$2X+1=0 \quad X+4=0$
$2X=-1 \quad X=-4$
$X=-\dfrac{1}{2}$

2. $2\left(-\frac{1}{2}\right)^2+9\left(-\frac{1}{2}\right)+4=0$

$2\left(\frac{1}{4}\right)-\frac{9}{2}+\frac{8}{2}=0$

$\frac{2}{4}-\frac{9}{2}+\frac{8}{2}=0$

$\frac{1}{2}-\frac{9}{2}+\frac{8}{2}=0$

$0=0$

$2(-4)^2+9(-4)+4=0$

$2(16)-36+4=0$

$32-36+4=0$

$0=0$

3. $X^2+13X-68=0$

$(X+17)(X-4)=0$

$X+17=0 \quad X-4=0$

$X=-17 \quad X=4$

4. $(-17)^2+13(-17)-68=0$

$289-221-68=0$

$0=0$

$(4)^2+13(4)-68=0$

$16+52-68=0$

$0=0$

5. $X^2-2X+5=8$

$X^2-2X+5-8=0$

$X^2-2X-3=0$

$(X-3)(X+1)=0$

$X-3=0 \quad X+1=0$

$X=3 \quad X=-1$

6. $(3)^2-2(3)+5=8$

$9-6+5=8$

$8=8$

$(-1)^2-2(-1)+5=8$

$1+2+5=8$

$8=8$

7. $X^2-8X+16=(X-4)(X-4)$

(X – 4)

(X – 4)

8.

$$\begin{array}{r} X-4 \\ \times X-4 \\ \hline -4X+16 \\ X^2+-4X \quad \\ \hline X^2-8X+16 \end{array}$$

9. $X^2-Y^2=(X-Y)(X+Y)$

10. $4X^2-4Y^2=(4)(X^2-Y^2)$

$=(4)(X-Y)(X+Y)$

or: $4X^2-4Y^2=(2X-2Y)(2X+2Y)$

$=(2)(X-Y)(2)(X+Y)$

$=(4)(X-Y)(X+Y)$

11. $-3^2-(2)^2=-(3\times3)-(2)(2)=$

$-9-4=-13$

12. $4^{-2}\times4^3=4^{-2+3}=4^1=4$

13. $(X^2)^3(X^{-2})^2=X^{2\times3}X^{-2\times2}=$

$X^6X^{-4}=X^{6+(-4)}=X^2$

14. $2B^2B^1-\frac{3B^{-1}}{B^{-4}}+\frac{5B^4}{B^{-1}}=$

$2B^{2+1}-3B^{-1}B^4+5B^4B^1=$

$2B^3-3B^{-1+4}+5B^{4+1}=$

$2B^3-3B^3+5B^5=$

$-B^3+5B^5$ or $5B^5-B^3$

15. $\frac{B}{4}=\frac{9}{25}$

$25B=4\times9$

$B=\frac{36}{25}=1\frac{11}{25}$

16. $\frac{3.4}{5}=\frac{R}{15}$

$5R=3.4\times15$

$5R=51$

$R=\frac{51}{5}=10\frac{1}{5}$ or 10.2

17. $520\div65=8$ hours

18. $240\div6=40$ mph

19. $X = 4$

$$Y + 2X = -2 \Rightarrow Y + 2(4) = -2$$
$$Y + 8 = -2$$
$$Y = -2 - 8$$
$$Y = -10$$

$$Y + 2X = -2 \Rightarrow (-10) + 2X = -2$$
$$2X = 8$$
$$X = 4$$

Alternately, you may take the value for X directly from equation 2.

20. $(3X+4)(X+2) = (3X)(X+2) + (4)(X+2)$
$= (3X^2 + 6X) + (4X + 8)$

Systematic Review 27E

1. $4X^2 + 8X + 3 = 0$
$(2X+1)(2X+3) = 0$

$2X + 1 = 0 \qquad 2X + 3 = 0$
$2X = -1 \qquad 2X = -3$
$X = -\frac{1}{2} \qquad X = -\frac{3}{2}$

2. $4\left(-\frac{1}{2}\right)^2 + 8\left(-\frac{1}{2}\right) + 3 = 0$
$4\left(\frac{1}{4}\right) - \frac{8}{2} + 3 = 0$
$1 - 4 + 3 = 0$
$0 = 0$

$4\left(-\frac{3}{2}\right)^2 + 8\left(-\frac{3}{2}\right) + 3 = 0$
$4\left(\frac{9}{4}\right) - \frac{24}{2} + 3 = 0$
$\frac{36}{4} - 12 + 3 = 0$
$9 - 12 + 3 = 0$
$0 = 0$

3. $X^2 + 7X + 12 = 0$
$(X+3)(X+4) = 0$

$X + 3 = 0 \qquad X + 4 = 0$
$X = -3 \qquad X = -4$

4. $(-3)^2 + 7(-3) + 12 = 0$
$9 - 21 + 12 = 0$
$0 = 0$

$(-4)^2 + 7(-4) + 12 = 0$
$16 - 28 + 12 = 0$
$0 = 0$

5. $X^2 + X + 1 = 13$
$X^2 + X + 1 - 13 = 0$
$X^2 + X - 12 = 0$
$(X+4)(X-3) = 0$

$X + 4 = 0 \qquad X - 3 = 0$
$X = -4 \qquad X = 3$

6. $(-4)^2 + (-4) + 1 = 13$
$16 - 4 + 1 = 13$
$13 = 13$

$(3)^2 + (3) + 1 = 13$
$9 + 3 + 1 = 13$
$13 = 13$

7. $(X-5)(X-5) = X^2 - 10X + 25$

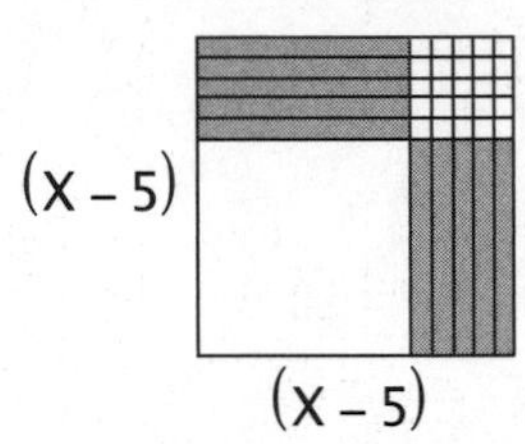

8.
$$\begin{array}{r} X - 5 \\ \times X - 5 \\ \hline -5X + 25 \\ X^2 - 5X \quad \\ \hline X^2 - 10X + 25 \end{array}$$

9. $16X^2 - 4 = (4)(4X^2 - 1) = (4)(2X-1)(2X+1)$

or: $16X^2 - 4 = (4X-2)(4X+2)$
$= (2)(2X-1)(2)(2X+1)$
$= (4)(2X-1)(2X+1)$

10. $X^2 - 100 = (X-10)(X+10)$

11. $(-3)^2 - (5)^2 = 9 - 25 = -16$

12. $2^{-4} \times 2^4 = 2^{-4+4} = 2^0 = 1$

13. $(X^2)^{-3}(X^3)^{-2} = X^{(2)(-3)}X^{(3)(-2)}$
$= X^{-6}X^{-6} = X^{-6+(-6)} = X^{-12}$

14. $5M^4N^2M^{-1} - \frac{2NM^4}{N^{-3}M} = 5M^{4+(-1)}N^2 - 2N^1M^4N^3M^{-1}$
$= 5M^3N^2 - 2N^{1+3}M^{4+(-1)}$
$= 5M^3N^2 - 2M^3N^4$

15. $\frac{5}{8} = \frac{G}{20}$
$5 \times 20 = 8G$
$100 = 8G$
$\frac{100}{8} = G = 12\frac{1}{2}$

16. $\frac{7}{2} = \frac{100}{T}$
$7T = 2 \times 100$
$7T = 200$
$T = \frac{200}{7} = 28\frac{4}{7}$

17. $N^2 + 2N - 2$

18. $N^2 + 2N - 2 = 22$
$N^2 + 2N - 2 - 22 = 0$
$N^2 + 2N - 24 = 0$
$(N+6)(N-4) = 0$
$N + 6 = 0 \quad N - 4 = 0$
$N = -6 \quad N = 4$

19. $(5Y - X = -6)(3) \Rightarrow \quad 15Y - 3X = -18$
$+ \quad 4Y + 3X = -20$
$19Y \quad = -38$
$Y = \frac{-38}{19} = -2$
$5Y - X = -6 \Rightarrow 5(-2) - X = -6$
$-10 - X = -6$
$-X = 4$
$X = -4$
$(-4, -2)$

20. $(X+2)(3X+1) =$
$(X)(3X+1) + (2)(3X+1) =$
$(3X^2 + X) + (6X + 2)$

Lesson Practice 28A

1. 1 foot = 12 inches
2. Feet in numerator to remain in final answer. Inches in denominator so they will cancel.
3. $84\ \text{in} \times \frac{1\ \text{ft}}{12\ \text{in}} = \frac{84\ \text{ft}}{12} = 7\ \text{ft}$
4. 3 feet = 1 yard
5. Yards in numerator to remain in final answer. Feet in denominator so they will cancel.
6. $63\ \text{ft} \times \frac{1\ \text{yd}}{3\ \text{ft}} = \frac{63\ \text{yd}}{3} = 21$ yards
7. 1 foot = 12 inches
8. Inches in numerator to remain in final answer. Feet in denominator so they will cancel.
9. $15\ \text{ft} \times \frac{12\ \text{in}}{1\ \text{ft}} = \frac{180\ \text{in}}{1} = 180\ \text{in}$
10. 4 quarts = 1 gallon
11. Quarts in numerator to remain in final answer. Gallons in denominator so they will cancel.
12. $25\ \text{gal} \times \frac{4\ \text{qt}}{\text{gal}} = \frac{100\ \text{qt}}{1} = 100\ \text{qt}$
13. 16 oz = 1 lb
14. pounds in numerator ounces in denominator
15. $272\ \text{oz} \times \frac{1\ \text{lb}}{16\ \text{oz}} = 17\ \text{lb}$
16. 4 qt = 1 gal
17. gallons in numerator quarts in denominator
18. $52\ \text{qt} \times \frac{1\ \text{gal}}{4\ \text{qt}} = 13\ \text{gal}$

Lesson Practice 28B

1. 1 meter = 100 centimeters
2. Centimeters in numerator to remain in final answer. Meters in denominator so they will cancel.
3. $14\ \cancel{m} \times \frac{100\ cm}{1\ \cancel{m}} =$ $\frac{1{,}400\ cm}{1} = 1{,}400\ cm$
4. 1 kilometer = 1,000 meters
5. Meters in numerator to remain in final answer. Kilometers in denominator so they will cancel.
6. $200\ \cancel{km} \times \frac{1{,}000\ m}{1\ \cancel{km}} =$ $\frac{200{,}000\ m}{1} = 200{,}000\ m$
7. 1 dekaliter = 10 liters
8. Dekaliters in numerator to remain in final answer. Liters in denominator so they will cancel.
9. $3{,}500\ \cancel{liters} \times \frac{1\ dkl}{10\ \cancel{liters}} =$ $\frac{3500\ dkl}{10} = 350\ dkl$
10. 1 liter = 1,000 milliliters
11. Liters in numerator to remain in final answer. Milliliters in denominator so they will cancel.
12. $67{,}000\ \cancel{ml} \times \frac{1\ liter}{1{,}000\ \cancel{ml}} =$ $\frac{67{,}000\ liters}{1{,}000} = 67\ liters$
13. 1 hectoliter = 100 liters
14. Liters in numerator to remain in final answer. Hectoliters in denominator so they will cancel.
15. $4.5\ \cancel{hl} \times \frac{100\ liters}{1\ \cancel{hl}} =$ $\frac{450\ liters}{1} = 450\ liters$
16. 1 gram = 10 decigrams
17. Grams in numerator to remain in final answer. Decigrams in denominator so they will cancel.
18. $790\ \cancel{dg} \times \frac{1\ g}{10\ \cancel{dg}} = \frac{790\ g}{10} = 79\ g$

Systematic Review 28C

1. 12 inches = 1 foot
2. Feet in numerator to remain in final answer. Inches in denominator so they will cancel.
3. $60\ \cancel{in} \times \frac{1\ ft}{12\ \cancel{in}} = \frac{60\ ft}{12} = 5\ ft$
4. 3 feet = 1 yard
5. Yards in numerator to remain in final answer. Feet in denominator so they will cancel.
6. $24\ \cancel{ft} \times \frac{1\ yd}{3\ \cancel{ft}} = \frac{24\ yd}{3} = 8\ yd$
7. 16 ounces = 1 pound
8. Pounds in numerator to remain in final answer. Ounces in denominator so they will cancel.
9. $32\ \cancel{oz} \times \frac{1\ lb}{16\ \cancel{oz}} = \frac{32\ lb}{16} = 2\ lb$
10. 4 quarts = 1 gallon
11. Gallons in numerator to remain in final answer. Quarts in denominator so they will cancel.
12. $28\ \cancel{qts} \times \frac{1\ gal}{4\ \cancel{qts}} = \frac{28\ gal}{4} = 7\ gal$

13.
$$\begin{array}{r} X-3\ R\ 4 \\ X-2\,\overline{)\,X^2-5X+10} \\ -(X^2-2X) \\ \hline -3X+10 \\ -(-3X+6) \\ \hline 4 \end{array}$$

14.
$$\begin{array}{r} X-2 \\ \times X-3 \\ \hline -3X+6 \\ X^2-2X \quad \\ \hline X^2-5X+6 \\ +4 \\ \hline X^2-5X+10 \end{array}$$

15. $3X^2+10X+3=0$

$(3X+1)(X+3)=0$

$3X+1=0 \quad X+3=0$

$3X=-1 \quad X=-3$

$X=-\frac{1}{3}$

16. $3\left(-\frac{1}{3}\right)^2+10\left(-\frac{1}{3}\right)+3=0$

$3\left(\frac{1}{9}\right)-\frac{10}{3}+\frac{9}{3}=0$

$\frac{3}{9}-\frac{10}{3}+\frac{9}{3}=0$

$\frac{1}{3}-\frac{10}{3}+\frac{9}{3}=0$

$0=0$

$3(-3)^2+10(-3)+3=0$

$3(9)-30+3=0$

$27-30+3=0$

$0=0$

17. $(10)(.2X^2-.6X+1=3)$

$2X^2-6X+10=30$

$2X^2-6X-20=0$

$2(X^2-3X-10)=0$

$2(X-5)(X+2)=0$

$X-5=0 \quad X+2=0$

$X=5 \quad X=-2$

18. $.2(5)^2-.6(5)+1=3$

$.2(25)-3+1=3$

$5-3+1=3$

$3=3$

$.2(-2)^2-.6(-2)+1=3$

$.2(4)+1.2+1=3$

$.8+1.2+1=3$

$3=3$

19. $\frac{Q}{.2}=\frac{25}{10}$

$10Q=.2\times25$

$10Q=5$

$Q=\frac{5}{10}=\frac{1}{2}$ or $.5$

20. $\frac{A}{B}=\frac{C}{D}$

$AD=BC$

$A=\frac{BC}{D}$

Systematic Review 28D

1. 12 inches = 1 foot
2. Inches in numerator to remain in final answer. Feet in denominator so they will cancel.
3. $4\text{ ft}\times\frac{12\text{ in}}{1\text{ ft}}=\frac{48\text{ in}}{1}=48\text{ in}$
4. 1 mile = 5,280 ft
5. Feet in numerator to remain in final answer. Miles in denominator so they will cancel.
6. $3\text{ mi}\times\frac{5{,}280\text{ ft}}{1\text{ mi}}=$ $\frac{15{,}840\text{ ft}}{1}=15{,}840\text{ ft}$
7. 2,000 lb = 1 ton
8. Pounds in numerator to remain in final answer. Tons in denominator so they will cancel.

9. $6 \text{ tons} \times \frac{2{,}000 \text{ lb}}{1 \text{ ton}} = \frac{12{,}000 \text{ lb}}{1} =$
   12,000 lb
10. 2 pints = 1 quart
11. Pints in numerator to remain in final answer.
    Quarts in denominator so they will cancel.
12. $2.5 \text{ qt} \times \frac{2 \text{ pt}}{1 \text{ qt}} = \frac{5 \text{ pt}}{1} = 5 \text{ pt}$
13. $$\begin{aligned} 2X^2 - 3X - 5 &= 0 \\ (2X-5)(X+1) &= 0 \end{aligned}$$
    $$2X - 5 = 0 \qquad X + 1 = 0$$
    $$2X = 5 \qquad X = -1$$
    $$X = \frac{5}{2}$$
14. $$\begin{aligned} 2\left(\frac{5}{2}\right)^2 - 3\left(\frac{5}{2}\right) - 5 &= 0 \\ 2\left(\frac{25}{4}\right) - \frac{15}{2} - \frac{10}{2} &= 0 \\ \frac{50}{4} - \frac{15}{2} - \frac{10}{2} &= 0 \\ \frac{25}{2} - \frac{15}{2} - \frac{10}{2} &= 0 \\ 0 &= 0 \end{aligned}$$
    $$\begin{aligned} 2(-1)^2 - 3(-1) - 5 &= 0 \\ 2(1) + 3 - 5 &= 0 \\ 2 + 3 - 5 &= 0 \\ 0 &= 0 \end{aligned}$$
15. $$\begin{aligned} 3X^2 + 8X + 4 &= 0 \\ (3X+2)(X+2) &= 0 \end{aligned}$$
    $$\begin{aligned} 3X + 2 &= 0 \\ 3X &= -2 \\ X &= -\frac{2}{3} \\ X &= -2 \end{aligned}$$
16. $$\begin{aligned} 3\left(-\frac{2}{3}\right)^2 + 8\left(-\frac{2}{3}\right) + 4 &= 0 \\ 3\left(\frac{4}{9}\right) - \frac{16}{3} + \frac{12}{3} &= 0 \\ \frac{12}{9} - \frac{16}{3} + \frac{12}{3} &= 0 \\ \frac{4}{3} - \frac{16}{3} + \frac{12}{3} &= 0 \\ 0 &= 0 \end{aligned}$$
    $$\begin{aligned} 3(-2)^2 + 8(-2) + 4 &= 0 \\ 3(4) - 16 + 4 &= 0 \\ 16 - 16 + 4 &= 0 \\ 0 &= 0 \end{aligned}$$
17. $$\begin{aligned} 3Y^2 - 12 &= 0 \\ (3)(Y^2 - 4) &= 0 \\ (3)(Y-2)(Y+2) &= 0 \end{aligned}$$
    $$Y - 2 = 0 \qquad Y + 2 = 0$$
    $$Y = 2 \qquad Y = -2$$
18. $$\begin{aligned} 3(2)^2 - 12 &= 0 \\ 3(4) - 12 &= 0 \\ 12 - 12 &= 0 \\ 0 &= 0 \end{aligned}$$
    $$\begin{aligned} 3(-2)^2 - 12 &= 0 \\ 3(4) - 12 &= 0 \\ 12 - 12 &= 0 \\ 0 &= 0 \end{aligned}$$
19. $X = \frac{65 \text{ mi}}{1 \text{ hr}} \times \frac{3 \text{ hr}}{1} = \frac{195 \text{mi}}{1} = 195 \text{mi}$
20. $X = \frac{45 \text{ mi}}{1 \text{ hr}} \times \frac{5 \text{ hr}}{1} = \frac{225 \text{mi}}{1} = 225 \text{mi}$

Systematic Review 28E

1. 12 inches = 1 foot
2. Inches in numerator to remain in final answer.
   Feet in denominator so they will cancel.
3. $7{,}920 \text{ ft} \times \frac{12 \text{ in}}{1 \text{ ft}} =$
   $\frac{95{,}040 \text{ in}}{1} = 95{,}040 \text{ in}$

4. 2,000 pounds = 1 ton
5. Tons in numerator to remain in final answer. Pounds in denominator so they will cancel.
6. $10{,}000\ \text{lb} \times \dfrac{1\ \text{ton}}{2{,}000\ \text{lb}} = \dfrac{10{,}000\ \text{tons}}{2{,}000} = 5\ \text{tons}$
7. 16 ounces = 1 pound
8. Ounces in numerator to remain in final answer. Pounds in denominator so they will cancel.
9. $5\ \text{lb} \times \dfrac{16\ \text{oz}}{1\ \text{lb}} = \dfrac{80\ \text{oz}}{1} = 80\ \text{oz}$
10. 2 pints = 1 quart
11. Pints in numerator to remain in final answer. Quarts in denominator so they will cancel.
12. $13\ \text{qt} \times \dfrac{2\ \text{pt}}{1\ \text{qt}} = \dfrac{26\ \text{pt}}{1} = 26\ \text{pt}$
13. $2X^2 + X - 6 = 0$

$(2X-3)(X+2) = 0$

$2X - 3 = 0 \quad X + 2 = 0$

$2X = 3 \quad X = -2$

$X = \dfrac{3}{2}$

14. $2\left(\dfrac{3}{2}\right)^2 + \left(\dfrac{3}{2}\right) - 6 = 0$

$2\left(\dfrac{9}{4}\right) + \dfrac{3}{2} - \dfrac{12}{2} = 0$

$\dfrac{18}{4} + \dfrac{3}{2} - \dfrac{12}{2} = 0$

$\dfrac{9}{2} + \dfrac{3}{2} - \dfrac{12}{2} = 0$

$0 = 0$

$2(-2)^2 + (-2) - 6 = 0$

$2(4) - 2 - 6 = 0$

$8 - 2 - 6 = 0$

$0 = 0$

15. $5B^2 - 125 = 0$

$(5)(B^2 - 25) = 0$

$(5)(B-5)(B+5) = 0$

$B - 5 = 0 \quad B + 5 = 0$

$B = 5 \quad B = -5$

16. $5(5)^2 - 125 = 0$

$5(25) - 125 = 0$

$125 - 125 = 0$

$0 = 0$

$5(-5)^2 - 125 = 0$

$5(25) \ 125 = 0$

$125 - 125 = 0$

$0 = 0$

17. $6X^2 - 6X + 18 = 90$

$6X^2 - 6X - 72 = 0$

$(6)(X^2 - X - 12) = 0$

$(6)(X-4)(X+3) = 0$

$X - 4 = 0 \quad X + 3 = 0$

$X = 4 \quad X = -3$

18. $6(4)^2 - 6(4) + 18 = 90$

$6(16) - 24 + 18 = 90$

$96 - 24 + 18 = 90$

$90 = 90$

19. $\dfrac{X\ \text{mi}}{1\ \text{hr}} = \dfrac{6\ \text{mi}}{.5\ \text{hr}}$

$.5X = 6$

$X = 6 \div .5 = 12\ \text{mi}$

rate = 12 mph

20. $R = \dfrac{10\ \text{mi}}{.8\ \text{hr}} = 12.5\ \text{mph}$

Lesson Practice 29A

1. $\dfrac{1\ \text{ft}^2}{1} \times \dfrac{12\ \text{in}}{1\ \text{ft}} \times \dfrac{12\ \text{in}}{1\ \text{ft}} = 144\ \text{in}^2$
2. $\dfrac{2\ \text{ft}^2}{1} \times \dfrac{12\ \text{in}}{1\ \text{ft}} \times \dfrac{12\ \text{in}}{1\ \text{ft}} = 288\ \text{in}^2$
3. $\dfrac{1\ \text{yd}^2}{1} \times \dfrac{3\ \text{ft}}{1\ \text{yd}} \times \dfrac{3\ \text{ft}}{1\ \text{yd}} = 9\ \text{ft}^2$

4. $\frac{1\text{ yd}^3}{1} \times \frac{36\text{ in}}{1\text{ yd}} \times \frac{36\text{ in}}{1\text{ yd}} \times \frac{36\text{ in}}{1\text{ yd}} = 46{,}656\text{ in}^3$

5. $\frac{2\text{ ft}^3}{1} \times \frac{12\text{ in}}{1\text{ ft}} \times \frac{12\text{ in}}{1\text{ ft}} \times \frac{12\text{ in}}{1\text{ ft}} = 3{,}456\text{ in}^3$

6. $\frac{8\text{ cm}^2}{1} \times \frac{10\text{ mm}}{1\text{ cm}} \times \frac{10\text{ mm}}{1\text{ cm}} =$ 800 mm^2

7. $\frac{9\text{ yd}^2}{1} \times \frac{36\text{ in}}{1\text{ yd}} \times \frac{36\text{ in}}{1\text{ yd}} = 11{,}664\text{ in}^2$

8. $\frac{1\text{ mi}^2}{1} \times \frac{5{,}280\text{ ft}}{1\text{ mi}} \times \frac{5{,}280\text{ ft}}{1\text{ mi}} =$ $27{,}878{,}400\text{ ft}^2$

9. $\frac{100\text{ ft}^2}{1} \times \frac{1\text{ yd}}{3\text{ ft}} \times \frac{1\text{ yd}}{3\text{ ft}} \approx 11.11\text{ yd}^2$

10. $\frac{.5\text{ yd}^2}{1} \times \frac{3\text{ ft}}{1\text{ yd}} \times \frac{3\text{ ft}}{1\text{ yd}} = 4.5\text{ ft}^2$

11. $\frac{300\text{ ft}^2}{1} \times \frac{1\text{ mi}}{5{,}280\text{ ft}} \times \frac{1\text{ mi}}{5{,}280\text{ ft}} \approx$ $.00001\text{ mi}^2$

12. $\frac{950\text{ cm}^2}{1} \times \frac{1\text{ m}}{100\text{ cm}} \times \frac{1\text{ m}}{100\text{ cm}} = .095\text{ m}^2$

13. $43{,}560\text{ ft}^2$

14. $4\text{ ft} \times 4\text{ ft} \times 8\text{ ft} = 128\text{ ft}^3$

15. $3\text{ ft} \times 3\text{ ft} \times 3\text{ ft} = 27\text{ ft}^3$

16. $3\text{ ft} \times 3\text{ ft} = 9\text{ ft}^2$

Lesson Practice 29B

1. $\frac{7\text{ ft}^2}{1} \times \frac{12\text{ in}}{1\text{ ft}} \times \frac{12\text{ in}}{1\text{ ft}} = 1{,}008\text{ in}^2$

2. $\frac{3\text{ m}^2}{1} \times \frac{100\text{ cm}}{1\text{ m}} \times \frac{100\text{ cm}}{1\text{ m}} = 30{,}000\text{ cm}^2$

3. $\frac{.8\text{ ft}^2}{1} \times \frac{12\text{ in}}{1\text{ ft}} \times \frac{12\text{ in}}{1\text{ ft}} = 115.2\text{ in}^2$

4. $\frac{1.5\text{ ft}^2}{1} \times \frac{12\text{ in}}{1\text{ ft}} \times \frac{12\text{ in}}{1\text{ ft}} = 216\text{ in}^2$

5. $\frac{8\text{ m}^3}{1} \times \frac{10\text{ dm}}{1\text{ m}} \times \frac{10\text{ dm}}{1\text{ m}} \times \frac{10\text{ dm}}{1\text{ m}} =$ $8{,}000\text{ dm}^3$

6. $\frac{3\text{ km}^3}{1} \times \frac{1{,}000\text{ m}}{1\text{ km}} \times \frac{1{,}000\text{ m}}{1\text{ km}} \times \frac{1{,}000\text{ m}}{1\text{ km}} =$ $3{,}000{,}000{,}000\text{ m}^3$

7. $\frac{5.6\text{ ft}^3}{1} \times \frac{12\text{ in}}{1\text{ ft}} \times \frac{12\text{ in}}{1\text{ ft}} \times \frac{12\text{ in}}{1\text{ ft}} =$ $9{,}676.8\text{ in}^3$

8. $\frac{2\text{ ft}^3}{1} \times \frac{12\text{ in}}{1\text{ ft}} \times \frac{12\text{ in}}{1\text{ ft}} \times \frac{12\text{ in}}{1\text{ ft}} = 3{,}456\text{ in}^3$

9. $\frac{7\text{ yd}^3}{1} \times \frac{36\text{ in}}{1\text{ yd}} \times \frac{36\text{ in}}{1\text{ yd}} \times \frac{36\text{ in}}{1\text{ yd}} =$ $326{,}592\text{ in}^3$

10. $\frac{4\text{ mi}^3}{1} \times \frac{5{,}280\text{ ft}}{1\text{ mi}} \times \frac{5{,}280\text{ ft}}{1\text{ mi}} \times \frac{5{,}280\text{ ft}}{1\text{ mi}} \times$ $\frac{12\text{ in}}{1\text{ ft}} \times \frac{12\text{ in}}{1\text{ ft}} \times \frac{12\text{ in}}{1\text{ ft}} \approx$ $1{,}017{,}400{,}000{,}000{,}000\text{ in}^3$

11. $\frac{370\text{ cm}^3}{1} \times \frac{1\text{ m}}{100\text{ cm}} \times \frac{1\text{ m}}{100\text{ cm}} \times \frac{1\text{ m}}{100\text{ cm}} =$ $.00037\text{ m}^3$

12. $\frac{18\text{ cm}^2}{1} \times \frac{1\text{ m}}{100\text{ cm}} \times \frac{1\text{ m}}{100\text{ cm}} = .0018\text{ m}^2$

13. $\frac{2\text{ acres}}{1} \times \frac{43{,}560\text{ ft}^2}{1\text{ acre}} = 87{,}120\text{ ft}^2$

14. $4\text{ ft} \times 4\text{ ft} \times 8\text{ ft} = 128\text{ ft}^3$

15. $\frac{2\text{ yards}}{1} \times \frac{27\text{ ft}^3}{1\text{ yard}} = 54\text{ ft}^3$

16. $\frac{2\text{ yards}}{1} \times \frac{9\text{ ft}^2}{1\text{ yard}} = 18\text{ ft}^2$

Systematic Review 29C

1. $\frac{1\text{ ft}^2}{1} \times \frac{12\text{ in}}{1\text{ ft}} \times \frac{12\text{ in}}{1\text{ ft}} = 144\text{ in}^2$

2. $\frac{1\text{ yd}^2}{1} \times \frac{3\text{ ft}}{1\text{ yd}} \times \frac{3\text{ ft}}{1\text{ yd}} = 9\text{ ft}^2$

3. $\frac{1\ \text{mi}^2}{1} \times \frac{5{,}280\ \text{ft}}{1\ \text{mi}} \times \frac{5{,}280\ \text{ft}}{1\ \text{mi}}$

$= 27{,}878{,}400\ \text{ft}^2$

4. $\frac{1\ \text{m}^2}{1} \times \frac{100\ \text{cm}}{1\ \text{m}} \times \frac{100\ \text{cm}}{1\ \text{m}} = 10{,}000\ \text{cm}^2$

5. $\frac{4\ \text{ft}^2}{1} \times \frac{12\ \text{in}}{1\ \text{ft}} \times \frac{12\ \text{in}}{1\ \text{ft}} = 576\ \text{in}^2$

6. $\frac{7\ \text{yd}^2}{1} \times \frac{3\ \text{ft}}{1\ \text{yd}} \times \frac{3\ \text{ft}}{1\ \text{yd}} = 63\ \text{ft}^2$

7. $\frac{3.2\ \text{mi}^2}{1} \times \frac{5{,}280\ \text{ft}}{1\ \text{mi}} \times \frac{5{,}280\ \text{ft}}{1\ \text{mi}} =$

$89{,}210{,}880\ \text{ft}^2$

8. $\frac{15.7\ \text{m}^2}{1} \times \frac{100\ \text{cm}}{1\ \text{m}} \times \frac{100\ \text{cm}}{1\ \text{m}} =$

$157{,}000\ \text{cm}^2$

9. $43{,}560\ \text{ft}^2$

10. $3\ \text{ft} \times 3\ \text{ft} = 9\ \text{ft}^2$

11.
$$3X^2 - 5X + 2 = 0$$
$$(3X-2)(X-1) = 0$$
$$3X - 2 = 0 \quad X - 1 = 0$$
$$3X = 2 \quad X = 1$$
$$X = \frac{2}{3}$$

12.
$$3\left(\frac{2}{3}\right)^2 - 5\left(\frac{2}{3}\right) + 2 = 0$$
$$3\left(\frac{4}{9}\right) - \frac{10}{3} + \frac{6}{3} = 0$$
$$\frac{12}{9} - \frac{10}{3} + \frac{6}{3} = 0$$
$$\frac{4}{3} - \frac{10}{3} + \frac{6}{3} = 0$$
$$0 = 0$$

$$3(1)^2 - 5(1) + 2 = 0$$
$$3(1) - 5 + 2 = 0$$
$$3 - 5 + 2 = 0$$
$$0 = 0$$

13.
$$2X^2 - 10X + 12 = 0$$
$$(2)(X^2 - 5X + 6) = 0$$
$$(2)(X-3)(X-2) = 0$$
$$0 = 0$$
$$X - 3 = 0 \quad X - 2 = 0$$
$$X = 3 \quad X = 2$$

14.
$$2(3)^2 - 10(3) + 12 = 0$$
$$2(9) - 30 + 12 = 0$$
$$18 - 30 + 12 = 0$$
$$0 = 0$$

$$2(2)^2 - 10(2) + 12 = 0$$
$$2(4) - 20 + 12 = 0$$
$$8 - 20 + 12 = 0$$
$$0 = 0$$

15. $(X-4)^2 = (X-4)(X-4) = X^2 - 8X + 16$

16. $35^2 > 32 \times 38$

$1{,}225 > 1{,}216$

17. $X^2 + 7X + 10 = (X+2)(X+5)$

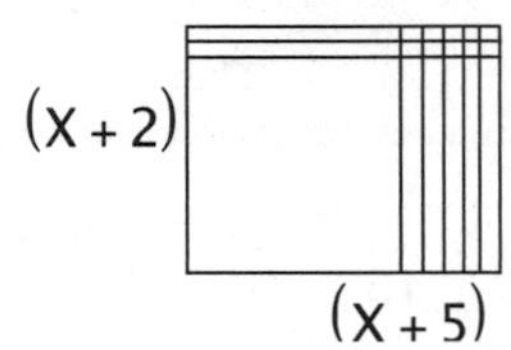

18. $(3X+2)(X+2) = 3X^2 + 8X + 4$

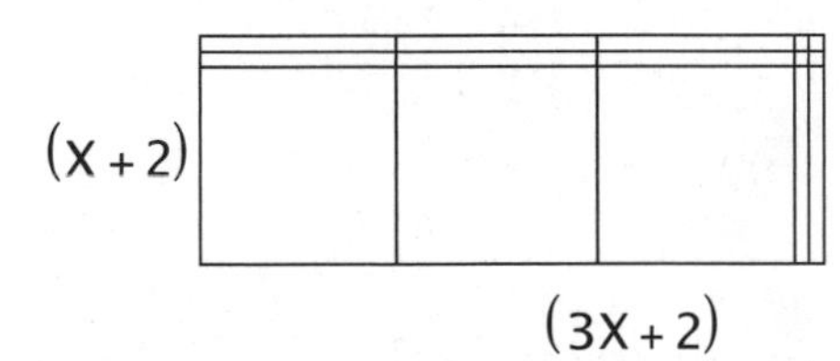

19. $WF \times 3\ \text{ft} = 1\ \text{ft}$

$WF = \frac{1}{3} = .33\bar{3} = 33\frac{1}{3}\%$

20. $WF \times 9\ \text{ft}^2 = 1\ \text{ft}^2$

$WF = \frac{1\ \text{ft}^2}{9\ \text{ft}^2} = \frac{1}{9}$

Systematic Review 29D

1. $\frac{9\ \text{ft}^2}{1} \times \frac{12\ \text{in}}{1\ \text{ft}} \times \frac{12\ \text{in}}{1\ \text{ft}} = 1{,}296\ \text{in}^2$

2. $\frac{5\ \text{yd}^2}{1} \times \frac{3\ \text{ft}}{1\ \text{yd}} \times \frac{3\ \text{ft}}{1\ \text{yd}} = 45\ \text{ft}^2$

3. $\frac{6\ \text{mi}^2}{1} \times \frac{5{,}280\ \text{ft}}{1\ \text{mi}} \times \frac{5{,}280\ \text{ft}}{1\ \text{mi}} =$
   $167{,}270{,}400\ \text{ft}^2$

4. $\frac{18\ \text{m}^2}{1} \times \frac{100\ \text{cm}}{1\ \text{m}} \times \frac{100\ \text{cm}}{1\ \text{m}} = 180{,}000\ \text{cm}^2$

5. $\frac{.75\ \text{ft}^2}{1} \times \frac{12\ \text{in}}{1\ \text{ft}} \times \frac{12\ \text{in}}{1\ \text{ft}} = 108\ \text{in}^2$

6. $\frac{1.3\ \text{yd}^2}{1} \times \frac{3\ \text{ft}}{1\ \text{yd}} \times \frac{3\ \text{ft}}{1\ \text{yd}} = 11.7\ \text{ft}^2$

7. $\frac{25\ \text{mi}^2}{1} \times \frac{5{,}280\ \text{ft}}{1\ \text{mi}} \times \frac{5{,}280\ \text{ft}}{1\ \text{mi}} =$
   $696{,}960{,}000\ \text{ft}^2$

8. $\frac{.67\ \text{m}^2}{1} \times \frac{100\ \text{cm}}{1\ \text{m}} \times \frac{100\ \text{cm}}{1\ \text{m}} = 6{,}700\ \text{cm}^2$

9. $\frac{5\ \text{acres}}{1} \times \frac{43{,}560\ \text{ft}^2}{1\ \text{acre}} = 217{,}800\ \text{ft}^2$

10. $\frac{2\ \text{cords}}{1} \times \frac{128\ \text{ft}^3}{1\ \text{cord}} = 256\ \text{ft}^3$

11. $3X^2 - 9X - 12 = 0$
    $(3X + 3)(X - 4) = 0$
    $3X = -3 \quad X - 4 = 0$
    $X = -1 \quad X = 4$

12. $3(-1)^2 - 9(-1) - 12 = 0$
    $3(1) + 9 - 12 = 0$
    $3 + 9 - 12 = 0$
    $0 = 0$

    $3(4)^2 - 9(4) - 12 = 0$
    $3(16) - 36 - 12 = 0$
    $48 - 36 - 12 = 0$
    $0 = 0$

13. $X^2 - 36 = 0$
    $(X - 6)(X + 6) = 0$
    $X - 6 = 0 \quad X + 6 = 0$
    $X = 6 \quad X = -6$

14. $(6)^2 - 36 = 0$
    $36 - 36 = 0$
    $0 = 0$

    $(-6)^2 - 36 = 0$
    $36 - 36 = 0$
    $0 = 0$

15. $(X - 5)^2 = (X - 5)(X - 5) = X^2 - 10X + 25$

16. $45^2 > 40 \times 50$
    $2{,}025 > 2{,}000$

17. $X^2 + 10X + 21 = (X + 7)(X + 3)$

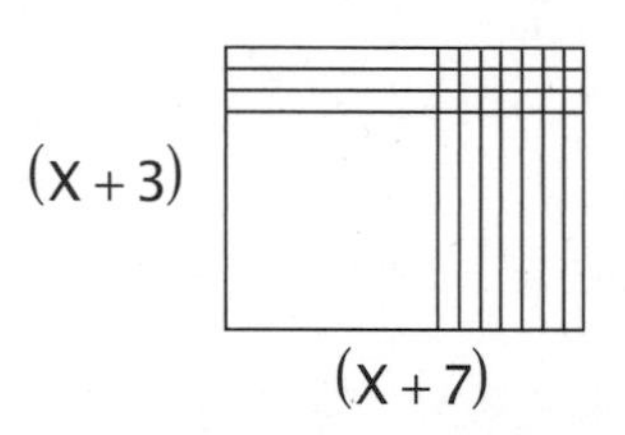

(X + 3)
(X + 7)

18. $(X + 3)(X - 9) = X^2 - 6X - 27$

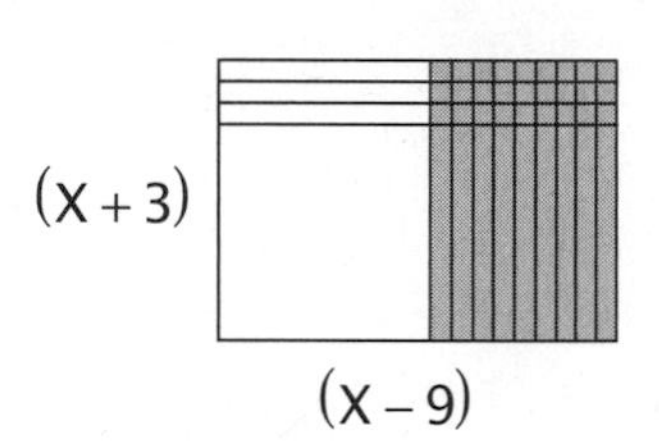

(X + 3)
(X − 9)

19. $WF \times 36\ \text{in} = 1\ \text{in}$
    $WF = \frac{1}{36} = .02\overline{7} \approx 2.8\%$

20. $WF \times 144\ \text{in}^2 = 1\ \text{in}^2$
    $WF = \frac{1}{144}$

Systematic Review 29E

1. $\frac{27\ \text{ft}^2}{1} \times \frac{1\ \text{yd}}{3\ \text{ft}} \times \frac{1\ \text{yd}}{3\ \text{ft}} = 3\ \text{yd}^2$

2. $\frac{3\ \text{yd}^2}{1} \times \frac{3\ \text{ft}}{1\ \text{yd}} \times \frac{3\ \text{ft}}{1\ \text{yd}} = 27\ \text{ft}^2$

3. $\frac{10{,}000\ \text{ft}^2}{1} \times \frac{1\ \text{mi}}{5{,}280\ \text{ft}} \times \frac{1\ \text{mi}}{5{,}280\ \text{ft}} \approx$
   $.00036\ \text{mi}^2$

4. $\frac{1{,}200\ \text{cm}^2}{1} \times \frac{1\ \text{m}}{100\ \text{cm}} \times \frac{1\ \text{m}}{100\ \text{cm}} = .12\ \text{m}^2$

5. $\frac{1\ \text{ft}^3}{1} \times \frac{12\ \text{in}}{1\ \text{ft}} \times \frac{12\ \text{in}}{1\ \text{ft}} \times \frac{12\ \text{in}}{1\ \text{ft}} = 1{,}728\ \text{in}^3$

6. $\frac{1\ \text{yd}^3}{1} \times \frac{3\ \text{ft}}{1\ \text{yd}} \times \frac{3\ \text{ft}}{1\ \text{yd}} \times \frac{3\ \text{ft}}{1\ \text{yd}} = 27\ \text{ft}^3$

7.

$\frac{1\ \text{mi}^3}{1} \times \frac{5{,}280\ \text{ft}}{1\ \text{mi}} \times \frac{5{,}280\ \text{ft}}{1\ \text{mi}} \times \frac{5{,}280\ \text{ft}}{1\ \text{mi}} \approx$

$147{,}000{,}000{,}000\ \text{ft}^3$

8. $\frac{3\ \text{m}^3}{1} \times \frac{100\ \text{cm}}{1\ \text{m}} \times \frac{100\ \text{cm}}{1\ \text{m}} \times \frac{100\ \text{cm}}{1\ \text{m}}$

$= 3{,}000{,}000\ \text{cm}^3$

9. $\frac{3\ \text{cords}}{1} \times \frac{128\ \text{ft}^3}{1\ \text{cord}} = 384\ \text{ft}^3$

10. $\frac{2\ \text{yards}}{1} \times \frac{27\ \text{ft}^3}{1\ \text{yard}} = 54\ \text{ft}^3$

11. $X^2 - 10X + 25 = 0$

$(X-5)(X-5) = 0$

$X - 5 = 0$

$X = 5$

12. $(5)^2 - 10(5) + 25 = 0$

$25 - 50 + 25 = 0$

$X = 5$

13. $X^2 - 12X + 35 = 0$

$(X-7)(X-5) = 0$

$X - 7 = 0 \quad X - 5 = 0$

$X = 7 \quad X = 5$

14. $(7)^2 - 12(7) + 35 = 0$

$49 - 84 + 35 = 0$

$0 = 0$

$(5)^2 - 12(5) + 35 = 0$

$25 - 60 + 35 = 0$

$0 = 0$

15. $(3X-1)(X-2) = 3X^2 - 7X + 2$

16. $73 \times 77 > 60 \times 80$

$5{,}621 > 4{,}800$

17. $WF \times 9\ \text{ft}^2 = 1\ \text{ft}^2$

$WF = \frac{1}{9} = .1\overline{1} = 11.\overline{1}\%$

18. $WF \times 27{,}878{,}400\ \text{ft}^2 = 43{,}560\ \text{ft}^2$

$WF = \frac{43{,}560}{27{,}878{,}400}$

$WF = \frac{1}{640}$

19. $R = \frac{100\ \text{yd}}{9\ \text{sec}} = 11.1\overline{1}\ \text{yd/sec}$

20. $R = \frac{200\ \text{mi}}{5\ \text{hr}} = 40\ \text{mph}$

Lesson Practice 30A

1. 2.5
2. .9
3. 1.6
4. 28
5. $\frac{10\ \text{km}}{1} \times \frac{.62\ \text{mi}}{1\ \text{km}} = 6.2\ \text{mi}$
6. $\frac{45\ \text{oz}}{1} \times \frac{28\ \text{g}}{1\ \text{oz}} = 1{,}260\ \text{g}$
7. $\frac{21\ \text{kg}}{1} \times \frac{2.2\ \text{lb}}{1\ \text{kg}} = 46.2\ \text{lb}$
8. $\frac{15\ \text{yd}}{1} \times \frac{.9\text{m}}{1\ \text{yd}} = 13.5\ \text{m}$
9. $\frac{15\ \text{cm}}{1} \times \frac{.4\ \text{in}}{1\ \text{cm}} = 6\ \text{in}$
10. $\frac{25\ \text{g}}{1} \times \frac{.035\ \text{oz}}{1\ \text{g}} = .875\ \text{oz}$
11. $\frac{5\ \text{qt}}{1} \times \frac{.95\ \text{liters}}{1\ \text{qt}} = 4.75\ \text{liters}$
12. $\frac{54\ \text{in}}{1} \times \frac{2.5\ \text{cm}}{1\ \text{in}} = 135\ \text{cm}$
13. $\frac{5\ \text{km}}{1} \times \frac{.62\ \text{mi}}{1\ \text{km}} = 3.1\ \text{mi}$
14. $\frac{45\ \text{lb}}{1} \times \frac{.45\ \text{kg}}{1\ \text{lb}} = 20.25\ \text{kg}$
15. $\frac{105\ \text{oz}}{1} \times \frac{28\ \text{g}}{1\ \text{oz}} = 2{,}940\ \text{g}$
16. $\frac{63\ \text{yd}}{1} \times \frac{.9\ \text{m}}{1\ \text{yd}} = 56.7\ \text{m}$

Lesson Practice 30B

1. .4
2. 1.1
3. 2.2
4. 1.06
5. $\frac{25\text{ cm}}{1} \times \frac{.4\text{ in}}{1\text{ cm}} = 10\text{ in}$
6. $\frac{36\text{ g}}{1} \times \frac{.035\text{ oz}}{1\text{ g}} = 1.26\text{ oz}$
7. $\frac{12\text{ qt}}{1} \times \frac{.95\text{ liters}}{1\text{ qt}} = 11.4\text{ liters}$
8. $\frac{110\text{ in}}{1} \times \frac{2.5\text{ cm}}{1\text{ in}} = 275\text{ cm}$
9. $\frac{36\text{ in}}{1} \times \frac{2.5\text{ cm}}{1\text{ in}} = 90\text{ cm}$
10. $\frac{75.5\text{ g}}{1} \times \frac{.035\text{ oz}}{1\text{ g}} = 2.64\text{ oz}$
11. $\frac{18.5\text{ yd}}{1} \times \frac{.9\text{ m}}{1\text{ yd}} = 16.65\text{ m}$
12. $\frac{55\text{ kg}}{1} \times \frac{2.2\text{ lbs}}{1\text{ kg}} = 121\text{ lb}$
13. $\frac{16.3\text{ mi}}{1} \times \frac{1.6\text{ km}}{1\text{ mi}} = 26.08\text{ km}$
14. $\frac{36\text{ liters}}{1} \times \frac{1.06\text{ qt}}{1\text{ liter}} = 38.16\text{ qt}$
15. $\frac{5.05\text{ oz}}{1} \times \frac{28\text{ g}}{1\text{ oz}} = 141.4\text{ g}$
16. $\frac{360.5\text{ cm}}{1} \times \frac{.4\text{ in}}{1\text{ cm}} = 144.2\text{ in}$

Systematic Review 30C

1. $\frac{5\text{ in}}{1} \times \frac{2.5\text{ cm}}{1\text{ in}} = 12.5\text{ cm}$
2. $\frac{3\text{ qt}}{1} \times \frac{.95\text{ liters}}{1\text{ qt}} = 2.85\text{ liters}$
3. $\frac{10\text{ oz}}{1} \times \frac{28\text{ g}}{1\text{ oz}} = 280\text{ g}$
4. $\frac{62\text{ lb}}{1} \times \frac{.45\text{ kg}}{1\text{ lb}} = 27.9\text{ kg}$
5. $$\begin{array}{r} 3X^2+X+4 \quad R-3 \\ 2X+1\,\overline{)\,6X^3+5X^2+9X+1} \\ -\underline{(6X^3+3X^2)} \\ 2X^2+9X \\ -\underline{(2X^2+X)} \\ 8X+1 \\ -\underline{(8X+4)} \\ -3 \end{array}$$
6. $$\begin{array}{r} 3X^2+X+4 \\ \times \quad 2X+1 \\ \hline 3X^2+X+4 \\ \underline{6X^3+2X^2+8X} \\ 6X^3+5X^2+9X+4 \\ -3 \\ \hline 6X^3+5X^2+9X+1 \end{array}$$
7. $\sqrt{9X^4} = 3X^2$
8. $\sqrt{9(10)^4} = 3(10)^2$
   $\sqrt{9(10{,}000)} = 3(100)$
   $\sqrt{90{,}000} = 300$
   $300 = 300$
9. $\frac{3.5\text{ acres}}{1} \times \frac{43{,}560\text{ ft}^2}{1\text{ acre}} =$
   $152{,}460\text{ ft}^2$
10. $\frac{1\text{ mi}^2}{1} \times \frac{5{,}280\text{ ft}}{1\text{ mi}} \times \frac{5{,}280\text{ ft}}{1\text{ mi}} =$
    $27{,}878{,}400\text{ ft}^2$
11. $13 \times 17 < 16 \times 14$
    $221 < 224$
12. $5\text{ yd}^2 = 50\text{ ft}^2$
    $\frac{5\text{ yd}^2}{1} \times \frac{3\text{ ft}}{1\text{ yd}} \times \frac{3\text{ ft}}{1\text{ yd}} = 45\text{ ft}^2$
    $5\text{ yd}^2 = 45\text{ ft}^2$
    $45\text{ ft}^2 < 50\text{ ft}^2$
13. $(X+6)^2 = (X+6)(X+6) =$
    $X^2+12X+36$

14. $X - 2Y = 4$

$-2Y = -X + 4$

$Y = \frac{1}{2}X - 2$

$m = \frac{1}{2}$

15. $X^2 + 5X + 6 = 20$

$X^2 + 5X - 14 = 0$

$(X + 7)(X - 2) = 0$

$X + 7 = 0 \qquad X - 2 = 0$

$X = -7 \qquad X = 2$

16. $(-7)^2 + 5(-7) + 6 = 20$

$49 - 35 + 6 = 20$

$20 = 20$

$(2)^2 + 5(2) + 6 = 20$

$4 + 10 + 6 = 20$

$20 = 20$

17. $(2X + 3)(X + 2) = (2X)(X + 2) + (3)(X + 2) =$

$(2X^2 + 4X + 3X + 6)$

18. $E = 400W - 100$

(E = earnings, W = weeks)

see graph

19. $300

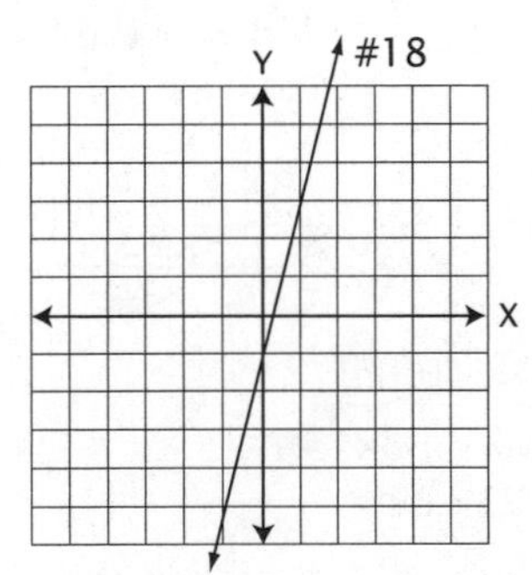

20. $E = 400W - 100$

$E = 400(30) - 100$

$E = 12{,}000 - 100$

$E = 11{,}900$

Systematic Review 30D

1. $\frac{7\ \text{mi}}{1} \times \frac{1.6\ \text{km}}{1\ \text{mi}} = 11.2\ \text{km}$

2. $\frac{8\ \text{lb}}{1} \times \frac{.45\ \text{kg}}{1\ \text{lb}} = 3.6\ \text{kg}$

3. $\frac{4\ \text{yd}}{1} \times \frac{.9\ \text{m}}{1\ \text{yd}} = 3.6\ \text{m}$

4. $\frac{2\ \text{qt}}{1} \times \frac{.95\ \text{liters}}{1\ \text{qt}} = 1.9\ \text{liters}$

5.

$$
\begin{array}{r}
2X^2 - 5X - 14 \text{ R } -21 \\
X - 2 \overline{)\,2X^3 - 9X^2 - 4X + 7} \\
-(2X^3 - 4X^2) \\
\hline
-5X^2 - 4X \\
-(-5X^2 + 10X) \\
\hline
-14X + 7 \\
-(-14X + 28) \\
\hline
-21
\end{array}
$$

6.

$$
\begin{array}{r}
2X^2 - 5X - 14 \\
\times \qquad X - 2 \\
\hline
-4X^2 + 10X + 28 \\
2X^3 - 5X^2 - 14X \\
\hline
2X^3 - 9X^2 - 4X + 28 \\
-21 \\
\hline
2X^3 - 9X^2 - 4X + 7
\end{array}
$$

7. $\sqrt{16X^2} = 4X$

8. $\sqrt{Y^2X^4} = YX^2$

9. $\frac{100{,}000\ \text{ft}^2}{1} \times \frac{1\ \text{acre}}{43{,}560\ \text{ft}^2}$

≈ 2.296 acres

10. $\frac{1.34\ \text{m}^2}{1} \times \frac{100\ \text{cm}}{1\ \text{m}} \times \frac{100\ \text{cm}}{1\ \text{m}}$

$= 13{,}400\ \text{cm}^2$

11. $82 \times 88 < 86 \times 84$

$7{,}216 < 7{,}224$

12. $7\text{ yd}^3 > 175\text{ ft}^3$

$$\frac{7\text{ yd}^3}{1} \times \frac{3\text{ ft}}{1\text{ yd}} \times \frac{3\text{ ft}}{1\text{ yd}} \times \frac{3\text{ ft}}{1\text{ yd}} = 189\text{ ft}^3$$

$189\text{ ft}^3 > 175\text{ ft}^3$

13. $(X-3)^2 = (X-3)(X-3) = X^2 - 6X + 9$

14. $X - 2Y = 4$

$-2Y = -X + 4$

$Y = \frac{1}{2}X - 2$

$m = -2\left(\text{negative reciprocal of } \frac{1}{2}\right)$

15. $X^2 - 12X + 35 = 15$

$X^2 - 12X + 20 = 0$

$(X-10)(X-2) = 0$

$X - 10 = 0 \qquad X - 2 = 0$

$X = 10 \qquad X = 2$

16. $(10)^2 - 12(10) + 35 = 15$

$100 - 120 + 35 = 15$

$15 = 15$

$(2)^2 - 12(2) + 35 = 15$

$4 - 24 + 35 = 15$

17. $(2X+7)(X+2) =$

$(2X)(X+2) + (7)(X+2) =$

$\left(2X^2 + 4X + 7X + 14\right)$

18. $D = RT$ divide both sides by R:

$\frac{D}{R} = T$

19. $T = \frac{D}{R}$

$T = \frac{(12)}{(6)} = 2$ hours

20. $2Y < 3X - 2$

$Y < \frac{3}{2}X - 1$

try $(0,0)$:

$(0) < \frac{3}{2}(0) - 1$

$0 < 0 - 1$

$0 < -1$; false

try $(2,-2)$:

$(-2) < \frac{3}{2}(2) - 1$

$-2 < 3 - 1$

$-2 < 2$; true

see graph

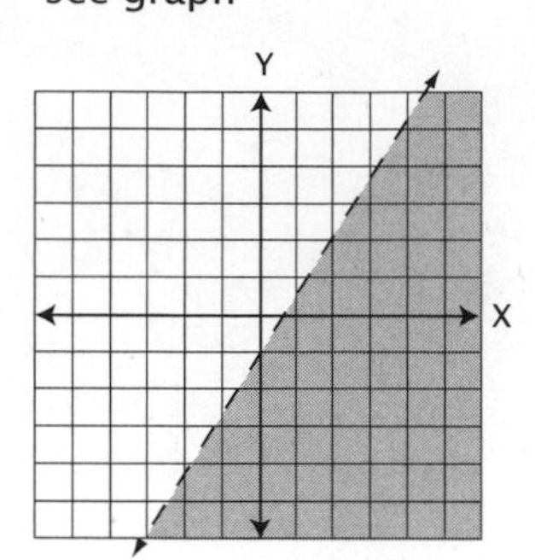

Systematic Review 30E

1. $\frac{25\text{ km}}{1} \times \frac{.62\text{ mi}}{1\text{ km}} = 15.5\text{ mi}$

2. $\frac{7\text{ m}}{1} \times \frac{1.1\text{ yd}}{1\text{ m}} = 7.7\text{ yd}$

3. $\frac{11\text{ kg}}{1} \times \frac{2.2\text{ lb}}{1\text{ kg}} = 24.2\text{ lb}$

4. $\frac{10\text{ liters}}{1} \times \frac{1.06\text{ qt}}{1\text{ liter}} = 10.6\text{ qt}$

5.

$$\begin{array}{r} 2X^2 + 3X + 8 \text{ R } 21 \\ 2X - 3 \overline{\left) 4X^3 + 0X^2 + 7X - 3 \right.} \\ -\left(4X^3 - 6X^2\right) \\ \hline 6X^2 + 7X \\ -\left(6X^2 - 9X\right) \\ \hline 16X - 3 \\ -(16X - 24) \\ \hline 21 \end{array}$$

6.
$$\begin{array}{r} 2X^2 + 3X + 8 \\ \times \quad 2X - 3 \\ \hline -6X^2 - 9X - 24 \\ 4X^3 + 6X^2 + 16X \quad \\ \hline 4X^3 + \quad 7X - 24 \\ + \quad 21 \\ \hline 4X^3 + \quad 7X - 3 \end{array}$$

7. $\sqrt{144} = 12$

8. $\sqrt{25X^6} = 5X^3$

9. $\frac{1.75 \not{yd}}{1} \times \frac{9 \text{ ft}^2}{1 \not{yd}} = 15.75 \text{ ft}^2$

10. $25 \times 18 = 450 \text{ ft}^2$

$\frac{450 \not{ft^2}}{1} \times \frac{1 \text{ yd}}{9 \not{ft^2}} = 50$ yards of carpet

11. $13 \times 17 < 16 \times 14$

$221 < 224$

12. $5 \text{ yd}^2 < 50 \text{ ft}^2$

$\frac{5 \not{yd^2}}{1} \times \frac{3 \text{ ft}}{\not{yd}} \times \frac{3 \text{ ft}}{\not{yd}} = 45 \text{ ft}^2$

$45 \text{ ft}^2 < 50 \text{ ft}^2$

13. $(X-2)^2 = (X-2)(X-2) =$

$X^2 - 4X + 4$

14. $(X+4)^2 = (X+4)(X+4) =$

$X^2 + 8X + 16$

15.
$$X^2 + 7X - 18 = 42$$
$$X^2 + 7X - 60 = 0$$
$$(X+12)(X-5) = 0$$

$X + 12 = 0 \qquad X - 5 = 0$

$X = -12 \qquad X = 5$

16.
$$(-12)^2 + 7(-12) - 18 = 42$$
$$144 - 84 - 18 = 42$$
$$42 = 42$$

$$(5)^2 + 7(5) - 18 = 42$$
$$25 + 35 - 18 = 42$$
$$42 = 42$$

17. $T = 4 \div 6 = \frac{4}{6} = \frac{2}{3}$ hour

To change to minutes:

$\frac{2}{3} = \frac{X}{60}$

(How many 60ths of an hour is $\frac{2}{3}$?)

$(2)(60) = 3X$

$120 = 3X$

$\frac{120}{3} = X = 40$ minutes

18. $T = 4 \div 3 = \frac{4}{3} = 1\frac{1}{3}$ hours

$\frac{4}{3} = \frac{X}{60}$

$(4)(60) = 3X$

$240 = 3X$

$\frac{240}{3} = X = 80$ minutes

or 1 hour and 20 minutes

19. $2P - 2 + 7 + P - P =$

$2P + P - P - 2 + 7 =$

$2P + 5$

20. $2P + 5 = 11$

$2P = 11 - 5$

$2P = 6$

$P = \frac{6}{2} = 3$

Lesson Practice 31A

1. $16^{\frac{3}{2}} = \left(\sqrt{16}\right)^3 = 4^3 = 64$

2. $2^{\frac{2}{1}} = 2^2 = 4$

3. $100^{\frac{1}{2}} = \sqrt{100} = 10$

4. $8^{\frac{2}{3}} = \sqrt[3]{8}^{\,2} = 2^2 = 4$

5. $\left(X^{\frac{5}{1}}\right)^{\frac{1}{10}} = X^{\left(\frac{5}{1}\right)\left(\frac{1}{10}\right)} = X^{\frac{5}{10}} = X^{\frac{1}{2}} = \sqrt{X}$

$\left(X^{\frac{5}{1}}\right)^{\frac{1}{10}} = X^{\frac{5}{10}} = X^{\frac{1}{2}} = \sqrt{X}$

6. $\left(Y^{\frac{1}{3}}\right)^{\frac{1}{5}} = Y^{\left(\frac{1}{3}\right)\left(\frac{1}{5}\right)} = Y^{\frac{1}{15}}$ or $\sqrt[15]{Y}$

7. $\left(Y^3 \cdot Y^5\right)^{\frac{1}{4}} = \left(Y^{3+5}\right)^{\frac{1}{4}} = \left(Y^8\right)^{\frac{1}{4}} =$
$Y^{(8)\left(\frac{1}{4}\right)} = Y^2$

8. $16^{\frac{3}{4}} = \sqrt[4]{16}^{\,3} = 2^3 = 8$

9. $\left(27^{\frac{1}{3}}\right)^{\frac{4}{1}} = 3^4 = 81$

10. $8^{\frac{1}{3}} \cdot 16 = \sqrt[3]{8} \cdot 16 = 2 \cdot 16 = 32$

11. $\left(64^{\frac{1}{2}}\right)^{\frac{2}{3}} = 64^{\left(\frac{1}{2}\right)\left(\frac{2}{3}\right)} = 64^{\frac{1}{3}} = \sqrt[3]{64} = 4$

12. $\left(X^5 \cdot X^7\right)^{\frac{1}{2}} = \left(X^{5+7}\right)^{\frac{1}{2}} = \left(X^{12}\right)^{\frac{1}{2}} =$
$X^{(12)\left(\frac{1}{2}\right)} = X^6$

13. $\left(M^{\frac{1}{2}} \cdot M^{\frac{2}{3}}\right)^{\frac{6}{1}} = \left(M^{\frac{1}{2}+\frac{2}{3}}\right)^{\frac{6}{1}} = \left(M^{\frac{3}{6}+\frac{4}{6}}\right)^{\frac{6}{1}} =$
$\left(M^{\frac{7}{6}}\right)^{\frac{6}{1}} = M^{\left(\frac{7}{6}\right)\left(\frac{6}{1}\right)} = M^7$

14. $\left[\left(X^3\right)^5 \cdot X^5\right]^{\frac{1}{2}} = \left[X^{(3)(5)} \cdot X^5\right]^{\frac{1}{2}} =$
$\left[X^{15}X^5\right]^{\frac{1}{2}} = \left[X^{15+5}\right]^{\frac{1}{2}} = \left[X^{20}\right]^{\frac{1}{2}} =$
$X^{(20)\left(\frac{1}{2}\right)} = X^{10}$

15. $\left[\left(X^5\right)^{\frac{2}{3}}\right]^{\frac{1}{6}} = X^{(5)\left(\frac{2}{3}\right)\left(\frac{1}{6}\right)} = X^{\frac{10}{18}} =$
$X^{\frac{5}{9}}$ or $\left(\sqrt[9]{X}\right)^5$

16. $\left[\left(M^8\right)^{\frac{1}{2}}\right]^{\frac{3}{4}} = M^{(8)\left(\frac{1}{2}\right)\left(\frac{3}{4}\right)} = M^{\frac{24}{8}} = M^3$

Lesson Practice 31B

1. $32^{\frac{2}{5}} = \left(\sqrt[5]{32}\right)^2 = 2^2 = 4$

2. $9^{\frac{3}{1}} = 9^3 = 729$

3. $81^{\frac{1}{2}} = \sqrt{81} = 9$

4. $625^{\frac{3}{4}} = \left(\sqrt[4]{625}\right)^3 = 5^3 = 125$

5. $\left(X^6\right)^{\frac{1}{3}} = X^{(6)\left(\frac{1}{3}\right)} = X^{\frac{6}{3}} = X^2$

6. $\left(Y^{\frac{1}{2}}\right)^{\frac{1}{7}} = Y^{\left(\frac{1}{2}\right)\left(\frac{1}{7}\right)} = Y^{\frac{1}{14}}$ or $\sqrt[14]{Y}$

7. $\left(Y^4 \cdot Y^6\right)^{\frac{1}{5}} = \left(Y^{4+6}\right)^{\frac{1}{5}} = \left(Y^{10}\right)^{\frac{1}{5}} =$
$Y^{(10)\left(\frac{1}{5}\right)} = Y^{\frac{10}{5}} = Y^2$

8. $27^{\frac{2}{3}} = \left(\sqrt[3]{27}\right)^2 = 3^2 = 9$

9. $\left(81^{\frac{1}{4}}\right)^{\frac{5}{1}} = 81^{\left(\frac{1}{4}\right)\left(\frac{5}{1}\right)} = 81^{\frac{5}{4}} = \left(\sqrt[4]{81}\right)^5 =$
$3^5 = 243$

10. $64^{\frac{1}{3}} \cdot 64^{\frac{1}{3}} = \sqrt[3]{64} \cdot \sqrt[3]{64} = 4 \cdot 4 = 16$

11. $\left(16^{\frac{1}{3}}\right)^{\frac{3}{4}} = 16^{\left(\frac{1}{3}\right)\left(\frac{3}{4}\right)} = 16^{\frac{1}{4}} = \sqrt[4]{16} = 2$

12. $\left(X^3 \cdot X^5\right)^{\frac{1}{4}} = \left(X^{3+5}\right)^{\frac{1}{4}} = \left(X^8\right)^{\frac{1}{4}} =$
$X^{(8)\left(\frac{1}{4}\right)} = X^{\frac{8}{4}} = X^2$

13. $\left(Y^{\frac{1}{2}} \cdot Y^{\frac{3}{4}}\right)^{\frac{4}{1}} = \left(Y^{\frac{1}{2}+\frac{3}{4}}\right)^{\frac{4}{1}} = \left(Y^{\frac{2}{4}+\frac{3}{4}}\right)^{\frac{4}{1}} =$
$\left(Y^{\frac{5}{4}}\right)^{\frac{4}{1}} = Y^{\left(\frac{5}{4}\right)\left(\frac{4}{1}\right)} = Y^5$

14. $\left[\left(X^3\right)^{\frac{1}{3}} \cdot X^4\right]^{\frac{1}{5}} = \left[X^{(3)\left(\frac{1}{3}\right)} \cdot X^4\right]^{\frac{1}{5}} =$
$\left[X^1 \cdot X^4\right]^{\frac{1}{5}} = \left[X^{1+4}\right]^{\frac{1}{5}} = \left[X^5\right]^{\frac{1}{5}} = X^{(5)\left(\frac{1}{5}\right)} = X^1 = X$

15. $\left[\left(X^4\right)^{\frac{3}{5}}\right]^{\frac{1}{6}} = X^{(4)\left(\frac{3}{5}\right)\left(\frac{1}{6}\right)} = X^{\frac{12}{30}} =$
$X^{\frac{2}{5}}$ or $\left(\sqrt[5]{X}\right)^2$

16. $\left(Y^6 \cdot Y^8\right)^{\frac{1}{2}} = \left(Y^{6+8}\right)^{\frac{1}{2}} = \left(Y^{14}\right)^{\frac{1}{2}} =$
$Y^{(14)\left(\frac{1}{2}\right)} = Y^7$

Systematic Review 31C

1. $8^{\frac{1}{3}} = \sqrt[3]{8} = 2$

2. $9^{\frac{1}{2}} = \sqrt{9} = 3$

3. $5^{\frac{3}{1}} = 5^3 = 125$

4. $1{,}000^{\frac{2}{3}} = \sqrt[3]{1{,}000}^{\,2} = 10^2 = 100$

5. $(X^2)^{\frac{3}{2}} = X^{(2)\left(\frac{3}{2}\right)} = X^{\frac{6}{2}} = X^3$

6. $2^{\frac{1}{3}} \cdot 4 = 2^{\frac{1}{3}} \cdot 2^2 = 2^{\frac{1}{3}+2} = 2^{\frac{1}{3}+\frac{6}{3}} = 2^{\frac{7}{3}}$ or $\left(\sqrt[3]{2}\right)^7$

7. $\left(Y^{\frac{2}{3}}\right)\left(Y^{\frac{1}{4}}\right) = Y^{\frac{2}{3}+\frac{1}{4}} = Y^{\frac{8}{12}+\frac{3}{12}} = Y^{\frac{11}{12}}$ or $\left(\sqrt[12]{Y}\right)^{11}$

8. $\left(5^{\frac{1}{4}}\right)\left(5^{\frac{2}{3}}\right) = 5^{\frac{1}{4}+\frac{2}{3}} = 5^{\frac{3}{12}+\frac{8}{12}} = 5^{\frac{11}{12}}$
or $\left(\sqrt[12]{5}\right)^{11}$

9. $\frac{8 \text{ in}}{1} \times \frac{2.5 \text{ cm}}{1 \text{ in}} = 20 \text{ cm}$

10. $\frac{30 \text{ qt}}{1} \times \frac{.95 \text{ liters}}{1 \text{ qt}} = 28.5 \text{ liters}$

11. $72 \times 78 < (75)(75)$
$5{,}616 < 5{,}625$

12. $2 \text{ mi}^2 > 1{,}200$ acres
$\frac{2 \text{ mi}^2}{1} \times \frac{5{,}280 \text{ ft}}{1 \text{ mi}} \times \frac{5{,}280 \text{ ft}}{1 \text{ mi}} =$
$55{,}756{,}800 \text{ ft}^2$
$\frac{1{,}200 \text{ acres}}{1} \times \frac{43{,}560 \text{ ft}^2}{1 \text{ acre}} = 52{,}272{,}000 \text{ ft}^2$
$55{,}756{,}800 \text{ ft}^2 > 52{,}272{,}000 \text{ ft}^2$
so $2 \text{ mi}^2 > 1{,}200$ acres

13. $(A+B)^2 = (A+B)(A+B) = A^2 + 2AB + B^2$

14. $(X-2)(X^2+2X+Y^2) =$
$(X)(X^2+2X+Y^2) + (-2)(X^2+2X+Y^2) =$
$X^3 + 2X^2 + XY^2 - 2X^2 - 4X - 2Y^2 =$
$X^3 + XY^2 - 4X - 2Y^2$

15. $X^2 + 11X + 24 = (X+8)(X+3)$

16. $-4Y - 4X = 20$
$Y + X = -5$
(divided both sides by -4)
$Y = -5 - X$
$5Y + 3X = 10 \Rightarrow 5(-5-X) + 3X = 10$
$-25 - 5X + 3X = 10$
$-5X + 3X = 10 + 25$
$-2X = 35$
$X = -\frac{35}{2}$ or $X = -17\frac{1}{2}$
$Y = -5 - X \Rightarrow Y = -5 - \left(-\frac{35}{2}\right)$
$Y = -\frac{10}{2} + \frac{35}{2}$
$Y = \frac{25}{2}$ or $12\frac{1}{2}$

17. Answers for the next two questions will vary. The example given is for the state of Pennsylvania.
$\frac{44{,}832 \text{ mi}^2}{1} \times \frac{5{,}280 \text{ ft}}{1 \text{ mi}} \times \frac{5{,}280 \text{ ft}}{1 \text{ mi}} \approx$
1,249,844,400,000 (rounded)

18. $1{,}249{,}844{,}400{,}000 \div$
$6{,}000{,}000{,}000 \approx 208 \text{ ft}^2$

19. $452 \times 62 = 28{,}024$ lb

20. $\frac{28{,}024 \text{ lb}}{1} \times \frac{1 \text{ ton}}{2{,}000 \text{ lb}} \approx 14$ tons

Systematic Review 31D

1. $4^{\frac{3}{2}} = \left(\sqrt[2]{4}\right)^3 = 2^3 = 8$

2. $81^{\frac{1}{2}} = \sqrt{81} = 9$

3. $7^{\frac{2}{1}} = 7^2 = 49$

4. $64^{\frac{1}{3}} = \sqrt[3]{64} = 4$

5. $\left(Y^{\frac{3}{2}}\right)^{\frac{1}{2}} = Y^{\left(\frac{3}{2}\right)\left(\frac{1}{2}\right)} = Y^{\frac{3}{4}}$ or $\left(\sqrt[4]{Y}\right)^3$

6. $10^{\frac{1}{3}} \cdot 1{,}000 = 10^{\frac{1}{3}} \cdot 10^3 = 10^{\frac{1}{3}+3} =$
$10^{\frac{1}{3}+\frac{9}{3}} = 10^{\frac{10}{3}}$ or $\left(\sqrt[3]{10}\right)^{10}$

7. $\left(A^{\frac{3}{4}}\right)\left(A^{\frac{1}{4}}\right) = A^{\frac{3}{4}+\frac{1}{4}} = A^{\frac{4}{4}} = A^1 = A$

8. $(X^2)^{\frac{3}{4}} = X^{(2)\left(\frac{3}{4}\right)} = X^{\frac{6}{4}} = X^{\frac{3}{2}}$ or $\sqrt[2]{X^3}$

9. $\frac{50 \text{ mi}}{1} \times \frac{1.6 \text{ km}}{1 \text{ mi}} = 80 \text{ km}$

10. $\frac{100 \text{ oz}}{1} \times \frac{28 \text{ g}}{1 \text{ oz}} = 2{,}800 \text{ g}$

11. $43 \times 47 < (45)(45)$
$2{,}021 < 2{,}025$

12. .25 acres < 12,000 sq ft
$\frac{.25 \text{ acres}}{1} \times \frac{43{,}560 \text{ ft}^2}{1 \text{ acre}} = 10{,}890 \text{ ft}^2$
$10{,}890 \text{ ft}^2 < 12{,}000 \text{ ft}^2$

13. $(X-A)^2 = (X-A)(X-A) =$
$X^2 - 2XA + A^2$

14. $(X+2)(X^2-2X+4) =$
$(X)(X^2-2X+4) + (2)(X^2-2X+4) =$
$X^3 - 2X^2 + 4X + 2X^2 - 4X + 8 = X^3 + 8$

15. $(X-1)(X-6) = X^2 - 7X + 6$

16. $X = -4$
$Y - X = 0 \Rightarrow Y - (-4) = 0$
$Y + 4 = 0$
$Y = -4$
$(-4, -4)$

17. $262{,}400 \text{ mi}^2$
$\frac{262{,}400 \text{ mi}^2}{1} \times \frac{5{,}280 \text{ ft}}{1 \text{ mi}} \times \frac{5{,}280 \text{ ft}}{1 \text{ mi}} \approx$
$7{,}315{,}292{,}160{,}000 \text{ ft}^2$

18. $7{,}315{,}292{,}160{,}000 \div 6{,}000{,}000{,}000$
$\approx 1{,}219 \text{ ft}^2$

19. $706 \times 62 = 43{,}772 \text{ lb}$

20. $\frac{43{,}772 \text{ lb}}{1} \times \frac{1 \text{ ton}}{2{,}000 \text{ lb}} =$
$\frac{43{,}772 \text{ ton}}{2{,}000} \approx 22 \text{ ton}$

Systematic Review 31E

1. $10^{\frac{4}{1}} = 10^4 = 10{,}000$

2. $25^{\frac{3}{2}} = \left(\sqrt{25}\right)^3 = 5^3 = 125$

3. $13^{\frac{4}{4}} = 13^1 = 13$

4. $16^{\frac{3}{2}} = \left(\sqrt{16}\right)^3 = 4^3 = 64$

5. $(A^3)^{\frac{1}{3}} = A^{(3)\left(\frac{1}{3}\right)} = A^{\frac{3}{3}} = A^1 = A$

6. $3^{\frac{1}{2}} \cdot 27 = 3^{\frac{1}{2}} \cdot 3^3 = 3^{\frac{1}{2}+3} = 3^{\frac{1}{2}+\frac{6}{2}} =$
$3^{\frac{7}{2}}$ or $\left(\sqrt{3}\right)^7$

7. $\left(X^{\frac{5}{6}}\right)\left(X^{\frac{1}{2}}\right) = X^{\frac{5}{6}+\frac{1}{2}} = X^{\frac{5}{6}+\frac{3}{6}} = X^{\frac{8}{6}} =$
$X^{\frac{4}{3}}$ or $\left(\sqrt[3]{X}\right)^4$

8. $\left(2^{\frac{1}{3}}\right)\left(2^{\frac{1}{2}}\right)\left(2^{\frac{7}{6}}\right) = 2^{\frac{1}{3}+\frac{1}{2}+\frac{7}{6}}$
$= 2^{\frac{2}{6}+\frac{3}{6}+\frac{7}{6}} = 2^{\frac{12}{6}} = 2^2 = 4$

9. $\frac{10 \text{ m}}{1} \times \frac{1.1 \text{ yd}}{1 \text{ m}} = 11 \text{ yd}$

10. $\frac{20 \text{ kg}}{1} \times \frac{2.2 \text{ lb}}{1 \text{ kg}} = 44 \text{ lb}$

11. $\frac{2 \text{ ft}^3}{1} \times \frac{12 \text{ in}}{1 \text{ ft}} \times \frac{12 \text{ in}}{1 \text{ ft}} \times \frac{12 \text{ in}}{1 \text{ ft}} = 3{,}456 \text{ in}^3$

12. $\frac{14 \text{ yd}^3}{1} \times \frac{3 \text{ ft}}{1 \text{ yd}} \times \frac{3 \text{ ft}}{1 \text{ yd}} \times \frac{3 \text{ ft}}{1 \text{ yd}} = 378 \text{ ft}^3$

13. $(5A+5B)^2 = (5A+5B)(5A+5B)$
$= 25A^2 + 50AB + 25B^2$

14. $(X-Y)(X^2+XY+Y^2) =$
$(X)(X^2+XY+Y^2) + (-Y)(X^2+XY+Y^2) =$
$X^3 + X^2Y + XY^2 - X^2Y - XY^2 - Y^3 = X^3 - Y^3$

15. $(X+1)(4X+6) = 4X^2 + 10X + 6$

16. $6 + Y = 2X \Rightarrow Y = 2X - 6$
$3Y - 4X = 2$
$(-3)(Y - 2X = -6) \Rightarrow -3Y + 6X = 18$
$2X = 20$
$X = \frac{20}{2}$
$X = 10$

$Y = 2X - 6 \Rightarrow Y = 2(10) - 6$
$Y = 20 - 6$
$Y = 14$
$(10, 14)$

17. 586,400 square miles

$\frac{586{,}400\ \text{mi}^2}{1} \times \frac{5{,}280\ \text{ft}}{1\ \text{mi}} \times \frac{5{,}280\ \text{ft}}{1\ \text{mi}} \approx$

$16{,}347{,}893{,}760{,}000\ \text{ft}^2$

18. $16{,}347{,}893{,}760{,}000 \div 6{,}000{,}000{,}000$

$\approx 2{,}724.6\ \text{ft}^2$

19. $100 \times 100 \times 50 = 500{,}000\ \text{ft}^3$

$500{,}000 \times 62 = 31{,}000{,}000\ \text{lb}$

20. $\frac{31{,}000{,}000\ \text{lb}}{1} \times \frac{1\ \text{ton}}{2{,}000\ \text{lb}} =$

$\frac{31{,}000{,}000\ \text{ton}}{2{,}000} = 15{,}500\ \text{tons}$

Lesson Practice 32A

1. $500{,}000 = 5 \times 10^5$
2. $356{,}000{,}000 = 3.56 \times 10^8$
3. $54{,}800{,}000 = 5.48 \times 10^7$
4. $.00096 = 9.6 \times 10^{-4}$
5. $.00468 = 4.68 \times 10^{-3}$
6. $.0000000913 = 9.13 \times 10^{-8}$
7. $200{,}000 \times 6{,}000{,}000 =$

$1{,}200{,}000{,}000{,}000$

$(1.9 \times 10^5)(6 \times 10^6) =$

$(1.9 \times 6)(10^5 \times 10^6) =$

$11.4 \times 10^{11} \approx 1 \times 10^{12}$ 1 SD (significant digit)

8. $200{,}000 \times 4{,}000{,}000{,}000 =$

$800{,}000{,}000{,}000{,}000$

$(1.815 \times 10^5)(4.16 \times 10^9) =$

$(1.815 \times 4.16)(10^5 \times 10^9) =$

$7.5504 \times 10^{14} \approx 7.55 \times 10^{14}$ (3 SD)

9. $900{,}000 \times 40{,}000{,}000 =$

$36{,}000{,}000{,}000{,}000$

$(8.6 \times 10^5)(3.64 \times 10^7) =$

$(8.6 \times 3.64)(10^5 \times 10^7) =$

$31.304 \times 10^{12} \approx 3.1 \times 10^{13}$ (2 SD)

10. $.00009 \times 9{,}000{,}000{,}000 = 810{,}000$

$(8.5 \times 10^{-5})(9 \times 10^9) =$

$(8.5 \times 9)(10^{-5} \times 10^9) = 76.5 \times 10^4 =$

$7.65 \times 10^5 \approx 8 \times 10^5$ (1 SD)

11. $.0009 \times 50{,}000 = 45$

$(9.3 \times 10^{-4})(5 \times 10^4) =$

$(9.3 \times 5)(10^{-4} \times 10^4) =$

$46.5 \times 10^0 = 4.65 \times 10^1 \approx 5 \times 10^1$ (1 SD)

12. $.002 \times .0004 = .0000008$

$(2.1 \times 10^{-3})(3.50 \times 10^{-4}) =$

$(2.1 \times 3.50)(10^{-3} \times 10^{-4}) =$

$7.35 \times 10^{-7} \approx 7.4 \times 10^{-7}$ (2 SD)

13. $600{,}000 \div 4{,}000{,}000{,}000 = .00015$

$(5.6 \times 10^5) \div (4 \times 10^9) =$

$(5.6 \div 4)(10^5 \div 10^9) = 1.4 \times 10^{-4} \approx 1 \times 10^{-4}$ (1 SD)

14. $\frac{10{,}000{,}000}{2{,}000{,}000} = 5$

$(9.8 \times 10^6) \div (2.45 \times 10^6) =$

$(9.8 \div 2.45)(10^6 \div 10^6) = 4.0 \times 10^0$ or 4.0 (2 SD)

15. $.004 \div .01 = .4$

$(3.6 \times 10^{-3}) \div (1.2 \times 10^{-2}) =$

$(3.6 \div 1.2)(10^{-3} \div 10^{-2}) = 3.0 \times 10^{-1}$ (2 SD)

Lesson Practice 32B

1. $600{,}000 = 6 \times 10^5$
2. $854{,}000{,}000 = 8.54 \times 10^8$
3. $62{,}800{,}000 = 6.28 \times 10^7$
4. $.000095 = 9.5 \times 10^{-5}$
5. $.00528 = 5.28 \times 10^{-3}$
6. $.000000921 = 9.21 \times 10^{-7}$
7. $200{,}000 \times 5{,}000{,}000 =$

$1{,}000{,}000{,}000{,}000$

$(1.8 \times 10^5)(5 \times 10^6) =$

$(1.8 \times 5)(10^5 \times 10^6) = 9 \times 10^{11}$ (1 SD)

8. $900{,}000 \times 3{,}000{,}000 = 2{,}700{,}000{,}000{,}000$

$(9.15 \times 10^5)(3 \times 10^6) = (9.15 \times 3)(10^5 \times 10^6) =$

$27.45 \times 10^{11} = 2.745 \times 10^{12} \approx 3 \times 10^{12}$ (1 SD)

9. $100{,}000 \times 40{,}000{,}000 =$
$4{,}000{,}000{,}000{,}000$
$(9.6 \times 10^4)(4.36 \times 10^7) =$
$(9.6 \times 4.36)(10^4 \times 10^7) =$
$41.856 \times 10^{11} = 4.1856 \times 10^{12}$
$\approx 4.2 \times 10^{12}$ (2 SD)

10. $.00008 \times 9{,}000{,}000{,}000 = 720{,}000$
$(7.5 \times 10^{-5})(9 \times 10^9) =$
$(7.5 \times 9)(10^{-5} \times 10^9) =$
$67.5 \times 10^4 = 6.75 \times 10^5 \approx 7 \times 10^5$ (1 SD)

11. $.00008 \times 60{,}000 = 4.8$
$(7.9 \times 10^{-5})(6.25 \times 10^4) =$
$49.375 \times 10^{-1} = 4.9375 \times 10^0 \approx$
4.9×10^0 or 4.9 (2 SD)

12. $.0003 \times .0000004 = .00000000012$
$(3.1 \times 10^{-4})(4 \times 10^{-7}) =$
$(3.1 \times 4)(10^{-4} \times 10^{-7}) =$
$12.4 \times 10^{-11} = 1.24 \times 10^{-10} \approx$
1×10^{-10} (1 SD)

13. $50{,}000 \div 40{,}000{,}000 = .00125$
$(5.2 \times 10^4) \div (4 \times 10^7) =$
$(5.2 \div 4)(10^4 \div 10^7) =$
$1.3 \times 10^{-3} \approx 1 \times 10^{-3}$ (1 SD)

14. $\dfrac{20{,}000{,}000}{60{,}000{,}000{,}000} = .000\overline{3}$
$(2.4 \times 10^7) \div (6 \times 10^{10}) =$
$(2.4 \div 6)(10^7 \div 10^{10}) =$
$.4 \times 10^{-3} = 4 \times 10^{-4}$ (1 SD)

15. $.0004 \div .007 = 0.0\overline{571428}$
$(3.5 \times 10^{-4}) \div (7 \times 10^{-3}) =$
$(3.5 \div 7)(10^{-4} \div 10^{-3}) = .5 \times 10^{-1} =$
5×10^{-2} (1 SD)

Systematic Review 32C

1. $700{,}000 = 7 \times 10^5$
2. $.0076 = 7.6 \times 10^{-3}$
3. $5{,}000 \times 8{,}000{,}000 =$
$40{,}000{,}000{,}000 = 4 \times 10^{10}$
$(5 \times 10^3)(8 \times 10^6)$
4. $(5 \times 8)(10^3 \times 10^6) = 40 \times 10^9$
5. 4×10^{10} (1 SD)
6. Check with calculator
7. $60{,}000 \div 100 = 600 = 6 \times 10^2$
$(6.13 \times 10^4) \div (1.2 \times 10^2)$
8. $(6.13 \div 1.2)(10^4 \times 10^2) = 5.108 \times 10^2$
9. 5.1×10^2 (2 SD)
10. Check with calculator
11. $1{,}000^{\frac{2}{3}} \cdot 10^2 \cdot 10^{-3} = \left(\sqrt[3]{1{,}000}\right)^2 \cdot 10^2 \cdot 10^{-3} =$
$10^2 \cdot 10^2 \cdot 10^{-3} = 10^{2+2+(-3)} = 10^1$ or 10
12. $8^{\frac{2}{3}} \cdot 4 = \left(\sqrt[3]{8}\right)^2 \cdot 2^2 = 2^2 \cdot 2^2 = 2^4 = 16$
13. $10^{\frac{1}{3}} \cdot 100^{\frac{3}{2}} \cdot 10^{-1} =$
$10^{\frac{1}{3}} \cdot \left(\sqrt{100}\right)^3 \times 10^{-1} =$
$10^{\frac{1}{3}} \cdot 10^3 \times 10^{-1} = 10^{\frac{1}{3}+3+(-1)} =$
$10^{\frac{1}{3}+\frac{9}{3}+\left(-\frac{3}{3}\right)} = 10^{\frac{7}{3}}$ or $\left(\sqrt[3]{10}\right)^7$
14. $A^5 A^{\frac{-1}{2}} A^{\frac{-3}{2}} = A^{\frac{10}{2}+\left(\frac{-1}{2}\right)+\left(\frac{-3}{2}\right)} =$
$A^{\frac{6}{2}} = A^3$
15. $\dfrac{10 \text{ km}}{1} \times \dfrac{1 \text{ mi}}{1.6 \text{ km}} = 6.25$ miles
16. $\dfrac{75 \text{ g}}{1} \times \dfrac{.035 \text{ oz}}{1 \text{ g}} = 2.625$ oz
17. $(3X - 3Y)^2 = (3X - 3Y)(3X - 3Y) =$
$9X^2 - 18XY + 9Y^2$
18. $(X + Y)(X^2 - XY + Y^2) =$
$(X)(X^2 - XY + Y^2) + (Y)(X^2 - XY + Y^2) =$
$X^3 - X^2Y + XY^2 + X^2Y - XY^2 + Y^3 = X^3 + Y^3$
19. $X(X + 4) + 5X + 3 = -17$
$X^2 + 4X + 5X + 3 + 17 = 0$
$X^2 + 9X + 20 = 0$
$(X + 4)(X + 5) = 0$

$X+4=0 \qquad X+5=0$

$X=-4 \qquad X=-5$

Check :

$(-4)((-4)+4)+5(-4)+3=-17$

$(-4)(0)-20+3=-17$

$0-20+3=-17$

$-17=-17$

$(-5)((-5)+4)+5(-5)+3=-17$

$(-5)(-1)-25+3=-17$

$5-25+3=-17$

$-17=-17$

20. $X(2X-9)=0$

$X=0 \qquad 2X-9=0$

$2X=9$

$X=\frac{9}{2}$

Check : $(0)(2(0)-9)=0$

$(0)(0-9)=0$

$(0)(-9)=0$

$0=0$

$\left(\frac{9}{2}\right)\left[2\left(\frac{9}{2}\right)-9\right]=0$

$\left(\frac{9}{2}\right)\left[\frac{18}{2}-9\right]=0$

$\left(\frac{9}{2}\right)(9-9)=0$

$\left(\frac{9}{2}\right)(0)=0$

$0=0$

Systematic Review 32D

1. $586{,}000{,}000=5.86\times10^{8}$
2. $.000595=5.95\times10^{-4}$
3. $20{,}000\times.007=140$
   $(1.8\times10^{4})(7.2\times10^{-3})$
4. $(1.8\times7.2)(10^{4}\times10^{-3})=12.96\times10^{1}$
5. $1.296\times10^{2}\approx1.3\times10^{2}$ (2 SD)
6. Check with calculator
7. $1{,}000{,}000\div300=3{,}333.\bar{3}$
   $(1.45\times10^{6})\div(2.9\times10^{2})$
8. $(1.45\div2.9)(10^{6}\div10^{2})=.5\times10^{4}$
9. 5.0×10^{3} (2 SD)
10. Check with calculator
11. $\left(5^{\frac{1}{2}}\right)^{-4}5^{0}\,5^{2}=5^{\left(\frac{1}{2}\right)(-4)}5^{0+2}=$
    $5^{-2}\,5^{2}=5^{-2+2}=5^{0}=1$
12. $9^{\frac{3}{2}}\cdot27\cdot81^{\frac{1}{4}}=\left(\sqrt{9}\right)^{3}\cdot3^{3}\cdot\left(\sqrt[4]{81}\right)^{1}=$
    $3^{3}\cdot3^{3}\cdot3^{1}=3^{3+3+1}=3^{7}$ or 2,187
13. $\frac{26\text{ mi}}{1}\times\frac{1.6\text{ km}}{1\text{ mi}}=41.6\text{ km}$
14. $\frac{500\text{ g}}{1}\times\frac{.035\text{ oz}}{1\text{ g}}=17.5\text{ oz}$
15. $(D-5)(D^{2}+5D+25)=$
    $(D)(D^{2}+5D+25)+(-5)(D^{2}+5D+25)=$
    $D^{3}+5D^{2}+25D-5D^{2}-25D-125=$
    $D^{3}-125$
16.

$$\begin{array}{r} A^{2}-AT+T^{2} \\ A+T\overline{\big)\,A^{3}+0A^{2}+0A+T^{3}} \\ -(A^{3}+A^{2}T) \\ \hline -A^{2}T+0A \\ -(-A^{2}T-AT^{2}) \\ \hline AT^{2}+T^{3} \\ -(AT^{2}+T^{3}) \\ \hline 0 \end{array}$$

17. $X(5X-10)=0$

$5X-10=0 \qquad X=0$

$5X=10$

$X=\frac{10}{5}=2$

$(0)(5(0)-10)=0 \qquad (2)(5(2)-10)=0$

$(0)(0-10)=0 \qquad (2)(10-10)=0$

$(0)(-10)=0 \qquad (2)(0)=0$

$0=0 \qquad 0=0$

18.
$$X^2+7X-18=42$$
$$X^2+7X-60=0$$
$$(X+12)(X-5)=0$$

$X+12=0$ $\quad$ $X-5=0$

$X=-12$ $\quad$ $X=5$

$$(-12)^2+7(-12)-18=42$$
$$144-84-18=42$$
$$42=42$$

$$(5)^2+7(5)-18=42$$
$$25+35-18=42$$
$$42=42$$

19.
$$10(N)+2(N+2)-4(N+4)+8=3(N+4)-11$$
$$10N+2N+4-4N-16+8=3N+12-11$$
$$10N+2N-4N-3N=12-11-4+16-8$$
$$5N=5$$
$$N=1$$

1; 3; 5

20.
$$(.10D+.05N=1.35)(100) \qquad 10D+5N=135$$
$$(D+N=16)(-5) \qquad -5D-5N=-80$$
$$5D=55$$
$$D=\frac{55}{5}$$
$$D=11$$

$D+N=16$ $\quad$ $(11)+N=16$
$$N=16-11$$
$$N=5$$

Systematic Review 32E

1. $23{,}800{,}000=2.38\times10^7$
2. $.000000112=1.12\times10^{-7}$
3. $.9\times600{,}000=540{,}000$
   $(9.2\times10^{-1})(6.4\times10^5)$
4. $(9.2\times6.4)(10^{-1}\times10^5)=58.88\times10^4$
5. $5.888\times10^5\approx5.9\times10^5$ (2 SD)
6. Check with calculator
7. $.4\times.3\div.001=120$
   $(4\times10^{-1})(2.5\times10^{-1})\div(1\times10^{-3})$
8. $(4\times2.5\div1)(10^{-1}\times10^{-1}\div10^{-3})=10\times10^1$
9. 1×10^2 (1 SD)
10. Check with calculator
11. $A^{\frac{3}{4}}A^{\frac{4}{3}}=A^{\frac{3}{4}+\frac{4}{3}}=A^{\frac{9}{12}+\frac{16}{12}}=A^{\frac{25}{12}}$
    or $\left(\sqrt[12]{A}\right)^{25}$
12. $9^{\frac{1}{2}}\cdot3^2\cdot27^{\frac{4}{3}}=\left(\sqrt{9}\right)\cdot3^2\cdot\left(\sqrt[3]{27}\right)^4=$
    $3^1\cdot3^2\cdot3^4=3^{1+2+4}=3^7$
13. $\frac{100\text{ m}}{1}\times\frac{1.1\text{ yd}}{1\text{ m}}=110\text{ yd}$
14. $\frac{2\text{ liters}}{1}\times\frac{1.06\text{ qt}}{1\text{ liter}}=2.12\text{ qt}$
15. $X^2-B^2=(X-B)(X+B)$
16. $4X^5-324X=(4X)(X^4-81)=$
    $(4X)(X^2-9)(X^2+9)=$
    $(4X)(X-3)(X+3)(X^2+9)$

17.
$$X^2+X-12=60$$
$$X^2+X-72=0$$
$$(X+9)(X-8)=0$$

$X+9=0$ $\quad$ $X-8=0$

$X=-9$ $\quad$ $X=8$

$(-9)^2+(-9)-12=60$ $\quad$ $(8)^2+(8)-12=60$

$81-9-12=60$ $\quad$ $64+8-12=60$

$60=60$ $\quad$ $60=60$

18.
$$4-A^2=0$$
$$(2-A)(2+A)=0$$

$2-A=0$ $\quad$ $2+A=0$

$2=A$ $\quad$ $A=-2$

$4-(2)^2=0$ $\quad$ $4-(-2)^2=0$

$4-4=0$ $\quad$ $4-4=0$

$0=0$

19. 9.5×10^6 square miles

$$\frac{9.5\times10^6\cancel{mi^2}}{1}\times\frac{5.28\times10^3\text{ ft}}{1\cancel{mi}}\times\frac{5.28\times10^3\text{ ft}}{1\cancel{mi}}=$$

$264.8448\times10^{12}\approx2.6\times10^{14}$ sq ft (2 SD) Your answer may be slightly different, depending on how many significant digits were given by the source of information that you used, and the point at which you rounded.

20.

$(2.65\times10^{14}\text{ sq ft})\div(6\times10^9\text{ people})=$

$(2.65\div6)(10^{14}\div10^9)=.44\times10^5=$

$4.4\times10^4\approx4\times10^4\text{ ft}^2$ per person

1 acre $\approx4.4\times10^4\text{ ft}^2$

so 1 acre per person

Your answer to this problem will be affected by the answer to the previous problem. As long as yours is close to the one given here, it can be counted correct.

Lesson Practice 33A

1. 3^3 is the largest power of $3\le80$

$3^3=27;\ 3^2=9;\ 3^1=3;\ 3^0=1$

$$\begin{array}{r}2\\27\overline{)80}\\\underline{54}\\26\end{array}\qquad\begin{array}{r}2\\9\overline{)26}\\\underline{18}\\8\end{array}\qquad\begin{array}{r}2\\3\overline{)8}\\\underline{6}\\2\end{array}\qquad\begin{array}{r}2\\1\overline{)2}\\\underline{2}\\0\end{array}$$

$2\times3^3+2\times3^2+2\times3^1+2\times3^0=2222_3$

2. 5^2 is the largest power of $5\le80$

$5^2=25;\ 5^1=5;\ 5^0=1$

$$\begin{array}{r}3\\25\overline{)80}\\\underline{75}\\5\end{array}\qquad\begin{array}{r}1\\5\overline{)5}\\\underline{5}\\0\end{array}\qquad\begin{array}{r}0\\1\overline{)0}\\\underline{0}\\0\end{array}$$

$3\times5^2+1\times5^1+0\times5^0=310_5$

3. 4^3 is the largest power of $4\le80$

$4^3=64;\ 4^2=16;\ 4^1=4;\ 4^0=1$

$$\begin{array}{r}1\\64\overline{)80}\\\underline{64}\\16\end{array}\qquad\begin{array}{r}1\\16\overline{)16}\\\underline{16}\\0\end{array}\qquad\begin{array}{r}0\\4\overline{)0}\\\underline{0}\\0\end{array}\qquad\begin{array}{r}0\\1\overline{)0}\\\underline{0}\\0\end{array}$$

$1\times4^3+1\times4^2+0\times4^1+0\times4^0=1100_4$

4. 6^2 is the largest power of $6\le100$

$6^2=36;\ 6^1=6;\ 6^0=1$

$$\begin{array}{r}2\\36\overline{)100}\\\underline{72}\\28\end{array}\qquad\begin{array}{r}4\\6\overline{)28}\\\underline{24}\\4\end{array}\qquad\begin{array}{r}4\\1\overline{)4}\\\underline{4}\\0\end{array}$$

$2\times6^2+4\times6^1+4\times6^0=244_6$

5. 8^3 is the largest power of $8\le1{,}352$

$8^3=512;\ 8^2=64;\ 8^1=8;\ 8^0=1$

$$\begin{array}{r}2\\512\overline{)1352}\\\underline{1024}\\328\end{array}\qquad\begin{array}{r}5\\64\overline{)328}\\\underline{320}\\8\end{array}\qquad\begin{array}{r}1\\8\overline{)8}\\\underline{8}\\0\end{array}\qquad\begin{array}{r}0\\1\overline{)0}\\\underline{0}\\0\end{array}$$

$2\times8^3+5\times8^2+1\times8^1+0\times8^0=2510_8$

6. 6^4 is the largest power of $6\le1{,}352$

$6^4=1{,}296;\ 6^3=216;\ 6^2=36;\ 6^1=6;\ 6^0=1$

$$\begin{array}{r}1\\1296\overline{)1352}\\\underline{1296}\\56\end{array}\qquad\begin{array}{r}0\\216\overline{)56}\\\underline{0}\\56\end{array}\qquad\begin{array}{r}1\\36\overline{)56}\\\underline{36}\\20\end{array}$$

$$\begin{array}{r}3\\6\overline{)20}\\\underline{18}\\2\end{array}\qquad\begin{array}{r}2\\1\overline{)2}\\\underline{2}\\0\end{array}$$

$1\times6^4+0\times6^3+1\times6^2+3\times6^1+2\times6^0=10132_6$

7. $563_7=5\times7^2+6\times7^1+3\times7^0=$

$5(49)+6(7)+3(1)=$

$245+42+3=290$

8. $441_5 = 4\times5^2+4\times5^1+1\times5^0 =$
$4(25)+4(5)+1(1) =$
$100+20+1=121$

9. $2121_3 = 2\times3^3+1\times3^2+2\times3^1+1\times3^0 =$
$2(27)+1(9)+2(3)+1(1) =$
$54+9+6+1=70$

10. $3421_5 = 3\times5^3+4\times5^2+2\times5^1+1\times5^0 =$
$3(125)+4(25)+2(5)+1(1) =$
$375+100+10+1=486$

11. $6A8_{12} = 6\times12^2+10\times12^1+8\times12^0 =$
$6(144)+10(12)+8(1) =$
$864+120+8=992$

12. $B81_{13} = 11\times13^2+8\times13^1+1\times13^0 =$
$11(169)+8(13)+1(1) =$
$1,859+104+1=1,964$

Lesson Practice 33B

1. 2^6 is the largest power of $2 \le 95$
$2^6=64;\ 2^5=32;\ 2^4=16;\ 2^3=8;$
$2^2=4;\ 2^1=2;\ 2^0=1$

$$\begin{array}{r}1\\64\overline{)95}\\64\\\hline31\end{array}\quad\begin{array}{r}0\\32\overline{)31}\\0\\\hline31\end{array}\quad\begin{array}{r}1\\16\overline{)31}\\16\\\hline15\end{array}$$

$$\begin{array}{r}1\\8\overline{)15}\\8\\\hline7\end{array}\quad\begin{array}{r}1\\4\overline{)7}\\4\\\hline3\end{array}\quad\begin{array}{r}1\\2\overline{)3}\\2\\\hline1\end{array}\quad\begin{array}{r}1\\1\overline{)1}\\1\\\hline0\end{array}$$

$1\times2^6+0\times2^5+1\times2^4+1\times2^3+1\times2^2+$
$1\times2^1+1\times2^0=1011111_2$

2. 5^2 is the largest power of $5 \le 95$
$5^2=25;\ 5^1=5;\ 5^0=1$

$$\begin{array}{r}3\\25\overline{)95}\\75\\\hline20\end{array}\quad\begin{array}{r}4\\5\overline{)20}\\20\\\hline0\end{array}\quad\begin{array}{r}0\\1\overline{)0}\\0\\\hline0\end{array}$$

$3\times5^2+4\times5^1+0\times5^0=340_5$

3. 7^2 is the largest power of $7 \le 95$
$7^2=49;\ 7^1=7;\ 7^0=1$

$$\begin{array}{r}1\\49\overline{)95}\\49\\\hline46\end{array}\quad\begin{array}{r}6\\7\overline{)46}\\42\\\hline4\end{array}\quad\begin{array}{r}4\\1\overline{)4}\\4\\\hline0\end{array}$$

$1\times7^2+6\times7^1+4\times7^0=164_7$

4. 8^2 is the largest power of $8 \le 100$
$8^2=64;\ 8^1=8;\ 8^0=1$

$$\begin{array}{r}1\\64\overline{)100}\\64\\\hline36\end{array}\quad\begin{array}{r}4\\8\overline{)36}\\32\\\hline4\end{array}\quad\begin{array}{r}4\\1\overline{)4}\\4\\\hline0\end{array}$$

$1\times8^2+4\times8^1+4\times8^0=144_8$

5. 12^2 is the largest power of $12 \le 1,352$
$12^2=144;\ 12^1=12;\ 12^0=1$

$$\begin{array}{r}9\\144\overline{)1352}\\1296\\\hline56\end{array}\quad\begin{array}{r}4\\12\overline{)56}\\48\\\hline8\end{array}\quad\begin{array}{r}8\\1\overline{)8}\\8\\\hline0\end{array}$$

$9\times12^2+4\times12^1+8\times12^0=948_{12}$

6. 9^3 is the largest power of $9 \le 1{,}352$

$9^3 = 729;\ 9^2 = 81;\ 9^1 = 9;\ 9^0 = 1$

$$\begin{array}{r} 1 \\ 729\overline{)1352} \\ \underline{729} \\ 623 \end{array} \qquad \begin{array}{r} 7 \\ 81\overline{)623} \\ \underline{567} \\ 56 \end{array}$$

$$\begin{array}{r} 6 \\ 9\overline{)56} \\ \underline{54} \\ 2 \end{array} \qquad \begin{array}{r} 2 \\ 1\overline{)2} \\ \underline{2} \\ 0 \end{array}$$

$1\times9^3+7\times9^2+6\times9^1+2\times9^0 = 1762_9$

7. $11001_2 =$

$1\times2^4+1\times2^3+0\times2^2+0\times2^1+1\times2^0 =$

$1(16)+1(8)+0(4)+0(2)+1(1) =$

$16+8+0+0+1 = 25$

8. $2121_7 =$

$2\times7^3+1\times7^2+2\times7^1+1\times7^0 =$

$2(343)+1(49)+2(7)+1(1) =$

$686+49+14+1 = 750$

9. $465_7 = 4\times7^2+6\times7^1+5\times7^0 =$

$4(49)+6(7)+5(1) =$

$196+42+5 = 243$

10. $3421_6 =$

$3\times6^3+4\times6^2+2\times6^1+1\times6^0 =$

$3(216)+4(36)+2(6)+1(1) =$

$648+144+12+1 = 805$

11. $26A_{12} = 2\times12^2+6\times12^1+10\times12^0 =$

$2(144)+6(12)+10(1) = 288+72+10 = 370$

12. $3B4_{20} = 3\times20^2+11\times20^1+4\times20^0 =$

$3(400)+11(20)+4(1) =$

$1{,}200+220+4 = 1{,}424$

Systematic Review 33C

1. 3^4 is the largest power of $3 \le 100$

$3^4 = 81;\ 3^3 = 27;\ 3^2 = 9;$

$3^1 = 3;\ 3^0 = 1$

$$\begin{array}{r} 1 \\ 81\overline{)100} \\ \underline{81} \\ 19 \end{array} \qquad \begin{array}{r} 0 \\ 27\overline{)19} \\ \underline{0} \\ 19 \end{array} \qquad \begin{array}{r} 2 \\ 9\overline{)19} \\ \underline{18} \\ 1 \end{array}$$

$$\begin{array}{r} 0 \\ 8\overline{)1} \\ \underline{0} \\ 1 \end{array} \qquad \begin{array}{r} 1 \\ 1\overline{)1} \\ \underline{1} \\ 0 \end{array}$$

$1\times3^4+0\times3^3+2\times3^2+0\times3^1+1\times3^0 = 10201_3$

2. 6^3 is the largest power of $6 \le 245$

$6^3 = 216;\ 6^2 = 36;\ 6^1 = 6;\ 6^0 = 1$

$$\begin{array}{r} 1 \\ 216\overline{)245} \\ \underline{216} \\ 29 \end{array} \qquad \begin{array}{r} 0 \\ 36\overline{)29} \\ \underline{0} \\ 29 \end{array}$$

$$\begin{array}{r} 4 \\ 6\overline{)29} \\ \underline{24} \\ 5 \end{array} \qquad \begin{array}{r} 5 \\ 1\overline{)5} \\ \underline{5} \\ 0 \end{array}$$

$1\times6^3+0\times6^2+4\times6^1+5\times6^0 = 1045_6$

3. $56_7 = 5\times7^1+6\times7^0 =$

$5(7)+6(1) = 35+6 = 41$

4. $173_8 = 1\times8^2+7\times8^1+3\times8^0 =$

$1(64)+7(8)+3(1) = 64+56+3 = 123$

5. $300\times7{,}000\times.8 =$

$(3\times10^2)(7\times10^3)(8\times10^{-1}) =$

$(3\times7\times8)(10^2\times10^3\times10^{-1}) =$

$168\times10^4 = 1.68\times10^6 \approx 2\times10^6$ (1 SD)

6. $60\times.05\times40{,}000 =$

$(6\times10^1)(5\times10^{-2})(4\times10^4) =$

$(6\times5\times4)(10^1\times10^{-2}\times10^4) =$

$120\times10^3 = 1.2\times10^5 \approx 1\times10^5$ (1 SD)

7. $\frac{9{,}000 \times .04}{300{,}000 \times .2} = \frac{(9\times10^{3})(4\times10^{-2})}{(3\times10^{5})(2\times10^{-1})} =$

$\frac{36\times10^{1}}{6\times10^{4}} = 6\times10^{-3}$

8. $\frac{1.4\times.005}{350{,}000} = \frac{(1.4)(5\times10^{-3})}{(3.5\times10^{5})} =$

$\frac{7\times10^{-3}}{3.5\times10^{5}} = 2\times10^{-8}$

9. $\left[(10^{2})^{\frac{1}{2}}\right]^{0} = 10^{(2)\left(\frac{1}{2}\right)(0)} = 10^{0} = 1$

10. $4^{\frac{3}{2}} \cdot 16^{2} \cdot 64^{\frac{2}{3}} = (\sqrt{4})^{3} \cdot (2^{4})^{2} \cdot (2^{6})^{\frac{2}{3}} =$

$2^{3} \cdot 2^{4\times2} \cdot 2^{(6)\left(\frac{2}{3}\right)} = 2^{3} \cdot 2^{8} \cdot 2^{4} =$

$2^{3+8+4} = 2^{15}$ or 32,768

11. $100^{\frac{3}{2}} \cdot (10^{2})^{4} = (10^{2})^{\frac{3}{2}} \cdot (10^{2})^{4} =$

$10^{(2)\left(\frac{3}{2}\right)} \cdot 10^{(2)(4)} = 10^{3} \cdot 10^{8} =$

$10^{3+8} = 10^{11}$

12. $D^{\frac{-1}{3}} \cdot D^{6} \cdot D^{\frac{2}{3}} = D^{\frac{-1}{3}+6+\frac{2}{3}} = D^{\frac{-1}{3}+\frac{18}{3}+\frac{2}{3}} = D^{\frac{19}{3}}$

13. $\frac{880 \text{ yd}}{1} \times \frac{.9 \text{ m}}{1 \text{ yd}} = 792 \text{ m}$

14. Answers will vary: multiply your weight by .45 kg

15. $\frac{4 \text{ qt}}{1} \times \frac{.95 \text{ liters}}{1 \text{ qt}} = 3.8 \text{ liters}$

16. Answers will vary: multiply your weight in kg (# 14) by 1,000 g

17.

$\frac{1.24\times10^{6} \text{ mi}^{2}}{1} \times \frac{5.28\times10^{3} \text{ ft}}{1 \text{ mi}} \times \frac{5.28\times10^{3} \text{ ft}}{1 \text{ mi}}$

$= 34.6\times10^{12} = 3.46\times10^{13} \text{ ft}^{2}$

18.

$(3.46\times10^{13} \text{ sq ft}) \div (6\times10^{9} \text{ people}) =$

$(3.46 \div 6)(10^{13} \div 10^{9}) = .577\times10^{4} =$

$5.77\times10^{3} \text{ ft}^{2}$ per person

19. $B^{2} - A^{2} = (B-A)(B+A)$

20. $C^{4} - D^{4} = (C^{2}-D^{2})(C^{2}+D^{2}) =$

$(C-D)(C+D)(C^{2}+D^{2})$

Systematic Review 33D

1. 7^{2} is the smallest power of $7 \le 100$

$7^{2} = 49;\ 7^{1} = 7;\ 7^{0} = 1$

$100 \div 49 = 2$, 98, remainder 2; $2 \div 7 = 0$, 0, remainder 2; $2 \div 1 = 2$, 2, remainder 0

$2\times7^{2} + 0\times7^{1} + 2\times7^{0} = 202_{7}$

2. 8^{2} is the smallest power of $8 \le 245$

$8^{2} = 64;\ 8^{1} = 8;\ 8^{0} = 1$

$245 \div 64 = 3$, 192, remainder 53; $53 \div 8 = 6$, 48, remainder 5; $5 \div 1 = 5$, 5, remainder 0

$3\times8^{2} + 6\times8^{1} + 5\times8^{0} = 365_{8}$

3. $2120_{3} =$

$2\times3^{3} + 1\times3^{2} + 2\times3^{1} + 0\times3^{0} =$

$2(27) + 1(9) + 2(3) + 0(1) =$

$54 + 9 + 6 + 0 = 69$

4. $3210_{4} =$

$3\times4^{3} + 2\times4^{2} + 1\times4^{1} + 0\times4^{0} =$

$3(64) + 2(16) + 1(4) + 0(1) =$

$192 + 32 + 4 + 0 = 228$

5. $.032\times8{,}000\times.7 =$

$(3.2\times10^{-2})(8\times10^{3})(7\times10^{-1}) =$

$(3.2\times8\times7)(10^{-2}\times10^{3}\times10^{-1}) =$

$179.2\times10^{0} = 1.792\times10^{2} \approx$

2×10^{2} (1 SD)

6. $.003\times500 = (3\times10^{-3})(5\times10^{2}) =$

$(3\times5)(10^{-3}\times10^{2}) = 15\times10^{-1} = 1.5\times10^{0} \approx$

2×10^{0} or 2 (1 SD)

7. $12{,}400 \div .04 = (1.24\times10^{4}) \div (4\times10^{-2}) =$

$(1.24 \div 4)(10^{4} \div 10^{-2}) = .31\times10^{6} =$

$3.1\times10^{5} \approx 3\times10^{5}$ (1 SD)

8. $1{,}000{,}000 \div 5{,}000{,}000 = (1\times10^{6}) \div (5\times10^{6}) =$

$(1 \div 5)(10^{6} \div 10^{6}) = .2\times10^{0} = 2\times10^{-1}$ (1 SD)

9. $8^{\frac{4}{3}} = \left(\sqrt[3]{8}\right)^4 = 2^4 = 16$

10. $\left(X^{\frac{4}{3}}\right)^{\frac{1}{2}} = X^{\left(\frac{4}{3}\right)\left(\frac{1}{2}\right)} = X^{\frac{4}{6}} = X^{\frac{2}{3}}$

11. $A^{-5}A^{4}A^{-3}A^{\frac{1}{2}} = A^{-5+4+(-3)+\frac{1}{2}} =$
$A^{-4+\frac{1}{2}} = A^{\frac{-8}{2}+\frac{1}{2}} = A^{\frac{-7}{2}}$

12. $\frac{B^5A^{-2}}{B^{-3}A^7} = B^5B^3A^{-2}A^{-7} = A^{-2+(-7)}B^{5+3} =$
$A^{-9}B^8$ or $\frac{B^8}{A^9}$

13. $\frac{2{,}000 \text{ lb}}{1} \times \frac{.45 \text{ kg}}{1 \text{ lb}} = 900 \text{ kg}$

14. $\frac{4 \text{ ft}^2}{1} \times \frac{12 \text{ in}}{1 \text{ ft}} \times \frac{12 \text{ in}}{1 \text{ ft}} = 576 \text{ in}^2$

15. $\frac{2.2}{11} = \frac{1.5}{E}$
$2.2E = 11 \times 1.5$
$2.2E = 16.5$
$E = \frac{16.5}{2.2} = 7.5$

16. $\frac{A}{B} = \frac{C}{D}$
$AD = BC$
$\frac{AD}{C} = B$

17. $\frac{3.69 \times 10^6 \text{ mi}^2}{1} \times \frac{5.28 \times 10^3}{1 \text{ mi}} \times \frac{5.28 \times 10^3}{1 \text{ mi}} =$
$103 \times 10^{12} = 1.03 \times 10^{14} \text{ ft}^2$

18. $\left(1.03 \times 10^{14} \text{ ft}^2\right) \div \left(6 \times 10^9 \text{ people}\right) =$
$(1.03 \div 6)\left(10^{14} \div 10^9\right) = .172 \times 10^5 =$
$1.72 \times 10^4 \text{ ft}^2$ per person

19. $\frac{1.72 \times 10^4}{4.4 \times 10^4} = (1.72 \div 4.4)\left(10^4 \div 10^4\right) =$
$.39 \times 10^0 = 3.9 \times 10^{-1}$
or .39 acres per person

20. $5Y + 4X \geq 10$
$5Y \geq -4X + 10$
$Y \geq -\frac{4}{5}X + \frac{10}{5}$
$Y \geq -\frac{4}{5}X + 2$
See graph.

Systematic Review 33E

1. 9^2 is the largest power of $9 \leq 100$
$9^2 = 81;\ 9^1 = 9;\ 9^0 = 1$

$100 \div 81 = 1$, remainder 19; $19 \div 9 = 2$, remainder 1; $1 \div 1 = 1$, remainder 0

$1 \times 9^2 + 2 \times 9^1 + 1 \times 9^0 = 121_9$

2. 4^3 is the largest power of $4 \leq 245$
$4^3 = 64;\ 4^2 = 16;\ 4^1 = 4;\ 4^0 = 1$

$245 \div 64 = 3$, remainder 53; $53 \div 16 = 3$, remainder 5; $5 \div 4 = 1$, remainder 1; $1 \div 1 = 1$, remainder 0

$3 \times 4^3 + 3 \times 4^2 + 1 \times 4^1 + 1 \times 4^0 = 3311_4$

3. $35AB_{12} = 3 \times 12^3 + 5 \times 12^2 + 10 \times 12^1 + 11 \times 12^0 =$
$3(1{,}728) + 5(144) + 10(12) + 11(1) =$
$5{,}184 + 720 + 120 + 11 = 6{,}035$

4. $404_5 = 4 \times 5^2 + 0 \times 5^1 + 4 \times 5^0 =$
$4(25) + 0(5) + 4(1) = 100 + 0 + 4 = 104$

5. $60{,}200{,}000 \times .507 =$
$\left(6.02 \times 10^7\right)\left(5.07 \times 10^{-1}\right) =$
$(6.02 \times 5.07)\left(10^7 \times 10^{-1}\right) =$
$30.5 \times 10^6 = 3.05 \times 10^7$ (3 SD)

6. $2{,}000 \times 5{,}000 \times 400 =$
$\left(2 \times 10^3\right)\left(5 \times 10^3\right)\left(4 \times 10^2\right) =$
$(2 \times 5 \times 4)\left(10^3 \times 10^3 \times 10^2\right) =$
$40 \times 10^8 = 4 \times 10^9$ (1 SD)

7. $90{,}000{,}000{,}000 \times .000021 =$
$(9 \times 10^{10})(2.1 \times 10^{-5}) =$
$(18.9 \times 10^{5}) = 1.89 \times 10^{6} \approx$
2×10^{6} (1 SD)

8. $40{,}000 \times 30{,}000 \div 60 =$
$(4 \times 10^{4} \times 3 \times 10^{4} \div 6 \times 10^{1}) =$
$(4 \times 3 \div 6)(10^{4} \times 10^{4} \div 10^{1}) =$
$(12 \div 6)(10^{7}) = 2 \times 10^{7}$ (1 SD)

9. $\left(X^{\frac{2}{5}}\right)\left(X^{\frac{1}{3}}\right) = X^{\frac{2}{5}+\frac{1}{3}} = X^{\frac{6}{15}+\frac{5}{15}} = X^{\frac{11}{15}}$

10. $\left(X^{\frac{2}{5}}\right)^{\frac{1}{3}} = X^{\left(\frac{2}{5}\right)\left(\frac{1}{3}\right)} = X^{\frac{2}{15}}$

11. $\left(X^{\frac{2}{3}}\right)\left(X^{\frac{-1}{5}}\right) = X^{\frac{2}{3}+\frac{-1}{5}} = X^{\frac{10}{15}+\frac{-3}{15}} = X^{\frac{7}{15}}$

12. $\frac{B^6BC^{-4}}{C^9C^{-4}} = B^6B^1C^{-4}C^{-9}C^4 =$
$B^{6+1}C^{-4+(-9)+4} = B^7C^{-9}$ or $\frac{B^7}{C^9}$

13. $\frac{100 \text{ mi}}{1} \times \frac{1.6 \text{ km}}{1 \text{ mi}} = 160 \text{ km}$

14. $\frac{14 \text{ yd}^3}{1} \times \frac{3 \text{ ft}}{1 \text{ yd}} \times \frac{3 \text{ ft}}{1 \text{ yd}} \times \frac{3 \text{ ft}}{1 \text{ yd}} = 378 \text{ ft}^3$

15. $\frac{.03}{2} = \frac{1.5}{W}$
$.03W = 2 \times 1.5$
$.03W = 3$
$W = \frac{3}{.03} = 100$

16. $\frac{X^2}{Y} = \frac{X^2Z^2}{A}$
$X^2A = YX^2Z^2$
$A = \frac{YX^2Z^2}{X^2}$
$A = YZ^2$

17. $3X^2 + 14X + 8 = 0$
$(3X + 2)(X + 4) = 0$
$X + 4 = 0$
$X = -4$
$3X + 2 = 0$
$3X = -2$
$X = -\frac{2}{3}$

$3\left(-\frac{2}{3}\right)^2 + 14\left(-\frac{2}{3}\right) + 8 = 0$
$3\left(\frac{4}{9}\right) - \frac{28}{3} + 8 = 0$
$\frac{12}{9} - \frac{28}{3} + \frac{24}{3} = 0$
$\frac{4}{3} - \frac{28}{3} + \frac{24}{3} = 0$
$0 = 0$

$3(-4)^2 + 14(-4) + 8 = 0$
$3(16) - 56 + 8 = 0$
$48 - 56 + 8 = 0$
$0 = 0$

18. $3Y - 2X = 9$
$3Y = 2X + 9$
$Y = \frac{2}{3}X + 3$
See graph.

19. $m = \frac{2}{3}$ See graph.

20. $m = -\frac{3}{2}$ See graph.

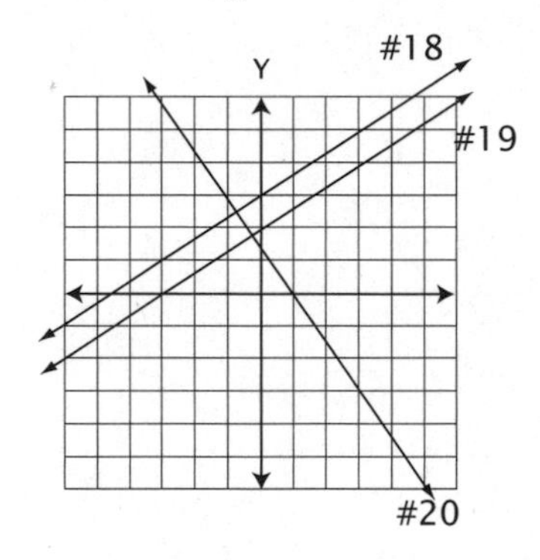

Lesson Practice 34A

1. $X^2 + Y^2 = 16$
$(0)^2 + Y^2 = 16$
$Y^2 = 16$
$Y = \pm 4$

2. $X^2 + Y^2 = 16$
$X^2 + (0)^2 = 16$
$X = \pm 4$

3. $(0,0)$

4. $\sqrt{16} = 4$

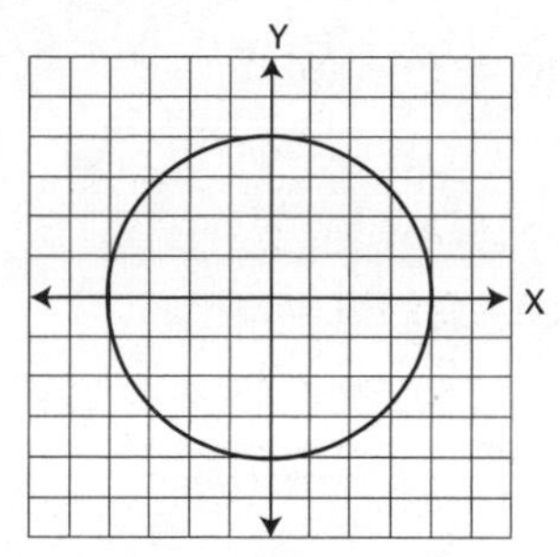

5. $(X-1)^2 + (Y-2)^2 = 9$
$((1)-1)^2 + (Y-2)^2 = 9$
$0^2 + (Y-2)^2 = 9$
$(Y-2)^2 = 9$
$Y - 2 = \pm 3$

$Y - 2 = 3$ $\quad$ $Y - 2 = -3$
$Y = 5$ $\quad$ $Y = -1$

5, −1

6. $(X-1)^2 + (Y-2)^2 = 9$
$(X-1)^2 + ((2)-2)^2 = 9$
$(X-1)^2 + 0^2 = 9$
$(X-1)^2 = 9$
$X - 1 = \pm 3$

$X - 1 = 3$ $\quad$ $X - 1 = -3$
$X = 4$ $\quad$ $X = -2$

4, −2

7. $(1,2)$

8. $\sqrt{9} = 3$

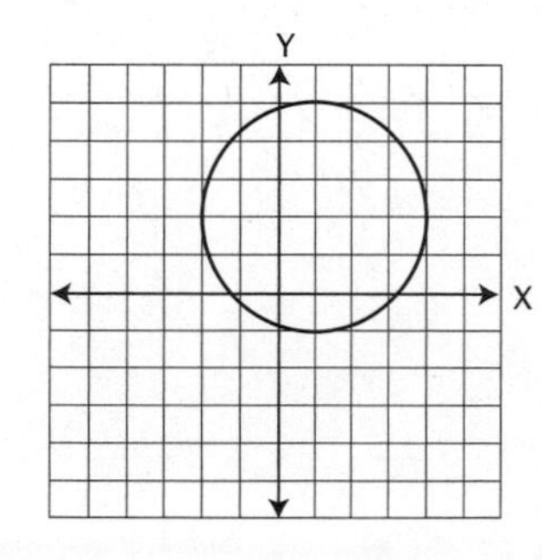

9. $4X^2 + Y^2 = 9$
$4(0)^2 + Y^2 = 9$
$Y^2 = 9$
$Y = \pm 3$

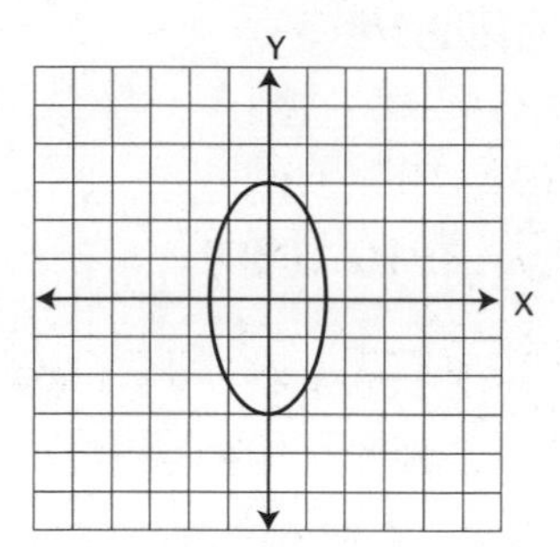

10. $4X^2 + Y^2 = 9$
$4X^2 + (0)^2 = 9$
$4X^2 = 9$
$X^2 = \frac{9}{4}$
$X = \pm \frac{3}{2}$

11. ellipse

12. $6X^2 + 4Y^2 = 12$
$6(0)^2 + 4Y^2 = 12$
$4Y^2 = 12$
$Y^2 = 3$
$Y = \pm\sqrt{3}$

13. $6X^2 + 4Y^2 = 12$
$6X^2 + 4(0)^2 = 12$
$6X^2 = 12$
$X^2 = 2$
$X = \pm\sqrt{2}$

14. ellipse

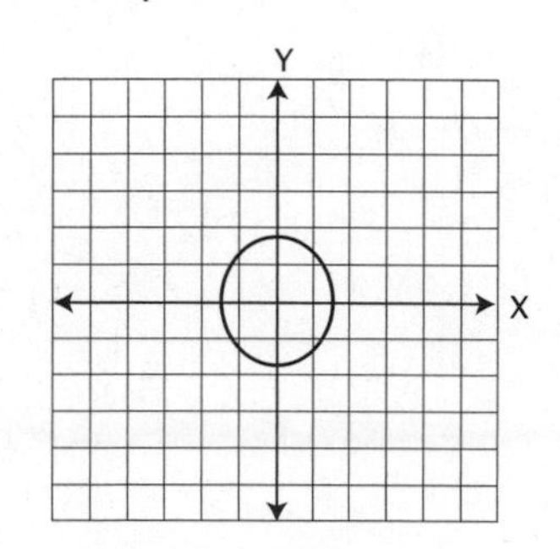

15. $X^2 + Y^2 = 25$

$(0)^2 + Y^2 = 25$

$Y^2 = 25$

$Y = \pm 5$

Points $(0, 5)$ and $(0, -5)$

$X^2 + Y^2 = 25$

$X^2 + (0)^2 = 25$

$X^2 = 25$

$X^2 = \pm 5$

Points $(5, 0)$ and $(-5, 0)$

see graph

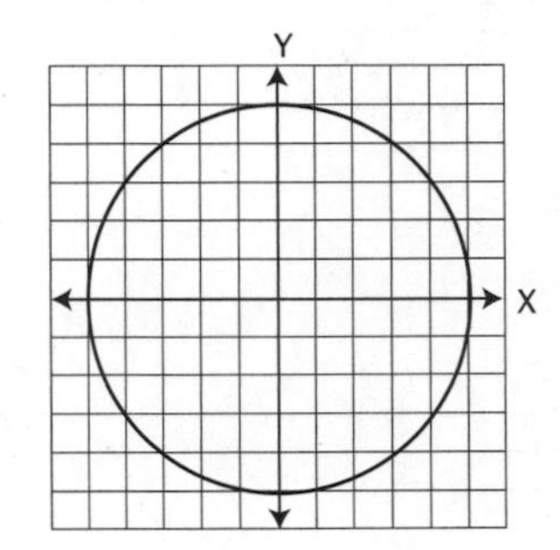

16. $(X+3)^2 + (Y-1)^2 = 4$

$((-3)+3)^2 + (Y-1)^2 = 4$

$(Y-1)^2 = 4$

$Y - 1 = \pm 2$

$Y - 1 = 2 \qquad Y - 1 = -2$

$Y = 3 \qquad Y = -1$

Points $(-3, 3)$ and $(-3, -1)$

$(X+3)^2 + (Y-1)^2 = 4$

$(X+3)^2 + (1-1)^2 = 4$

$(X+3)^2 = 4$

$X + 3 = \pm 2$

$X + 3 = 2 \qquad X + 3 = -2$

$X = -1 \qquad X = -5$

Points $(-1, 1)$ and $(-5, 1)$

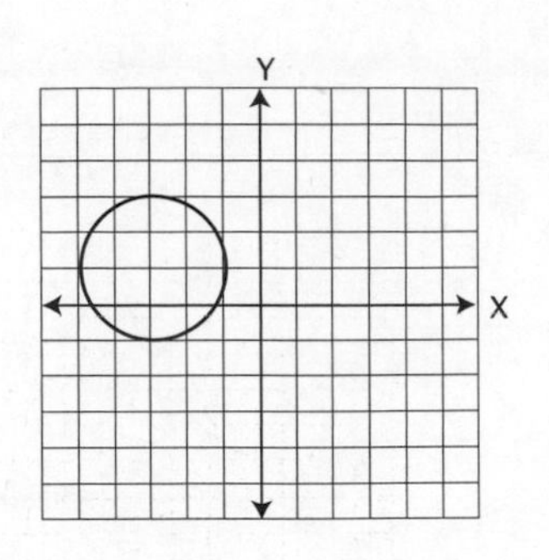

Lesson Practice 34B

1. $X^2 + Y^2 = 4$

$(0)^2 + Y^2 = 4$

$Y^2 = 4$

$Y = \pm 2$

2. $X^2 + Y^2 = 4$

$X^2 + (0)^2 = 4$

$X^2 = 4$

$X = \pm 2$

3. $(0,0)$

4. $\sqrt{4} = 2$

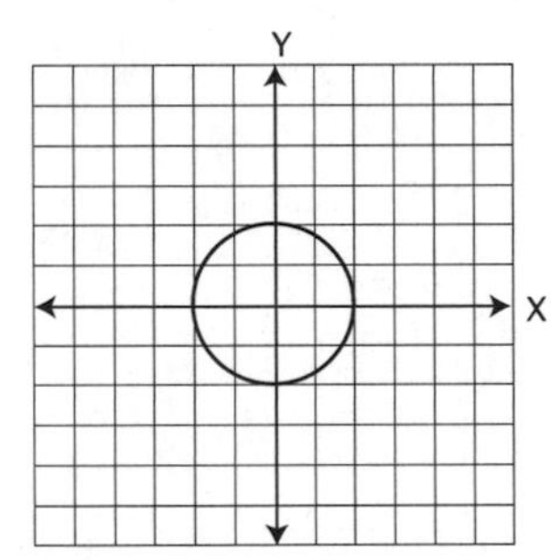

5. $(X+3)^2 + (Y-4)^2 = 9$

$((-3)+3)^2 + (Y-4)^2 = 9$

$(Y-4)^2 = 9$

$Y - 4 = \pm 3$

$Y - 4 = 3 \qquad Y - 4 = -3$

$Y = 7 \qquad Y = 1 \qquad 7,1$

6. $(X+3)^2 + (Y-4)^2 = 9$

$(X+3)^2 + ((4)-4)^2 = 9$

$(X+3)^2 = 9 \qquad X + 3 = 3$

$X + 3 = \pm 3 \qquad X = 0$

$X+3=-3$ 0, −6
$X=-6$

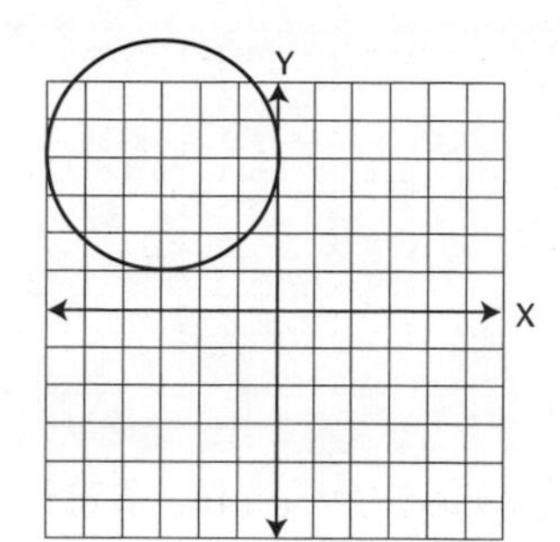

7. $(-3,4)$
8. $\sqrt{9}=3$
9. $3X^2+2Y^2=12$
$3(0)^2+2Y^2=12$
$2Y^2=12$
$Y^2=6$
$Y=\pm\sqrt{6}$
10. $3X^2+2Y^2=12$
$3X^2+2(0)^2=12$
$3X^2=12$
$X^2=4$
$X=\pm\sqrt{4}$
$X=\pm 2$
11. ellipse

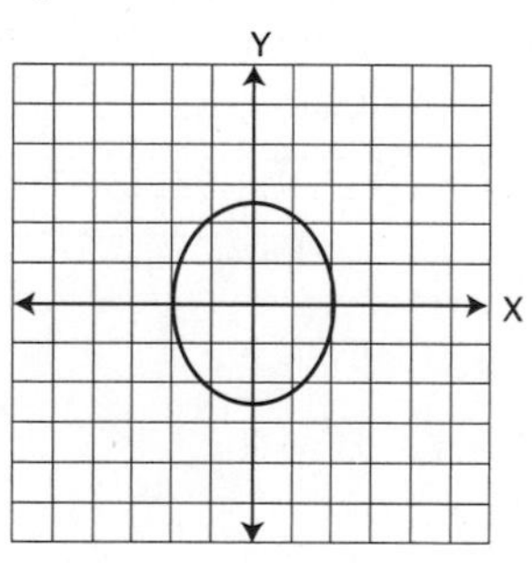

12. $5X^2+3Y^2=15$
$5(0)^2+3Y^2=15$
$3Y^2=15$
$Y^2=5$
$Y=\pm\sqrt{5}$
13. $5X^2+3Y^2=15$
$5X^2+3(0)^2=15$
$5X^2=15$
$X^2=3$
$X=\pm\sqrt{3}$
14. ellipse

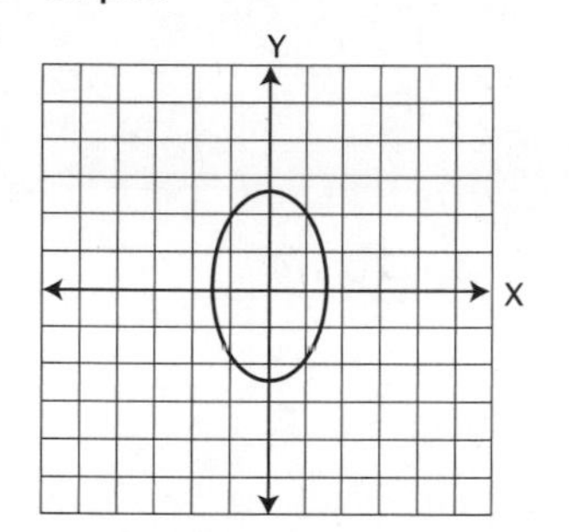

15. $X^2+5Y^2=20$
$(0)^2+5Y^2=20$
$5Y^2=20$
$Y^2=4$
$Y=\pm 2$
Points $(0,2)$ and $(0,-2)$
$X^2+5Y^2=20$
$X^2+5(0)^2=20$
$X^2=20$
$X=\pm\sqrt{20}\approx\pm 4.5$
Points $(4.5,0)$ and $(-4.5,0)$
see graph

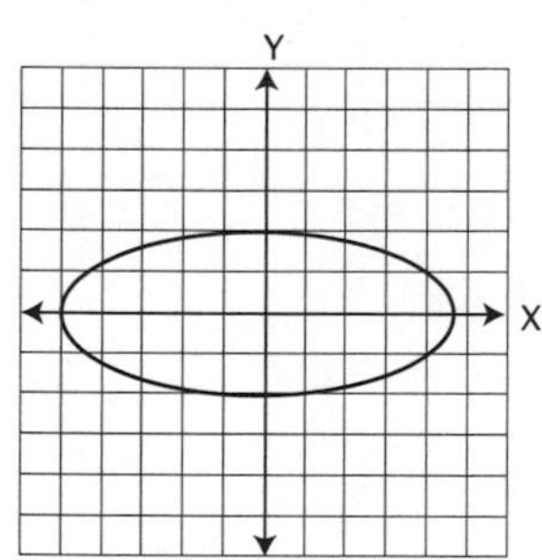

16. $(X+4)^2+(Y-4)^2=16$

$((-4)+4)^2+(Y-4)^2=16$

$(Y-4)^2=16$

$Y-4=\pm4$

$Y-4=4$

$Y=8$

$Y-4=-4$

$Y=0$

Points $(-4, 8)$ and $(-4, 0)$

$(X+4)^2+(Y-4)^2=16$

$(X+4)^2+((4)-4)^2=16$

$(X+4)^2=16$

$X+4=\pm4$

$X+4=4$

$X=0$

$X+4=-4$

$X=-8$

Points $(0, 4)$ and $(-8, 4)$

see graph

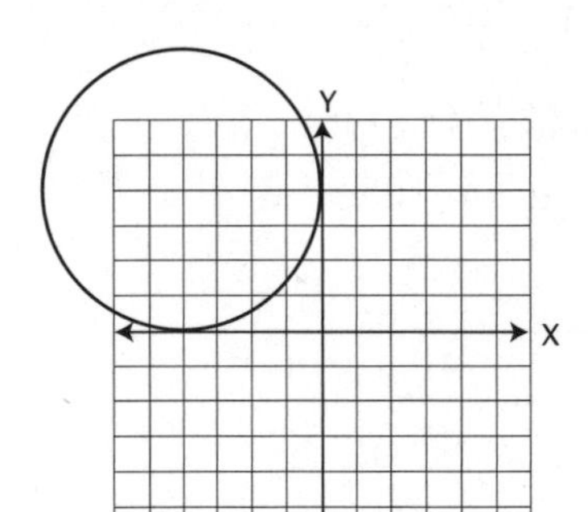

Systematic Review 34C

1. $X^2+Y^2=9$

$(0)^2+Y^2=9$

$Y^2=9$

$Y=\pm3$

2. $X^2+Y^2=9$

$X^2+(0)^2=9$

$X^2=9$

$X=\pm3$

3. $(0,0)$

4. $\sqrt{9}=3$

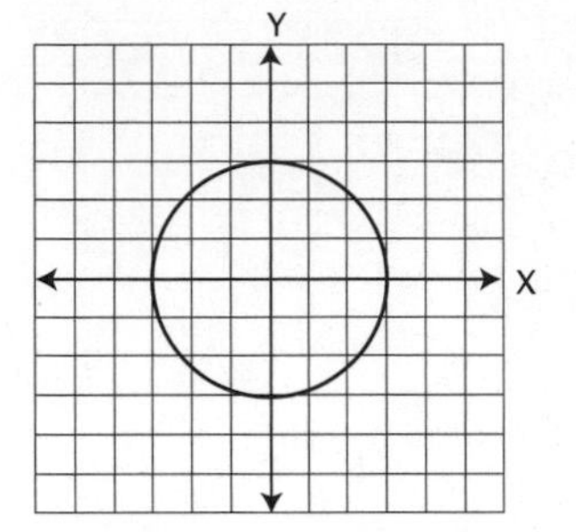

5. $(X-1)^2+(Y-2)^2=9$

$((1)-1)^2+(Y-2)^2=9$

$(Y-2)^2=9$

$Y-2=\pm3$

$Y-2=3$ $\quad$ $Y-2=-3$

$Y=5$ $\quad$ $Y=-1$

6. $(X-1)^2+(Y-2)^2=9$

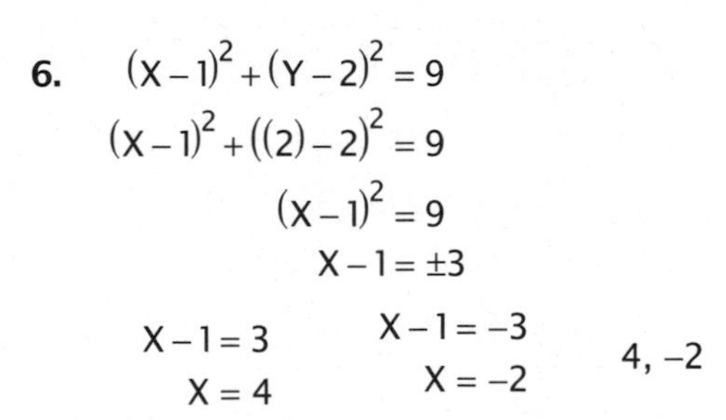

$(X-1)^2+((2)-2)^2=9$

$(X-1)^2=9$

$X-1=\pm3$

$X-1=3$ $\quad$ $X-1=-3$

$X=4$ $\quad$ $X=-2$ $\quad$ 4, –2

7. $(1,2)$

8. $\sqrt{9}=3$

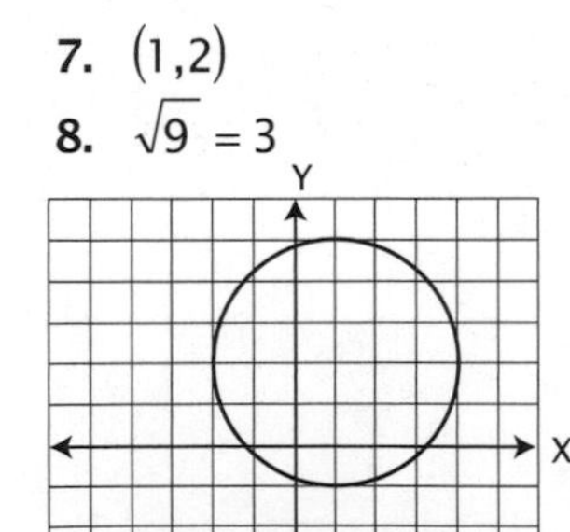

9. $4X^2 + 9Y^2 = 36$

$4(0)^2 + 9Y^2 = 36$

$9Y^2 = 36$

$Y^2 = 4$

$Y = \pm 2$

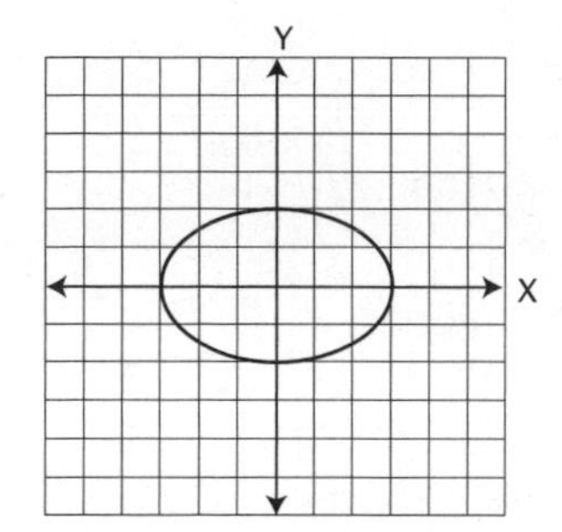

10. $4X^2 + 9Y^2 = 36$

$4X^2 + 9(0)^2 = 36$

$4X^2 = 36$

$X^2 = 9$

$X = \pm 3$

11. 8^3 is the largest power of $8 \le 1{,}721$

$8^3 = 512;\ 8^2 = 64;\ 8^1 = 8;\ 8^0 = 1$

$\begin{array}{r} 3 \\ 512\overline{)1721} \\ 1536 \\ \hline 185 \end{array} \quad \begin{array}{r} 2 \\ 64\overline{)185} \\ 128 \\ \hline 57 \end{array}$

$\begin{array}{r} 7 \\ 8\overline{)57} \\ 56 \\ \hline 1 \end{array} \quad \begin{array}{r} 1 \\ 1\overline{)1} \\ 1 \\ \hline 0 \end{array}$

$3 \times 8^3 + 2 \times 8^2 + 7 \times 8^1 + 1 \times 8^0 = 3271_8$

12. 5^5 is the largest power of $5 \le 3{,}125$

$5^5 = 3125;\ 5^4 = 625;\ 5^3 = 125;$
$5^2 = 25;\ 5^1 = 5;\ 5^0 = 1$

$\begin{array}{r} 1 \\ 3125\overline{)4090} \\ 3125 \\ \hline 965 \end{array} \quad \begin{array}{r} 1 \\ 625\overline{)965} \\ 625 \\ \hline 340 \end{array} \quad \begin{array}{r} 2 \\ 125\overline{)340} \\ 250 \\ \hline 90 \end{array}$

$\begin{array}{r} 3 \\ 25\overline{)90} \\ 75 \\ \hline 15 \end{array} \quad \begin{array}{r} 3 \\ 5\overline{)15} \\ 15 \\ \hline 0 \end{array} \quad \begin{array}{r} 0 \\ 1\overline{)0} \\ 0 \\ \hline 0 \end{array}$

$1 \times 5^5 + 1 \times 5^4 + 2 \times 5^3 + 3 \times 5^2 +$
$3 \times 5^1 + 0 \times 5^0 = 112330_5$

13. $654_7 = 6 \times 7^2 + 5 \times 7^1 + 4 \times 7^0 =$
$6(49) + 5(7) + 4(1) =$
$294 + 35 + 4 = 333$

14. $8B0_{12} = 8 \times 12^2 + 11 \times 12^1 + 0 \times 12^0 =$
$8(144) + 11(12) + 0(1) =$
$1{,}152 + 132 + 0 = 1{,}284$

15. $1{,}000 \times 500 \times 70{,}000 =$
$(1 \times 10^3)(5 \times 10^2)(7 \times 10^4) =$
$(1 \times 5 \times 7)(10^3 \times 10^2 \times 10^4) =$
$35 \times 10^9 = 3.5 \times 10^{10} \approx 4 \times 10^{10}$ (1 SD)

16. $.000058 \times .0023 =$
$(5.8 \times 10^{-5})(2.3 \times 10^{-3}) =$
$(5.8 \times 2.3)(10^{-5} \times 10^{-3}) =$
$13.34 \times 10^{-8} =$
$1.334 \times 10^{-7} \approx 1.3 \times 10^{-7}$ (2 SD)

17. $Y = 2X + 2$

$Y + 4X = -4 \Rightarrow (2X + 2) + 4X = -4$

$6X + 2 = -4$

$6X = -6$

$X = \frac{-6}{6}$

$X = -1$

$Y = 2X + 2 \Rightarrow Y = 2(-1) + 2$

$Y = -2 + 2$

$Y = 0$

$(-1, 0)$

18. $3Y - 2X = -1 \Rightarrow 3Y = 2X - 1$

$\left(Y = \frac{2}{3}X + 1\right)(-3) \Rightarrow -3Y = -2X - 3$

$0 = 0 - 4$

$0 = -4$

When a result like this is obtained, it means that there is no solution for this set of problems. Another way of arriving at this conclusion is to put both of the equations into the Y-intercept form, and take note of the fact that they have the same slope but different intercepts. Since parallel lines never cross, there is no solution.

19. $2Y^5 - 162Y = (2Y)(Y^4 - 81) =$
$(2Y)(Y^2 - 9)(Y^2 + 9) =$
$(2Y)(Y - 3)(Y + 3)(Y^2 + 9)$

20. $Y^8 - 1 = (Y^4 - 1)(Y^4 + 1) =$
$(Y^2 - 1)(Y^2 + 1)(Y^4 + 1) =$
$(Y - 1)(Y + 1)(Y^2 + 1)(Y^4 + 1)$

Systematic Review 34D

1. $2X^2 + 2Y^2 = 8$
$2(0)^2 + 2Y^2 = 8$
$2Y^2 = 8$
$Y^2 = 4$
$Y = \pm 2$

2. $2X^2 + 2Y^2 = 8$
$2X^2 + 2(0)^2 = 8$
$2X^2 = 8$
$X^2 = 4$
$X = \pm 2$

3. $(0,0)$

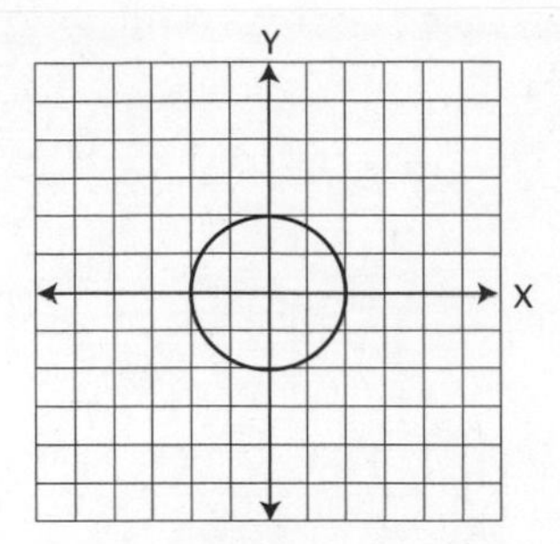

4. Note that this can be simplified:
$2X^2 + 2Y^2 = 8$
$X^2 + Y^2 = 4$
(dividing both sides by 2)
$r = \sqrt{4} = 2$

5. $(X+1)^2 + (Y+3)^2 = 4$
$((-1)+1)^2 + (Y+3)^2 = 4$
$(Y+3)^2 = 4$
$Y + 3 = \pm 2$

$Y + 3 = 2$ $\quad$ $Y + 3 = -2$
$Y = -1$ $\quad$ $Y = -5$

$-1, -5$

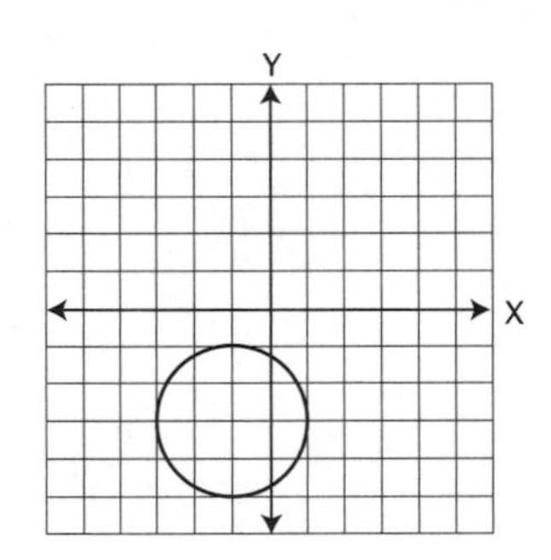

6. $(X+1)^2 + (Y+3)^2 = 4$
$(X+1)^2 + ((-3)+3)^2 = 4$
$(X+1)^2 = 4$
$X + 1 = \pm 2$

$X + 1 = 2$ $\quad$ $X + 1 = -2$
$X = 1$ $\quad$ $X = -3$

$1, -3$

7. $(-1,-3)$

8. $\sqrt{4} = 2$

9. $$\begin{aligned} 9X^2 + 4Y^2 &= 36 \\ 9(0)^2 + 4Y^2 &= 36 \\ 4Y^2 &= 36 \\ Y^2 &= 9 \\ Y &= \pm 3 \end{aligned}$$

10. $$\begin{aligned} 9X^2 + 4Y^2 &= 36 \\ 9X^2 + 4(0)^2 &= 36 \\ 9X^2 &= 36 \\ X^2 &= 4 \\ X &= \pm 2 \end{aligned}$$

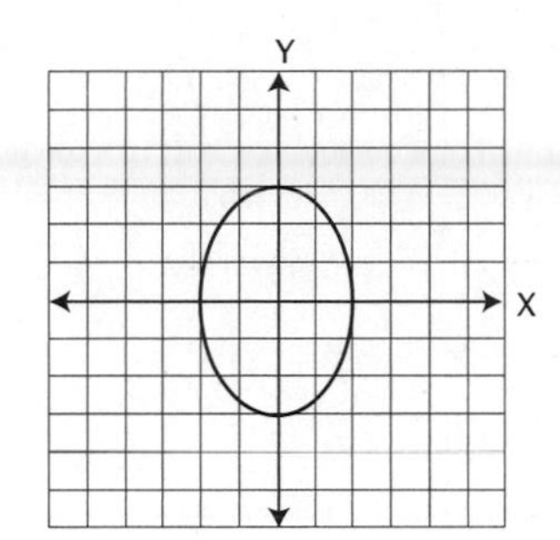

11. $$\begin{array}{r} X^2 - 2X + 1 \\ X - 1 \overline{)X^3 - 3X^2 + 3X - 1} \\ -(X^3 - X^2) \\ \hline -2X^2 + 3X \\ -(-2X^2 + 2X) \\ \hline X - 1 \\ -(X - 1) \\ \hline 0 \end{array}$$

12. $$\begin{array}{r} 8X^2 + 12X + 29 \text{ R } 59 \\ X - 2 \overline{)8X^3 - 4X^2 + 5X + 1} \\ -(8X^3 - 16X^2) \\ \hline 12X^2 + 5X \\ -(12X^2 - 24X) \\ \hline 29X + 1 \\ -(29X - 58) \\ \hline 59 \end{array}$$

13. 4^4 is the largest power of $4 \le 371$

$4^4 = 256$; $4^3 = 64$; $4^2 = 16$; $4^1 = 4$; $4^0 = 1$

$$\begin{array}{r} 1 \\ 256\overline{)371} \\ 256 \\ \hline 115 \end{array} \quad \begin{array}{r} 1 \\ 64\overline{)115} \\ 64 \\ \hline 51 \end{array} \quad \begin{array}{r} 3 \\ 16\overline{)51} \\ 48 \\ \hline 3 \end{array}$$

$$\begin{array}{r} 0 \\ 4\overline{)3} \\ 0 \\ \hline 3 \end{array} \quad \begin{array}{r} 3 \\ 1\overline{)3} \\ 3 \\ \hline 0 \end{array}$$

$1 \times 4^4 + 1 \times 4^3 + 3 \times 4^2 +$
$0 \times 4^1 + 3 \times 4^0 = 11303_4$

14. 8^2 is the largest power of $8 \le 215$

$8^2 = 64$; $8^1 = 8$; $8^0 = 1$

$$\begin{array}{r} 3 \\ 64\overline{)215} \\ 192 \\ \hline 23 \end{array} \quad \begin{array}{r} 2 \\ 8\overline{)23} \\ 16 \\ \hline 7 \end{array} \quad \begin{array}{r} 7 \\ 1\overline{)7} \\ 7 \\ \hline 0 \end{array}$$

$3 \times 8^2 + 2 \times 8^1 + 7 \times 8^0 = 327_8$

15. $406_7 = 4 \times 7^2 + 0 \times 7^1 + 6 \times 7^0 =$
$4(49) + 0(7) + 6(1) =$
$196 + 0 + 6 = 202$

16. $100_4 = 1 \times 4^2 + 0 \times 4^1 + 0 \times 4^0 =$
$1(16) + 0(4) + 0(1) =$
$16 + 0 + 0 = 16$

17. $(3 \times 10^{-5})(2 \times 10^{-2}) =$
$(3 \times 2)(10^{-5} \times 10^{-2}) =$
6×10^{-7}

18. $(4 \times 10^{-5})(5 \times 10^2) \div (2 \times 10^3) =$
$(4 \times 5 \div 2)(10^{-5} \times 10^2 \div 10^3) =$
$10 \times 10^{-6} = 1 \times 10^{-5}$

19. $Y = 3X - 1$

$$4Y = -3X - 19 \Rightarrow \quad 4(3X-1) = -3X - 19$$
$$12X - 4 = -3X - 19$$
$$12X + 3X = -19 + 4$$
$$15X = -15$$
$$X = -1$$

$$Y = 3X - 1 \Rightarrow \quad Y = 3(-1) - 1$$
$$Y = -3 - 1$$
$$Y = -4$$

$(-1, -4)$

20.
$$21 + 12M - 3M = 15M - 9$$
$$12M - 3M - 15M = -9 - 21$$
$$-6M = -30$$
$$M = 5$$

Systematic Review 34E

1.
$$3X^2 + 3Y^2 = 48$$
$$3(0)^2 + 3Y^2 = 48$$
$$3Y^2 = 48$$
$$Y^2 = 16$$
$$Y = \pm 4$$

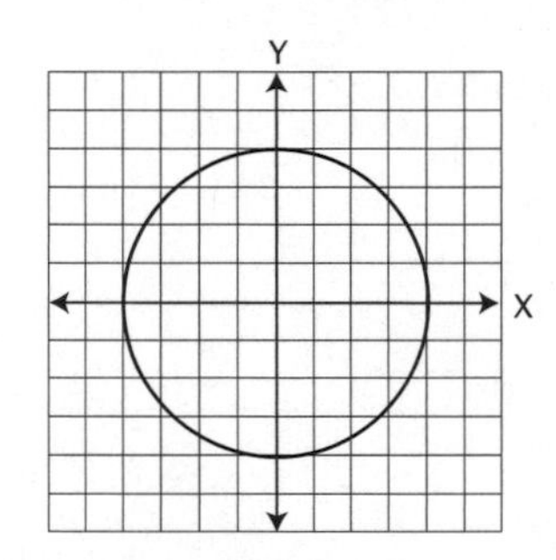

2.
$$3X^2 + 3Y^2 = 48$$
$$3X^2 + 3(0)^2 = 48$$
$$3X^2 = 48$$
$$X^2 = 16$$
$$X = \pm 4$$

3. $(0,0)$

4.
$$3X^2 + 3Y^2 = 48$$
$$X^2 + Y^2 = 16$$
$$r = \sqrt{16} = 4$$

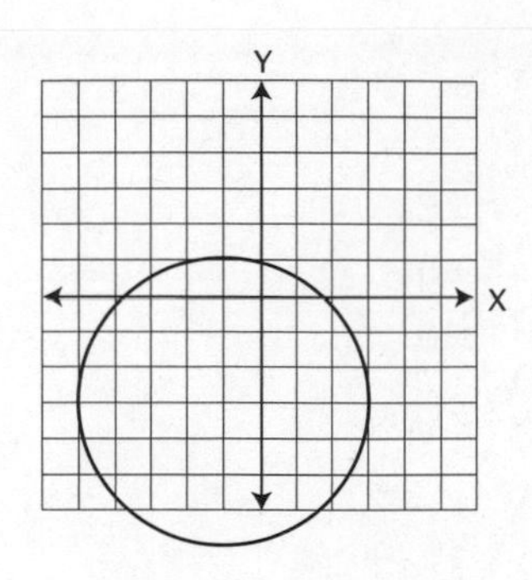

5.
$$3(X+1)^2 + 3(Y+3)^2 = 48$$
$$3((-1)+1)^2 + 3(Y+3)^2 = 48$$
$$3(Y+3)^2 = 48$$
$$(Y+3)^2 = 16$$
$$Y + 3 = \pm 4$$

$$Y + 3 = -4 \qquad Y + 3 = 4$$
$$Y = -7 \qquad Y = 1$$

6.
$$3(X+1)^2 + 3(Y+3)^2 = 48$$
$$3(X+1)^2 + 3((-3)+3)^2 = 48$$
$$3(X+1)^2 = 48$$
$$(X+1)^2 = 16$$
$$X + 1 = \pm 4$$

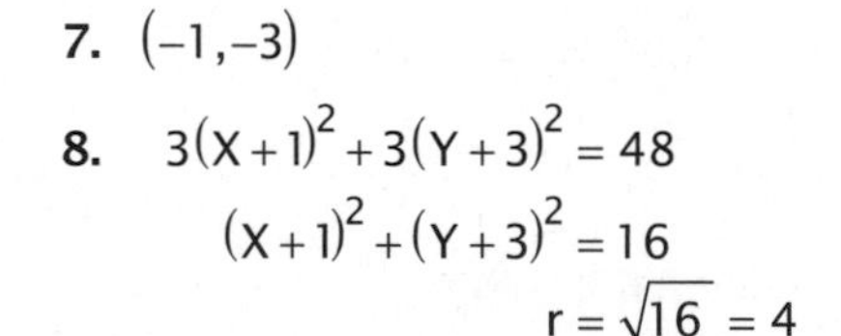

$$X + 1 = 4 \qquad X + 1 = -4$$
$$X = 3 \qquad X = -5$$

7. $(-1, -3)$

8.
$$3(X+1)^2 + 3(Y+3)^2 = 48$$
$$(X+1)^2 + (Y+3)^2 = 16$$
$$r = \sqrt{16} = 4$$

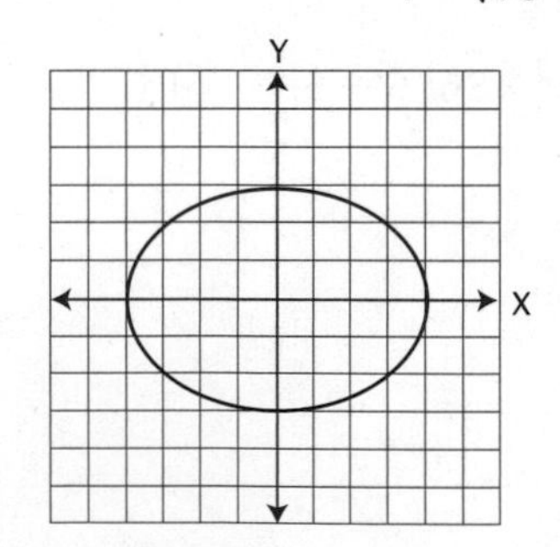

9.
$$9X^2+16Y^2=144$$
$$9(0)^2+16Y^2=144$$
$$16Y^2=144$$
$$Y^2=9$$
$$Y=\pm 3$$

10.
$$9X^2+16Y^2=144$$
$$9X^2+16(0)^2=144$$
$$9X^2=144$$
$$X^2=16$$
$$X=\pm 4$$

11.
$$\begin{array}{r} X^2+2X+1 \\ X+1\overline{)X^3+3X^2+3X+1} \\ -(X^3+X^2) \\ \hline 2X^2+3X \\ -(2X^2+2X) \\ \hline X+1 \\ -(X+1) \\ \hline 0 \end{array}$$

12.
$$\begin{array}{r} X^2+2X+3 \\ X+2\overline{)X^3+4X^2+7X+6} \\ -(X^3+2X^2) \\ \hline 2X^2+7X \\ -(2X^2+4X) \\ \hline 3X+6 \\ -(3X+6) \\ \hline 0 \end{array}$$

13. 6^3 is the largest power of $6 \le 1,054$

$6^3=216;\ 6^2=36;\ 6^1=6;\ 6^0=1$

$$\begin{array}{r} 4 \\ 216\overline{)1054} \\ 864 \\ \hline 190 \end{array} \qquad \begin{array}{r} 5 \\ 36\overline{)190} \\ 180 \\ \hline 10 \end{array}$$

$$\begin{array}{r} 1 \\ 6\overline{)10} \\ 6 \\ \hline 4 \end{array} \qquad \begin{array}{r} 4 \\ 1\overline{)4} \\ 4 \\ \hline 0 \end{array}$$

$$4\times6^3+5\times6^2+1\times6^1\times4\times6^0=4514_6$$

14. $101111_2=$

$1\times2^5+0\times2^4+1\times2^3+1\times2^2+1\times2^1+1\times2^0=$

$1(32)+0(16)+1(8)+1(4)+1(2)+1(1)=$

$32+0+8+4+2+1=47$

15. $50\times60<55\times55$

$3,000<3,025$

16. $\frac{4\text{ acres}}{1}\times\frac{43,560\text{ ft}^2}{1\text{ acre}}=174,240\text{ ft}^2$

$174,240\text{ ft}^2<200,000\text{ ft}^2$

17. $(4.2\times10^4)\div(6\times10^{-3})=$

$(4.2\div6)(10^4\div10^{-3})=.7\times10^7=7\times10^6$

18. $[(7\times10^8)(8\times10^0)]\div[(4\times10^3)(1.4\times10^5)]=$

$[(7\times8)\div(4\times1.4)][(10^8\times10^0)\div(10^3\times10^5)]=$

$(56\div5.6)(10^8\div10^8)=10\times10^0=1\times10^1$ or 10

19. $X^4-16=(X^2-4)(X^2+4)=$

$(X-2)(X+2)(X^2+4)$

20.
$$1.25+.8A-1=.3$$
$$125+80A-100=30$$
$$80A=30-125+100$$
$$80A=5$$
$$A=\frac{1}{16}$$

Lesson Practice 35A

1.

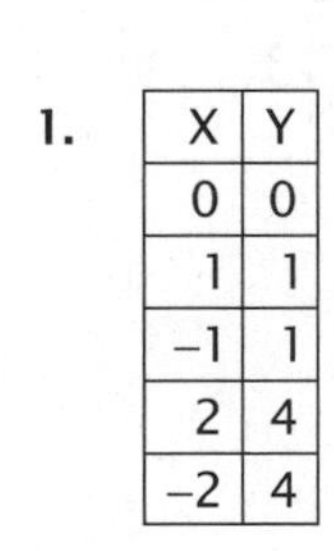

X	Y
0	0
1	1
–1	1
2	4
–2	4

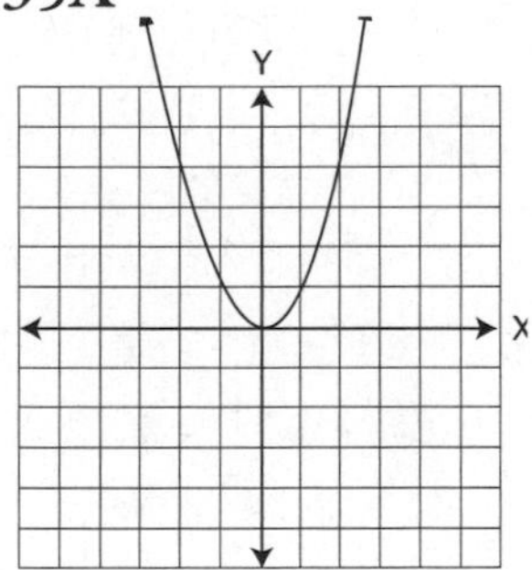

2.

X	Y
2	3
–2	–3
3	2
–3	–2
1	6
–1	–6
6	1
–6	–1

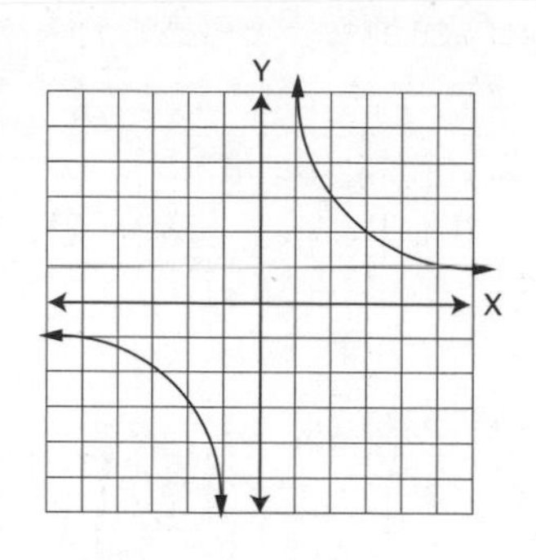

3.

X	Y
0	0
1	2
2	8
3	18
–1	2
–2	8

4.

X	Y
1	–2
–1	2
2	–1
–2	1
$\frac{1}{2}$	–4
$-\frac{1}{2}$	4
4	$-\frac{1}{2}$
–4	$\frac{1}{2}$

5.

X	Y
0	–3
1	–2
–1	–2
2	1
–2	1

6.

X	Y
2	4
–2	–4
4	2
–4	–2
$1\frac{1}{3}$	6
$-1\frac{1}{3}$	–6
6	$1\frac{1}{3}$
–6	$-1\frac{1}{3}$

Lesson Practice 35B

1.

X	Y
0	0
1	$\frac{1}{2}$
2	2
3	$\frac{9}{2}$
–1	$\frac{1}{2}$
–2	2

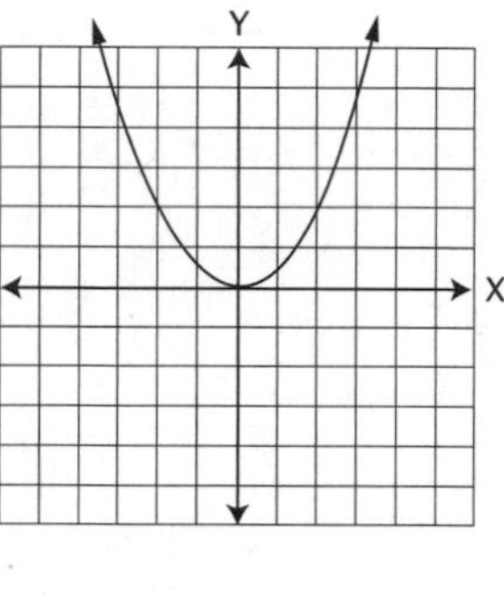

2.

X	Y
–1	3
1	–3
–3	1
3	–1
$1\frac{1}{2}$	–2
$-1\frac{1}{2}$	2
–2	$1\frac{1}{2}$
2	$-1\frac{1}{2}$

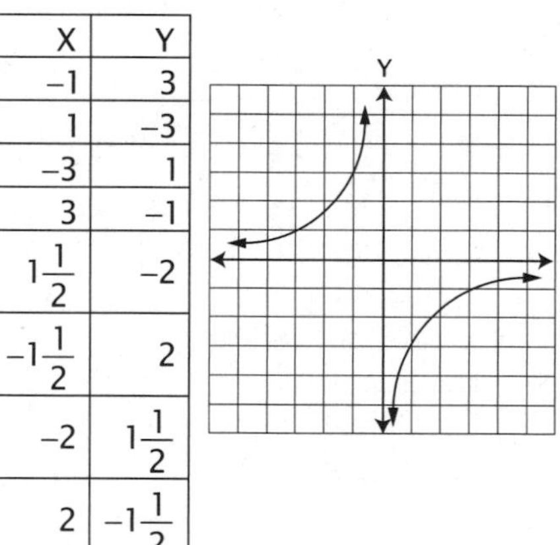

3.

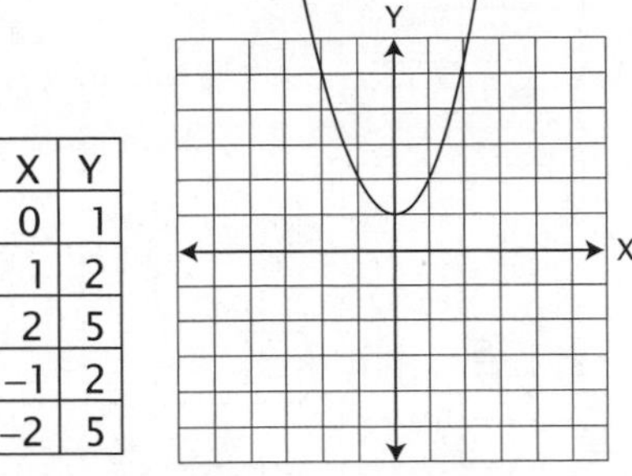

X	Y
0	1
1	2
2	5
–1	2
–2	5

4.

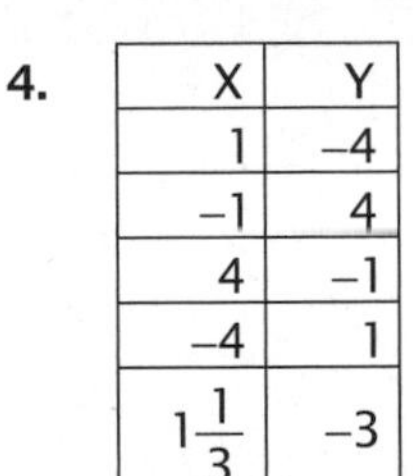

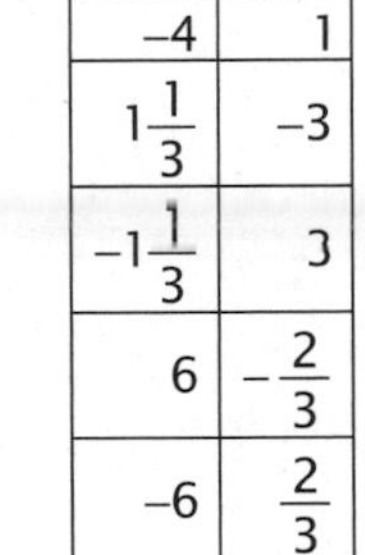

X	Y
1	–4
–1	4
4	–1
–4	1
$1\frac{1}{3}$	–3
$-1\frac{1}{3}$	3
6	$-\frac{2}{3}$
–6	$\frac{2}{3}$

5.

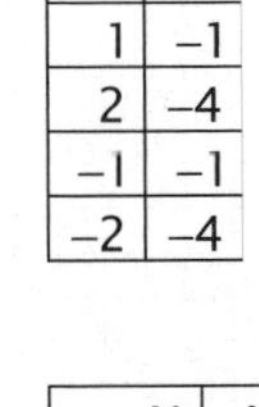

X	Y
0	0
1	–1
2	–4
–1	–1
–2	–4

6.

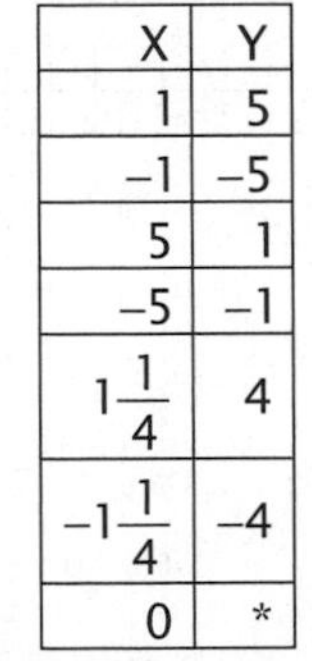

X	Y
1	5
–1	–5
5	1
–5	–1
$1\frac{1}{4}$	4
$-1\frac{1}{4}$	–4
0	*

*no possible value

Systematic Review 35C

1.

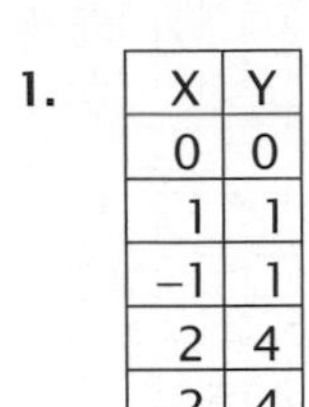

X	Y
0	0
1	1
–1	1
2	4
–2	4

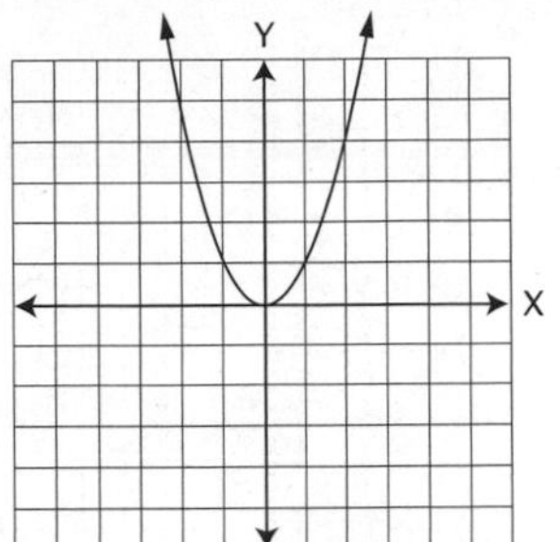

2.

X	Y
1	1
–1	–1
3	$\frac{1}{3}$
–3	$-\frac{1}{3}$
5	$\frac{1}{5}$
–5	$-\frac{1}{5}$

3. line
4. circle
5. ellipse
6. line
7. hyperbola
8. parabola
9. $1793_{12} = 1\times12^3+7\times12^2+9\times12^1+3\times12^0 =$
   $1(1{,}728)+7(144)+9(12)+3(1) =$
   $1{,}728+1{,}008+108+3 = 2{,}847$
10. 5^3 is the largest power of 5 less than 131

    $5^3 = 125;\ 5^2 = 25;\ 5^1 = 5;\ 5^0 = 1$

    $125\overline{)131}$ = 1, 125, remainder 6
    $25\overline{)6}$ = 0, 0, remainder 6
    $5\overline{)6}$ = 1, 5, remainder 1
    $1\overline{)1}$ = 1, 1, remainder 0

    $1\times5^3+0\times5^2+1\times5^1+1\times5^0 = 1011_5$

11. $1111_2 = 1\times2^3+1\times2^2+1\times2^1+1\times2^0 =$
$1(8)+1(4)+1(2)+1(1) = 8+4+2+1 = 15$
$202_3 = 2\times3^2+0\times3^1+2\times3^0 =$
$2(9)+0(3)+2(1) = 18+0+2 = 20$
$15 < 20$

12. $\frac{2\ \text{ft}^2}{1}\times\frac{12\ \text{in}}{1\ \text{ft}}\times\frac{12\ \text{in}}{1\ \text{ft}} = 288\ \text{in}$
$288 < 289$

13. $\left(7\times10^{-8}\right)\div\left(1.4\times10^{6}\right) =$
$\left(7\div1.4\right)\left(10^{-8}\div10^{6}\right) = 5\times10^{-14}$

14.
$\left[\left(2.4\times10^{-4}\right)\left(2.6\times10^{5}\right)\right]\div\left[\left(6\times10^{-5}\right)\left(5.2\times10^{-7}\right)\right] =$
$\left[\left(2.4\times2.6\right)\div\left(6\times5.2\right)\right]\left[\left(10^{-4}\times10^{5}\right)\div\left(10^{-5}\times10^{-7}\right)\right]$
$\left(6.24\div31.2\right)\left(10^{1}\div10^{-12}\right) =$
$.2\times10^{13} = 2\times10^{12}$

15. $\frac{2}{3}X+\frac{4}{5} = -\frac{17}{30}$
multiply each term by 30,
to eliminate fractions:
$\frac{30}{1}\cdot\frac{2}{3}X+\frac{30}{1}\cdot\frac{4}{5} = \frac{30}{1}\cdot-\frac{17}{30}$
$20X+24 = -17$
$20X = -17-24$
$20X = -41$
$X = \frac{-41}{20} = -2\frac{1}{20}$

16. $\frac{5}{6}-\frac{1}{3}X+\frac{4}{7} = 0$
multiply each term by 42:
$\frac{42}{1}\cdot\frac{5}{6}+\frac{42}{1}\cdot-\frac{1}{3}X+\frac{42}{1}\cdot\frac{4}{7} = 42(0)$
$35-14X+24 = 0$
$35+24 = 14X$
$59 = 14X$
$\frac{59}{14} = X = 4\frac{3}{14}$

17. $(78)(72) = 5{,}616$

18. $Y^3-Y = (Y)\left(Y^2-1\right) =$
$(Y)(Y-1)(Y+1)$

19. on graph

20. $m = \frac{(-4)-1}{1-(-1)} = -\frac{5}{2}$
$Y = mX+b$
$(1) = \left(-\frac{5}{2}\right)(-1)+b$
$1 = \frac{5}{2}+b$
$\frac{2}{2}-\frac{5}{2} = b;\ b = \frac{-3}{2}$
$Y = -\frac{5}{2}X-\frac{3}{2}$ or $2Y+5X = -3$

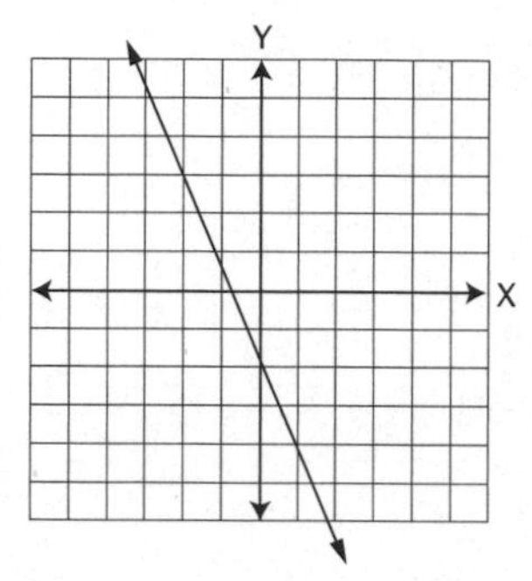

Systematic Review 35D

1.

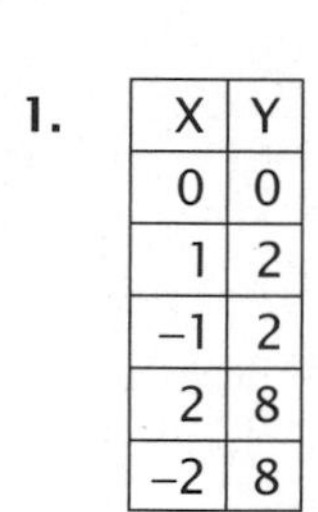

X	Y
0	0
1	2
–1	2
2	8
–2	8

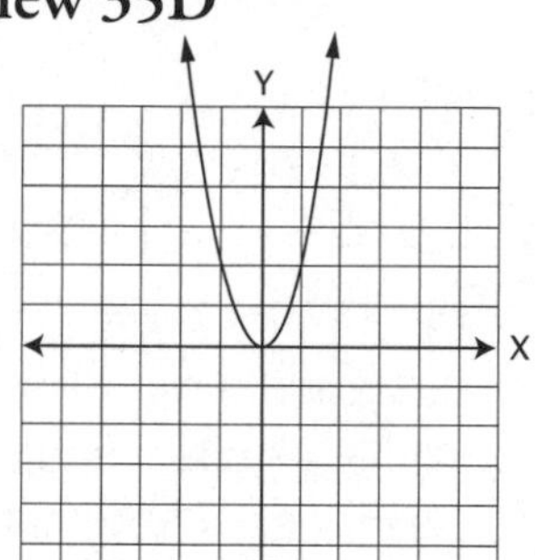

2.

X	Y
.5	12
–.5	–12
1	6
–1	–6
2	3
–2	–3
3	2
–3	–2

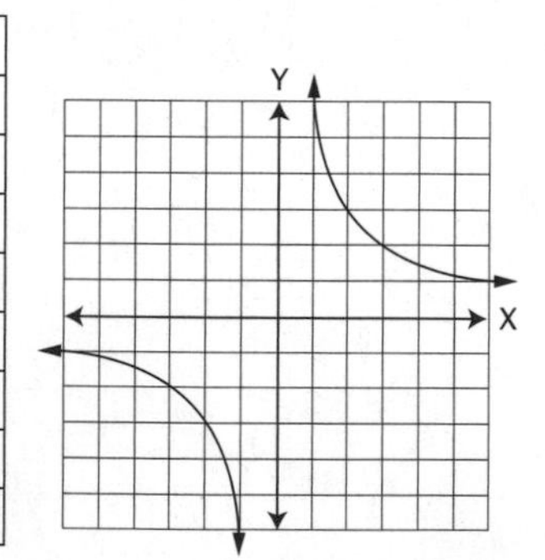

3. parabola
4. circle
5. hyperbola
6. parabola
7. ellipse
8. line
9. $132_4 = 1\times4^2+3\times4^1+2\times4^0 =$
   $1(16)+3(4)+2(1) = 16+12+2 = 30$
10. 8^3 is the largest power of $8 < 2{,}348$
    $8^3 = 512;\ 8^2 = 64;\ 8^1 = 8;\ 8^0 = 1$

$$512\overline{)2348}\ \to 4,\ 2348-2048=300;\quad 64\overline{)300}\ \to 4,\ 300-256=44;\quad 8\overline{)44}\ \to 5,\ 44-40=4;\quad 1\overline{)4}\ \to 4,\ 4-4=0$$

$$4\times8^3+4\times8^2+5\times8^1+4\times8^0 = 4454_8$$

11. $|17-3|\cdot(-2) > -5^2-4$
    $|14|\cdot-2 > -(5\times5)-4$
    $14\cdot-2 > -25-4$
    $-28 > -29$
12. $47\cdot43 < 45^2$
    $2{,}021 < 2{,}025$
13. $(6\times10^7)(2.5\times10^{-9}) = (6\times2.5)(10^7\times10^{-9}) =$
    $15\times10^{-2} = 1.5\times10^{-1}$
    Or 2×10^{-1} if the student took significant digits into account. Either answer is acceptable.
14. $[(1.1\times10^{-9})(1.5\times10^8)]\div[(5\times10^1)(3\times10^{-6})] =$
    $[(1.1\times1.5)\div(5\times3)][(10^{-9}\times10^8)\div(10^1\times10^{-6})] =$
    $(1.65\div15)(10^{-1}\div10^{-5}) =$
    $.11\times10^4 = 1.1\times10^3$
    Or 1×10^3 if the student took significant digits into account. Either answer is acceptable.
15. $Y^{-2}\div Y^{-6} = Y^{-2-(-6)} = Y^4$
16. $\frac{.25\ \text{mi}}{1}\times\frac{5{,}280\ \text{ft}}{1\ \text{mi}} = 1{,}320$ ft
17. $\frac{C^6D^3C^2}{D^{-9}D^{-2}C^8} = C^6D^3C^2D^9D^2C^{-8} =$
    $C^{6+2+(-8)}D^{3+9+2} = C^0D^{14} = D^{14}$
18. $3X^{-2}Y^2+\frac{4Y^4Y^0Y^{-2}}{X^{-1}} =$
    $\frac{3Y^2}{X^2}+\frac{4Y^{4+0+(-2)}}{X^{-1}} =$
    $\frac{3Y^2}{X^2}+\frac{4X^1Y^2}{1}$
    find common denominators:
    $\frac{3Y^2}{X^2}+\frac{X^2}{X^2}\cdot\frac{4XY^2}{1} =$
    $\frac{3Y^2}{X^2}+\frac{4X^3Y^2}{X^2} =$
    $\frac{3Y^2+4X^3Y^2}{X^2}$
    $3X^{-2}Y^2+4XY^2$ is also acceptable
19. see graph
20. see graph $\left(m = \frac{1}{4}\right)$

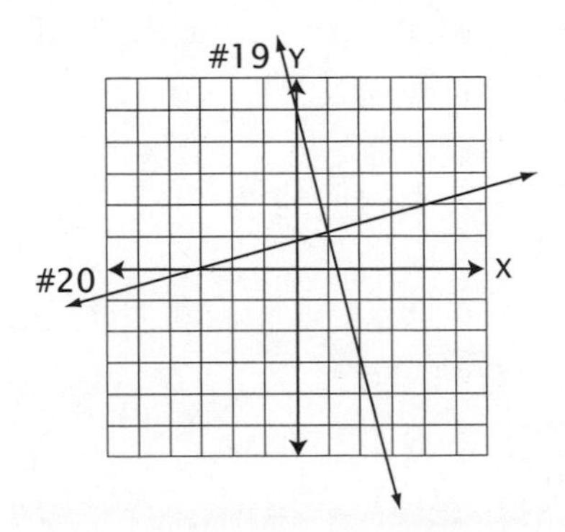

Systematic Review 35E

1.

X	Y
0	0
1	3
–1	3
2	12
–2	12

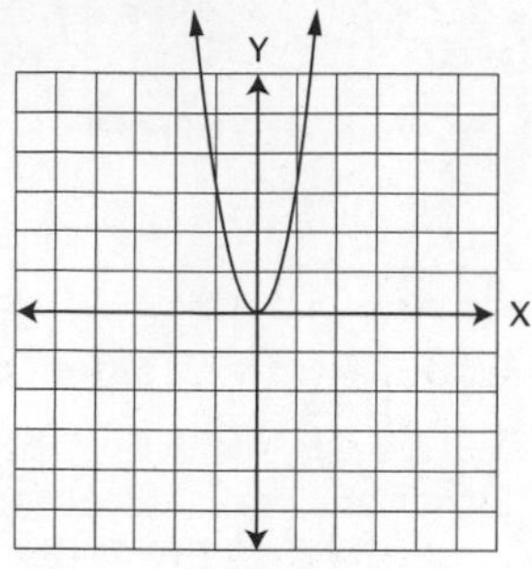

2.

X	Y
1	–10
–1	10
2	–5
–2	5
5	–2
–5	2

3. ellipse

4. hyperbola

5. line

6. circle

7. line

8. circle

9. $151_7 = 1\times7^2+5\times7^1+1\times7^0 =$
$1(49)+5(7)+1(1)=$
$49+35+1=85$

10. 4^4 is the largest power of $4 \le 291$

$4^4 = 256;\ 4^3 = 64;\ 4^2 = 16;$
$4^1 = 4;\ 4^0 = 1$

$256\overline{)291}$ quotient 1; -256; remainder 35

$64\overline{)35}$ quotient 0; -0; remainder 35

$16\overline{)35}$ quotient 2; -32; remainder 3

$4\overline{)3}$ quotient 0; -0; remainder 3

$1\overline{)3}$ quotient 3; -3; remainder 0

$1\times4^4+0\times4^3+2\times4^2+0\times4^1+3\times4^0 = 10203_4$

11. $|3\cdot2\cdot(-2)| > 24\div-3$
$|-12| > -8$
$12 > -8$

12. $(3^2)^{\frac{1}{3}} < (287)^2$
$(\sqrt[3]{9}) < 82{,}369$

13. $93{,}000{,}000 = 9.3\times10^7$

14. $.038 = 3.8\times10^{-2}$

15. $\frac{900\text{ g}}{1}\times\frac{.035\text{ oz}}{1\text{ g}} = 31.5\text{ oz}$

16. $\frac{1\text{ yd}^2}{1}\times\frac{36\text{ in}}{1\text{ yd}}\times\frac{36\text{ in}}{1\text{ yd}} = 1{,}296\text{ in}^2$

17. $A^3-25A = (A)(A^2-25) =$
$(A)(A-5)(A+5)$

18. $243-3X^4 = (3)(81-X^4) =$
$(3)(9-X^2)(9+X^2) =$
$(3)(3-X)(3+X)(9+X^2)$

19. $X+(X+4)+5-1-2X = 2X$
$4+5-1 = 2X$
$8 = 2X$
$4 = X$

20. $5D+(5)(4) = 45$
$5D+20 = 45$
$5D = 25$
$D = 5$

Honors Solutions

Honors Lesson 1H

1. $1\frac{1}{2} = \frac{3}{2}$

   $\frac{2}{3} \times \frac{3}{2} = \frac{6}{6} = 1$ cup shortening

   $\frac{3}{4} \times \frac{3}{2} = 1\frac{1}{8}$ cup of sugar

   $1 \times \frac{3}{2} = 1\frac{1}{2}$ eggs (rounds to 2)

   $1 \times \frac{3}{2} = 1\frac{1}{2}$ tablespoons of milk

   $1 \times \frac{3}{2} = 1\frac{1}{2}$ teaspoons vanilla

   $\frac{7}{4} \times \frac{3}{2} = \frac{21}{8} = 2\frac{5}{8}$ teaspoons of baking powder

   $\frac{1}{2} \times \frac{3}{2} = \frac{3}{4}$ teaspoon of salt

   $\frac{3}{4} \times \frac{3}{2} = \frac{9}{8} = 1\frac{1}{8}$ cups of rolled oats

   $\frac{1}{4} \times \frac{3}{2} = \frac{3}{8}$ cup of dried fruit

2. original recipe made three dozen, or 36 cookies

   $\frac{36}{1} \times \frac{3}{2} = \frac{108}{2} = 54$ cookies from the larger recipe

   $54 \div 2 = 27$ cookies from each bowl

3. Total of bills:

   $35.92 + $25.26 + $255.10 + $798.53 + $20.00 + $116.48 + $398.19 = $1,649.48

   $1,609.00 − 1649.48 = $ − 40.48

   The negative number indicates that Daniel is "in the hole" or owes that amount.

4. $(-3) \times 6 = -18$ in

   $(-18) + 5 = -13$ inches from starting level

5. 1st option

   $10 + 20 + 40 + 80 = \$150$

   2nd option

   $5 + 25 + 625 + 390{,}625 = \$391{,}280$

   The second option is definitely the better choice.

6. $4 \times 3 = 12$

   $3 \times 4 = 12$

   commutative

7. $5 + 6 = 11; 11 + 8 = 19$

   $6 + 8 = 14; 14 + 5 = 19$

   associative

8. $8 \div 4 = 2$ pizzas per person

   $4 \div 8 = \frac{1}{2}$ pizza per person

   division is not commutative

Honors Lesson 2H

1. There are 14. They are: 32, 33, 34, 35, 36, 38, 39, 40, 42, 44, 45, 46, 48, and 49.
2. 1, 17, 289
3. $\frac{18}{15} = 1\frac{3}{15} = 1\frac{1}{5} = \$1.20$
4. $75.78 − $45.78 = $30.00 labor

   $30.00 ÷ 1.5 = $20.00/hour

5. $\frac{1}{4} + \frac{7}{12} = \frac{3}{12} + \frac{7}{12} = \frac{10}{12} = \frac{5}{6}$
6. $\frac{5}{6} \times \frac{1}{10} = \frac{5}{60} = \frac{1}{12}$ of a tank used

   $\frac{5}{6} - \frac{1}{12} = \frac{10}{12} - \frac{1}{12} = \frac{9}{12} = \frac{3}{4}$

   $\frac{3}{4} \times 24 = 18$ gallons left

7. First, figure out how long it would take for him to do the whole job. 30 minutes is $\frac{3}{5}$ of the total time. In equation form:

   $30 = \frac{3}{5}T$

   $150 = 3T$

   $50 = T$, so 50 min for the whole job

   $\frac{1}{5}$ of $50 = 10$ min

   $\frac{1}{2}$ of $50 = 25$ min

8. $9 + 19 = 28$

   $28 \div 4 = 7$

   $7 + 5 = 12$

9. $5 \times 4 = 20$

   $20 - 1 = 19$

   $19 + 8 = 27$ yards

10. $\$-20.00+\$35.00=\$15.00$
$\$15+\$70.00=\$85.00$
$\$85.00-\$10.00=\$75.00$
$\$75.00-\$22.50=\$52.50$

Honors Lesson 3H

1. yes
2. rational
3. rational
4. $A=\frac{1}{2}bh$
$12=\frac{1}{2}(6)h$
$12=3h$
$4\text{ in}=h$
5. $P=2L+2W$
$30=2(10)+2W$
$30=20+2W$
$10=2W$
$5\text{ cm}=W$
6. $d=rt$
$11\frac{1}{4}=4\frac{1}{2}(t)$
$\frac{45}{4}=\frac{18}{4}(t)$
$45=18t$
$t=\frac{45}{18}=2\frac{1}{2}$ hours
using decimals: $11.25=4.5t$
$2.5=t$
7. $p=0.433d$
$43.3=.433d$
$43300=433d$
$d=100\text{ ft}$

Honors Lesson 4H

1. 2
2. 18
3. 1
4. 4
5. 9
6. Test 2
7. Test 4
8. $100-75=25$
9. John: $95+90+95+93+97=470$
$470\div5=94$
David: $98+90+90+75+100=453$
$453\div5=90.6$
You may have slightly different results depending on how you estimated the scores.
John had the highest average score.
10. Joe sold 25.
Jeff sold 20.
The graphs agree.
11. Jeff probably drew the first graph: it is unlikely that he would have presented the data in a way that made it look like he had only sold a fraction of what Joe sold.
Joe probably drew the sec ond graph.
12. Answers will vary.

Honors Lesson 5H

1. (6 steps, east)
2. $6+(-10)=-4$, so (4 steps, west)
3. (4 paces, south)
4. (2, south)
5. (6, north)
6. $A^2+B^2=C^2$
$4^2+4^2=C^2$
$32=C^2$
$C=\sqrt{32}$
7. ($\sqrt{32}$, northeast)
8. $8^2+8^2=C^2$
$128=C^2$
$C=\sqrt{128}$
($\sqrt{128}$, southwest)

Honors Lesson 6H

1\.

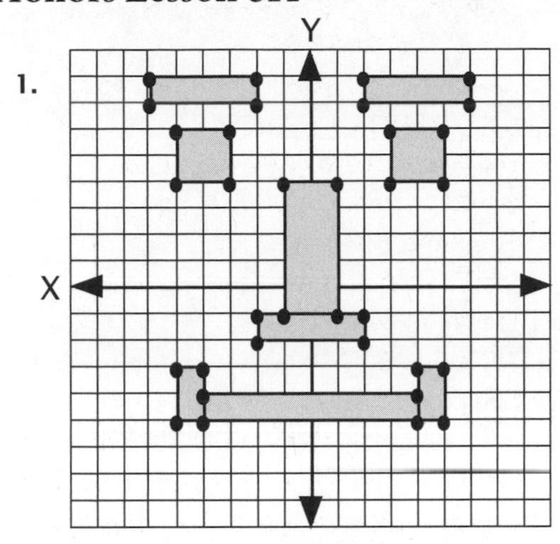

2\.

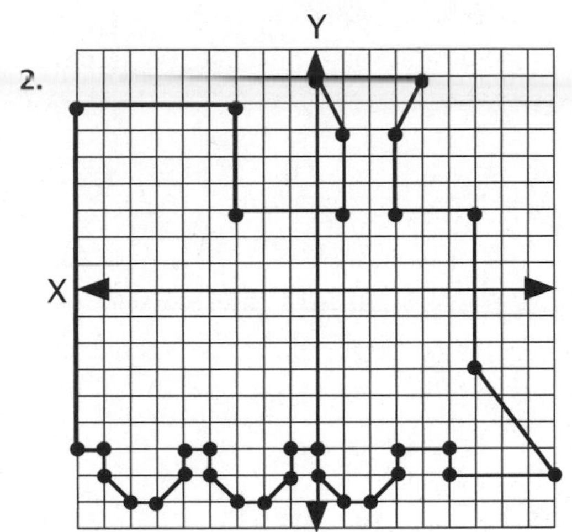

Honors Lesson 7H

1. done
2. done
3. slope is negative; less steep than 1;
   Y-intercept is 1.
   Line a is the best choice.
4. slope is positive; steeper than 1;
   Y-intercept is 1.
   Line c is the best choice.
5. slope is positive; less steep than 1;
   Y-intercept is -1.
   Line b is the best choice.
6. slope is positive; steeper than 1;
   Y-intercept is 0.
   Line d is the best choice.
7. slope is positive; steeper than 1;
   Y-intercept is 0.
   Line h is the best choice.
8. slope is positive; less steep than 1;
   Y-intercept is 3.
   Line f is the best choice.
9. slope is positive; equal to 1;
   Y-intercept is 0.
   Line g is the best choice.
10. slope is negative; equal to 1;
    Y-intercept is –3.
    Line e is the best choice.

Honors Lesson 8H

1. done
2. $\frac{4}{12}$ or $\frac{1}{3}$
3. $\frac{10}{15}$ or $\frac{2}{3}$
4. $\frac{9}{6}$ or $\frac{3}{2}$
5. done
6. $\frac{8}{24}$ or $\frac{1}{3}$
7. $\frac{10}{20}$ or $\frac{1}{2}$
8. $\frac{18}{24}$ or $\frac{3}{4}$

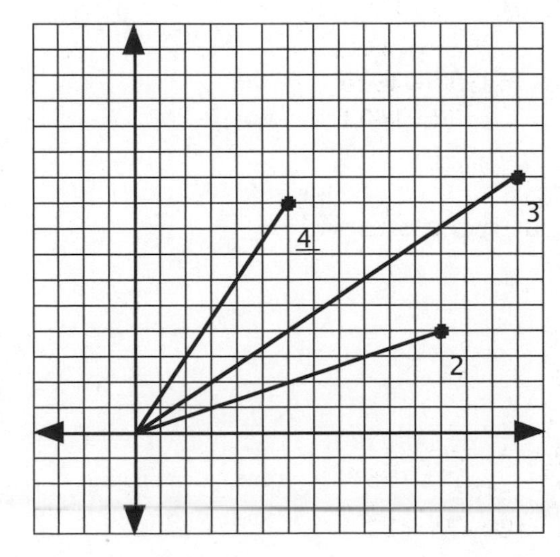

Honors Lesson 9H

1. X is greater than 11 and less than 20,

   so $C = 2.75X$
   $C = 2.75(12)$
   $C = \$33.00$

2. X is greater than 0 and less than 10, so

   so $C = 3X$
   $C = 3(5)$
   $C = \$15.00$

3. Cost of 10 reams $= \$3.00 \times 10 = \$30.00$
   Cost of 20 reams $= \$2.50 \times 20 = \$50.00$
   This shows that we can use the lowest price category.
   $\$50.00 \div \$2.50 = 20$ reams

4. Let C = service Charge, and B = Balance

   $C = .008B$ if $B > 1000$
   $C = .012B$ if $1000 \geq B \geq 50$
   $C = 1$ if $50 > B > 0$
   $C = 0$ if $B = 0$

5. $C = .012B$
   $C = .012(\$600)$
   $C = \$7.20$

6. The lowest possible charge if the balance is over \$1000 is $\$1000.01 \times .008 = \$8.00$ (rounded). If the balance were under \$50, the charge would have been \$1.00, so it must have been between \$50.00 and \$1000.00.

   $\$7.00 = .012 \times B$
   $\$7.00 \div .012 = B$
   $B = \$583.33$ (rounded)

7. Let P = Pay and H = Hours.

   $P = 10H$ for all hours under 40.
   $P = 15H$ for all hours over 40.
   $P = 20H$ for holiday hours.

8. $P = 10H$
   $P = 10(40)$
   $P = \$400$

9. $40(10) + 5(15) + 6(20) =$
   $400 + 75 + 120 = \$595$

10. $\$580 - \$400 = \$180$ in overtime pay.
    $\$180 \div \$15 = 12$ hours overtime;
    $12 + 40$ regular $= 52$ hours worked.

Honors Lesson 10H

1.

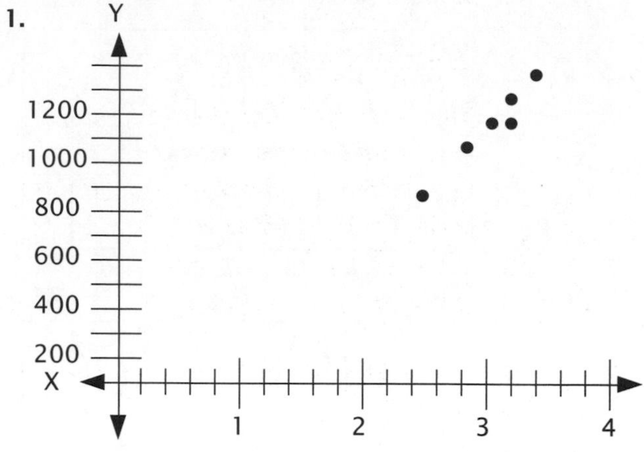

2.

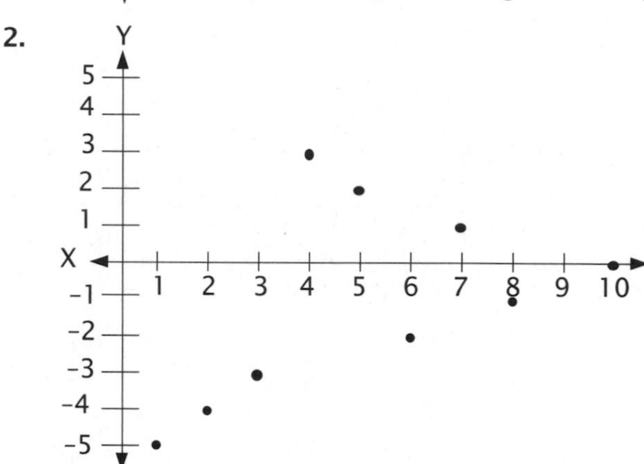

Honors Lesson 11H

1.

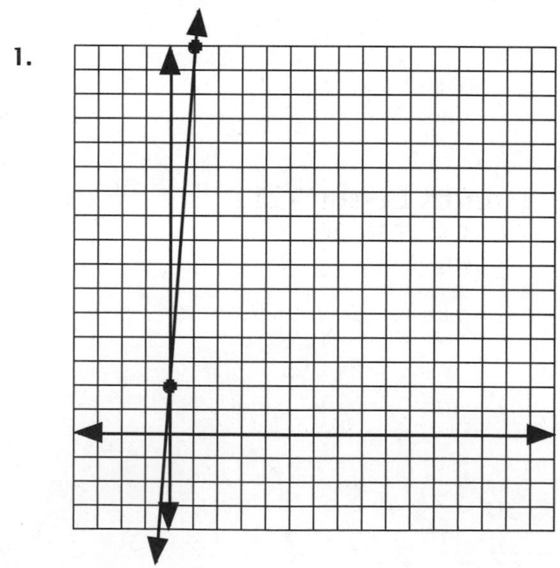

2. slope $= \dfrac{Y_2 - Y_1}{X_2 - X_1}$

$\dfrac{400-50}{5-0} = \dfrac{350}{5} = 70$

This is not the slope that you will get from a quick observation of the graph. Remember that you used two different scales for X- and Y-axes.

3. $Y = mX + b$
$50 = 70(0) + b$
$50 = b$
$Y = 70X + 50$

4. $G = 70T + 50$

5. $V = 70(30) + 50$
$V = 2{,}100 + 50$
$V = \$2{,}150$

6. start with points $(10, 50)$ and $(15, 80)$

slope $= \dfrac{80-50}{15-10} = \dfrac{30}{5} = 6$
$Y = mX + b$
$(50) = 6(10) + b$
$50 = 60 + b$
$-10 = b$
$Y = 6X - 10$
$G = 6T - 10$

7. $G = 6(12) - 10$
$G = 72 - 10$
$G = 62$

8. $90 = 6T - 10$
$100 = 6T$
$T = 16.67$ (rounded)

Honors Lesson 12H

1. $C = .15M + 20$
2. $C = 0M + 30$ or $C = 30$
3. plan 1: $C = .15(80) + 40(2 \text{ days})$
$C = 12 + 40$
$C = \$52$
plan 2: $C = \$60(2 \text{ days})$
Plan 1 is cheaper.
4. $\dfrac{5}{100} = \dfrac{1}{20}$

5. $\dfrac{Y_2 - Y_1}{X_2 - X_1}$
$\dfrac{2-0}{X-0} = \dfrac{1}{20}$
$2(20) = 1(X)$
$X = 40$ ft

Honors Lesson 13H

1. $X - Y = 2$
$-Y = -X + 2$
$Y = X - 2$
(Try a sample set of points to see which side of the line to shade.)

2. $X + Y = 6$
$Y = -X + 6$

3. yes

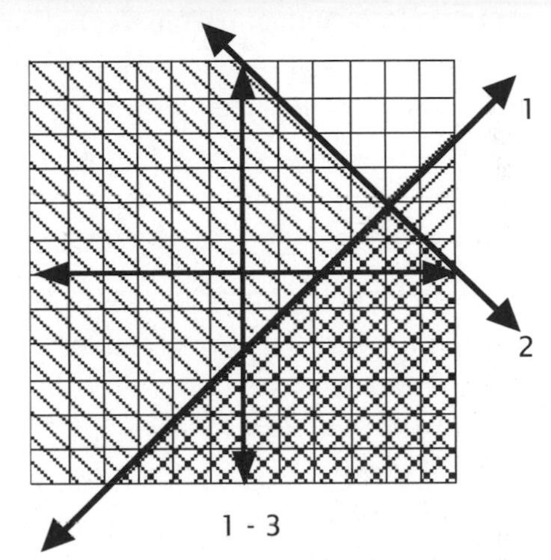

1 - 3

4. $2X - Y = 2$
$-Y = -2X + 2$
$Y = 2X - 2$
Original problem was inequality only, so line is dotted.

5. $3X + Y = 6$
$Y = -3X + 6$

6. no

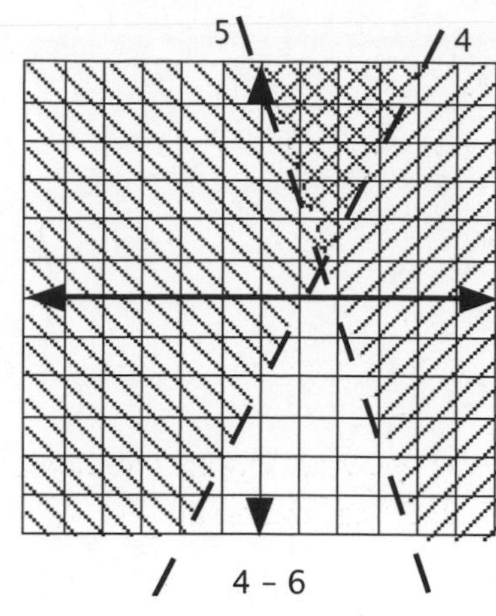

7. $A \ge 2B$
$200A + 500B \le 10{,}000$
$A \ge 5$
$B \ge 2$

8. See graph; only the final answer has been shaded here. The shaded side of each line is indicated by the small arrows.

9. 20 A's and 5 B's is one possible solution. Answers will vary.

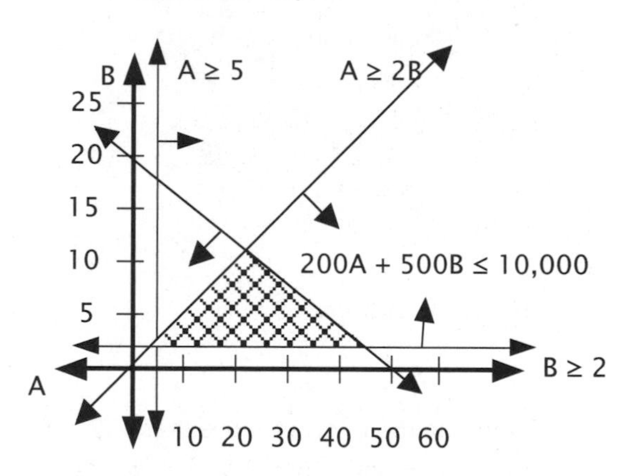

Honors Lesson 14H

1. 8 ft = 96 in $\frac{1}{12}$ can be written as $\frac{8}{96}$.
$\frac{9}{96} > \frac{8}{96}$, so no.

2. $\frac{10}{X} = \frac{1}{12}$
$X(1) = (10)(12)$
$X = 120$ in or 10 ft

3. $L = W + 3$
$2L + 2W = 4.5W$
$2(W+3) + 2W = 4.5W$
$2W + 6 + 2W = 4.5W$
$6 = .5W$
$12 = W$
$L = (12) + 3$
$L = 15$

4. $B = 12T + 42$
$B = 12(4) + 42$
$B = 48 + 42$
$B = 90$
\$90,000

5. $B = 12T + 42$
$B = 12(5) + 42$
$B = 60 + 42$
$B = 102$
\$102,000

6. $B = 12T + 42$
$126 = 12T + 42$
$84 = 12T$
$2000 + 7 = 2007$

7.

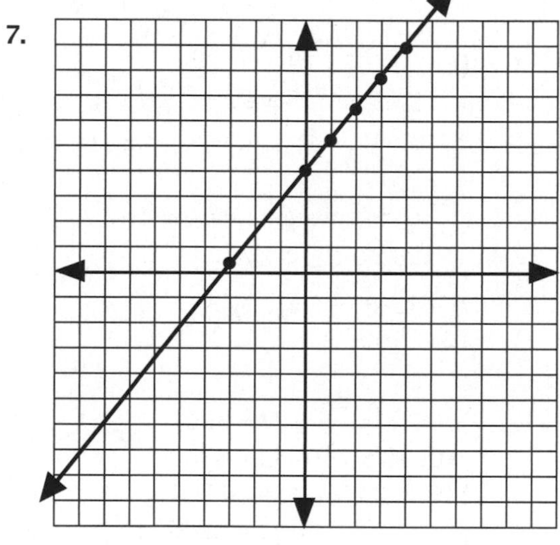

8. Any answer close to \$5,000 would be ok.

Honors Lesson 15H

1. $T + S = 62.48$
   $2T + S = 87.98$

   $S = 62.48 - T$
   $2T + (62.48 - T) = 87.98$
   $T + 62.48 = 87.98$
   $T = 25.50$

2. $10C + 5P = 85$
   $20C + 8P = 158$

   $5P = 85 - 10C$
   $P = 17 - 2C$
   $20C + 8(17 - 2C) = 158$
   $20C + 136 - 16C = 158$
   $4C = 22$
   $C = \$5.50$ per bag

3. $N + P = 2$
   $N = 2 - P$
   $2(2 - P) + 3P = 4.75$
   $4 - 2P + 3P = 4.75$
   $4 + P = 4.75$
   $P = \$.75$ per pen

4. $180 - 3W = 150 - 2W$
   $30 = W$ (weeks)

5. L = number of people working two years or less
   M = number of people working more than two years

   $L + M = 700$
   $10L + 15M = 8500$

   $M = 700 - L$
   $10L + 15(700 - L) = 8500$
   $10L + 10500 - 15L = 8500$
   $-5L = -2000$
   $L = 400$

6. $3X = \left(\frac{1}{4}\right)2Y$
   $12X = 2Y$
   $6X = Y$

   4, 24
   5, 30
   6, 36

   (Answers will vary. The second number will be six times the first.)

Honors Lesson 16H

1. $C = .12N + \$2000$
2. $R = .62N$
3. $C = R$
   $.12N + 2000 = .62N$
   $2000 = .62N - .12N$
   $2000 = .50N$
   $4{,}000 = N$
4. Plan 1: 19.95 a month for any number of hours

   Plan 2: $4.95 + 2 \times 2 = \$8.95$
   $4.95 + 2 \times 6 = \$16.95$
   $4.95 + 2 \times 10 = \$24.95$
   $4.95 + 2 \times 14 = \$32.95$
5. $C = \$19.95$; $C = 4.95 + 2H$
6. $19.95 = 4.95 + 2H$
   $19.95 - 4.95 = 2H$
   $15 = 2H$
   $7.5 = H$

   If you use the Internet more than 7.5 hours per month, then Plan 1 is better.

Honors Lesson 17H

1. Hometown: $F = 10 + .10(C - 50)$
   AmeriBank: $F = 8 + .12(C - 50)$
2. $10 + .10(C - 50) = 8 + .12(C - 50)$
   $10 + .10C - 5 = 8 + .12C - 6$
   $5 + .10C = 2 + .12C$
   $3 = .02C$
   $150 = C$

3. Hometown: $F = 10 + .10(60)$
$F = \$16$

AmeriBank: $F = 8 + .12(60)$
$F = \$15.20$

AmeriBank's program would be cheaper.

4. $C = 30{,}000 + 75S$

5. $C = 30{,}000 + 75(2000)$
$C = 30{,}000 + 150{,}000$
$C = 180{,}000$

6. $150{,}000 = 30{,}000 + 75S$
$120{,}000 = 75S$
$1{,}600 = S$

Honors Lesson 18H

1. D = number of dimes
$3D$ = number of nickels
$3D + 4$ = number of quarters

$D + 3D + 3D + 4 = 18$
$7D + 4 = 18$
$7D = 14$
$D = 2$ dimes

$3(2) = 6$ nickels
$(6) + 4 = 10$ quarters

2. C = number of children
$2C$ = number of adults
$4C$ = number of seniors

$4(C) + 8(2C) + 5(4C) = 1120$
$4C + 16C + 20C = 1120$
$40C = 1120$
$C = 28$ children

$2(28) = 56$ adults
$4(28) = 112$ seniors
$28 + 56 + 112 = 196$ people

3. number of business rooms = B
number of coupons rooms = $B + 8$
number of standard rooms = $(B + 8)10 = 10B + 80$
number of senior rooms = $(10B + 80) - 10 = 10B + 70$

$45(B) + 40(B + 8) + 50(10B + 80) + 35(10B + 70) = 8640$
$45B + 40B + 320 + 500B + 4000 + 350B + 2450 = 8640$
$935B + 6770 = 8640$
$935B = 1870$
$B = 2$ business

$(2) + 8 = 10$ coupon
$10(2) + 80 = 100$ standard
$10(2) + 70 = 90$ senior
$2 + 10 + 100 + 90 = 202$ rooms occupied
$250 - 202 = 48$ empty rooms

4. T = number of 20¢ stamps
$T + 5$ = number of 37¢ stamps
$10(T + 5)$ = number of 1¢ stamps

$.20T + .37(T + 5) + .01(10(T + 5)) = 5.70$
$20T + 37(T + 5) + 10(T + 5) = 570$
$20T + 37T + 185 + 10T + 50 = 570$
$67T + 235 = 570$
$67T = 335$

T = five 20¢ stamps
$T + 5$ = ten 37¢ stamps
$10(T + 5) = 100$ one-cent stamps

5. W = number of women
$W + 1$ = number of men
C = number of children

$8(W) + 10(W + 1) + 5C = 112$
$8W + 10W + 10 + 5C = 112$
$18W + 5C = 102$

$W + W + 1 + C = 15$
$2W + 1 + C = 15$
$2W + C = 14$
$C = 14 - 2W$

Substitute $14 - 2W$ for C in 1st equation:

$18W + 5(14 - 2W) = 102$
$18W + 70 - 10W = 102$
$8W = 32$
$W = 4$ women

$W + 1 = 5$ men
$15 - (4 + 5) = 15 - 9 = 6$ children

6. Let X = 1st digit, and Y = 2nd

$$X + Y = 10$$
$$10Y + X = 36 + 10X + Y$$
$$9Y = 36 + 9X$$
$$-9X + 9Y = 36$$
$$+(9X + 9Y = 90) \quad \text{1st eq. multiplied by 9}$$
$$18Y = 126$$
$$Y = 7(\text{second digit})$$
$$10 - 7 = 3(\text{first digit})$$
number is 37
$$73 - 37 = 36$$

Honors Lesson 19H

1. t = hours; b = bacteria in thousands

t	0	3	6	9	12	15	18	21	24
b	1	2	4	8	16	32	64	128	256

2.

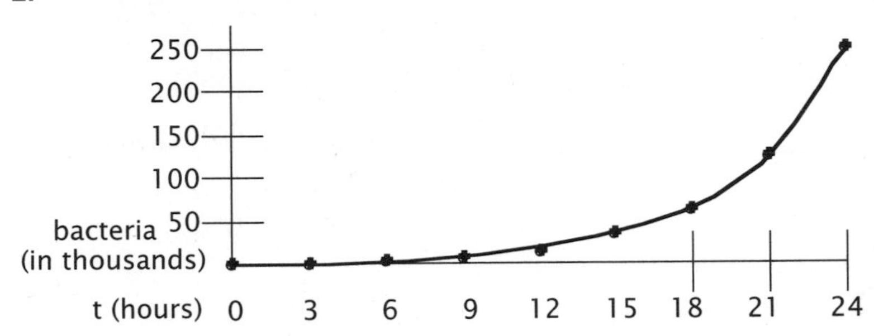

3. t = hours; b = bacteria in thousands

t	0	1	2	3	4	5	6	7	8	9	10	11	12
b	1	2	4	8	16	32	64	128	256	512	1,024	2,048	4,096

4.

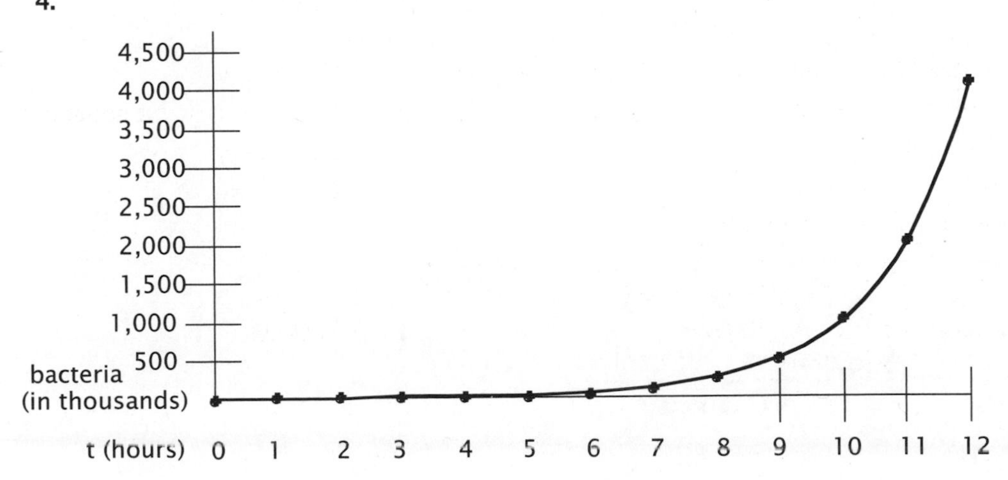

5. The rate of increase increases over time.

Honors Lesson 20H

1. x = # of months; m = mass in grams

x	0	1	2	3	4
m	200	100	50	25	12.5

2. 200 g
3. 1 month
4. 2 months
5. 12.5 g

6.

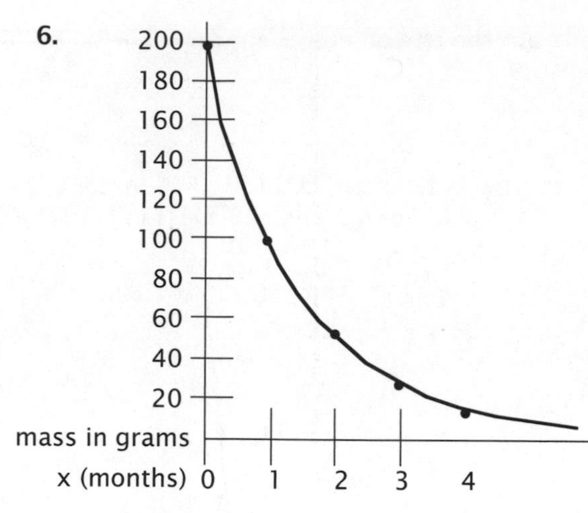

7. $m = 200(.5)^x$

$m = 200(.5)^6$

$m = 200(.0156) = 3.125 \text{ g}$

Honors Lesson 21H

1. done

2. $B = (A)2^{\frac{x}{D}}$

$B = 10\left(2^{\frac{30}{5}}\right)$

$B = \left(2^6\right)$

$B = 10(64) = 640$

3. $B = 10\left(2^{\frac{60}{5}}\right)$

$B = 10\left(2^{12}\right)$

$B = 10(4096) = 40{,}960$

4.

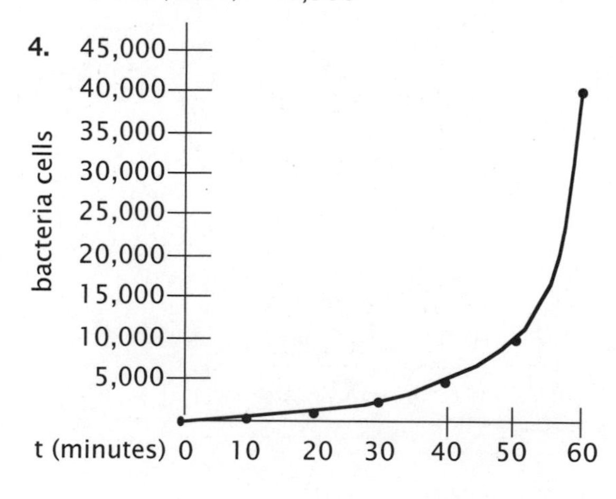

Honors Lesson 22H

1. never true: $3^{-2} = \frac{1}{9}$
2. sometimes true: $\left(\frac{1}{3}\right)^{-2} = 9$
3. never true: $\frac{1}{7} \neq \frac{1}{2} + \frac{1}{5}$
4. never true: $\frac{2}{X^0} = \frac{2}{1} = 2$
5. always true: $1 - 1 = 0$
6. never true: $8^{-1} = \frac{1}{8}$
7. always true: a number multiplied by its reciprocal always equals 1.
8. always true: When raising a power to a power, you multiply exponents.
9.

X	Y
0	4
1	5
2	7
−1	3.5
−2	3.25

10.

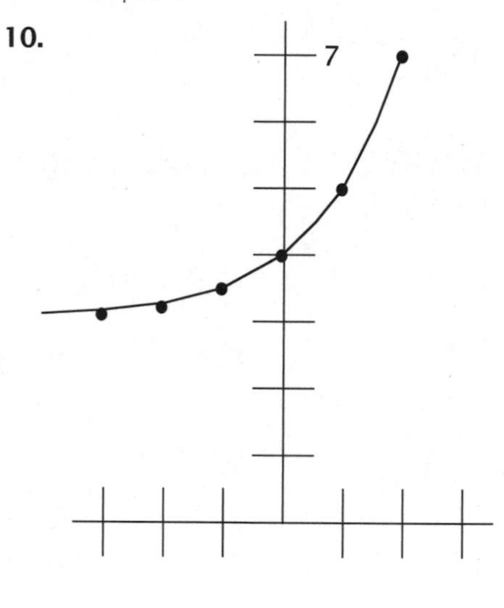

Honors Lesson 23H

1.

X	Y
0	2
1	3
2	5
3	9
–1	1.5
–2	1.25
–3	1.125

2.

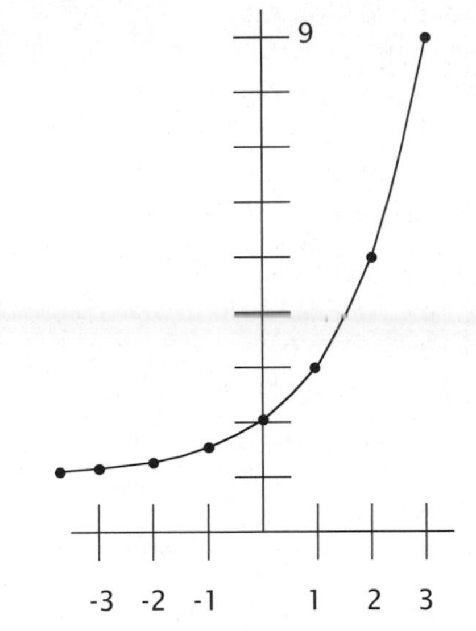

3. They get smaller.
4. They get larger.
5.

X	Y
0	1
1	3
2	9
3	27
4	81
–1	$\frac{1}{3}$
–2	$\frac{1}{9}$

6.

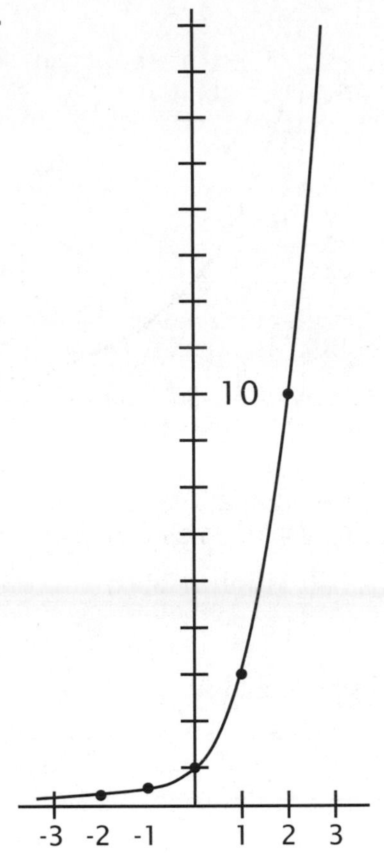

Honors Lesson 24H

1. $A = (2X)(12 + 2X)$

   $A = 4X^2 + 24X$

2. $4(10)^2 + 24(10) = 400 + 240 = 640 \text{ ft}^2$

3. $A = \frac{1}{2}bh$

   $A = \frac{1}{2}(Y+1)(2Y-1)$

   $A = \frac{1}{2}(2Y^2 - Y + 2Y - 1)$

   $A = \frac{1}{2}(2Y^2 + Y - 1)$

4.
$$\begin{array}{r} 2X^5 - 3X^4 + 7X \\ X^8 \quad\quad - 2X^4 + 3X \\ \hline X^8 + 2X^5 - 5X^4 + 10X \end{array}$$

5.
$$\begin{array}{r} 6X^4 \quad\quad\quad\quad +8 \\ \underline{2X^9-3X^5-7X^4+4X-2} \\ 2X^9-3X^5-1X^4+4X+6 \end{array}$$

6.
$$\begin{array}{r} 5X^3-7 \\ \underline{6X^4-5X^3+7} \\ 6X^4 \end{array}$$

7. $(2X^2+7)(3X^3+X)=$
$6X^5+2X^3+21X^3+7X=6X^5+23X^3+7X$

8. $(4X^5+3)(X^2-2)=4X^7-8X^5+3X^2-6$

9. $8X^4(7X^5-2X^3+3)=56X^9-16X^7+24X^4$

Honors Lesson 25H

1. $12,200 ÷ 800 = $15.25 profit per gun

2. $P=\$19.50(2{,}000)-3400$
$P=39{,}000-3400$
$P=\$35{,}600$

3. $35,600 ÷ 2,000 = $17.80 per gun
As long as fixed costs remain the same, selling more items means more profit per item

4. fixed costs = 1,500 rent + 1,600 equipment + (100 × 4) electricity = $3,500
$P=19.50-3{,}500$
$P=19.50(800)-3{,}500$
$P=\$12{,}100$

5. $C=3N+500$
$R=5N$

6. $P=5N-(3N+500)$
$P=2N-500$

7. $P=2(500)-500$
$P=1{,}000-500=\$500$

8. $P=2N-500$
$P=2(2{,}000)-500$
$P=4{,}000-500=\$3{,}500$

9. $0=2N-500$
$500=2N$
$250=N$
250 boxes of candy must be sold in order to break even.

Honors Lesson 26H

1. $P=100N-(65N+18{,}000)$
$P=35N-18{,}000$

2. $P=35(1{,}000)-18{,}000$
$P=35{,}000-18{,}000$
$P=\$17{,}000$

3. $17,000 ÷ 1,000 = $17 per item

4. $P=35(2{,}000)-18{,}000$
$P=70{,}000-18{,}000$
$52,000

5. $52,000 ÷ 2,000 = $26 per item

6. $0=35N-18{,}000$
$18{,}000=35N$
514.29 (rounded)
515 items is break – even point

7. $P=50N-(30N+10{,}000)$
$P=20N-10{,}000$
$P=20(1{,}000)-10{,}000$
$P=20{,}000-10{,}000=\$10{,}000$
$10,000 ÷ 1,000 = $10 per case

8. $P=50N-(30N+10{,}000)$
$P=20N-10{,}000$
$P=20(2{,}000)-10{,}000$
$P=40{,}000-10{,}000=\$30{,}000$
$30,000 ÷ 2,000 = $15 per case
It is more.

9. $R=50N$
$C=30N+10{,}000$
R will equal C when: $50N=30N+10{,}000$
$20N=10{,}000$
$N=500$ cases

Honors Lesson 27H

1. $2(2A^2+1)$
2. not factorable
3. not factorable
4. $8B(B^3+4)$
5. not factorable
6. $(X+3)(2X+1)$

7.

X	Y
0	1
1	.33
2	.11
–1	3
–2	9

8.

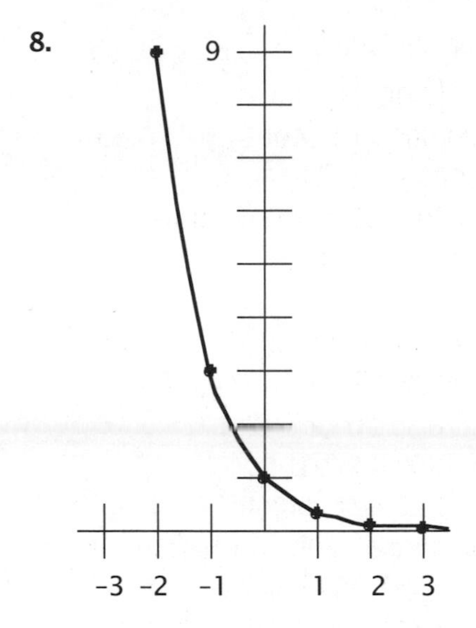

9. They get smaller. They get larger.

10.

X	Y
0	3
1	2.5
2	2.25
3	2.125
–1	4
–2	6
–3	10

11.

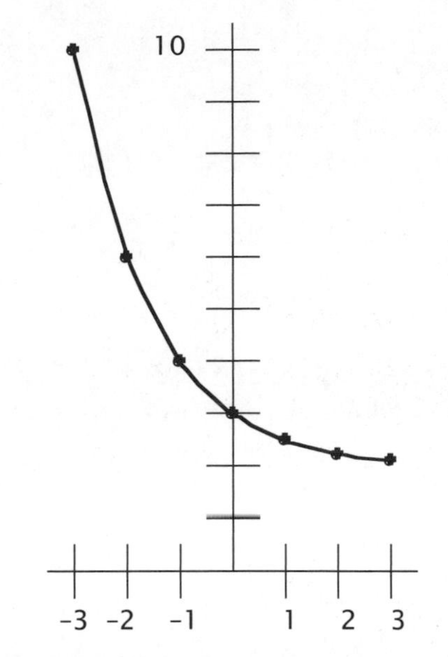

12. They get smaller. They get larger.

Honors Lesson 28H

1. $2X^3 - X^2 + 4X - 2 =$
   $X^2(2X - 1) + 2(2X - 1) = (2X - 1)(X^2 + 2)$
2. $3A^3 - 6A^2 - A + 2$
   $-3A^2(-A + 2) + 1(-A + 2) = (-A + 2)(-3A^2 + 1)$
3. $2B^3 + 3B^2 + 2B + 3 =$
   $B^2(2B + 3) + (2B + 3) = (B^2 + 1)(2B + 3)$
4. $2X^4 + 4X^3 - 3X - 6 =$
   $2X^3(X + 2) - 3(X + 2) = (2X^3 - 3)(X + 2)$
5. $4Y^2 + 6Y - 2Y - 3 =$
   $4Y^2 - 2Y + 6Y - 3 =$
   $2Y(2Y - 1) + 3(2Y - 1) =$
   $(2Y + 3)(2Y - 1)$
6. $6P^4 - 6P^3 + 14P^2 - 14P =$
   $6P^3(P - 1) + 14P(P - 1) =$
   $(6P^3 + 14P)(P - 1) = 2P(3P^2 + 7)(P - 1)$

7. $\frac{X^3+X^2-2X}{X^2+2X-3} \cdot \frac{X+3}{X^2+2X} =$
$\frac{X(X^2+X-2)}{(X+3)(X-1)} \cdot \frac{X+3}{X(X+2)} =$
$\frac{(X^2+X-2)}{(X-1)(X+2)} =$
$\frac{(X+2)(X-1)}{(X-1)(X+2)} = \frac{1}{1} = 1$

8. $\frac{X+5}{X^2-3X+2} \div \frac{X^2+3X}{X^3-X^2-6X} =$
$\frac{X+5}{(X+1)(X+2)} \cdot \frac{X(X^2-X-6)}{X(X-3)} =$
$\frac{X+5}{(X+1)(X+2)} \cdot \frac{(X+2)(X-3)}{X-3} =$
$\frac{(X+5)(X-3)}{(X+1)(X-3)} = \frac{(X+5)}{(X+1)}$

Honors Lesson 29H

1. $d = vt + 16t^2$
$96 = 16t + 16t^2$
$12 = 2t + 2t^2$
$2t^2 + 2t - 12 = 0$
$(2t-4)(t+3) = 0$
$t = 2, t = -3$
-3 makes no sense, so $t = 2$ seconds
2. $77 + 3 = 80$, so the rock was dropped from 80 ft above the water
$d = vt + 16t^2$
$80 = 8t + 16t^2$
$10 = t + 2t^2$
$2t^2 + t - 10 = 0$
$(2t+5)(t-2) = 0$
$t = -2.5, 2$
-2.5 makes no sense, so $t = 2$ seconds
3. $2000 - 80 = 1920$
so distance was 1,920 ft
$d = vt^2 + 16t^2$
$1920 = 32t + 16t^2$
$120 = 2t + t^2$
$t^2 + 2t - 120 = 0$
$(t+12)(t-10) = 0$
$t = -12, 10$
-12 makes no sense, so $t = 10$ seconds
4. $d = vt + 16t^2$
$d = 10(4) + 16(4)^2$
$d = 40 + 16(16)$
$d = 40 + 256$
$d = 296$ ft

Honors Lesson 30H

You may also use the unit multiplier method to get your answer. Either method is fine.

1. $300 \times 18 = 5{,}400$ in
$5{,}400 \div 12 = 450$ ft
This can also be figured by writing 18 in as 1.5 ft and multiplying.
2. 50×1.5 ft $= 75$ ft
30×1.5 ft $= 45$ ft
length from #1 = 450 ft
$450 \times 75 \times 45 = 1{,}518{,}750$ ft^3
3. 1 pace = 5 Roman feet
$5 \times 1000 = 5000$ Roman feet in Roman mile
It is shorter than a modern mile.
4. $5{,}280 \times 5{,}280 = 27{,}878{,}400$
28,000,000 ft^2 (rounded)
one acre = 43,560 ft^2 (from text)
$28{,}000{,}000 \div 43{,}560 = 643$ mornings (rounded)
5. a yard
6. 18×2 in $= 36$ in
This is the number of inches short his measure is.
36 in = 3 ft
$18 - 3 = 15$ ft = actual length of room
7. Answers will vary.

Honors Lesson 31H

You may also have used the unit multipler method to get your answer. Either method is fine unless the directions specified using unit multipliers.

1. $20{,}000 \times 3 = 60{,}000$ mi
   $60{,}000 \times 8 = 480{,}000$ furlongs
2. $1{,}920 \div 4 = 480$ chains
   $480 \div 10 = 48$ furlongs
   $48 \div 8 = 6$ mi
3. 1 furlong = 10 chains
   $10 \times 4 = 40$ rods
4. $1 \text{ mi} \times \frac{8 \text{ furlongs}}{1 \text{ mi}} \times \frac{10 \text{ chains}}{1 \text{ furlong}} \times \frac{22 \text{ yd}}{1 \text{ chain}} \times \frac{3 \text{ ft}}{1 \text{ yd}}$
   $= 5{,}280$ feet
5. 14 pounds in a stone
   $14 \times 2 = 28$ pounds in a quarter
   $28 \times 4 = 112$ pounds in a hundredweight
   $112 \times 20 = 2{,}240$ pounds in a ton
   heavier than an American ton
6. $6{,}400$ lb $\div 8 = 800$ gallons
   800 gallons $\div 2 = 400$ pecks
   400 pecks $\div 4 = 100$ bushels
7. $6{,}400 \div 2{,}240 = 2.86$ tons(rounded)
8. $\frac{1}{2}$ bushel = 2 pecks
   2 pecks $\times 2 = 4$ gal
   4 gal $\times 8 = 32$ lb

Honors Lesson 32H

1. $2(1.008) + 16 = 18.02$ amu(rounded)
2. $23 + 1.008 + 12 + 3(16) =$
   84 amu(rounded)
3. $1.008 + 35.5 = 36.5$ amu(rounded)
4. $12 + 2(16) = 44$ amu
5. hydrogen chloride:
   $36.5 \times 1.67 \times 10^{-24} = 6.10 \times 10^{-23}$ g
   carbon dioxide:
   $44 \times 1.67 \times 10^{-24} = 7.35 \times 10^{-23}$ g
6. $4(12) + 10(1.008) + 16 = 74.08$ amu
   $74.08 \times 1.67 \times 10^{-24} = 1.24 \times 10^{-22}$ g

Honors Lesson 33H

1. $225 \div 16 = 14$ remainder 1
   E1
2. $888 \div 256 = 3$ remainder 120
   $120 \div 16 = 7$ remainder 8
   378
3. $5 \times 256 = 1{,}280$
4. $7 \times 256 + 5 \times 16 = 1{,}872$
5. A little bit of red, a little bit of green, and a lot of blue: since the amounts of red and green are insignificant, the result is blue.
6. FFFFFF
   Remember that we are mixing light, not paint, so white is all colors mixed together.
7. blue – green

Honors Lesson 34H

Each step was rounded using significant digits.

1. $P^2 = 11.8^3$
   $P^2 = 1640$
   $P = \sqrt{1640} = 40.5$ A.U.
   $40.5 \times 365 = 14{,}782.5 =$
   14,800 days using sig. digits
2. Mercury.
   A planet with an AU less than 1 is closer to the sun than the earth is.
3. $P^2 = 1.88^3$
   $P^2 = 6.64$
   $P = \sqrt{6.64} = 2.58$ A.U.
   $2.58 \times 365 = 942$ days using sig. digits
4. $P = 100 \div 365 \approx .274$
   $.274^2 = A^3$
   $\sqrt[3]{.075} = A$
   $A = .422$ A.U.
   $.422 \times 9.3 \times 10^7 = 3.925 \times 10^7$ mi
   3.9×10^7 miles using sig. digits

Honors Lesson 35H

1. $2L+3W=1{,}200$

$$2L=1{,}200-3W$$

$$L=\frac{1{,}200-3W}{2}$$

$$A=\left(\frac{1{,}200-3W}{2}\right)(W)$$

$$=\frac{-3W^2+1{,}200W}{2}$$

$$=-\frac{3}{2}W^2+600W$$

$$h=\frac{-600}{(2)\left(-\frac{3}{2}\right)}=\frac{-600}{-3}=200$$

$$k=\frac{-3(200)^2+1{,}200(200)}{2}$$

$$=\frac{-120{,}000+240{,}000}{2}$$

$$=\frac{120{,}000}{2}=60{,}000\text{ ft}^2$$

$60{,}000 \div 200 = 300$ ft

dimensions with maximum area: 200 ft $\times$ 300 ft

2. $2L+2W=1{,}200$

$$2L=1{,}200-2W$$

$$L=600-W$$

$$A=(600-W)(W)=-W^2+600W$$

$$h=\frac{-600}{2(-1)}=300$$

$$k=-(300)^2+600(300)$$

$$=-90{,}000+180{,}000=90{,}000\text{ ft}^2$$

$90{,}000 \div 300 = 300$ ft

dimensions with maximum area: 300 ft $\times$ 300 ft

Test Solutions

Test 1

1. A: addition
2. A: addition
3. C: both associative and commutative properties
4. B: -8
5. D: -2
6. D: 20
7. A: -16
8. B: 3
9. A: -5
10. E: $2A+5C$
11. E: $5X+Y$
12. B: $10A-2B$
13. A: $2\times2\times2\times2\times2$
14. C: $2\times2\times5\times5$
15. D: $36=\underline{2}\times2\times\underline{3}\times3$
    $42=\underline{2}\times\underline{3}\times7$
    $2\times3=6$

Test 2

1. A: $3+5+2^2=3+5+4=12$
2. E: $6^2+4\div4=36+1=37$
3. A: $10^2\times(1+2)-1=100\times(3)-1=300-1=299$
4. B: $3\times1^2-1^2=3\times1-1=3-1=2$
5. D: $16\div2-1\times3=8-3=5$
6. E: $A=5+25\div5-2^2=5+5-4=6$
    $B=(5+25)\div5-2^2=30\div5-4=6-4=2$
    $C=(5+25)\div(5-2^2)=30\div(5-4)=30\div1=30$
    $D=5+25\div5+2^2=5+5+4=14$
    $E=5+25\times5-2^2=5+125-4=126$
7. C: $A=-3^2\div3+6=-9\div3+6=-3+6=3$
    $B=(3)^2\div3+5=9\div3+5=3+5=8$
    $C=-(3)^2\div3+5=-9\div3+5=-3+5=2$
    $D=(-3)^2\div3+4=9\div3+4=3+4=7$
    $E=3\div3+5=1+5=6$
8. B: $4A-5B+3C$
9. A: $|0-4|=|-4|=4$
10. E: $|6-10-2|=|-6|=6$
11. C: $\left|(2-8)^2\right|+\left|2-8^2\right|=\left|(-6)^2\right|+|2-64|$
    $=|36|+|-62|$
    $=36+62=98$
12. A: $A=|5-6\times5|=|5-30|=|-25|=25$
    $B=|(5-6)\times5|=|-1\times5|=|-5|=5$
    $C=|(5\times5)-6|=|25-6|=|19|=19$
    $D=|6-6\times5|=|6-30|=|-24|=24$
    $E=|5\times(5-6)|=|5\times-1|=|-5|=5$
13. E: $8=2\times2\times2$
    $10=2\times5$
    $LCM=2\times2\times2\times5=40$
14. A
15. D

Test 3

1. B: $-3X+2+5X-3=8+9$
    $2X-1=17$
    $2X=18$
    $X=9$
2. A: $3D-3+8+D-D=9+9-1$
    $3D+5=17$
    $3D=12$
    $D=4$
3. C: $-6+2+3B+4=2(4+1)-1$
    $3B=9$
    $B=3$
4. B: $-2(5)+2+5(5)+8=$
    $-10+2+25+8=25$
5. D: $((3)+7)\times\left((3)^2-10\right)=$
    $(10)(9-10)=$
    $(10)(-1)=-10$
6. B: $5Q-9-6=-1\times25$
    $5Q-15=-25$
    $5Q=-10$
    $Q=\frac{-10}{5}$
    $Q=-2$

7. E: $-3+Y+Y-6+2=6+7$
$$2Y-7=13$$
$$2Y=20$$
$$Y=\frac{20}{2}$$
$$Y=10$$

8. E: A. $X-3=9;\ X=9+3=12$
B. $X+3=9;\ X=9-3=6$
C. $3X=9;\ X=3$
D. $X+1=9;\ X=9-1=8$
E. $X-1=12;\ X=12+1=13$

9. E: A. $R+2R=15;\ 3R=15;\ R=5$
B. $2R+3=15;\ 2R=12;\ R=6$
C. $R+2R=18;\ 3R=18;\ R=6$
D. $R+5R=15;\ 6R=15;\ R=\frac{15}{6}=2\frac{1}{2}$
E. $R+5R=6;\ 6R=6;\ R=1$

10. A: I. $3Q-4=20;\ 3Q=24;\ Q=8$
II. $4Q-3=17;\ 4Q=20;\ Q=5$
III. $4Q+3=23;\ 4Q=20;\ Q=5$
IV. $4Q-3Q=21;\ Q=21$

11. A: $5+P-3=3(6)+5P$
$$P+2=18+5P$$
$$2-18=5P-P$$
$$-16=4P$$
$$-4=P$$

12. D: $^{3}\cancel{(12)}\frac{3}{\cancel{4}}+^{6}\cancel{(12)}\frac{1}{\cancel{2}}=^{4}\cancel{(12)}\frac{2}{\cancel{3}}X$
$$9+6=8X$$
$$15=8X$$
$$X=\frac{15}{8}=1\frac{7}{8}$$

13. B: $^{3}\cancel{(15)}\frac{3}{\cancel{5}}Y-^{5}\cancel{(15)}\frac{1}{\cancel{3}}=^{3}\cancel{(15)}\frac{1}{\cancel{5}}$
$$9Y-5=3$$
$$9Y=8$$
$$Y=\frac{8}{9}$$

14. D: $100(.09X)-100(1.8)=100(2.25)$
$$9X-180=225$$
$$9X=405$$
$$X=\frac{405}{9}=45$$

15. B: $10(.6A)+10(15)=10(7.2)$
$$6A+150=72$$
$$6A=72-150$$
$$6A=-78$$
$$A=\frac{-78}{6}=-13$$

Test 4

1. B: $14=2\times7$
$16=2\times2\times2\times2$
$28=2\times2\times7$
$GCF=2$

2. B: $14=2\times7$
$16=2\times2\times2\times2$
$24=2\times2\times2\times3$
$GCF=2$

3. A: $24=2\times2\times2\times3$
$36=2\times2\times3\times3$
$40=2\times2\times2\times5$
$GCF=2\times2=4$

4. E: $26=2\times13$
$52=2\times2\times13$
$65=5\times13$
$GCF=13$

5. E: $3(A+B+6)=$
$3A+3B+3(6)=3A+3B+18$

6. B: $6(X-2Y+3+Z)=6X-12Y+18+6Z$

7. D: $2(3T-5+4T+3)=$
$6T-10+8T+6=14T-4$

8. A: $A(B+4Q+1)=AB+4AQ+A$

9. B: $10B=2\times5\times B$ GCF = 5
$15B=3\times5\times B$
$40=2\times2\times5$

10. E: $36X=2\times2\times3\times3\times X$
$12Y=2\times2\times3\times Y$
$24Z=2\times2\times2\times3\times Z$
$GCF=2\times2\times3=12$

11. C: $60A=2\times2\times3\times5\times A$
$30D=2\times3\times5\times D$
$90=2\times3\times3\times5$
$GCF=2\times3\times5=30$

12. C: $18A+24B=30$ GCF is 6
$6(3A+4B)=6(5)$

13. B: $15P-25R=35T$
$5(3P-5R)=5(7T)$

14. E: $4G+16H-8J=32$
$4(G+4H-2J)=4(8)$
$G+4H-2J=8$

15. B: $9X+27Y=3Z$
$3(3X+9Y)=3(Z)$
$3X+9Y=Z$

Test 5

1. E: $(1, 1)$
2. B: $(-1, -2)$
3. C: $(-3, 1)$
4. A: $(3, -4)$
5. C: $(3, 0)$
6. D: quadrant IV
7. A: quadrant I
8. E: the origin
9. C: Descartes
10. D: 0
11. C: quadrant III
12. D: 0
13. C: algebra and geometry
14. A: They form a straight line.
15. D: They cannot be connected with a straight line.

Test 6

1. A: $G=D+3$
2. C: $S=2W+5$
3. B: $G=2W+2$

4. C: $C=2D+5$
$C=2(6)+5$
$C=12+5$
$C=17$

5. E: $M=10D+8$
$M=10(12)+8$ (12 days worked)
$M=120+8$
$M=\$128$

6. E: $Y=4X-1$
$Y=4(3)-1$
$Y=12-1$
$Y=11$

7. B: $A=6B+4$
$A=6(0)+4$
$A=0+4$
$A=4$

8. E: $R=T-5$
$R=(2)-5$
$R=-3$

9. A: Use trial and error to check all answers. Answer A is the one that yields a true statement:

$Y=3X+1$
$(13)=3(4)+1$
$13=12+1$
$13=13$

10. D: Use the same process as in #9.

$Y=X-4$
$(-6)=(-2)-4$
$-6=-6$

11. B:

$(0, 1)$:
$Y=3X+1$
$(1)=3(0)+1$
$1=1$

$(-1, -2)$:
$Y=3X+1$
$(-2)=3(-1)+1$
$-2=-3+1$
$-2=-2$

Both points are tested because it takes 2 points to define a line.

12. E: The Y-axis includes all points where $X=0$.

13. C: Any 2 points from line S can be chosen. We show $(0, 2)$ and $(-2, 0)$ here:

$(0, 2)$:
$Y=X+2$
$(2)=(0)+2$
$2=2$

$(-2, 0)$:
$Y=X+2$
$(0)=(-2)+2$
$0=0$

The student can try just one point on the possible answers given until one works, then the other point can be tested.

14. B: Use the same process as #11:

$(0, 4)$: $Y = 2X + 4$; $(4) = 2(0) + 4$; $4 = 4$

$(-1, 2)$: $Y = 2X + 4$; $(2) = 2(-1) + 4$; $2 = -2 + 4$; $2 = 2$

15. E: The X-axis includes all points where Y = 0.

Test 7

1. C
2. D
3. A
4. C
5. D
6. C: slope is rise over run or $\frac{3}{4}$
7. A: $\frac{\text{rise}}{\text{run}} = \frac{2}{2} = 1$
8. D
9. B: $\frac{\text{rise}}{\text{run}} = \frac{2}{3}$
10. A: $\frac{\text{rise}}{\text{run}} = \frac{1}{-1} = -1$
11. E
12. C: $\frac{\text{rise}}{\text{run}} = \frac{3}{1} = 3$
13. E (No line crosses the Y-axis at 3.)
14. A (The line crosses the Y-axis at -2.)
15. C (The line crosses the Y-axis at -3.)

Test 8

1. A: $P = (-1)W + 3$
2. D: $R = -10W + 50$
3. E: $M = -5D + 4$
4. C: $M = 4D - 5$
5. D: $T = -3W - 4$
6. A: 4
7. C: -2
8. E: $\frac{1}{2}$
9. B: Since the y-intercept is the point where the X-coordinate is 0, $(0, 2)$ from the points given is the y-intercept. The Y-coordinate of that point is 2.
10. A: slope $= \frac{3}{1} = 3$

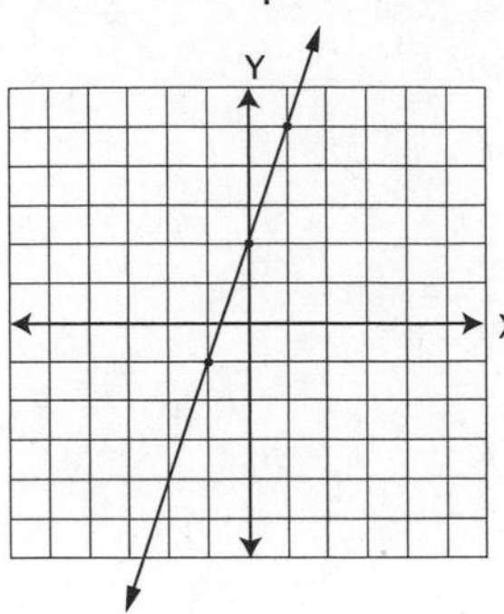

11. C: slope = 3; Y-intercept = 2

$Y = 3X + 2$

12. B: slope $= \frac{1}{1} = 1$
13. B: slope $= \frac{-2}{1} = -2$
14. A: slope $= \frac{0}{\text{any X}}$
15. A: slope $= \frac{-1}{1} = -1$

Test 9

1. A: slope
2. C: parallel

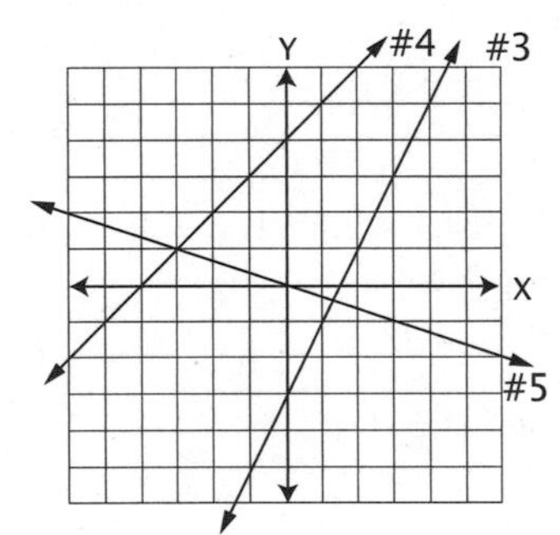

3. D: slope $= \frac{2}{1} = 2$
4. C: slope $= \frac{1}{1} = 1$
5. B: slope $= \frac{-1}{3}$
6. B: $2Y = 6X + 4$

$Y = 3X + 2$

(divide both sides by 2)

slope = 3

7. A: $3Y = 6X + 3$
$Y = 2X + 1$
(divide both sides by 3)
slope = 2
8. A: $3X + 2Y = 3$
9. E: 6
10. E: $Y = 2X + 6$
$-2X + Y = 6$
$2X - Y = -6$
11. D: 2
12. D: $Y = 2X - 4$
13. A: line F; intercept is 0, and slope is -1
14. E: -3
15. C: 0

Test 10

1. C: perpendicular
2. E: none of the above
correct answer would be negative reciprocal

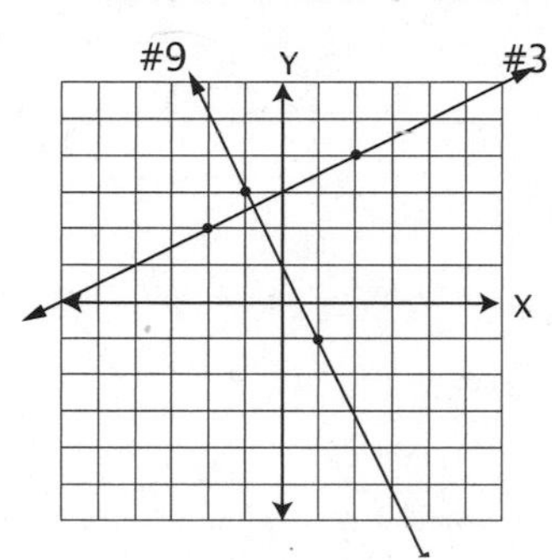

3. B: slope $= \frac{2}{4} = \frac{1}{2}$
4. D: 3
5. D: slope $= \frac{1}{2}$; y-intercept = 3
$Y = \frac{1}{2}X + 3$
6. A: $Y = \frac{1}{2}X + 3$
$-\frac{1}{2}X + Y = 3$
$-X + 2Y = 6$
7. C: -2 (negative reciprocal of $\frac{1}{2}$)
8. C: slope $= \frac{-4}{2} = -2$
9. E: Graph the points and connect them, then note where the line crosses the Y-axis. (0, 1)
10. A: slope = -2; y-intercept = 1
$Y = -2X + 1$
11. B: $Y = -2X + 1$
$2X + Y = 1$
12. D: $\frac{1}{2}$
13. E: $Y = -3X + 6$ slope must be -3
14. E: $Y = \frac{1}{4}X - 1$ slope must be $\frac{1}{4}$
15. C: $3Y + 6X = 12$
$3Y = -6X + 12$
$Y = -2X + 4$
negative reciprocal of -2 is $\frac{1}{2}$

Test 11

1. D: slope $= \frac{3-1}{5-2} = \frac{2}{3}$
2. B: slope $= \frac{1-0}{-2-4} = \frac{1}{-6} = -\frac{1}{6}$
3. E: slope $= \frac{8-(-2)}{4-(-3)} = \frac{8+2}{4+3} = \frac{10}{7}$
4. E: $2X + Y = 13$
$Y = -2X + 13$
5. B: $-3X + 4Y = 8$
$4Y = 3X + 8$
$Y = \frac{3}{4}X + 2$
6. C: $2X - 2Y - 6 = 0$
$-2Y = -2X + 6$
$Y = X - 3$
7. A: 6
8. B: $2X + 3Y = 4$
$3Y = -2X + 4$
$Y = -\frac{2}{3}X + \frac{4}{3}$
slope $= -\frac{2}{3}$
9. D: $-4X + 2Y = 16$
$2Y = 4X + 16$
$Y = 2X + 8$
slope = 2
10. C: a point and the slope or two points

11. B: $Y = mX + b$
$(1) = (3)(2) + b$
$1 = 6 + b$
$1 - 6 = b;\ b = -5$

12. A: $Y = mX + b$
$(-2) = (-1)(-2) + b$
$-2 = 2 + b$
$-2 - 2 = b;\ b = -4$

13. E: slope $= \frac{3-1}{6-4} = \frac{2}{2} = 1$
$Y = mX + b$
$(3) = (1)(6) + b$
$3 = 6 + b$
$3 - 6 = b = -3$
OR $Y = mX + b$
$1 = (1)(4) + b$
$1 = 4 + b$
$1 - 4 = b = -3$
$Y = X - 3$

14. B: slope $= \frac{6-0}{-4-1} = \frac{6}{-5} = -\frac{6}{5}$
$Y = mX + b$
$(0) = \left(-\frac{6}{5}\right)(1) + b$
$\frac{6}{5} = b$
OR $Y = mX + b$
$(6) = \left(-\frac{6}{5}\right)(-4) + b$
$\frac{30}{5} = \frac{24}{5} + b$
$\frac{30}{5} - \frac{24}{5} = b = \frac{6}{5}$
$Y = -\frac{6}{5}X + \frac{6}{5}$

15. A: Slope is -1; this eliminates all but A and C.
using $(2, 3)$:
$(3) = (-1)(2) + 5$
$3 = -2 + 5$
$3 = 3$
(Either point could have been used.)

Unit Test I

I

1. $-3^2 = -(3 \times 3) = -9$
2. $-2 + 3^2 - 1 \times 4 = -2 + 9 - 4 = 3$
3. $|3-2| - |1-4| = |1| - |-3| = 1 - 3 = -2$

II

1. $3X - 2 + 2X = 4 - X$
$3X + 2X + X = 4 + 2$
$6X = 6$
$X = 1$

2. $\frac{1}{2}B + \frac{1}{3} = \frac{2}{9}$
$\frac{18}{1} \cdot \frac{1}{2}B + \frac{18}{1} \cdot \frac{1}{3} = \frac{18}{1} \cdot \frac{2}{9}$
$9B + 6 = 4$
$9B = -2$
$B = -\frac{2}{9}$

3. $.03Y + 1 = 4.3$
$(100)(.03Y) + (100)(1) = (100)(4.3)$
$3Y + 100 = 430$
$3Y = 330$
$Y = \frac{330}{3} = 110$

III

1. Point B

IV

1. associative
2. distributive
3. commutative

V

1. see graph
2. see graph

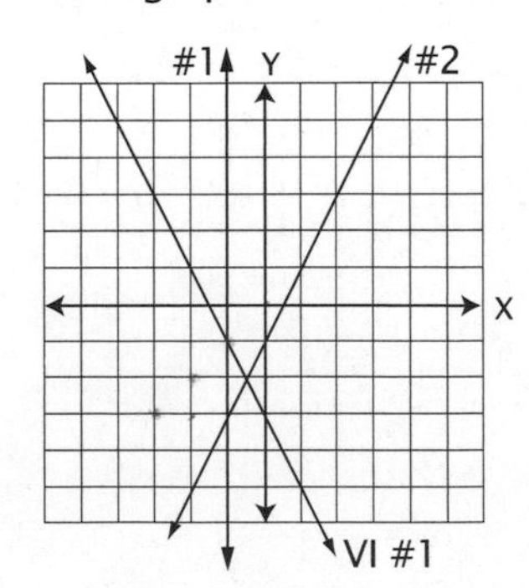

VI

1. $Y + 3 = -2X$
   $Y = -2X - 3$
   slope is -2
   Y-intercept is -3
   see graph

VII

1. $M = 3D - 2$

VIII

1. perpendicular, so slope is $-\frac{1}{3}$

$Y = mX + b$

$(1) = \left(-\frac{1}{3}\right)(2) + b$

$\frac{3}{3} = -\frac{2}{3} + b$

$\frac{3}{3} + \frac{2}{3} = b = \frac{5}{3}$

$Y = -\frac{1}{3}X + \frac{5}{3}$ (Y-intercept form)

$\frac{1}{3}X + Y = \frac{5}{3}$

$X + 3Y = 5$ (standard form)

see graph

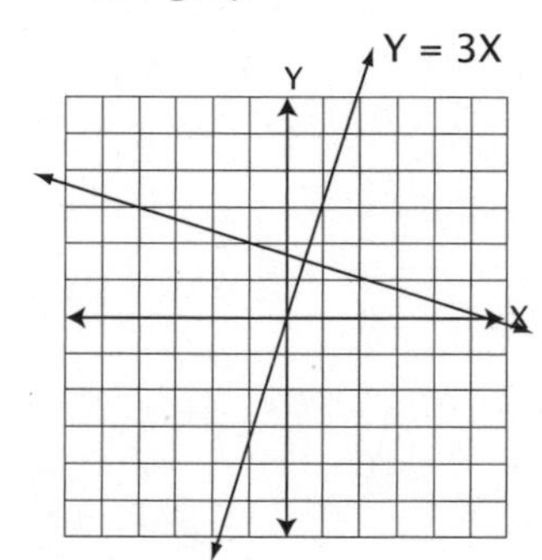

IX

1. $m = \frac{4-1}{0-2} = \frac{3}{-2} = -\frac{3}{2}$

$Y = mX + b$

$(4) = -\frac{3}{2}(0) + b$

$4 = b$

$Y = -\frac{3}{2}X + 4$ (Y-intercept form)

$\frac{3}{2}X + Y = 4$

$3X + 2Y = 8$ (standard form)

Either the Y-intercept form or the standard form may be considered correct.

X

a. $m = 3$
b. $m = 3$
c. $Y + 3X = 7$
   $Y = -3X + 7$
   $m = -3$
d. $3X = Y + 12$
   $-Y = -3X + 12$
   $Y = 3X - 12$
   $m = 3$

a, b and d all have a slope of 3, so they are parallel

Test 12

1. B: $2Y > X + 4$ because of the > sign
2. E: $-2Y > 4X + 8$; divide both sides by -2:
   $Y < -2X - 4$
   dividing by a negative number changes the direction of the inequality
3. B: $3Y < 6X - 6$
   $Y < 2X - 2$
   dividing by a positive number does not change the direction of the inequality
4. C: sketch graph to determine location
   slope is 1 and Y-intercept is -4
5. C: II and V
6. B: $2Y > 6X + 2$
   $Y > 3X + 1$ (see #3)

7. D: slope is 3; Y-intercept is 1; dotted line
8. E: $-4Y+8X>16$
$-4Y>-8X+16$
$Y<2X-4$ (see #2)
9. C: slope is 2; Y-intercept is -4; dotted line
10. D: $-3Y+9X\le 12$
$-3Y\le -9X+12$
$Y\ge 3X-4$ (see #2)
11. A: slope is 3; Y-intercept is -4; solid line
12. B: $3Y-9X\le 12$
$3Y\le 9X+12$
$Y\le 3X+4$ (see #3)
13. E: slope is 3; Y-intercept is 4; solid line
14. A: $-2Y>6X+2$
$Y<-3X-1$ (see #2)
$Y\le 3X+4$ (see #3)
15. B: slope is -3; Y-intercept is -1; dotted line

Test 13

1. A: infinite
2. B: 1
3. D: the lines intersect
4. E: "(3, 4) will not satisfy either line" is a false statement
5. A: slope $=-4$; Y – intercept $=-2$
6. B: slope $=1$; Y – intercept $=3$
7. C: from the graph
8. E: slope $=4$; Y – intercept $=2$
9. B: from the graph
10. B: slope $=-1$; Y – intercept $=4$
11. A: slope $=1$; Y – intercept $=-4$
12. C: slope $=4$; Y – intercept $=1$
13. D: slope $=-4$; Y – intercept $=1$
14. E: from the graph
15. B: from the graph

Test 14

1. D: substitution
2. B: $Y-2$
3. E: $Y+X=8 \Rightarrow Y+(Y-2)=8$
$2Y-2=8$
$2Y=10$
$Y=5$

$Y+X=8 \Rightarrow (5)+X=8$
$X=8-5$
$X=3$

(3, 5)
4. E: $8Y+X=2$
$X=-8Y+2$
5. A: $Y+5=X \Rightarrow Y+5=-8Y+2$
$Y+8Y=2-5$
$9Y=-3$
$Y=\frac{-3}{9}=-\frac{1}{3}$

$Y+5=X \Rightarrow \left(-\frac{1}{3}\right)+5=X$
$-\frac{1}{3}+\frac{15}{3}=X$
$\frac{14}{3}=X=4\frac{2}{3}$

$\left(4\frac{2}{3},-\frac{1}{3}\right)$
6. D: substitute to find the other variable
7. E: the answer may be an estimate
8. B: $X+Y=8 \Rightarrow X=-Y+8$
$2X-Y=7 \Rightarrow 2(-Y+8)-Y=7$
$-2Y+16-Y=7$
$-2Y-Y=7-16$
$-3Y=-9$
$Y=\frac{-9}{-3}=3$

$X+Y=8 \Rightarrow X+(3)=8$
$X=8-3$
$X=5$

(5, 3)

9. A:
$$Y = X + 2$$
$$2X + Y = 14 \Rightarrow 2X + (X+2) = 14$$
$$3X + 2 = 14$$
$$3X = 12$$
$$X = \frac{12}{3} = 4$$
$$Y = X + 2 \Rightarrow Y = (4) + 2$$
$$Y = 6$$
$(4, 6)$

10. B:
$$Y - X = -1 \Rightarrow Y = X - 1$$
$$2X + Y = 2 \Rightarrow 2X + (X - 1) = 2$$
$$3X - 1 = 2$$
$$3X = 3$$
$$X = 1$$
$$Y - X = -1 \Rightarrow Y - (1) = -1$$
$$Y = -1 + 1$$
$$Y = 0$$
$(1, 0)$

11. B: $X = 2Y + 3$
$$Y = 6X + 4 \Rightarrow Y = 6(2Y+3) + 4$$
$$Y = 12Y + 18 + 4$$
$$Y - 12Y = 22$$
$$-11Y = 22$$
$$Y = \frac{22}{-11} = -2$$
$$X = 2Y + 3 \Rightarrow X = 2(-2) + 3$$
$$X = -4 + 3$$
$$X = -1$$
$(-1, -2)$

12. E: $470 \div 55 = 8.5$ hours
6:00 AM + 8.5 hours = 2:30 PM

13. D: $470 \div 23.5 = 20$ mpg

14. D: 36 The numbers are squares of consecutive counting numbers.

15. C:

A
$$4(0) - 2Y = 12$$
$$-2Y = 12$$
$$Y = -6$$

B
$$4(-10) - 2Y = 12$$
$$-40 - 2Y = 12$$
$$-2Y = 52$$
$$Y = -26$$

C
$$4(3) - 2Y = 12$$
$$12 - 2Y = 12$$
$$-2Y = 0$$
$$Y = \frac{0}{-2} = 0$$

D
$$4(1) - 2Y = 12$$
$$4 - 2Y = 12$$
$$-2Y = 8$$
$$Y = -4$$

E
$$4(2) - 2Y = 12$$
$$8 - 2Y = 12$$
$$-2Y = 4$$
$$Y = -2$$

Test 15

1. D: graphing, substitution, or elimination
2. C: Make sure both are in the same form.
3. B: -2
4. E: 4
5. C:
$$\begin{aligned} X - 3Y &= 6 \\ X + 3Y &= 12 \\ \hline 2X \quad &= 18 \\ X &= 9 \end{aligned}$$
6. B: $X - 3Y = 6 \Rightarrow (9) - 3Y = 6$
$$-3Y = 6 - 9$$
$$-3Y = -3$$
$$Y = 1$$
7. D: $(9, 1)$
8. E:
$$\begin{aligned} 2X + 4Y &= 13 \\ (3X + Y = 2)(-4) \Rightarrow -12X - 4Y &= -8 \\ \hline -10X \quad &= 5 \\ X &= -\frac{1}{2} \end{aligned}$$
9. A: $2X + 4Y = 13 \Rightarrow 2\left(-\frac{1}{2}\right) + 4Y = 13$
$$-1 + 4Y = 13$$
$$4Y = 14$$
$$Y = \frac{14}{4} = 3\frac{1}{2}$$
10. A: $\left(-\frac{1}{2}, 3\frac{1}{2}\right)$

11. C:

$$\begin{aligned} X+6Y&=12 \\ (X-2Y=4)(3) \Rightarrow \quad 3X-6Y&=12 \\ \hline 4X \quad &=24 \\ X&=6 \end{aligned}$$

$$\begin{aligned} X+6Y=12 \Rightarrow \quad (6)+6Y&=12 \\ 6Y&=12-6 \\ 6Y&=6 \\ Y&=1 \end{aligned}$$

$(6, 1)$

12. D:

$$\begin{aligned} -3X-Y&=9 \\ 2X+Y&=3 \\ \hline -X \quad &=12 \\ X&=-12 \end{aligned}$$

$$\begin{aligned} 2X+Y=3 \Rightarrow \quad 2(-12)+Y&=3 \\ -24+Y&=3 \\ Y&=27 \end{aligned}$$

$(-12, 27)$

13. E: $6N-4N=8N \div 4$

14. B: 100

The numbers are squares of consecutive counting numbers.

15. D:

A
$$\begin{aligned} 2(0)+3Y&=6 \\ 3Y&=6 \\ Y&=2 \end{aligned}$$

B
$$\begin{aligned} 2(1)+3Y&=6 \\ 2+3Y&=6 \\ 3Y&=4 \\ Y&=\frac{4}{3}=1\frac{1}{3} \end{aligned}$$

C
$$\begin{aligned} 2(5)+3Y&=6 \\ 10+3Y&=6 \\ 3Y&=-4 \\ Y&=\frac{-4}{3}=-1\frac{1}{3} \end{aligned}$$

D
$$\begin{aligned} 2(-5)+3Y&=6 \\ -10+3Y&=6 \\ 3Y&=16 \\ Y&=\frac{16}{3}=5\frac{1}{3} \end{aligned}$$

E
$$\begin{aligned} 2(8)+3Y&=6 \\ 16+3Y&=6 \\ 3Y&=-10 \\ Y&=\frac{-10}{3}=-3\frac{1}{3} \end{aligned}$$

Test 16

1. A: $P+5N=28$

2. B:

$$\begin{aligned} (N+D=7)(-10) \Rightarrow \quad -10N-10D&=-70 \\ (.05N+.10D=.50)(100) \Rightarrow \quad 5N+10D&=50 \\ \hline -5N \quad &=-20 \\ N&=\frac{-20}{-5} \\ N&=4 \end{aligned}$$

3. A: $N+D=7 \Rightarrow (4)+D=7$

$D=7-4=3$

4. D:

$$\begin{aligned} (N+D=13)(-5) \Rightarrow \quad -5N-5D&=-65 \\ (.05N+.10D=1.10)(100) \Rightarrow \quad 5N+10D&=110 \\ \hline 5D&=45 \\ D&=9 \end{aligned}$$

5. A:

$$\begin{aligned} N+D=13 \Rightarrow \quad N+(9)&=13 \\ N&=13-9 \\ N&=4 \end{aligned}$$

6. D:

$$\begin{aligned} (N+Q=13)(-5) \Rightarrow \quad -5N-5Q&=-65 \\ (.05N+.25Q=1.85)(100) \Rightarrow \quad 5N+25Q&=185 \\ \hline 20Q&=120 \\ Q&=6 \end{aligned}$$

7. E:

$$\begin{aligned} N+Q=13 \Rightarrow \quad N+(6)&=13 \\ N&=13-6 \\ N&=7 \end{aligned}$$

8. A: elimination

9. B:

$$\begin{aligned} (D+Q=10)(-25) \Rightarrow \quad -25D-25Q&=-250 \\ (.10D+.25Q=2.05)(100) \Rightarrow \quad 10D+25Q&=205 \\ \hline -15D \quad &=-45 \\ D&=3 \end{aligned}$$

10. C:

$$\begin{aligned} D+Q=10 \Rightarrow \quad (3)+Q&=10 \\ Q&=10-3 \\ Q&=7 \end{aligned}$$

11. E:

$$\begin{aligned} (A+B=5)(-30) \Rightarrow \quad -30A-30B&=-150 \\ (.30A+.75B=2.40)(100) \Rightarrow \quad 30A+75B&=240 \\ \hline 45B&=90 \\ B&=2 \end{aligned}$$

12. B: $A+B=5 \Rightarrow A+(2)=5$
$$A=5-2$$
$$A=3$$

13. B: $Y=15X+50$

14. A: $Y=15X+50 \Rightarrow Y=15(10)+50$
$$Y=150+50$$
$$Y=\$200$$

15. E: $Y=20X+50 \Rightarrow Y=20(10)+50$
$$Y=200+50$$
$$Y=\$250$$

Test 17

1. B: $N,\ N+1,\ N+2$

2. C: $N,\ N+2,\ N+4$

3. C: $N,\ N+2,\ N+4$

4. D. $6N-2(N+1)=(N+2)+5$

5. C: $6N-2(N+1)=(N+2)+5$
$$6N-2N-2=N+2+5$$
$$6N-2N-N=2+5+2$$
$$3N=9$$
$$N=3$$

6. A: $6N+5(N+2)+3(N+2)=10(N+1)+10$
$$6N+5N+10+3N+6=10N+10+10$$
$$6N+5N+3N-10N=10+10-10-6$$
$$4N=4$$
$$N=1$$

7. B: $N+2 \Rightarrow (1)+2=3$

8. C: $3N+(N+2)+2=3(N+4)$
$$3N+N+2+2=3N+12$$
$$3N+N-3N=12-2-2$$
$$N=8$$

9. E: $N+2 \Rightarrow (8)+2=10$

10. D: $10N+10(N+2)=10(N+4)+10$
$$10N+10N+20=10N+40+10$$
$$10N+10N-10N=40+10-20$$
$$10N=30$$
$$N=3$$
second integer $=N+2=(3)+2=5$

11. A: 3

12. A: $3N-2(N+2)+13=-3[N+(N+4)]$
$$3N-2N-4+13=-3[2N+4]$$
$$3N-2N-4+13=-6N-12$$
$$3N-2N+6N=-12+4-13$$
$$7N=-21$$
$$N=-3$$

13. C: $N+4 \Rightarrow (-3)+4=1$

14. C: $4N+2(N+1)=4(N+2)$
$$4N+2N+2=4N+8$$
$$4N+2N-4N=8-2$$
$$2N=6$$
$$N=3$$

15. C: $N+2 \Rightarrow (3)+2=5$

Test 18

1. E: $(-6)^2=(-6\times -6)=36$

2. D: $-6^2=-(6\times 6)=-36$

3. A: $R^2\times R^4=R^{2+4}=R^6$

4. B: $R^4\div R^2=R^{4-2}=R^2$

5. E: $R^8\div R^2=R^{8-2}=R^6$

6. B: $A^{5X}\cdot A^{3X}=A^{5X+3X}=A^{8X}$

7. C: $C^4C^3D^2D^1=C^{4+3}D^{2+1}=C^7D^3$

8. C: $\sqrt{144}=12$

9. A: $4^8\div 4^2=4^{8-2}=4^6$

10. C: $X^2Y^3X^4Y=X^{2+4}Y^{3+1}=X^6Y^4$

11. B: $-\sqrt{A^2}=-A$

12. C: $\sqrt{81B^2}=9B$

13. C: $\sqrt{2^2 2^2}=\sqrt{4\cdot 4}=\sqrt{16}=4$

14. B: subtracted

15. A: added

Test 19

1. C: $\frac{1}{X^{-3}}=X^3$

2. D: $\frac{1}{X^3X^4}=X^{-3}X^{-4}=X^{-7}$

3. E: $X^{-4}=\frac{1}{X^4}$

4. E: $5^{-5}=\frac{1}{5^5}$

5. A: $8^{-2} \cdot 8^{-2} = 8^{-2+(-2)} = 8^{-4}$
6. C: $7^{-5} \div 7^{3} = 7^{-5-3} = 7^{-8}$
7. E: $X^{8} \div X^{2} = X^{8-2} = X^{6}$
8. A: $X^{-2}X^{-3} = X^{-2+(-3)} = X^{-5} = \frac{1}{X^{5}}$
9. B: $X^{0} = 1$;
   Any number raised to the 0 power equals 1.
10. E: $X^{-2}Y^{6}X^{-3}Y = X^{-2+(-3)}Y^{6+1} = X^{-5}Y^{7}$
11. B: $A^{-1}A^{-8}B^{7}B^{2} = A^{-1+(-8)}B^{7+2} = A^{-9}B^{9}$
12. E: $\frac{B^{4}B^{2}}{B^{-3}} = B^{4}B^{2}B^{3} = B^{4+2+3} = B^{9}$
13. A: $\frac{P^{3}N^{-2}}{N^{2}P^{4}} = P^{3}N^{-2}N^{-2}P^{-4} = P^{3+(-4)}N^{-2+(-2)} = P^{-1}N^{-4}$
14. D: $\left(9^{2}\right)^{5} = 9^{2 \cdot 5} = 9^{10}$
15. C: $\left(X^{A}\right)^{B} = X^{AB}$

Test 20

1. A: equation is a specific kind of polynomial called a trinomial
2. D: $$\begin{array}{r} X^2+3X+2 \\ +\ X^2+4X+5 \\ \hline 2X^2+7X+7 \end{array}$$
3. A: $$\begin{array}{r} X^2+X+10 \\ +\ X^2-2X+4 \\ \hline 2X^2-X+14 \end{array}$$
4. E: $$\begin{array}{r} X^2+8X+6 \\ +\ X^2-3X-1 \\ \hline 2X^2+5X+5 \end{array}$$
5. E: $$\begin{array}{r} X^2-5X-2 \\ +\ X^2-4X-3 \\ \hline 2X^2-9X-5 \end{array}$$
6. C: $$\begin{array}{r} 2X+3 \\ +\ 4X-5 \\ \hline 6X-2 \end{array}$$
7. B: $$\begin{array}{r} 2X^2-9X+5 \\ +\ X^2+4X-1 \\ \hline 3X^2-5X+4 \end{array}$$
8. C: $$\begin{array}{r} 4X+3 \\ \times\ X+1 \\ \hline 4X+3 \\ 4X^2+3X \\ \hline 4X^2+7X+3 \end{array}$$
9. B: $$\begin{array}{r} X+3 \\ \times\ X+2 \\ \hline 2X+6 \\ X^2+3X \\ \hline X^2+5X+6 \end{array}$$
10. A: $$\begin{array}{r} X+4 \\ \times\ X-2 \\ \hline -2X-8 \\ X^2+4X \\ \hline X^2+2X-8 \end{array}$$
11. C: $$\begin{array}{r} X+1 \\ \times\ X+5 \\ \hline 5X+5 \\ X^2+X \\ \hline X^2+6X+5 \end{array}$$
12. D: $$\begin{array}{r} X-3 \\ \times\ X-6 \\ \hline -6X+18 \\ X^2-3X \\ \hline X^2-9X+18 \end{array}$$
13. B: Multiplying the two first terms:
    $7X \cdot X = 7X^2$
14. B: Multiplying the two first terms:
    $2X \cdot X = 2X^2$
15. B: trinomial

Test 21

1. E: $$\begin{array}{r} X+A \\ \times\ X+B \\ \hline BX+AB \\ X^2+AX \\ \hline X^2+(A+B)X+AB \end{array}$$
2. B: $(A+B)X$
3. B: $(X+1)(X+2)$
4. E: $(X+3)(X+5)$
5. B: $(X+6)(X+6)$

6. B: $(X+2)(X+10)$
7. C: $(X+3)(X+8)$
8. D: $(X+1)(X+5)$
9. A: $(X+7)(X+7)$
10. B: $(X+1)(X+10)$
11. D:
$$\begin{array}{r} A+B \\ \times\ A+B \\ \hline AB+B^2 \\ A^2+AB \\ \hline A^2+2AB+B^2 \end{array}$$
12. E:
$$\begin{array}{r} X+BY \\ \times\ X+BY \\ \hline BYX+(BY)^2 \\ X^2+BYX \\ \hline X^2+2BYX+(BY)^2 \end{array}$$
13. C: $(X+R)(X+T)$
14. B: $(X+R)(X+R)$
15. B: factors

12. B:
$$\begin{array}{r} A+C \\ \times\ 2A+B \\ \hline BA+BC \\ 2A^2+\quad 2CA \\ \hline 2A^2+(2C+B)A+BC \end{array}$$
13. A: $(3X+R)(X+R)$
14. E: $(2A+B)(A+3B)$
15. D:
$$\begin{array}{r} 5X+Y \\ \times\ X+Y \\ \hline 5XY+Y^2 \\ 5X^2+XY \\ \hline 5X^2+6XY+Y^2 \end{array}$$
first and second term are affected by the coefficient "5"

Test 22

1. B:
$$\begin{array}{r} 2X+A \\ \times\ X+A \\ \hline 2AX+A^2 \\ 2X^2+AX \\ \hline 2X^2+3AX+A^2 \end{array}$$
final term is A^2
2. E: 3AX
3. B: $(2X+1)(X+2)$
4. E: $(3X+2)(X+4)$
5. D: $(2X+3)(X+3)$
6. A: $(4X+6)(X+1)$
7. B: $(2X+1)(X+1)$
8. A: $(3A+4)(A+2)$
9. E: $(2Y+6)(Y+3)$
10. E: $(5B+2)(B+2)$
11. D:
$$\begin{array}{r} 2A+B \\ \times\ A+B \\ \hline 2AB+B^2 \\ 2A^2+AB \\ \hline 2A^2+3AB+B^2 \end{array}$$

Test 23

1. C: the last term
2. D:
$$\begin{array}{r} X-A \\ \times\ X-B \\ \hline -BX+AB \\ X^2-AX \\ \hline X^2-(A+B)X+AB \end{array}$$
first negative, second positive
3. A:
$$\begin{array}{r} X-2 \\ \times\ X-3 \\ \hline -3X+6 \\ X^2-2X \\ \hline X^2-5X+6 \end{array}$$
4. D:
$$\begin{array}{r} X-2 \\ \times\ X+3 \\ \hline 3X-6 \\ X^2-2X \\ \hline X^2+X-6 \end{array}$$
5. C:
$$\begin{array}{r} X+2 \\ \times\ X-3 \\ \hline -3X-6 \\ X^2+2X \\ \hline X^2-X-6 \end{array}$$
6. B: $(X-1)(X+2)$
7. E: $(X-4)(X+1)$
8. B: $(X-3)(X-2)$

9. A: $(A+3)(A-4)$
10. B: $(A-3)(A+4)$
11. E:

$$\begin{array}{r} X-Y \\ \times\ X-Y \\ \hline -XY+Y^2 \\ X^2-XY \\ \hline X^2-2XY+Y^2 \end{array}$$

12. D:

$$\begin{array}{r} X+Y \\ \times\ X-Y \\ \hline -XY-Y^2 \\ X^2+XY \\ \hline X^2-Y^2 \end{array}$$

13. D: $(X-R)(X-R)$
14. E: they are either both negative or both positive
15. D: the second term of either factor with the largest value

Unit Test II

I

1. $5^2 \cdot 5^3 = 5^{2+3} = 5^5$ or 3,125
2. $(-3)^3 = (-3)(-3)(-3) = -27$
3. $(2^{-2})^2 = 2^{(-2)(2)} = 2^{-4}$ or $\frac{1}{16}$
4. $3^{10} \div 3^2 = 3^{10-2} = 3^8$ or 6,561
5. $A^2B^3AB^4 = A^{2+1}B^{3+4} = A^3B^7$

6.

$$\begin{array}{r} 3X+2 \\ \times\ X-1 \\ \hline -3X-2 \\ 3X^2+2X \\ \hline 3X^2-X-2 \end{array}$$

II

1.

$$(.05N+.10D=1.10)(100) \Rightarrow 5N+10D=110$$
$$(N+D=16)(-5) \Rightarrow -5N-5D=-80$$
$$5D=30$$
$$D=6$$

III

1. using elimination:

$$(2X-Y=-1)(2) \Rightarrow 4X-2Y=-2$$
$$2Y=6$$
$$4X=4$$
$$X=1$$

$2Y=6 \Rightarrow Y=3$

$(1, 3)$

using substitution:

$2Y=6 \Rightarrow Y=3$

$2X+1=Y \Rightarrow 2X+1=(3)$

$2X=2$

$X=1$

$(1, 3)$

using graphing: $(1, 3)$

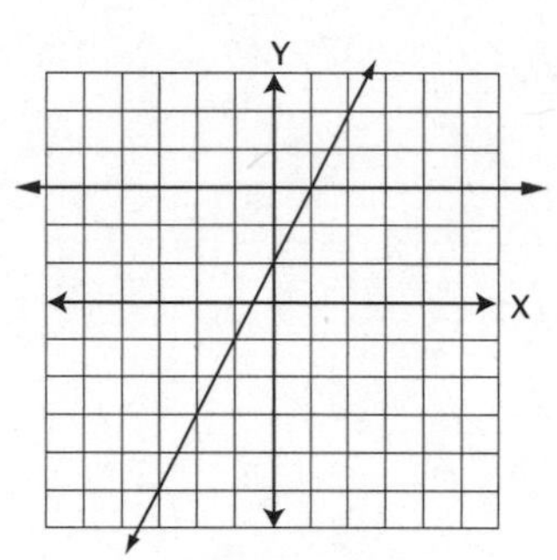

2. using elimination:

$Y-3=X+2$

$-X-3-2=-Y$

$-X-5=-Y$

$$(-X-5=-Y)(2) \Rightarrow -2X-10=-2Y$$
$$2X-1=Y$$
$$-11=-Y$$
$$Y=11$$

$2X-1=Y \Rightarrow 2X-1=11$

$2X=12$

$X=6$

$(6, 11)$

using substitution:

$2X - 1 = Y$

$$\begin{aligned} Y - 3 = X + 2 \Rightarrow \quad (2X - 1) - 3 &= X + 2 \\ 2X - 4 &= X + 2 \\ 2X - X &= 2 + 4 \\ X &= 6 \end{aligned}$$

$$\begin{aligned} 2X - 1 = Y \Rightarrow \quad 2(6) - 1 &= Y \\ 12 - 1 &= Y \\ 11 &= Y \end{aligned}$$

using graphing: We can only estimate, but (6, 11) looks reasonable.

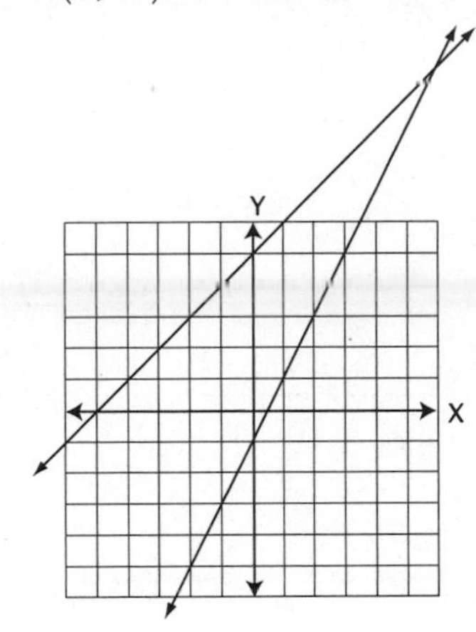

IV

1. $$\begin{aligned} 2N + 1 &= N + 4 \\ 2N - N &= 4 - 1 \\ N &= 3 \end{aligned}$$

   3, 5, 7

V

1. $2X^2 + 28 = 2(X^2 + 14)$
2. $2X^2 + 8X + 6 = (2)(X^2 + 4X + 3)$
   $= (2)(X + 3)(X + 1)$
3. $3X^2 + 19X + 20 = (3X + 4)(X + 5)$

VI

$2Y \le 4X - 8$

$Y = 2X - 4$

$2(0) \le 4(0) - 8$

$0 \le -8$ false

$2(-3) \le 4(4) - 8$

$-6 \le 8$ true

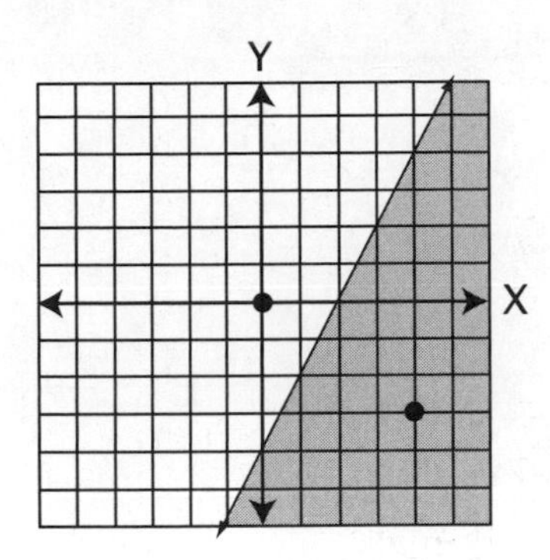

Test 24

1. A: $X + 3$
2. E: $X + 4$
3. B: $X + 1$
4. C: $X + 6$
5. B: $X + 2$
6. E: $X + 5$
7. A:

$$\begin{array}{r|l} & X + 1 \\ \hline X + 2 & X^2 + 3X + 2 \\ & -(X^2 + 2X) \\ \hline & \quad\quad X + 2 \\ & \quad -(X + 2) \\ \hline \end{array}$$

8. D:

$$\begin{array}{r|l} & X + 6 \quad r\ 2 \\ \hline X + 3 & X^2 + 9X + 20 \\ & -(X^2 + 3X) \\ \hline & \quad\quad 6X + 20 \\ & \quad -(6X + 18) \\ \hline & \quad\quad\quad\quad 2 \end{array}$$

9. E:

$$\begin{array}{r|l} & X \quad r\ -5 \\ \hline X + 4 & X^2 + 4X - 5 \\ & -(X^2 + 4X) \\ \hline & \quad\quad -5 \end{array}$$

10. B:

$$\begin{array}{r} X-7 \\ X+3\overline{\smash{)}X^2-4X-21} \\ \underline{-(X^2+3X)} \\ -7X-21 \\ \underline{-(-7X-21)} \end{array}$$

11. C:

$$\begin{array}{r} X+3 \\ X+5\overline{\smash{)}X^2+8X+15} \\ \underline{-(X^2+5X)} \\ 3X+15 \\ \underline{-(3X+15)} \end{array}$$

12. B:

$$\begin{array}{r} X+4 \quad r\,2 \\ X+2\overline{\smash{)}X^2+6X+10} \\ \underline{-(X^2+2X)} \\ 4X+10 \\ \underline{-(4X+8)} \\ 2 \end{array}$$

13. A:

$$\begin{array}{r} X+5 \\ \underline{\times X+5} \\ 5X+25 \\ \underline{X^2+5X} \\ X^2+10X+25 \end{array}$$

14. D:

$$\begin{array}{r} X+7 \\ \underline{\times X+7} \\ 7X+49 \\ \underline{X^2+7X} \\ X^2+14X+49 \end{array}$$

15. B:

$$\begin{array}{r} 2X+7 \\ \underline{\times X+5} \\ 10X+35 \\ \underline{2X^2+7X} \\ 2X^2+17X+35 \end{array}$$

Test 25

1. E:

$$\begin{array}{r} X+B \\ \underline{\times X-B} \\ -BX-B^2 \\ \underline{X^2+BX} \\ X^2-B^2 \end{array}$$

2. C: $(R+4)(R-4)$

3. D: $(S+5)(S-5)$

4. A:

$$\begin{array}{r} R+T \\ \underline{\times R-T} \\ -RT-T^2 \\ \underline{R^2+RT} \\ R^2-T^2 \end{array}$$

5. A: $(X-6)(X+6)$

6. E: none of the above

7. E: none of the above

8. C: $(X+10)(X-10)$

9. C:

$$\begin{array}{r} X+8 \\ \underline{\times X-8} \\ -8X-64 \\ \underline{X^2+8X} \\ X^2-64 \end{array}$$

10. A:

$$\begin{array}{r} X+3 \\ \underline{\times X-3} \\ -3X-9 \\ \underline{X^2+3X} \\ X^2-9 \end{array}$$

11. D:

$$\begin{array}{r} X+7 \\ \underline{\times X-7} \\ -7X-49 \\ \underline{X^2+7X} \\ X^2-49 \end{array}$$

12. C: add to 10

13. A: be the same

14. B: $4\times5=20$
$5\times5=25$
2,025

15. D: $6\times7=42$
$3\times7=21$
4,221

Test 26

1. D: $X^4-81=(X^2+9)(X^2-9)$
$=(X^2+9)(X-3)(X+3)$

2. D: $X^4-9=(X^2+3)(X^2-3)$

3. E: $A^4-16=(A^2+4)(A^2-4)$
$=(A^2+4)(A+2)(A-2)$

4. D: $4X^4$

5. B: $5X$

6. E: 1,000 is not a perfect square

7. D: $B^4-10{,}000=(B^2+100)(B^2-100)$
$=(B^2+100)(B+10)(B-10)$

8. C: $X^4-Y^4=(X^2+Y^2)(X^2-Y^2)$

9. B: $X^4-Y^4=(X^2+Y^2)(X^2-Y^2)$
$=(X^2+Y^2)(X+Y)(X-Y)$

10. D: $2X^3+16X^2+24X=(2X)(X^2+8X+12)$
$=(2X)(X+2)(X+6)$

11. E: $4X^3-64X=(4X)(X^2-16)$
$=(4X)(X+4)(X-4)$

12. A: $3X^3-12X^2-15X=(3X)(X^2-4X-5)$
$=(3X)(X-5)(X+1)$

13. C: $8X^3-72X=(8X)(X^2-9)$
$=(8X)(X+3)(X-3)$

14. B: $480\div 60=8$ hours

15. D: $5\times 65=325$ miles

Test 27

1. C: subtract 2 from each side

2. E: Each value of X must make at least one term equal 0.

3. A: divide each term by 2

4. B: factor out X

5. A: Each value of X must make at least one term equal 0.

6. C: $X^2+11X+30=0$
$(X+5)(X+6)=0$
$X=-5$ or $X=-6$

7. A: $2X^2+7X+6=0$
$(2X+3)(X+2)=0$

$2X+3=0$ $\quad$ $X+2=0$
$2X=-3$ $\quad$ $X=-2$
$X=-\frac{3}{2}$

8. E: $2X^2-7X+6=0$
$(2X-3)(X-2)=0$

$2X-3=0$ $\quad$ $X-2=0$
$2X=3$ $\quad$ $X=2$
$X=\frac{3}{2}$

9. B: $X^2+9X+20=0$
$(X+4)(X+5)=0$

$X+4=0$ $\quad$ $X+5=0$
$X=-4$ $\quad$ $X=-5$

10. D: $3X^2-3X-18=0$
$(3)(X^{?}-X-6)=0$
$(3)(X-3)(X+2)=0$

$X-3=0$ $\quad$ $X+2=0$
$X=3$ $\quad$ $X=-2$

11. C: $X^2-8X+16=1$
$X^2-8X+15=0$
$(X-3)(X-5)=0$

$X-3=0$ $\quad$ $X-5=0$
$X=3$ $\quad$ $X=5$

12. E: $2X^2-2X-4=20$
$2X^2-2X-24=0$
$(2)(X^2-X-12)=0$
$(X-4)(X+3)=0$

$X-4=0$ $\quad$ $X+3=0$
$X=4$ $\quad$ $X=-3$

13. D: $3X^2+9X=12$
$3X^2+9X-12=0$
$(3)(X^2+3X-4)=0$
$(X+4)(X-1)=0$

$X+4=0$ $\quad$ $X-1=0$
$X=-4$ $\quad$ $X=1$

14. B: $X^2-10X+25=0$
$(X-5)(X-5)=0$

$X-5=0$
$X=5$

15. A: $X^2+(R+S)X+RS=0$
$(X+R)(X+S)=0$

$X+R=0$ $\quad$ $X+S=0$
$X=-R$ $\quad$ $X=-S$

Test 28

1. B: 12 inches = 1 foot
   feet in denominator to cancel
   inches in numerator to remain in answer
2. D: 4 quarts = 1 gallon
   quarts in denominator
   gallons in numerator
3. C: 3 feet = 1 yard
   feet in denominator
   yards in numerator
4. A: 16 ounces = 1 pound
   pounds in denominator
   ounces in numerator
5. D: 2,000 pounds = 1 ton
   pounds in denominator
   tons in numerator
6. B: 2 pints = 1 quart
   pints in denominator
   quarts in numerator
7. B: $\frac{48\text{ in}}{1} \times \frac{1\text{ ft}}{12\text{ in}} = 4\text{ ft}$
8. C: $\frac{16\text{ gal}}{1} \times \frac{4\text{ qt}}{1\text{ gal}} = 64\text{ qt}$
9. A: $\frac{10\text{ lb}}{1} \times \frac{16\text{ oz}}{1\text{ lb}} = 160\text{ oz}$
10. D: $\frac{6\text{ qt}}{1} \times \frac{2\text{ pt}}{1\text{ qt}} = 12\text{ pt}$
11. B: $\frac{8{,}000\text{ lb}}{1} \times \frac{1\text{ ton}}{2{,}000\text{ lb}} = 4\text{ tons}$
12. C: $\frac{80\text{ oz}}{1} \times \frac{1\text{ lb}}{16\text{ oz}} = 5\text{ lb}$
13. D: $\frac{6\text{ yd}}{1} \times \frac{36\text{ in}}{1\text{ yd}} = 216\text{ in}$
14. C: the name of the desired answer
15. E: equal to 1

Test 29

1. B: 2
2. C: 3
3. C: 12 inches = 1 foot
   5,280 feet = 1 mile
4. B: 1 quart = 2 pints
   1 gallon = 4 quarts
5. E: $\frac{2\text{ yd}^2}{1} \times \frac{3\text{ ft}}{1\text{ yd}} \times \frac{3\text{ ft}}{1\text{ yd}} = 18\text{ ft}^2$
6. D: $\frac{6\text{ yd}^3}{1} \times \frac{3\text{ ft}}{1\text{ yd}} \times \frac{3\text{ ft}}{1\text{ yd}} \times \frac{3\text{ ft}}{1\text{ yd}} = 162\text{ ft}^3$
7. C: $\frac{8\text{ yd}^2}{1} \times \frac{3\text{ ft}}{1\text{ yd}} \times \frac{3\text{ ft}}{1\text{ yd}} = 72\text{ ft}^2$
   $\frac{72\text{ ft}^2}{1} \times \frac{12\text{ in}}{1\text{ ft}} \times \frac{12\text{ in}}{1\text{ ft}} = 10{,}368\text{ in}^2$
8. B: $\frac{87{,}120\text{ ft}^2}{1} \times \frac{1\text{ acre}}{43{,}560\text{ ft}^2} =$
   $\frac{87{,}120\text{ acres}}{43{,}560} = 2\text{ acres}$
9. C: $4\text{ ft} \times 4\text{ ft} \times 16\text{ ft} = 256\text{ ft}^3$
   $\frac{256\text{ ft}^3}{1} \times \frac{1\text{ cord}}{128\text{ ft}^3} = \frac{256\text{ cords}}{128} = 2\text{ cords}$
10. E: $\frac{6\text{ m}^3}{1} \times \frac{100\text{ cm}}{1\text{ m}} \times \frac{100\text{ cm}}{1\text{ m}} \times \frac{100\text{ cm}}{1\text{ m}} =$
    $6{,}000{,}000\text{ cm}^3$
11. B: $\frac{1\text{ yd}^2}{1} \times \frac{3\text{ ft}}{1\text{ yd}} \times \frac{3\text{ ft}}{1\text{ yd}} = 9\text{ ft}^2$
    $3\text{ ft}^2 < 9\text{ ft}^2$
12. C: $9\text{ ft}^2 = 9\text{ ft}^2$
13. A: $\frac{1\text{ mi}^2}{1} \times \frac{5{,}280\text{ ft}}{1\text{ mi}} \times \frac{5{,}280\text{ ft}}{1\text{ mi}} = 27{,}878{,}400\text{ ft}^2$
    $27{,}878{,}400\text{ ft}^2 > 43{,}560\text{ ft}^2$
14. D: Cannot be determined, because we don't know the relationship between X and Y.
15. B: $\frac{8\text{ yd}^3}{1} \times \frac{3\text{ ft}}{1\text{ yd}} \times \frac{3\text{ ft}}{1\text{ yd}} \times \frac{3\text{ ft}}{1\text{ yd}} = 216\text{ ft}^3$
    $215\text{ ft}^3 < 216\text{ ft}^3$

Test 30

1. D: $\frac{5\text{ km}}{1} \times \frac{.62\text{ mi}}{1\text{ km}} = 3.1\text{ mi}$
2. B: $\frac{6\text{ kg}}{1} \times \frac{2.2\text{ lb}}{1\text{ kg}} = 13.2\text{ lb}$

3. A: $\frac{3\,\cancel{yd}}{1} \times \frac{.9\text{ m}}{1\,\cancel{yd}} = 2.7\text{ m}$

4. B: $\frac{6\,\cancel{mi}}{1} \times \frac{1.6\text{ km}}{1\text{ mi}} = 9.6\text{ km}$

5. D: $\frac{8\,\cancel{lb}}{1} \times \frac{.45\text{ kg}}{1\,\cancel{lb}} = 3.6\text{ kg}$

6. E: $\frac{6\,\cancel{m}}{1} \times \frac{1.1\text{ yd}}{1\,\cancel{m}} = 6.6\text{ yd}$

7. A: $\frac{9\,\cancel{cm}}{1} \times \frac{.4\text{ in}}{1\,\cancel{cm}} = 3.6\text{ in}$

8. C: $\frac{9\,\cancel{in}}{1} \times \frac{2.5\text{ cm}}{1\,\cancel{in}} = 22.5\text{ cm}$

9. E: $\frac{3\,\cancel{\text{liters}}}{1} \times \frac{1.06\text{ qt}}{1\,\cancel{\text{liters}}} = 3.18\text{ qt}$

10. B: $\frac{1\,\cancel{gal}}{1} \times \frac{4\,\cancel{qt}}{1\,\cancel{gal}} \times \frac{.95\text{ liters}}{1\,\cancel{qt}} = 3.8\text{ liters}$

11. B: $\frac{2\,\cancel{qt}}{1} \times \frac{.95\text{ liters}}{1\,\cancel{qt}} = 1.9\text{ liters}$

    $1.9\text{ liters} < 2\text{liters}$

12. A: $\frac{5\,\cancel{mi}}{1} \times \frac{1.6\text{ km}}{1\,\cancel{mi}} = 8\text{ km}$

    $8\text{ km} > 5\text{ km}$

13. B: $\frac{X\,\cancel{lb}}{1} \times \frac{1\text{ kg}}{2.2\,\cancel{lb}} = \frac{X\text{ kg}}{2.2}$

    $= \frac{1}{2.2}X\text{ kg} = .45X\text{ kg}$

    $.45X\text{ kg} < 1X\text{ kg}$

14. C: $\frac{2\,\cancel{oz}}{1} \times \frac{28\text{ g}}{1\,\cancel{oz}} = 56\text{ g}$

    $56\text{ g} = 56\text{ g}$

15. D: Cannot be determined from the information given: The relationship between centimeters and inches is known, but the relationship between X and Y is not known.

Test 31

1. B: radical
2. A: add the exponents
3. C: $27^{\frac{1}{3}} = \sqrt[3]{27} = 3$
4. E: $\left(X^3\right)^{\frac{1}{3}} = X^{(3)\left(\frac{1}{3}\right)} = X^1 = X$
5. B: $125^{\frac{2}{3}} = \sqrt[3]{125}^{\,2} = 5^2 = 25$
6. A: $X^{\frac{2}{3}} = \sqrt[3]{X}^{\,2}$
7. D: $2\cdot 16^{\frac{1}{4}} = 2\cdot\sqrt[4]{16} = 2\cdot 2 = 4$
8. B: $\left(Y^{\frac{1}{6}}\right)\left(Y^{\frac{2}{5}}\right) = Y^{\frac{1}{6}+\frac{2}{5}} = Y^{\frac{5}{30}+\frac{12}{30}} = Y^{\frac{17}{30}}$
9. D: $\left(X^2\cdot X^4\right)^{\frac{1}{2}} = \left(X^{2+4}\right)^{\frac{1}{2}}$

   $= \left(X^6\right)^{\frac{1}{2}} = X^{(6)\left(\frac{1}{2}\right)} = X^3$
10. C: $10^{\frac{2}{3}}\cdot 1{,}000 = 10^{\frac{2}{3}}\cdot 10^3$

    $= 10^{\frac{2}{3}+3} = 10^{\frac{2}{3}+\frac{9}{3}} = 10^{\frac{11}{3}}$
11. A: $2^2 = 2\times 2 = 4$

    $\left(2^{\frac{1}{2}}\right)^2 = 2^{\left(\frac{1}{2}\right)(2)} = 2^1 = 2$

    $4 > 2$
12. B: $X^3X^3X^{\frac{1}{3}} = X^{3+3+\frac{1}{3}} = X^{6+\frac{1}{3}} = X^{\frac{19}{3}}$

    $X^{\frac{19}{3}} < X^{19}$
13. C: $\left(3^{\frac{1}{3}}\right)^3 = 3^{\left(\frac{1}{3}\right)(3)} = 3^1 = 3$

    $\left|\sqrt{9}\right| = |3| = 3$

    $3 = 3$
14. C: $10^2\cdot 1{,}000 = 10^2\cdot 10^3 =$

    $10^{2+3} = 10^5$

    $10^5 = 10^5$
15. A: $\left[\left(B^3B^5\right)^{\frac{1}{2}}\right]^2 = \left[\left(B^{3+5}\right)^{\frac{1}{2}}\right]^2$

    $= \left[\left(B^8\right)^{\frac{1}{2}}\right]^2$

    $= B^{(8)\left(\frac{1}{2}\right)(2)} = B^8$

    $B^8 > B^4$

Test 32

1. A: To simplify computations with very large or very small numbers
2. B: $6{,}300{,}000 = 6.3\times 10^6$

3. D: $543{,}000 = 5.43 \times 10^5$

4. B: $.00065 = 6.5 \times 10^{-4}$

5. E: $.0000781 = 7.81 \times 10^{-5}$

6. C: $(10)^7(10)^7 = 10^{7+7} = 10^{14}$

7. C: $(10^1)(10^{-6}) = 10^{1+(-6)} = 10^{-5}$

8. A: $12{,}000 \times .006 = (1.2 \times 10^4)(6 \times 10^{-3})$

9. C: $3.6 \times 10^{-6} = .0000036$

10. E: $1.02 \times 10^5 = 102{,}000$

11. C: $.25 \times 130{,}000 = (2.5 \times 10^{-1})(1.3 \times 10^5)$
$= (2.5 \times 1.3)(10^{-1} \times 10^5)$
$= 3.25 \times 10^4$

12. D: $50{,}000{,}000 \times .610 = (5 \times 10^7)(6.1 \times 10^{-1})$
$= (5 \times 6.1)(10^7 \times 10^{-1})$
$= 30.5 \times 10^6$
$= 3.05 \times 10^7$

13. C: $2.4 \times 3.06 = 7.344 \approx 7.3$ (2 SD)

14. D: $(1.24 \times 10^6)(4.7 \times 10^3)$
$= (1.24 \times 4.7)(10^6 \times 10^3)$
$= 5.828 \times 10^9$
$\approx 5.8 \times 10^9$ (2 SD)

15. B: $(6.25 \times 10^8) \div (3.241 \times 10^4)$
$= (6.25 \div 3.241)(10^8 \div 10^4)$
$= 1.9284 \times 10^4$ (rounded)
$\approx 1.93 \times 10^4$ (3 SD)

Test 33

1. C: 10

2. C: 10

3. C: Write the number in exponential notation

4. E: 4_{10}

5. D: 10_4

6. C: $5^3 = 125$, and is the largest power of 5 less than 300.

7. A: $4^3 = 64$, and is the largest power of 4 less than 95.

8. C: 4^2 is the largest power of 4 less than 34
$4^2 = 16;\ 4^1 = 4;\ 4^0 = 1$

$$\begin{array}{r} 2 \\ 16\overline{)34} \\ 32 \\ \hline 2 \end{array} \qquad \begin{array}{r} 0 \\ 4\overline{)2} \\ 0 \\ \hline 2 \end{array} \qquad \begin{array}{r} 2 \\ 1\overline{)2} \\ 2 \\ \hline 0 \end{array}$$

$2 \times 4^2 + 0 \times 4^1 + 2 \times 4^0 = 202_4$

9. E: 2^5 is the largest power of 2 less than 45.
$2^5 = 32;\ 2^4 = 16;\ 2^3 = 8;\ 2^2 = 4;\ 2^1 = 2;\ 2^0 = 1$

$$\begin{array}{r} 1 \\ 32\overline{)45} \\ 32 \\ \hline 13 \end{array} \quad \begin{array}{r} 0 \\ 16\overline{)13} \\ 0 \\ \hline 13 \end{array} \quad \begin{array}{r} 1 \\ 8\overline{)13} \\ 8 \\ \hline 5 \end{array} \quad \begin{array}{r} 1 \\ 4\overline{)5} \\ 4 \\ \hline 1 \end{array} \quad \begin{array}{r} 0 \\ 2\overline{)1} \\ 0 \\ \hline 1 \end{array} \quad \begin{array}{r} 1 \\ 1\overline{)1} \\ 1 \\ \hline 0 \end{array}$$

$1 \times 2^5 + 0 \times 2^4 + 1 \times 2^3 + 1 \times 2^2 + 0 \times 2^1 + 1 \times 2^0 = 101101_2$

10. B: 12^2 is the largest power of 12 less than 356.
$12^2 = 144;\ 12^1 = 12;\ 12^0 = 1$

$$\begin{array}{r} 2 \\ 144\overline{)356} \\ 288 \\ \hline 68 \end{array} \qquad \begin{array}{r} 5 \\ 12\overline{)68} \\ 60 \\ \hline 8 \end{array} \qquad \begin{array}{r} 8 \\ 1\overline{)8} \\ 8 \\ \hline 0 \end{array}$$

$2 \times 12^2 + 5 \times 12^1 + 8 \times 12^0 = 258_{12}$

11. E: $122_6 = 1 \times 6^2 + 2 \times 6^1 + 2 \times 6^0$
$= 1(36) + 2(6) + 2(1) = 36 + 12 + 2 = 50_{10}$

12. B: $4B3_{12} = 4 \times 12^2 + 11 \times 12^1 + 3 \times 12^0$
$= 4(144) + 11(12) + 3(1)$
$= 576 + 132 + 3 = 711_{10}$

13. A: $52A_{12} = 5 \times 12^2 + 2 \times 12^1 + 10 \times 12^0$
$= 5(144) + 2(12) + 10(1)$
$= 720 + 24 + 10 = 754_{10}$

14. E: $122_3 = 1 \times 3^2 + 2 \times 3^1 + 2 \times 3^0$
$= 1(9) + 2(3) + 2(1) = 9 + 6 + 2 = 17_{10}$

15. B: $1000_7 = 1 \times 7^3 + 0 \times 7^2 + 0 \times 7^1 + 0 \times 7^0$
$= 1(343) + 0(49) + 0(7) + 0(1) = 343_{10}$

Test 34

1. A: circle
2. C: ellipse
3. B: the coordinates of the center of a circle
4. C: P is the radius of a circle
5. E: Only ellipse has different X and Y coefficients
6. C: $(0, 0)$
7. A: radius $= \sqrt{9} = 3$
8. E: $-(-3) = 3;\ -(-4) = 4:\ (3, 4)$
9. D: center at $(-3, -3)$; radius $= \sqrt{4} = 2$
10. A: center at $(3, 2)$; radius $= \sqrt{1} = 1$
11. E: center at $(-3, 3)$; radius $= \sqrt{1} = 1$
    not on graph
12. D: center at $(3, -3)$; radius $= \sqrt{2} = 4$
13. C: center at $(0, 0)$

    When X is 0: $X^2 + 4Y^2 = 4$
    $$(0)^2 + 4Y^2 = 4$$
    $$4Y^2 = 4$$
    $$Y^2 = 1$$
    $$Y = \pm 1$$
    When Y is 0: $X^2 + 4Y^2 = 4$
    $$X^2 + 4(0)^2 = 4$$
    $$X^2 = 4$$
    $$X = \pm 2$$
    Points: $(0, 1); (0, -1); (-2, 0); (2, 0)$
14. E: center at $(0, 0)$

    When X is 0: $4X^2 + Y^2 = 4$
    $$4(0)^2 + Y^2 = 4$$
    $$Y^2 = 4$$
    $$Y = \pm 2$$
    When Y is 0: $4X^2 + Y^2 = 4$
    $$4X^2 + (0)^2 = 4$$
    $$4X^2 = 4$$
    $$X^2 = 1$$
    $$X = \pm 1$$
    Points: $(0, 2); (0, -2); (1, 0); (-1, 0)$
    not on graph
15. C: center at $(0, 0)$; radius $= \sqrt{1} = 1$

Test 35

1. C: ellipse
2. E: hyperbola
3. B: circle
4. A: line
5. D: parabola
6. B: I and III
   When a pair of factors multiply to equal a positive number, they will always be both positive or both negative.
7. C: II and IV
   When a pair of factors multiply to equal a negative number, one will be positive, and the othter negative.
8. D: The parabola gets narrower.
9. A: $XY = 12 \Rightarrow (-4)Y = 12$
   $$Y = \frac{12}{-4} = -3$$
10. E: hyperbola
    make chart and use test points
11. D: hyperbola
    make chart and use test points
12. B: hyperbola
    make chart and use test points
13. A: parabola
    make chart and use test points
14. C: parabola
    make chart and use test points
15. B: parabola
    make chart and use test points

Unit Test III

I

1.
$$\begin{array}{r} 2X \ +1 \\ X+2\,\overline{)\,2X^2+5X\ +2} \\ -(2X^2+4X) \\ \hline X+2 \\ -(X+2) \\ \hline \end{array} \qquad \begin{array}{r} 2X+1 \\ \times X+2 \\ \hline 4X+2 \\ 2X^2+\ X\ \ \\ \hline 2X^2+5X+2 \end{array}$$

2.

$$\begin{array}{r} X^2+5X+1 \\ X-2\overline{)X^3+3X^2-9X-2} \\ -(X^3-2X^2) \\ \hline 5X^2-9X \\ -(5X^2-10X) \\ \hline X-2 \\ X-2 \\ \hline \end{array}$$

$$\begin{array}{r} X^2+5X+1 \\ \times X-2 \\ \hline -2X^2-10X-2 \\ X^3+5X^2+X \\ \hline X^3+3X^2-9X-2 \end{array}$$

II

1. $3X^2-12=(3)(X^2-4)$
   $=(3)(X-2)(X+2)$
2. $Q^2-R^2=(Q-R)(Q+R)$
3. $2X^2-4X-30=(2)(X^2-2X-15)$
   $=(2)(X+3)(X-5)$

III

1.

$X^2+5X+16=10$
$X^2+5X+6=0$
$(X+2)(X+3)=0$

$X+2=0$ $\quad$ $X+3=0$
$X=-2$ $\quad$ $X=-3$

$X^2+5X+16=10$
$(-2)^2+5(-2)+16=10$
$4-10+16=10$
$10=10$

$X^2+5X+16=10$
$(-3)^2+5(-3)+16=10$
$9-15+16=10$
$10=10$

2.

$2X^3-18X=0$
$(2X)(X^2-9)=0$
$(2X)(X-3)(X+3)=0$

$2X=0$ $\quad$ $X-3=0$ $\quad$ $X+3=0$
$X=0$ $\quad$ $X=3$ $\quad$ $X=-3$

$2X^3-18X=0$
$2(0)^3-18(0)=0$
$0-0=0$
$0=0$

$2X^3-18X=0$
$2(3)^3-18(3)=0$
$2(27)-54=0$
$54-54=0$
$0=0$

$2X^3-18X=0$
$2(-3)^3-18(-3)=0$
$2(-27)+54=0$
$-54+54=0$
$0=0$

IV

1. $\frac{100\text{ oz}}{1}\times\frac{28\text{ g}}{1\text{ oz}}=2{,}800\text{ g}$
2. $\frac{6\text{ km}}{1}\times\frac{.62\text{ mi}}{1\text{ km}}=3.72\text{ mi}$

V

1. $456{,}700{,}000=4.567\times10^8$
2. $.0260=2.6\times10^{-2}$

VI

1. $.0003\times4.2=(3\times10^{-4})(4.2\times10^0)$
   $=(3\times4.2)(10^{-4}\times10^0)$
   $=12.6\times10^{-4}$
   $=1.26\times10^{-3}$ or 1×10^{-3}
   with significant digits taken into account: either answer is acceptable.
2. $\frac{6{,}800{,}000}{200{,}000}=(6.8\times10^6)\div(2\times10^5)$
   $=(6.8\div2)(10^6\div10^5)$
   $=3.4\times10^1$ or 3×10^1
   with significant digits taken into account: either answer is acceptable.

VII

1. $\sqrt{196} = 14$
2. $\sqrt{100A^2} = 10A$
3. $\sqrt{X^2 + 18X + 81} = X + 9$

VIII

1. 7^2 is the largest power of $7 < 70$

   $7^2 = 49;\ 7^1 = 7;\ 7^0 = 1$

   $70 \div 49 = 1$, remainder $70 - 49 = 21$; $21 \div 7 = 3$, remainder $21 - 21 = 0$; $0 \div 1 = 0$, remainder $0 - 0 = 0$

   $1 \times 7^2 + 3 \times 7^1 + 0 \times 7^0 = 130_7$

2. $$\begin{aligned} 2210_3 &= 2 \times 3^3 + 2 \times 3^2 + 1 \times 3^1 + 0 \times 3^0 \\ &= 2(27) + 2(9) + 1(3) + 0(1) \\ &= 54 + 18 + 3 + 0 = 75_{10} \end{aligned}$$

IX

1. $16^{\frac{1}{2}} = \sqrt{16} = 4$
2. $(1{,}000)^{\frac{2}{3}} = \sqrt[3]{1{,}000}^{\,2} = 10^2 = 100$

X

1. hyperbola

X	Y
3	$-\frac{1}{3}$
−3	$\frac{1}{3}$
$\frac{1}{3}$	−3
$-\frac{1}{3}$	3
2	$-\frac{1}{2}$
−2	$\frac{1}{2}$
$\frac{1}{2}$	−2
$-\frac{1}{2}$	2
1	−1
−1	1

2. circle: center at $(0, 0)$;

   radius $= \sqrt{4} = 2$

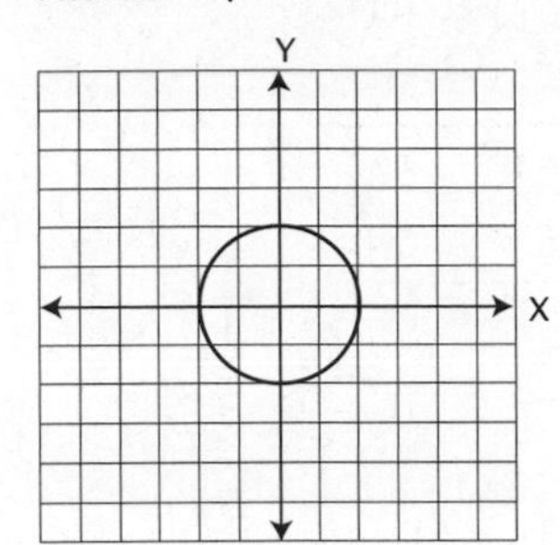

Final Exam

I

1. $$\begin{aligned} \left(-\frac{1}{2}\right)^2 + (ab)^0 - 3^2 &= \frac{1}{4} + 1 - 9 \\ &= \frac{1}{4} - 8 \\ &= \frac{1}{4} - \frac{32}{4} \\ &= -\frac{31}{4} = -7\frac{3}{4} \end{aligned}$$
2. $\sqrt{16X^2} = 4X$
3. $(2^2)^3(2^2) = 4^3 \cdot 4 = 64 \cdot 4 = 256$
4. $|6 - 8| = |-2| = 2$
5. $\sqrt{X^2 + 4X + 4} = X + 2$
6. $81^{\frac{1}{2}} = 9$
7. $\frac{3X^2}{X^{-4}} + \frac{5X}{X^{-1}} = 3X^2X^4 + 5XX = 3X^6 + 5X^2$

II

1. $3X^2 - 27 = (3)(X^2 - 9) = (3)(X - 3)(X + 3)$
2. $5X^2 - 9X - 2 = (5X + 1)(X - 2)$
3. $$\begin{aligned} X^3 + 5X^2 + 6X &= (X)(X^2 + 5X + 6) \\ &= (X)(X + 2)(X + 3) \end{aligned}$$
4. $$\begin{aligned} 14Y^2 - 7Y - 42 &= (7)(2Y^2 - Y - 6) \\ &= (7)(2Y + 3)(Y - 2) \end{aligned}$$

III.

1. $$\begin{aligned} 10^6 &= 10^{(3)(X)} \\ 10^{(3)(2)} &= 10^{(3)(X)} \\ X &= 2 \end{aligned}$$

IV.

1. $3X^2 - 6X = 0$

$(3X)(X-2) = 0$

$3X = 0 \qquad X - 2 = 0$

$X = 0 \qquad X = 2$

2. $\frac{1}{6}X - \frac{1}{2} = \frac{2}{3}$

$(30)\left(\frac{1}{6}\right)X - (30)\left(\frac{1}{2}\right) = (30)\left(\frac{2}{3}\right)$

$\frac{30}{6}X - \frac{30}{2} = \frac{60}{3}$

$5X - 15 = 20$

$5X = 35$

$X = 7$

3. $X + .25X = .4$

$(100)(X + .25X = .4)$

$100X + 25X = 40$

$125X = 40$

$X = \frac{40}{125} = \frac{8}{25}$ or .32

V.

1. parabola

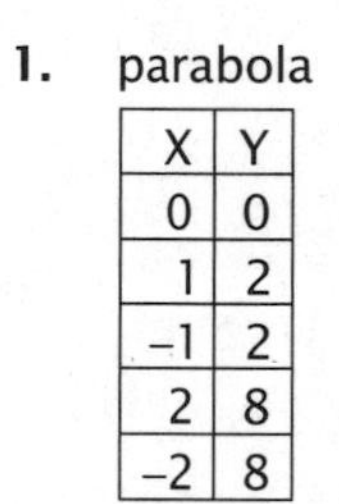

X	Y
0	0
1	2
–1	2
2	8
–2	8

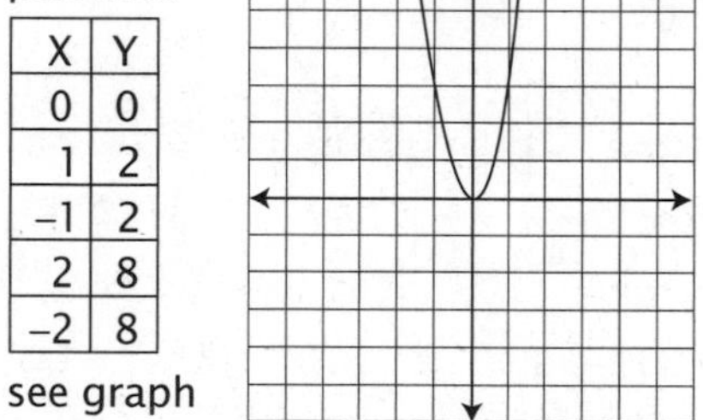

see graph

2. ellipse

When $X = 0$:

$4X^2 + Y^2 = 16$

$4(0)^2 + Y^2 = 16$

$Y^2 = 16$

$Y = \pm 4$

When $Y = 0$:

$4X^2 + Y^2 = 16$

$4X^2 + (0)^2 = 16$

$4X^2 = 16$

$X^2 = 4$

$X = \pm 2$

Points: $(0, 4); (0, -4); (+2, 0); (-2, 0)$

see graph

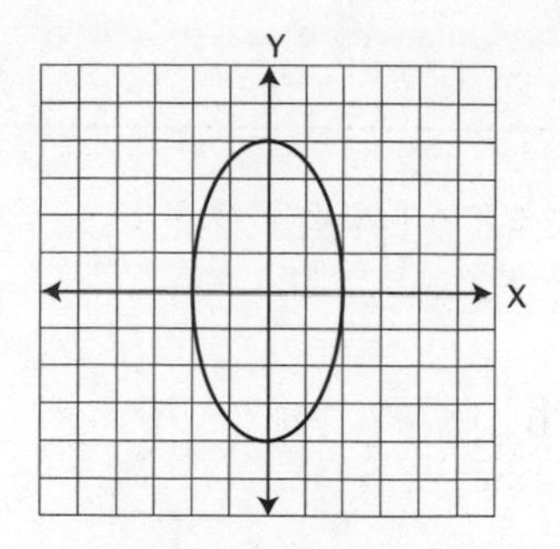

3. y – intercept = 1; slope = 3

see graph

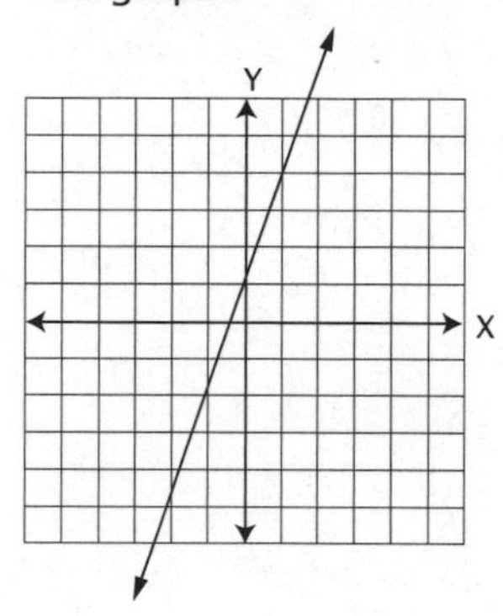

Symbols & Tables

U. S. CUSTOMARY MEASURE

2 cups = 1 pint (pt)
2 pints = 1 quart (qt)
4 quarts = 1 gallon (gal)
8 pints = 1 gallon
12 inches (in) = 1 foot (ft)
3 feet = 1 yard (yd)
5,280 feet = 1 mile (mi)
16 ounces (oz) = 1 pound (lb)
1 ton = 2,000 pounds

1 acre of land = 43,560 ft^2
1 cord of wood = $4' \times 4' \times 8' = 128$ ft^3
1 yard of carpet = $3' \times 3' = 1$ yd^2
1 yard of concrete = $3' \times 3' \times 3' = 1$ yd^3

U. S. CUSTOMARY TO METRIC

1 inch ≈ 2.5 centimeters
1 yard ≈ .9 meter
1 mile ≈ 1.6 kilometers
1 ounce ≈ 28 grams
1 pound ≈ .45 kilogram
1 quart ≈ .95 liter

METRIC TO U. S. CUSTOMARY

1 centimeter ≈ .4 inch
1 meter ≈ 1.1 yards
1 kilometer ≈ .62 mile
1 gram ≈ .035 ounce
1 kilogram ≈ 2.2 pounds
1 liter ≈ 1.06 quarts

SYMBOLS

Symbol	Meaning
$<$	less than
$>$	greater than
$\leq$	less than or equal to
$\geq$	greater than or equal to
$=$	equal in numerical value
$\neq$	not equal
$\approx$	approximately equal
$\sqrt{\ }$	square root (radical sign)
%	percent
$\perp$	perpendicular
$\parallel$	parallel
1'	1 foot
1"	1 inch

$|X| = X$ absolute value of X equals X
$|-X| = X$ absolute value of –X equals X

FORMULAS

Standard Form of Equations

line	$AX + BY = C$
circle	$AX^2 + AY^2 = R^2$
ellipse	$AX^2 + BY^2 = C$
parabola	$Y = X^2$
hyperbola	$XY = C$

Slope-Intercept Form of the Equation of a Line

$$Y = mX + b$$

where m = slope
b = Y-intercept

Slope of a line $= \dfrac{\text{up}}{\text{over}} = \dfrac{\text{rise}}{\text{run}} = \dfrac{Y_2 - Y_1}{X_2 - X_1}$

PROPERTIES

Associative Property

$(X + B) + C = A + (B + C)$

$(A \times B) \times C = A \times (B \times C)$

Commutative Property

$A + B = B + A$

$A \times B = B \times A$

Distributive Property

$A(B + C) = AB + AC$

MISCELLANEOUS

Exponents

$X^a \cdot X^b = X^{a+b}$

$X^a \div X^b = X^{a-b}$

$X^{-a} = \frac{1}{X^a}$

$(X^a)^b = X^{ab}$

$X^{\frac{1}{a}} = \sqrt[a]{X}$

Order of Operations

Parachute Expert My Dear Aunt Sally

1) parentheses
2) exponents
3) multiplication and division
4) addition and subtraction

Rules for Divisibility

Number is divisible by

2 if it ends in even number

3 if digits add to multiple of 3

9 if digits add to multiple of 9

5 if it ends in 5 or 0

METRIC MEASURES

1,000 millimeters (mm) - 1 meter (m)
100 centimeters (cm) - 1 meter
10 decimeters (dm) - 1 meter
10 meters = 1 dekameter (dam or dkm)
100 meters = 1 hectometer (hm)
1,000 meters = 1 kilometer (km)

1,000 milligrams (mg) - 1 gram (g)
100 centigrams (cg) - 1 gram
10 decigrams (dg) - 1 gram
10 grams = 1 dekagram (dag or dkg)
100 grams = 1 hectogram (hg)
1,000 grams = 1 kilogram (kg)

1,000 milliliters (ml) - 1 liter (l or L)
100 centiliters (cl) - 1 liter
10 deciliters (dl) - 1 liter
10 liters = 1 dekaliter (dal or dkl)
100 liters = 1 hectoliter (hl)
1,000 liters = 1 kiloliter (kl)

Glossary

A

absolute value - the value of a number without its sign, or the difference between a number and zero expressed as a positive number

additive inverse - the number that, when added to another, results in a sum of 0

algebra - a branch of mathematics that deals with numbers, which may be represented by letters or symbols

analytic geometry - a branch of mathematics that applies algebraic principles to geometry

area - the measure of the space covered by a plane shape, expressed in square units

Associative Property - a property that states that the way terms are grouped in an addition or multiplication expression does not affect the result

B

base - a number that is raised to a power; 2. the number that is the foundation in a given number system; for example, the decimal system describes numbers in relation to powers of 10, such as 100, 101, 102, and so on

binomial - an algebraic expression with two terms

break–even point - in a business, the point where revenue and costs are the same

C

Cartesian coordinate system - a system of representing points in space by using axes and coordinates, named for René Descartes

circle - the plane figure made up of all points equidistant from a given center

coefficient - a quantity placed before and multiplying the variable in an algebraic expression

Commutative Property - a property that states that the order in which numbers are added or multiplied does not affect the result

composite numbers - may be factored in more than one way; not prime

cone - a solid with a circular base and a curved surface that rises to a point

conic section - a curve that results when a cone is intersected by a plane

consecutive integers - integers that follow each other in order, ex., -2, -1, 0, 1, 2, ...

constant - a fixed, unchanging value

conversion factor - a ratio equal to one that is used to convert measures; also called *unit multiplier*

counting numbers - whole numbers from 1 to infinity; also called *natural numbers*

cube - a number multiplied by itself three times

cube root - the number which, when multiplied by itself three times, produces a given number

D - E

dimension - a measurement in a particular direction (length, width, height, depth)

Distributive Property of Multiplication over Addition - a property for multiplying a sum by a given factor

elimination - a method of solving simultaneous equations by using additive inverses

ellipse - a regular oval created by moving a point around two foci

exponent - a raised number that indicates the number of times a factor is multiplied by itself; also called *power*

F - G

factor - (n.) a whole number that multiplies with another to form a product; (v.) to find the factors of a given product

formula - a mathematical expression that shows a relationship, often used to solve a problem

greatest common factor (GCF) - the greatest number that will divide evenly into two or more numbers

H - L

hyperbola - a conic section that forms two congruent open curves facing in opposite directions on a graph

inequality - a mathematical statement showing that two expressions have different values

integer - a non-fractional number that can be positive, negative, or zero

inverse - opposite or reverse

irrational numbers - numbers that cannot be written as fractions and form non-repeating, non-terminating decimals

least common multiple (LCM) - the least number that is a multiple of two or more other numbers

like terms - terms in an algebraic expression that have the same variables and the same powers of those variables

line - in geometry, a set of connected points that extends infinitely in two directions

linear equation - an equation that creates a straight line when graphed

M - O

multiplicative inverse - the number that, when multiplied by a given number, has a product of 1; also called *reciprocal*

natural numbers - whole numbers from 1 to infinity; also called counting numbers

negative number - a number less than zero

origin - on a coordinate grid, the point at the intersection of the axes, generally identified by the ordered pair (0, 0)

P - Q

parabola - a conic section that forms a symmetrical curve on a graph

parallel lines - lines in the same plane that do not intersect

perpendicular lines - lines that form right angles when they intersect

perimeter - the distance around the outside edge of a plane shape

polynomial - an algebraic expression with more than one term

positive numbers - numbers greater than zero

power - another name for an *exponent*; indicates the number of times a factor is multiplied by itself

Pythagorean theorem - states that the square of the length of the hypotenuse of a right triangle is equal to the sum of the squares of the lengths of the other sides

quadrant - one of the four sections formed by the axes of a Cartesian coordinate grid

R

radical - an expression containing a root

radius - the distance from the center of a circle to its edge; in a regular polygon, the distance from the center to any vertex; in a sphere, the distance from the center to any point on the surface; plural is *radii*

rational numbers - numbers that can be written as ratios or fractions, including decimals

ray - a geometric figure that starts at a definite point (called the origin) and extends infinitely in one direction

real numbers - numbers that can be written as decimals, including rational and irrational numbers

reciprocal - the number that, when multiplied by a given number, has a product of 1; also called *multiplicative inverse*

rise - the vertical distance of a line for a specified horizontal distance; divided by the run to determine slope

run - the horizontal distance of a line for a specified vertical distance; divided into the rise to determine slope

S

scientific notation - a way to write numbers using the product of a base and a power of ten

significant digits - digits that indicate the accuracy of a measurement

simultaneous equations - a pair of equations with two unknown variables that must be solved at the same time

slope - the steepness of a line, found by determining the ratio of the rise to the run

slope-intercept form - the equation of a line written in the form $y = mx + b$, where m indicates the slope and b indicates the y-intercept

square - a number multiplied by itself

square root - a number that can be multiplied by itself to form a specified product

substitution - a method of solving simultaneous equations by expressing one unknown in terms of the other

system of equations - two or more equations with the same set of unknowns

T - Z

term - a part of an algebraic expression which may be a number, a variable, or a product

trinomial - an algebraic expression with three terms

undefined - having no mathematical meaning

unit - the place in a place-value system representing numbers less than the base; a quantity used as a standard of measure

unit multiplier - a ratio equal to one that is used to convert measures; also called *conversion factor*

unknown - a specific quantity that has not yet been determined

variable - a value that is not fixed or determined, often representing a range of possible values

vertex - the highest or lowest point of a parabola; the endpoint shared by two rays, line segments, or edges; plural is *vertices*

volume - the number of cubic units that can be contained in a solid

whole numbers - counting numbers from zero to infinity, excluding fractions

X-axis - the vertical number line used as a reference on a Cartesian coordinate system

Y-axis - the horizontal number line used as a reference on a Cartesian coordinate system

Y-intercept - the coordinate of a point at which a line, curve, or surface intersects the Y-axis

Secondary Levels Master Index

This index lists the levels at which main topics are presented in the instructions for Pre-Algebra through PreCalculus. For more detail, see the description of each level at MathUSee.com

Algebra 1 Index